D1748423

Magnetic Nanoparticles

Edited by
Sergey P. Gubin

Related Titles

Vedmedenko, E.

Competing Interactions and Patterns in Nanoworld

215 pages
2007
Hardcover
ISBN: 978-3-527-40484-1

Rao, C. N. R., Müller, A., Cheetham, A. K. (eds.)

The Chemistry of Nanomaterials

Synthesis, Properties and Applications

761 pages in 2 volumes with 310 figures and 20 tables
2004
Hardcover
ISBN: 978-3-527-30686-2

Schmid, G. (ed.)

Nanoparticles

From Theory to Application

444 pages with 257 figures
2004
Hardcover
ISBN: 978-3-527-30507-0

Poole, Jr., C. P., Owens, F. J.

Introduction to Nanotechnology

approx. 400 pages
2003
Hardcover
ISBN: 978-0-471-07935-4

Magnetic Nanoparticles

Edited by
Sergey P. Gubin

WILEY-VCH Verlag GmbH & Co. KGaA

The Editors

Prof. Sergey P. Gubin
Russian Academy of Sciences
General/Inorganic
31 Leninsky Pr.
119991 Moscow
QRF

■ All books published by Wiley-VCH are carefully produced. Nevertheless, authors, editors, and publisher do not warrant the information contained in these books, including this book, to be free of errors. Readers are advised to keep in mind that statements, data, illustrations, procedural details or other items may inadvertently be inaccurate.

Library of Congress Card No.:
applied for

British Library Cataloguing-in-Publication Data
A catalogue record for this book is available from the British Library.

Bibliographic information published by the Deutsche Nationalbibliothek
The Deutsche Nationalbibliothek lists this publication in the Deutsche Nationalbibliografie; detailed bibliographic data are available on the Internet at http://dnb.d-nb.de.

© 2009 WILEY-VCH Verlag GmbH & Co. KGaA, Weinheim

All rights reserved (including those of translation into other languages). No part of this book may be reproduced in any form – by photoprinting, microfilm, or any other means – nor transmitted or translated into a machine language without written permission from the publishers. Registered names, trademarks, etc. used in this book, even when not specifically marked as such, are not to be considered unprotected by law.

Cover Illustration by Spieszdesign Neu-Ulm, Germany
Typesetting Laserwords Private Ltd, Chennai, India
Printing Strauss GmbH, Mörlenbach
Binding Litges & Dopf Buchbinderei GmbH, Heppenheim

Printed in the Federal Republic of Germany
Printed on acid-free paper

ISBN: 978-3-527-40790-3

Contents

Preface *XI*

List of Contributors *XIII*

1 **Introduction** *1*
Sergey P. Gubin
1.1 Some Words about Nanoparticles *1*
1.2 Scope *3*
1.2.1 Magnetic Nanoparticles Inside Us and Everywhere Around Us *4*
1.3 The Most Extensively Studied Magnetic Nanoparticles and Their Preparation *5*
1.3.1 Metals *6*
1.3.2 Nanoparticles of Rare Earth Metals *9*
1.3.3 Oxidation of Metallic Nanoparticles *10*
1.3.4 Magnetic Alloys *11*
1.3.4.1 Fe–Co Alloys *11*
1.3.5 Magnetic Oxides *13*
1.3.6 Final Remarks *19*

2 **Synthesis of Nanoparticulate Magnetic Materials** *25*
Vladimir L. Kolesnichenko
2.1 What Makes Synthesis of Inorganic Nanoparticles Different from Bulk Materials? *25*
2.2 Synthesis of Magnetic Metal Nanoparticles *28*
2.2.1 Reduction of Metal Salts in Solution *29*
2.2.1.1 Electron Transfer Reduction *29*
2.2.1.2 Reduction via Intermediate Complexes *32*
2.2.2 Thermal Decomposition Reactions *36*
2.2.2.1 Decomposition of Metal Carbonyls *36*
2.2.2.2 Decomposition of Metal Alkene and Arene Complexes *38*
2.2.3 Combination Methods Used for Synthesis of Alloy Nanoparticles *39*

Magnetic Nanoparticles. Sergey P. Gubin
© 2009 WILEY-VCH Verlag GmbH & Co. KGaA, Weinheim
ISBN: 978-3-527-40790-3

2.3	Synthesis of Magnetic Metal Oxide Nanoparticles	40
2.3.1	Reactions of Hydrolysis	40
2.3.1.1	Hydrolysis in Aqueous Solutions	40
2.3.1.2	Hydrolysis in Nonaqueous Solutions	42
2.3.2	Oxidation Reactions	46
2.3.3	Thermal Decomposition of Metal Complexes with O-Donor Ligands	47
2.4	Technology of the Preparation of Magnetic Nanoparticles	49
2.4.1	Stabilizing Agents in Homogeneous Solution Techniques	50
2.4.2	Heterogeneous Solution Techniques	51
2.5	Conclusions	53

3 Magnetic Metallopolymer Nanocomposites: Preparation and Properties 59
Gulzhian I. Dzhardimalieva, Anatolii D. Pomogailo, Aleksander S. Rozenberg and Marcin Leonowicz

3.1	Introduction	59
3.2	The General Methods of Synthesis and Characterization of Magnetic Nanoparticles in a Polymer Matrix	60
3.2.1	Magnetic Nanoparticles in Inorganic Matrices	60
3.2.2	Magnetic Nanoparticles in Polymer Matrices	60
3.2.3	Preparation of Magnetic Polymer Nanocomposites in Magnetic Fields	63
3.2.4	Peculiarities of Magnetic Behavior of Metallic Nanoparticles in Polymer Matrix	63
3.3	Magnetic Metal Nanoparticles in Stabilizing the Polymer Matrix *In Situ* via Thermal Transformations of Metal-Containing Monomers	67
3.3.1	The Kinetics of Thermolysis of Metal-Containing Monomers	68
3.3.1.1	Dehydration	68
3.3.1.2	Polymerization	69
3.3.1.3	Kinetics of Decarboxylation	69
3.3.2	The Topography and Structure of Magnetic Metallopolymer Nanocomposites	75
3.3.3	The Magnetic Properties of the Metallopolymer Nanocomposites	78
3.4	Conclusion	82
	Acknowledgments	83

4 Magnetic Nanocomposites Based on the Metal-Containing (Fe, Co, Ni) Nanoparticles Inside the Polyethylene Matrix 87
Gleb Yu Yurkov, Sergey P. Gubin, and Evgeny A. Ovchenkov

4.1	Introduction	87
4.2	Experimental Details	88
4.2.1	Synthesis	88

4.2.2	Composition and Structure of Magnetic Nanometallopolymers	92
4.3	Magnetic Properties of Metal-Containing Nanoparticles	93
4.3.1	Iron Containing Nanoparticles	93
4.3.2	Iron Oxide Nanoparticles	101
4.3.3	Cobalt Nanoparticles	104
4.3.4	Co@Fe_2O_3 Particles	106
4.4	FMR Investigations of Nanocomposites	108
4.5	Conclusions	112
	Acknowledgments	114

5	**Organized Ensembles of Magnetic Nanoparticles: Preparation, Structure, and Properties**	**117**
	Gennady B. Khomutov and Yury A. Koksharov	
5.1	Introduction	117
5.2	Two-Dimensional Systems: Layers and Nanofilms	119
5.2.1	Self-Assembled Nanostructures	121
5.2.2	Langmuir–Blodgett Technique	124
5.2.3	Layer-by-Layer Assembly	130
5.2.3.1	Supported Films	131
5.2.3.2	Nanofilm Capsules	134
5.2.3.3	Free-Standing and Free-Floating Films	135
5.2.4	Bulk Phase Self-Assembled Sheetlike Nanofilms	137
5.3	Anisotropic and Quasilinear (1D) Systems	145
5.3.1	Synthesis under Applied Magnetic Field	146
5.3.2	Ligand Effects and Synthesis in Nanostructured Media	148
5.3.3	Quasilinear Nanoparticulate Structures	153
5.3.4	Templated Structures	156
5.3.5	Nanotubes	160
5.4	Patterned, Self-Organized, Composite, and Other Complex Magnetic Nanoparticulate Nanostructures	163
5.5	Bioinorganic Magnetic Nanostructures	170
5.6	Magnetic Properties of Organized Ensembles of Magnetic Nanoparticles	174
5.7	Conclusions and Perspectives	182
	Acknowledgments	183

6	**Magnetism of Nanoparticles: Effects of Size, Shape, and Interactions**	**197**
	Yury A. Koksharov	
6.1	Introduction	197
6.2	Magnetism of Nanoparticles in the View of Atomic and Solid State Physics	198
6.3	Magnetic Finite-Size Effects and Characteristic Magnetic Lengths. Single-Domain Particles	199
6.4	Shape Effects	208

6.5	Superparamagnetism	210
6.6	Surface Effects	214
6.7	Matrix Effects	220
6.8	Interparticle Interaction Effects	223
6.9	Nanoparticles of Typical Magnetic Materials: Illustrative Examples	227
6.10	Antiferromagnetic Nanoparticles	232
6.11	Semiconductor Magnetic Nanoparticles	233
6.11.1	Magnetism of Intrinsic and Diluted Magnetic Semiconductors	233
6.11.2	Unusual Magnetism of Magnetic Semiconductors and Role of Nanosize Effects	235
6.12	Some Applications of Magnetic Nanoparticles	238
6.12.1	High-Density Information Magnetic Storage	238
6.12.2	Traditional and New Applications of Ferrofluids	240
6.12.3	Magnetic Nanoparticles and Spintronics	244
6.13	Final Remarks	246
7	**Electron Magnetic Resonance of Nanoparticles: Superparamagnetic Resonance**	**255**
	Janis Kliava	
7.1	Introduction	255
7.2	Superparamagnetic Resonance Spectrum in a Disordered System	257
7.3	Resonance Magnetic Field	259
7.4	Resonance Lineshapes	263
7.4.1	Damped Gyromagnetic Precession	263
7.4.1.1	Definitions	263
7.4.1.2	The Bloch–Bloembergen Equation	265
7.4.1.3	The Modified Bloch Equation	268
7.4.1.4	The Gilbert Equation	269
7.4.1.5	The Landau–Lifshitz Equation	269
7.4.1.6	The Callen Equation	271
7.4.2	Linewidths and Apparent Shift of the Resonance Field	272
7.4.3	Angular Dependence of the Linewidth	274
7.5	Superparamagnetic Narrowing of the Resonance Spectra	275
7.5.1	De Biasi and Devezas model	275
7.5.2	Raikher and Stepanov Model	278
7.6	Nanoparticle Size and Shape Distribution	279
7.6.1	Distribution of Diameters	279
7.6.2	Nonsphericity of Nanoparticles: Distribution of Demagnetizing Factors	280
7.6.3	Joint Distribution of Diameters and Demagnetizing Factors	285
7.7	Superparamagnetic Resonance in Oxide Glasses: Some Experimental Results	287

7.7.1	Lithium Borate Glass	*287*
7.7.2	Sol–Gel Silica Glass	*292*
7.7.3	Potassium-Alumino-Borate Glass	*295*
7.7.4	Gadolinium-Containing Multicomponent Oxide Glass	*297*
7.8	Conclusions and Prospective	*297*

8 Micromagnetics of Small Ferromagnetic Particles *303*
Nickolai A. Usov and Yury B. Grebenshchikov

8.1	Introduction	*303*
8.2	Particle Morphology and Single-Domain Radius	*306*
8.2.1	Quasiuniform States	*306*
8.2.1.1	Particles of Perfect Geometrical Shape	*306*
8.2.1.2	Particle of Quasiellipsoidal Shape	*310*
8.2.2	Nonuniform States	*313*
8.2.2.1	Spherical Particle	*314*
8.2.2.2	Ellipsoidal Particle	*317*
8.2.2.3	Cylindrical Particle	*319*
8.2.2.4	Cubic Particle	*321*
8.2.3	Effective Single-Domain Radius	*322*
8.3	Surface and Interface Effects	*322*
8.3.1	Brown's Surface Anisotropy	*324*
8.3.2	Surface Spin Disorder	*327*
8.3.3	Interface Boundary Condition	*330*
8.4	Thermally Activated Switching	*332*
8.4.1	Analytical Estimates of the Relaxation Time	*333*
8.4.2	Stochastic Langevin Equation	*334*
8.4.3	Simple Examples	*335*
8.4.3.1	Uniaxial Anisotropy	*335*
8.4.3.2	Nonsymmetrical Case	*336*
8.4.3.3	Cubic Anisotropy	*339*
8.4.4	Nonuniform Modes	*341*
8.5	Conclusions	*343*

9 High-Spin Polynuclear Carboxylate Complexes and Molecular Magnets with VII and VIII Group 3d-Metals *349*
Igor L. Eremenko, Aleksey A. Sidorov, and Mikhail A. Kiskin

9.1	Introduction	*349*
9.2	High-Spin 3d-Metal Pivalate Polymers as a Good Starting Spin Materials	*351*
9.3	Chemical Design of High-Spin Polynuclear Structures with Different Magnetic Properties	*365*
9.4	Pivalate-Bridged Heteronuclear Magnetic Species	*374*
9.5	Pivalate-Based Single Molecular Magnets	*385*
9.6	Conclusions	*388*

10	**Biomedical Applications of Magnetic Nanoparticles** *393*
	Vladimir N. Nikiforov and Elena Yu. Filinova
10.1	Introduction *393*
10.2	Biocompatibility of Magnetic Nanoparticles *397*
10.3	Magnetic Separation for Purification and Immunoassay *401*
10.3.1	Cell Labeling for Separation *402*
10.3.2	Magnetic Separator Design *404*
10.3.3	The Biomedicine Applications *405*
10.3.4	The Immunomagnetic Separation *405*
10.3.5	Basic Principles of Magnetic Separation of Proteins and Peptides *408*
10.3.6	Examples of Magnetic Separations of Proteins and Peptides *410*
10.3.6.1	Perspectives *412*
10.4	Magnetic Nanoparticles in Cancer Therapy *414*
10.4.1	Is a Heat Suitable for Health? *415*
10.4.2	Basics of Hyperthermia *419*
10.4.3	Magnetic Hyperthermia in Cancer Treatment *422*
10.4.4	Prospective of Clinical Applications *425*
10.4.5	Conclusions *428*
10.5	Targeted Drug and Gene Delivery *429*
10.5.1	Magnetic Targeting Local Hyperthermia *431*
10.5.2	Magnetic Targeting Applications *431*
10.5.3	Targeted Liposomal Drug and Gene Delivery *432*
10.5.4	Magnetic Targeting of Radioactivity *436*
10.5.5	Other Magnetic Targeting Applications *436*
10.5.6	MRI Contrast Enhancement *437*
10.6	Prospective of MRI *443*
10.7	Problems and Perspectives *444*

Index *457*

Preface

Nanoparticles are essential part of our natural environment, modern science, and high technology. Magnetic nanoparticles are most common and very promising for applications. Although the magnetism of fine particles has been studied for almost 60 years, there is the rich variety of phenomena which remain to be understood. However, the meaningful progress in this field, including theory and in particular – practice, is rather recent and particularly remarkable for the last two decades. The major significance of magnetic nanoparticles is attributed to the uniformity in magnetic properties of individual particles in the real dispersion system, which allows us to directly correlate the magnetic properties of a whole material with those of each particle and facilitates theoretical approaches.

One of my main aims in preparing this book is to build bridge between theory of nanomagnetism and the study of synthetic materials with isolated magnetic nanoparticles. This book is not a comprehensive review of the many studies concerned with magnetic nanoparticles; instead, we concentrate our attention on giving a broad overview and key examples and attempt to motivate a deeper than usual examination of forefront fundamental developments in this field.

This book provides a forum for critical reviews on many aspects of nanoparticle magnetism, which are at the forefront of nanoscience today. The chapters do not cover the whole spectrum of nanomagnetism, which would be limitless, but present highlights especially in the domains of interest of the authors and editor.

I hope that this book, probably the first book dealing with general aspects of magnetic nanoparticles, will serve as a guide to the magnetic nanotechnology for wide audience: from senior- and graduate-level students up to advanced specialists in both academic and industrial centers.

The editor gratefully acknowledges the contributing authors of these chapters, who are world renowned experts in this burgeoning field of nanoscience. I thank all the colleagues who spend considerable time and effort in writing these high-level contributions.

Moscow, Russia
July 2008

S.P. Gubin

Magnetic Nanoparticles. Sergey P. Gubin
© 2009 WILEY-VCH Verlag GmbH & Co. KGaA, Weinheim
ISBN: 978-3-527-40790-3

List of Contributors

Gulzhian I. Dzhardimalieva
Institute of Problems
of Chemical Physics
Russian Academy of Sciences
Chernogolovka Moscow Region,
142432
Russian Federation

Igor L. Eremenko
N.S. Kurnakov Institute of General
and Inorganic Chemistry
Russian Academy of Sciences
31 Leninsky prosp.
119991 Moscow
Russian Federation

Elena Yu. Filinova
N.N. Blokhin Cancer Research Center
Russian Academy of Medical Sciences
Kashirskoe Shosse, 24
Moscow, 115478
Russian Federation

Yury B. Grebenshchikov
Institute of Terrestrial Magnetism
Ionosphere and Radio Wave
Propagation (IZMIRAN)
Russian Academy of Sciences
142190, Troitsk
Moscow
Russian Federation

Sergey Gubin
Kurnakov Institute of General
an Inorganic Chemistry
Russian Academy of Science
Leninskii prosp., 31
119071 Moscow
Russian Federation

Mikhail A. Kiskin
N.S. Kurnakov Institute of General
and Inorganic Chemistry
Russian Academy of Sciences
31 Leninsky prosp.
119991 Moscow
Russian Federation

Gennady B. Khomutov
Faculty of Physics
M.V. Lomonosov Moscow
State University
Leninskie Gory
Moscow 119991
Russian Federation

Janis Kliava
Centre de Physique Moléculaire
Optique et Hertzienne (CPMOH)
Université Bordeaux 1
351 cours de la Libération
33405 Talence Cedex
France

Magnetic Nanoparticles. Sergey P. Gubin
© 2009 WILEY-VCH Verlag GmbH & Co. KGaA, Weinheim
ISBN: 978-3-527-40790-3

List of Contributors

Yury A. Koksharov
Faculty of Physics
M.V. Lomonosov Moscow
State University
Leninskie Gory
Moscow 119991
Russian Federation

Vladimir L. Kolesnichenko
Xavier University of LA
Inorganic Chemistry
1, Drexel Drive New Orleans
LA, 70125
USA

Marcin Leonowicz
Faculty of Materials Science
and Engineering
Warsaw University of Technology
Pl. Politechniki 1
00-661 Warsaw
Poland

Vladimir N. Nikiforov
Faculty of Physics
M.V. Lomonosov Moscow
State University
Leninskie Gory
Moscow, 119991
Russian Federation

Evgeny A. Ovchenkov
Physics Faculty
M.V. Lomonosov Moscow
State University
Leninskie Gory
Moscow, 119992
Russian Federation

Anatolii D. Pomogailo
Institute of Problems
of Chemical Physics
Russian Academy of Sciences
Chernogolovka Moscow Region,
142432
Russian Federation

Aleksander S. Rozenberg
Institute of Problems
of Chemical Physics
Russian Academy of Sciences
Chernogolovka Moscow Region,
142432
Russian Federation

Aleksey A. Sidorov
N.S. Kurnakov Institute of General
and Inorganic Chemistry
Russian Academy of Sciences
31 Leninsky prosp.
119991 Moscow
Russian Federation

Nikolaj A. Usov
Institute of Terrestrial Magnetism
Ionosphere and Radio Wave
Propagation (IZMIRAN)
Russian Academy of Sciences
142190, Troitsk
Moscow
Russian Federation Ltd.
"Magnetic and Cryogenic Systems"
142190, Troits
Moscow
Russian Federation

Gleb Yu. Yurkov
Institution of Russian Academy
of Science A.A. Baikov Institute
of Metallurgy and Materials Science
RAS (A.A. Baikov IMET RAS)
Moscow, 119991
Russian Federation

1
Introduction
Sergey P. Gubin

1.1
Some Words about Nanoparticles

First of all, it is necessary to consider the general concepts related to the nanosized objects. A nanoobject is a physical object differing appreciably in properties from the corresponding bulk material and having at least 1 nm dimension (not more than 100 nm). When dealing with nanoparticles, magnetic properties (and other physical ones) are size dependent to a large extent. Therefore, particles whose sizes are comparable with (or lesser than) the sizes of magnetic domains in the corresponding bulk materials are the most interesting from a magnetism scientist viewpoint.

Nanotechnology is the technology dealing with both single nanoobjects and materials, and devices based on them, and with processes that take place in the nanometer range. Nanomaterials are those materials whose key physical characteristics are dictated by the nanoobjects they contain. Nanomaterials are classified into compact materials and nanodispersions. The first type includes so-called nanostructured materials [1], i.e., materials isotropic in the macroscopic composition and consisting of contacting nanometer-sized units as repeating structural elements [2]. Unlike nanostructured materials, nanodispersions include a homogeneous dispersion medium (vacuum, gas, liquid, or solid) and nanosized inclusions dispersed in this medium and isolated from each other. The distance between the nanoobjects in these dispersions can vary over broad limits from tens of nanometers to fractions of a nanometer. In the latter case, we are dealing with nanopowders whose grains are separated by thin (often monoatomic) layers of light atoms, which prevent them from agglomeration. Materials containing magnetic nanoparticles, isolated in nonmagnetic matrices at the distances longer than their diameters, are most interesting for magnetic investigations.

A nanoparticle is a quasi-zero-dimensional (0D) nanoobject in which all characteristic linear dimensions are of the same order of magnitude (not more than 100 nm). Nanoparticles can basically differ in their properties from larger

Magnetic Nanoparticles. Sergey P. Gubin
© 2009 WILEY-VCH Verlag GmbH & Co. KGaA, Weinheim
ISBN: 978-3-527-40790-3

particles, for example, from long- and well-known ultradispersed powders with a grain size above 0.5 μm. As a rule, nanoparticles are shaped like spheroids. Nanoparticles with a clearly ordered arrangement of atoms (or ions) are called nanocrystallites. Nanoparticles with a clear-cut discrete electronic energy levels are often referred to as "quantum dots" or "artificial atoms"; most often, they have compositions of typical semiconductor materials, but not always. Many magnetic nanoparticles have the same set of electronic levels.

Nanoparticles are of great scientific interest because they represent a bridge between bulk materials and molecules and structures at an atomic level. The term "cluster," which has been widely used in the chemical literature in previous years, is currently used to designate small nanoparticles with sizes less than 1 nm. Magnetic polynuclear coordination compounds (magnetic molecular clusters) belong to the special type of magnetic materials often with unique magnetic characteristics. Unlike nanoparticles, which always have the distributions in sizes, molecular magnetic clusters are the fully identical small magnetic nanoparticles. Their magnetism is usually described in terms of exchange-modified paramagnetism.

Nanorods and nanowires, as shown in Figure 1.1, are quasi-one-dimensional (1D) nanoobjects. In these systems, one dimension exceeds by an order of magnitude the other two dimensions, which are in the nanorange.

The group of two-dimensional objects (2D) includes planar structures – nanodisks, thin-film magnetic structures, magnetic nanoparticle layers, etc., in which two dimensions are an order of magnitude greater than the third one, which is in the nanometer range. The nanoparticles are considered by many authors as giant pseudomolecules having a core and a shell and often also external functional groups. The unique magnetic properties are usually

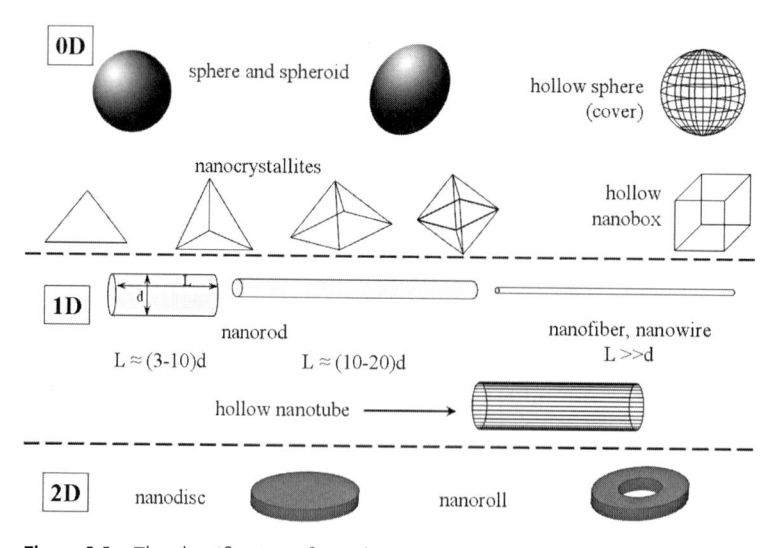

Figure 1.1 The classification of metal containing nanoparticles by the shape.

inherent in the particles with a core size of 2–30 nm. For magnetic nanoparticles, this value coincides (or less) with the size of a magnetic domain in most bulk magnetic materials. Methods of synthesis and properties of nanoparticles were considered in the books and reports [3].

1.2
Scope

Among many of known nanomaterials, the special position belong to those, in which isolated magnetic nanoparticles (magnetic molecular clusters) are divided by dielectric nonmagnetic medium. These nanoparticles present giant magnetic pseudoatoms with the huge overall magnetic moment and "collective spin." In this regard nanoparticles fundamentally differ from the classic magnetic materials with their domain structure. As a result of recent investigations, the new physics of magnetic phenomena – nanomagnetism – was developed. Nanomagnetism advances include superparamagnetism, ultrahigh magnetic anisotropy and coercive force, and giant magnetic resistance. The fundamental achievement of the last time became the development of the solution preparation of the objects with advanced magnetic parameters.

Currently, unique physical properties of nanoparticles are under intensive research [4]. A special place belongs to the magnetic properties in which the difference between a massive (bulk) material and a nanomaterial is especially pronounced. In particular, it was shown that magnetization (per atom) and the magnetic anisotropy of nanoparticles could be much greater than those of a bulk specimen, while differences in the Curie or Néel temperatures between nanoparticle and the corresponding microscopic phases reach hundreds of degrees. The magnetic properties of nanoparticles are determined by many factors, the key of these including the chemical composition, the type and the degree of defectiveness of the crystal lattice, the particle size and shape, the morphology (for structurally inhomogeneous particles), the interaction of the particle with the surrounding matrix and the neighboring particles. By changing the nanoparticle size, shape, composition, and structure, one can control to an extent the magnetic characteristics of the material based on them. However, these factors cannot always be controlled during the synthesis of nanoparticles nearly equal in size and chemical composition; therefore, the properties of nanomaterials of the same type can be markedly different.

In addition, magnetic nanomaterials were found to possess a number of unusual properties – giant magnetoresistance, abnormally high magnetocaloric effect, and so on.

Nanomagnetism usually considers so-called single-domain particles; typical values for the single-domain size range from 15 to 150 nm. However, recently the researchers focused their attention on the particles, whose sizes are smaller than the domain size range; a single particle of size comparable to the minimum domain size would not break up into domains; there is a reason to call these particles domain free magnetic nanoparticles (DFMN). Each such particle behaves like a giant paramagnetic atom and shows superparamagnetic behavior when the temperature is above the so-called blocking temperature. The experiment shows that the last one can vary in wide diapasons, from few kelvins till higher than room temperature.

Thus, this is a book describing what we need to know to perform nanoscale magnetism – the magnetism of single nanoparticles, well dispersed and isolated one from another. It is important to mention that the intensity of interparticle interactions can dramatically affect the magnetic behavior of their macroscopic ensemble.

Now it became possible to prepare individual nanometer metal or oxide particles not only as ferromagnetic fluids (whose preparation was developed back in the 1960s) [5] but also as single particles covered by ligands or as particles included into "rigid" matrices (polymers, zeolites, etc.).

The purpose of this book is to survey the state-of-the-art views on physics, chemistry and methods of preparation and stabilization of magnetic nanoparticles used in nanotechnology for the design of new instruments and devices.

Let us list the most important applications of magnetic nanoparticles: ferrofluids for seals, bearings and dampers in cars and other machines, magnetic recording industry, magnetooptic recording devices, and giant magnetoresistive devices. In recent years, there has been an increasing interest to use magnetic nanoparticles in biomedical applications. Examples of the exciting and broad field of magnetic nanoparticles applications include drug delivery, contrast agents, magnetic hyperthermia, therapeutic *in vivo* applications of magnetic carriers, and *in vitro* magnetic separation and purification, molecular biology investigations, immunomagnetic methods in cell biology and cell separation and in pure medical applications. All of these topics are described to some extent in the following chapters of the book.

1.2.1
Magnetic Nanoparticles Inside Us and Everywhere Around Us

Interstellar space, lunar samples, and meteorites have inclusive magnetic nanoparticles. The geomagnetic navigational aids in all migratory birds, fishes and other animals contain magnetic nanoparticles. The most common iron storage protein ferritin ($[FeOOH]_n$ containing magnetic nanoparticle) is present in almost every cell of plants and animals including humans. The human brain contains over 10^8 magnetic nanoparticles of magnetite–maghemite per gram of tissue [6]. Denis G. Rancourt has written a nice survey of magnetism of Earth, planetary and environmental nanomaterials [6].

Readers who are interested in more detailed information about the physical properties, magnetic behavior, chemistry, or biomedical applications of magnetic nanoparticles are referred to specific reviews [7].

1.3
The Most Extensively Studied Magnetic Nanoparticles and Their Preparation

A series of general methods for the nanoparticle synthesis have now been developed [8] Most of them can also be used for the preparation of magnetic particles. An essential feature of their synthesis is the preparation of particles of a specified size and shape; at least, the dispersity should be small, 5%–10%, and controllable, since the blocking temperature (and other magnetic characteristics) depends on the particle size. The shape control and the possibility of synthesis of anisotropic magnetic structures are especially important. In order to eliminate (or substantially decrease) the interparticle interactions, magnetic nanoparticles often need to be isolated from one another by immobilization on a substrate surface or in the bulk of a stabilizing matrix or by surfacing with long chain ligands. It is important that the distance between the particles in the matrix should be controllable. Finally, the synthetic procedure should be relatively simple, inexpensive and reproducible.

The development of magnetic materials is often faced with the necessity of preparing nanoparticles of a complex composition, namely, ferrites, FePt, NdFeB or $SmCo_5$ alloys, etc. In these cases, the range of synthetic approaches substantially narrows down. For example, the thermal evaporation of compounds with a complex elemental composition is often accompanied by a violation of the stoichiometry in the vapor phase, resulting in the formation of other substances, while the atomic beam synthesis does not yield a homogeneous distribution of elements in the substrate. The mechanochemical methods of powder dispersion also violate (in some cases, substantially) the phase composition: in particular, ferrites do not retain the homogeneity and oxygen stoichiometery. Furthermore, there is a difficulty of synthesis of the heteroelement precursors required composition. For example, no precursors for $SmCo_5$ with a Sm atom bonded to five Co atoms are known; the maximum chemically attainable element ratio in $Sm[Co(CO)_4]_3$ is $1:3$. It is even more difficult to propose a stoichiometeric precursor for the synthesis of NdFeB nanoparticles. The overview of general aspects of nanoalloys preparation and characterization and resulting difficulties is presented in [19].

The physical characteristics of nanoparticles are known to be substantially dependent on their dimensions. Unfortunately, most of the currently known methods of synthesis afford nanoparticles with rather broad size distributions (dispersion > 10%). The thorough control of reaction parameters (time, temperature, stirring velocity, and concentrations of reactants and stabilizing ligands) does not always allow one to narrow down this distribution to the required range. Therefore, together with the development of methods for synthesis of nanoparticles with a narrow size distribution, the techniques of separation of nanoparticles into rather monodisperse fractions are perfected. This is done using controlled precipitation of particles from surfactant-stabilized solutions followed by centrifugation. The process is repeated until nanoparticle fractions with specified sizes and dispersion degrees are obtained.

The methods of nanoparticle preparation cannot be detached from stabilization methods. For 1–10 nm particles with a high surface energy, it is difficult to select a really inert medium [10], because the surface of each nanoparticle bears the products of its chemical modification, which affect appreciably the nanomaterial properties. This is especially important for magnetic nanoparticles in which the modified surface layer may possess magnetic characteristics markedly differing from those of the particle core. Nevertheless, the general methods for nanoparticle synthesis are not related directly to the stabilization and the special methods exist where the nanoparticle formation is accompanied by stabilization (in matrices, by encapsulation, etc.).

We do not consider in detail the common methods of magnetic nanoparticles preparation and stabilization. One can find it in the reviews, and partly, in the subsequent chapters of the book.

Among a wide range of the magnetic nanomaterials, nanoparticles of magnetic metals, simple and complex magnetic oxides, and alloys may be separated for detailed analysis.

1.3.1
Metals

The metallic nanoparticles have larger magnetization compared to metal oxides, which is interesting for many applications. But metallic magnetic nanoparticles are not air stable, and are easily oxidized, resulting in the change or loss (full or partially) of their magnetization.

Fe

Iron is a ferromagnetic material with high magnetic moment density (about 220 emu/g) and is magnetically soft. Iron nanoparticles in the size range below 20 nm are superparamagnetic.

Procedures leading to monodisperse Fe nanoparticles have been well documented [11]. Nevertheless, the preparation of nanoparticles consisting of pure iron is a complicated task, because they often contain oxides, carbides and other impurities. A sample containing pure iron as nanoparticles (10.5 nm) can be obtained by evaporation of the metal in an Ar atmosphere followed by deposition on a substrate [12]. When evaporation took place in a helium atmosphere, the particle size varied in the range of 10–20 nm [13]. Relatively small (100–500 atoms) Fe nanoparticles are formed in the gas phase on laser vaporization of pure iron [14].

The common chemical methods used for the preparations include thermal decomposition of $Fe(CO)_5$ (the particles so prepared are extremely reactive), reductive decomposition of some iron(II) salts, or reduction of iron(III) acetylacetonate; there is a chemical reduction with TOPO capping [15]. A sonochemical method for the synthesis of amorphous iron was developed in [16]. The method of reducing metal salts by NaBH$_4$ has been widely used to synthesize iron-containing nanoparticles in organic solvents [17]. Normally, reductive synthesis of Fe nanoparticles in an aqueous solution with NaBH$_4$ yields a mixture including FeB [18]. Well-dispersed colloidal iron is required for applications in biological systems such as MRI contrast enhancement and biomaterials separation. Nevertheless, the syntheses have as yet a difficulty in producing stable Fe nanoparticle dispersions, especially aqueous dispersions, for potential biomedical applications.

The phase composition of the obtained nanoparticles was not always determined reliable. The range of specific methods was proposed to prepare nanoparticles of the defined phase composition. Thus, the α-Fe nanoparticles with a body-centered cubic (bcc) lattice and an average size of $\sim$10 nm were prepared by grinding a high-purity (99.999%) Fe powder for 32 h [19]. With face-centered cubic (fcc) Fe (γ-Fe) the situation is more complex. In the phase diagram of a bulk Fe, this phase exists at the ambient pressure in the temperature range of 1183–1663 K, i.e., above the Curie point (1096 K). In some special alloys, this phase, which exhibits antiferromagnetic properties (the Néel temperature is in the 40–67 K range), was observed at room temperature [20]. However, a Mössbauer spectroscopy study [21] has shown that the fcc-Fe nanoparticles (40 nm) remain paramagnetic down to 4.2 K. Some publications dealing with the synthesis of Fe nanoparticles present substantial reasons indicating that these nanoparticles had an fcc structure. Apparently, the nanoparticles containing γ-Fe were first obtained by Majima et al. [22]. These particles contained substantial amounts of carbon (up to 14 mass%) and had an austenite fcc structure analogous to γ-Fe. However, later, evidence for the existence of the γ-phase in the Fe nanoparticles that do not contain substantial amounts of carbon has been obtained. Nanoparticles ($\sim$8 nm) consisting, according to powder X-ray diffraction and Mössbauer spectroscopy, of γ-Fe (30 at.%), α-Fe (25 at.%), and iron oxides (45 at.%), were synthesized [23] by treatment of Fe(CO)$_5$ with a CO$_2$ laser. The content of the γ-phase in the nanoparticles did not change for several years; the particles remained nonmagnetic down to helium temperatures.

Sometimes the determination of phase composition as-synthesized nanoparticles is made difficult. Thus, Fe particles (8.5 nm) were obtained by thermal decomposition of Fe(CO)$_5$ in decalin (460 K) in the presence of surfactants [24]. The X-ray diffraction pattern of the powder formed did not display any sharp maxima, indicating the absence of a crystalline phase. It was assumed that amorphization was due to the high content of carbon (>11 mass%) in the nanoparticles studied. Similarly, on ultrasonic treatment of Fe(CO)$_5$ in the gas phase nanoparticles ($\sim$30 nm) were obtained which consisted of >96 mass% of Fe, <3 mass% of C, and 1 mass% of O [25]. Differential thermal analysis of the powder showed an exothermal transition around 585 K, which corresponded, in the author's opinion, to crystallization of the amorphous iron. As-synthesized particles were pyrophoric due to the large surface area. They were exposed to air which resulted in a thin layer of surface oxidation which also provides passivation. To prevent the iron nanoparticles from agglomerating, dispersing agents were added during synthesis, as a rule poly(vinylpyrolidone) (PVP). The size dispersion of the nanoparticles produced using physical methods is broader than that in nanoparticles synthesized by chemical methods like reverse micelle, coprecipitation, etc. However, chemical methods yield as a rule only limited quantities of materials.

Co

Co nanoparticles depending upon the synthetic route are observed in at least three crystallographic phases: typical for bulk Co hcp, ε-Co cubic [26], or multiply twinned fcc-based icosahedral [27]. Conditions of synthesis reactions is influence on the final product structure; in rare cases of the determined phase nanoparticles can be obtained. Often a size and phase selection was required to obtain Co nanocrystals with a specific size and even shape. Methods for the synthesis and magnetic properties of cobalt nanoparticles' different structures have been described in detail in a review [8c].

A popular approach is to synthesize colloidal particles by inversed micelle synthesis; the inverse micelles are defined as a microreactor [28]. In order to obtain stable cobalt nanoparticles with a narrow size distribution, $Co(AOT)_2$ reverse micelles are used; their reduction is obtained by using $NaBH_4$ as a reducing agent. Such particles are stabilized by surfactants and are often monodispersed in size, but are also unstable unless kept in a solution. Nevertheless, the chemical surface treatment by lauric acid highly improves the stability and cobalt nanoparticles could be stored without aggregation or oxidation for at least one week [29] In many instances it is possible to obtain Co nanoparticles coated by other ligands, which can be either dispersed in a solvent or deposited on a substrate; in the latter case, self-organized monolayers having a hexagonal structure can be obtained.

In some instances of reduction with $NaBH_4$ it is possible to obtain Co–B nanoparticles. The size, composition, and structure of this kind of nanoparticles strongly depend on the concentration of the solution, pH, and the mixing procedure [30]. It is well known that the presence of oxides in magnetic materials, which form spontaneously when the metallic surface is in contact with oxygen, drastically changes the magnetic behavior of the particles. An enhanced magnetoresistance, arising from the uniform Co core size and CoO shell thickness, has been reported [31]. This effect is caused by the strong exchange coupling between the ferromagnetic Co core and the antiferromagnetic CoO layer. However, up to now the understanding of this effect has not been well understood.

Ni

In contrast to cobalt and iron, relatively few reports have been dealing with the physical properties and synthesis of nickel particles. However, the nano-sized ferromagnetic Ni is also being widely studied as it presents both an interest for fundamental sciences and an interest for applications such as magnetic storage, ferrofluids, medical diagnosis, multilayer capacitors, and especially catalysis. Because these properties and applications can be tuned by manipulating the size and structure of the particles, the development of flexible and precise synthetic routes has been an active area of research. A wide variety of techniques have been used to produce nickel nanoparticles: thermal decomposition [32], sol–gel [33], spray pyrolysis [34], sputtering [35], and high-energy ball milling [36]. The organometallic precursors such as $Ni(CO)_4$, $Ni(COD)_2$, and $Ni(Cp)_2$ have also been used for the synthesis and spectroscopic studies of nickel nanoparticles [37, 38]. At present, Ni nanoparticles are generally prepared by microemulsion techniques, using cetyltrimethylammonium bromide (CTAB) [39] or by reduction of Ni ions in the presence of alkyl amines or trioctylphosphine oxide (TOPO) [40]. Some authors showed that the surface of Ni-nanoparticles was readily oxidized to NiO. On the basis of this discovery, they envisioned that the synthesis of large-sized Ni nanoparticles and their subsequent oxidation would provide an NiO shell having high affinity for biomolecules.

1.3.2
Nanoparticles of Rare Earth Metals

Six of the nine rare earth elements (REE) are ferromagnetic. The magnetic nanomaterials based on these REE occupy a special place, as they can be used in magnetic cooling systems [41]. However, REE nanoparticles (of both metals and oxides) are still represented by only a few examples, most of all, due to the high chemical activity of highly dispersed REE. A synthesis of coarse (95 × 280 nm) spindle-shaped ferromagnetic EuO nanocrystals suitable for the design of optomagnetic materials has been reported [42]. The EuS nanocrystals were prepared by passing H_2S through a solution of europium in liquid ammonia [43]. The size of the EuS magnetic nanoparticles formed can be controlled (to within 20–36 nm) by varying the amount of pyridine added to the reaction medium [43].

Gadolinium nanoparticles (12 nm) were prepared by reduction of gadolinium chloride by Na metal in THF. They proved to be extremely reactive and pyrophoric, which, however, did not prevent characterization of these particles and measuring their magnetic parameters [44]. The Gd, Dy, and Tb nanoparticles with an average size of 1.5–2.1 nm and an about 20% degree of dispersion were obtained in a titanium matrix by ion beam sputtering [45]. At 4.5 K, the coercive forces for ∼10 nm Tb and Gd nanoparticles were 22 and 1 kG, respectively. As the particle size decreases (<10 nm), the Hc value rapidly diminishes to zero, which is related, in the researchers' opinion [46], to the decrease in the Curie temperature for small nanoparticles.

1.3.3
Oxidation of Metallic Nanoparticles

Magnetic properties of metallic nanoparticles are dependent on the degree of oxidation of the surface. Therefore, the true knowledge of the degree of nanoparticle oxidation is necessary for the forecasting of magnetic characteristics of the obtained samples. However, as the experiments have shown, it was often difficult. It should be noted that the oxidation of magnetic metal nanoparticles during their synthesis cannot be avoided completely. Thorough mass-spectroscopic analysis of Fe nanoparticles obtained by laser vaporization of the metal in a pure He medium showed that at least 5% of particles contain at least one oxygen atom [47]. If the deposition of oxygen present in the gas phase in trace amounts on the nanoparticle surface cannot be avoided even under these "exceptional" conditions, it is evident that under "usual" conditions, the nanoparticles of magnetic metals would always contain some amounts of oxides or sub-oxides on the surface. It can be plainly seen in the HRTEM micrograph of Fe nanoparticles (20 nm) synthesized by laser pyrolysis of $Fe(CO)_5$ under inert atmosphere that the particles are coated with a (3.5 nm) layer of iron oxide (the content of the chemically bound oxygen is 14.4 mass%) [48]. At the same time the oxidation of amorphous $Fe_{1-x}C_x$ nanoparticles obtained by thermal decomposition of $Fe(CO)_5$ in decalin in the presence of oleic acid for several weeks in air has shown that the particles (6.9 nm) having a spherical shape and a very narrow size distribution consist of α- and γ-Fe_2O_3 [49]. However, passivation of nanocrystalline ($\sim$25 nm) Fe particles obtained by metal evaporation in a helium stream results in only a thin (1–2 nm) film of an antiferromagnetic oxide (apparently, FeO) forming on the surface [50].

The magnetic properties of cobalt nanoparticles, which were obtained by vacuum evaporation on the LiF substrate and then oxidized by exposure to air for a week, have been studied [51]. According to electron diffraction for two samples differing in the particle size (2.3 and 3.0 nm), the intensity of the HCP–Co reflections decreased after oxidation to become $\sim$1/3 of the CoO–HCP line intensity. Hence, a small stable core of unoxidized cobalt remains in all particles after oxidation. Comparative X-ray diffraction studies of the samples consisting of Co nanoparticles distributed in poly-vinylpyridine stored under Ar and in air (the storage time was not indicated) revealed no significant differences [52]. Therefore, the authors considered a low degree of oxidation for cobalt.

In a more comprehensive study [53], ^{57}Co-enriched cobalt nanoparticles were subjected to oxidation directly in a Mössbauer spectrometer. For this purpose, argon containing $\sim$80 ppm of O_2 was passed through the sample at 300 K for 18 h. Analysis of the emission Mössbauer spectra showed that oxidation results in a fairly well-organized CoO layer on the surface of Co particles. Passing pure oxygen through this gently oxidized sample for 1 h at 300 K did not induce any spectral changes; this is indicative of complete passivation of Co particles at the first oxidation stage. It is thought that such gentle surface oxidation of the Co nanoparticles was always necessary to obtain stable magnetic nanoparticles [54].

In some studies, the preparation of Fe nanoparticles was also followed by their passivation, for example, by keeping for several hours in an atmosphere of oxygen-diluted argon [55]. This procedure prevented further spontaneous particle aggregation. The structure and the magnetic characteristics of such passivated nanoparticles (15–40 nm) have been described in detail [56]. The continuous oxide layers that coat the metallic nanoparticle can be clearly seen in TEM images reported in this study. The interaction of the ferromagnetic core and the oxide shell, which resembles in the magnetic characteristics the interaction of magnetic moments in spin glass, was studied.

The data on the reactivity of Fe nanoparticles with respect to oxidation reported in the literature are contradictory. Thus rather large (~40 nm) nanoparticles of pure Fe obtained by thermal vaporization contained less than 8 mass% of the oxide after exposure to air for three months [57].

In the last few years, for oxidation as-synthesized Fe nanoparticles soft oxidizers such as $(CH_3)_3NO$ were often used.

1.3.4
Magnetic Alloys

1.3.4.1 Fe–Co Alloys

It is well known that Co and Fe form a body-centered-cubic solid solution (Co_xFe_{100-x}) over an extensive range. The ordered Co–Fe alloys are excellent soft magnetic materials with negligible magnetocrystalline anisotropy [58]. The saturation magnetization of Fe–Co alloys reaches a maximum at a Co content of 35 at.%; other magnetic characteristics of these metals also increase when they are mixed. Therefore, FeCo nanoparticles attract considerable attention. Thus Fe, Co, and Fe–Co (20 at.%, 40 at.%, 60 at.%, 80 at.%) nanoparticles (40–51 nm) with a structure similar to the corresponding bulk phases have been prepared in a stream of hydrogen plasma [59]. The Fe–Co particles reach a maximum saturation magnetization at 40 at.% of Co, and a maximum coercive force is attained at 80 at.% of Co. Chemical reduction by $NaBH_4$ is also used for the preparation of FeCo nanoparticles [60]. X-ray data show that the ratio of Co to Fe is around 30:70 in the prepared nanoparticles.

Fe-Ni

The bulk samples of the iron–nickel alloys are either nonmagnetic or magnetically soft ferromagnets (for example, permalloys containing >30% of Ni and various doping additives). The Fe–Ni nanoparticles have a much lower saturation magnetization than the corresponding bulk samples over the whole concentration range [61]. An alloy containing

37% of Ni has a low Curie point and an fcc structure. It consists of nanoparticles (12–80 nm) superparamagnetic over a broad temperature range [62]. Theoretical calculations predict a complex magnetic structure for these Fe–Ni particles [63].

Fe–Pt

Nanoparticles of this composition have received much attention in recent years due to the prospects for a substantial increase in the information recording density for materials based on them [64]. The face-centered tetragonal (fct) (also known as L10 phase) FePt alloy possesses a very high uniaxial magnetocrystalline anisotropy of ca. 6×10^6 J/m^3, which is more than 10 times as high as that of the currently utilized CoCr-based alloys and, thus, exhibit large coercivity at room temperature, even when their size is as small as several nanometers [65]. These unique properties make them possible candidates for the next generation of magnetic storage media and high-performance permanent magnets [66]. To realize these potentials it is important to develop synthetic methods that yield magnetic nanoparticles of tunable size, shape, dispersity and composition. For these syntheses the most commonly used is the thermal decomposition of organometallic precursors or reduction of metal salts in the presence of long-chain acid or amine and phosphine or phosphine oxide ligands [67]. The as-synthesized FePt nanoparticles possess an fcc structure and are superparamagnetic at room temperature. Thermal annealing converts the fcc FePt to fct FePt (L10), yielding nanocrystalline materials with sufficient coercivity [68]. L10 FePt nanoparticles can be synthesized directly using a polyol reduction method at high temperatures or annealed as-prepared chemically disordered face-centered cubic (fcc) to the chemically ordered L10 phase [69]. There are a number of requirements for such transformations into the L10 phase (except avoiding severe sintering and aggregation): the atomic composition of each nanoparticles should be within 40–60% Fe; the diameter should be larger than the superparamagnetic limit (ca. 3.3 nm); the size distribution should generally be below 10% [70].

The FePt nanoparticles (6 nm) with a narrow size distribution were prepared by joint thermolysis of Fe(CO)$_5$ and Pt(acac)$_2$ in the presence of oleic acid and oleylamine. Further heating resulted in the formation of a protective film from the products of thermal decomposition of the surfactant on the nanoparticle surface, which does not change significantly the particle size. These particles can be arranged to form regular films and so-called colloid crystals. For many practical applications, magnetic nanoparticles larger than 6 nm are preferred because coercivity and the saturation magnetization of the nanoparticles are closely related to the size of magnetic nanoparticles [71]. Later it has been however found that most FePt nanoparticles have the broad composition distribution: approximately 40% and 30% of the nanoparticles were Pt-rich and Fe-rich, respectively. Chemists are in general agreement that to obtain high-quality FePt nanoparticles via this synthetic route and to further control the size, composition, and size distribution, a better understanding of the reaction mechanism is required. It is believed that the progress in the chemical synthesis of these nanoparticles makes it possible to utilize FePt for large data storage capacities. An excellent review of the synthesis and properties of FePt nanoparticle materials is available [72]. The reaction of FePt nanoparticles with Fe_3O_4 followed by heating of the samples at 650 °C in an $Ar + 5\%H_2$ stream resulted in the FePt–Fe_3Pt nanocomposite with unusual magnetic characteristics [73].

Co–Pt
The impressive magnetic properties of CoPt nanoparticles according to their size, form, and crystal structure render them as important materials for high-density information storage [74], because they are chemically stable and have very high magnetocrystalline anisotropy $\sim 4 \times 10^6$ J/m^3 [75]. A general method applied for the synthesis of CoPt nanoparticles involves the reduction of Pt(acac)$_2$ by 1,2-hexadecanediol, with the simultaneous thermal decomposition of an organometallic cobalt source in dioctyl ether in the presence of oleic acid and oleyl amine [76]. An alternative way to produce metallic nanoparticles avoiding the use of organometallic precursors is the polyol method.

Using the synthesis of CoPt$_3$ nanoparticles as an example, the mechanism of homogeneous nucleation has been studied. It allowed us to deliberately and reproducibly obtain nanoparticles of fixed composition with a narrow size distribution in the 3–18 nm range [77].

The bimetallic particles are not always appropriately termed "alloys." For example, using the same initial compounds, $Co_2(CO)_8$ and Pt(hfac)$_2$, two types of Co–Pt nanoparticles with the same composition and different structures have been obtained [78], namely, particles with a uniform distribution of Co and Pt atoms and particles with a cobalt core and a platinum shell, Pt @ Co. In the latter type of particles, mixing of the atoms of two metals is possible only at the interface. The desired synthesis of such core/shell CoPt nanoparticles has been described [79]. The researchers first obtained Pt nanoparticles of diameter 2.5 nm and then coated them with a controlled amount of Co layers. This resulted in Co–Pt nanoparticles with a diameter of 7.6 nm.

1.3.5
Magnetic Oxides

Iron Oxides
Iron oxides have received increasing attention due to their extensive applications, such as magnetic recording media, catalysts, pigments, gas sensors, optical devices, and electromagnetic devices [80]. They exist in a rich variety of structures (polymorphs) and hydration states; therefore until recently, knowledge of the structural details, thermodynamics and reactivity of iron oxides

has been lacking. Furthermore, physical (magnetic) and chemical properties commonly change with particle size and degree of hydration. By definition, superparamagnetic iron oxide particles are generally classified with regard to their size into superparamagnetic iron oxide particles (SPIO), displaying hydrodynamic diameters larger than 30 nm, and ultrasmall superparamagnetic iron oxide particles (USPIO), with hydrodynamic diameters smaller than 30 nm. USPIO particles are now efficient contrast agents used to enhance relaxation differences between healthy and pathological tissues, due to their high saturation magnetization, high magnetic susceptibility, and low toxicity. The biodistribution and resulting contrast of these particles are highly dependent on their synthetic route, shape, and size [81]. There has been much interest in the development of synthetic methods to produce high-quality iron oxide systems. The synthesis of controlled size magnetic nanoparticles is described in multiple publications. High-quality iron oxide nanomaterials have been generated using high-temperature solution phase methods similar to those used for semiconductor quantum dots. Other synthesis methods such as polyol-mediated, sol–gel [82] and sonochemical [83] were also proposed. The effectiveness of the nonaqueous routes for the production of well-calibrated iron oxide nanoparticles was shown in [84]. The magnetite nanocrystals (and other) were easily purified using standard methods also developed for quantum dots.

For the variety of magnetic nanomaterials properties the different morphologies including spheres, rods, tubes, wires, belts, cubes, starlike, flowerlike, and other hierarchical architectures were fabricated by various approaches. Finally, some bacteria couple the reduction of Fe(III) with the metabolism of organic materials, which can include anthropogenic contaminants, or simply use iron oxides as electron sinks during respiration [85].

Fe_2O_3

Among several crystalline modifications of anhydrous ferric oxides there are two magnetic phases, namely, rhombohedral hematite (α-Fe_2O_3) and cubic maghemite (γ-Fe_2O_3), and the less common ε-Fe_2O_3 phases. In the α-structure, all Fe^{3+} ions have an octahedral coordination, whereas in γ-Fe_2O_3 having the structure of a cation-deficient AB_2O_4 spinel, the metal atoms A and B occur in tetrahedral and octahedral environments, respectively. The oxide α-Fe_2O_3 is antiferromagnetic at temperatures below 950 K, while above the Morin point (260 K) it exhibits so-called weak ferromagnetism. Hematite, the thermodynamically stable crystallographic phase of iron oxide with a band gap of 2.2 eV, is a very attractive material because of its wide applications, except magnetic recording materials, also in catalysis, as a gas sensors, pigments, and paints. Its nontoxicity is, attractive features for these applications.

The α-Fe_2O_3 and FeOOH (goethite) nanoparticles are obtained by controlled hydrolysis of Fe^{3+} salts [86]. In order to avoid the formation of other phases, a solution of ammonia is added to a boiling aqueous solution of $Fe(NO_3)_3$ with intensive stirring. After boiling for 2.5 h and treating with ammonium oxalate (to remove the impurities of other oxides), the precipitate forms a red powder containing α-Fe_2O_3 nanoparticles (20 nm) [87]. These nanoparticles are also formed on treatment of solutions of iron salts (Fe^{2+}: Fe^{3+} = 1 : 2) with an aqueous solution of ammonium hydroxide in air [88]. The synthesis of regularly arranged α-Fe_2O_3 nanowires with a diameter of 2–5 nm and a length of 20–40 nm has been described [89].

A bulk γ-Fe_2O_3 sample is a ferrimagnet below 620 °C. The γ-Fe_2O_3 nanoparticles (4–16 nm) with a relatively narrow size distribution have been obtained by mild oxidation (Me_3NO) of preformed metallic nanoparticles [90]. The same result can be attained by direct introduction of $Fe(CO)_5$ into a heated solution of Me_3NO. The oxidation with air is also used to prepare γ-Fe_2O_3 nanoparticles. For this purpose, the Fe_3O_4 nanoparticles (9 nm) are boiled in water at pH 12–13 [91]. The kinetics of this process was studied.

The most popular route to γ-Fe_2O_3 nanoparticles is thermal decomposition of Fe^{3+} salts in various media. Rather exotic groups are used in some cases as anions. For example, good results have been obtained by using iron complexes with cupferron [92]. A mechanochemical synthesis of γ-Fe_2O_3 has been described [93]. An iron powder was milled in a planetary mill with water; this is a convenient one-stage synthesis of maghemite nanoparticles (15 nm).

Additionally, nonspherical Fe_2O_3 nanoparticals, such as nanorods, nanowires, nanobelts, and nanotubes, have also been synthesized and used for investigating their peculiar magnetic properties [94].

Fe_3O_4 (Magnetite)

Among all iron oxides, magnetite Fe_3O_4 possess the most interesting properties because of the presence of iron cations in two valence states, Fe^{2+} and Fe^{3+}, in the inverse spinel structure. The cubic spinel Fe_3O_4 is ferrimagnetic at temperatures below 858 K. The route to these particles used most often involves treatment of a solution of a mixture of iron salts (Fe^{2+} and Fe^{3+}) with a base under an inert atmosphere. For example, the addition of an aqueous solution of ammonia to a solution of $FeCl_2$ and $FeCl_3$ (1 : 2) yields nanoparticles, which are transferred into a hexane solution by treatment with oleic acid [95]. The repeated selective precipitation gives Fe_3O_4 nanoparticles with a rather narrow size distribution. The synthesis can be performed starting only from $FeCl_2$, but in this case, a specified amount of an oxidant ($NaNO_2$) should be added to the aqueous solution apart from alkali. This method allows one to vary both the particle size (6.5–38 nm) and (to a certain extent) the particle shape [96].

In some cases, thermal decomposition of compounds containing Fe^{3+} ions under oxygen-deficient conditions is accompanied by partial reduction of Fe^{3+} to Fe^{2+}. Thus thermolysis of $Fe(acac)_3$ in diphenyl ether in the presence of small amounts of hexadecane-l,2-diol (probable reducer of a part of Fe^{3+} ions to Fe^{2+}) gives very fine Fe_3O_4 nanoparticles (about 1 nm), which can be enlarged by adding excess $Fe(acac)_3$ into the reaction mixture [97]. Fe_3O_4

nanoparticles can be also prepared in uniform sizes of about 9 nm by autoclave heating the mixture, consisting of FeCl$_3$, ethylene glycol, sodium acetate, and polyethylene glycol [98]. For partial reduction of Fe^{3+} ions, hydrazine has also been recommended [99]. The reaction of Fe(acac)$_3$ with hydrazine is carried out in the presence of a surfactant. This procedure resulted in superparamagnetic magnetite nanoparticles with controlled sizes, 8 and 11 nm.

The so-called dry methods are used alongside with the solution ones. Thus, Fe$_3$O$_4$ nanoparticles with an average size of 3.5 nm have been prepared by thermal decomposition of Fe$_2$(C$_2$O$_4$)$_3 \cdot$ 5H$_2$O at $T > 400\,°C$ [100]. Furthermore, the controlled reduction of ultradispersed α-Fe$_2$O$_3$ in a hydrogen stream at 723 K (15 min) is a more reliable method of synthesis of Fe$_3$O$_4$ nanoparticles. Particles with $\sim$13 nm size were prepared in this way [101].

The stabilization in the water media is interesting for bioapplications, but at the same time a problem also. For solving it cyclodextrin was used to transfer obtained organic ligand stabilized iron oxide nanoparticles to aqueous phase via forming an inclusion complex between surface-bound surfactants and cyclodextrin [102].

In contrast, higher nanoparticles (20 nm $< d <$ 100 nm) are of great interest, mainly for hyperthermia because of their ferrimagnetic behavior at room temperature. However, there are some difficulties encountered when obtaining a monodisperse magnetite particle of size larger than 20 nm and controlling the stoichiometery.

Ferrites

Microcrystalline ferrites form the basis of materials currently used for magnetic information recording and storage. To increase the recorded information density, it seems reasonable to obtain nanocrystalline ferrites and to prepare magnetic carriers based on them. Grinding of microcrystalline ferrite powders to reach the nanosize of grains is inefficient, as this gives particles with a broad size distribution, the content of the fraction with the optimal particle size (30–50 nm) being relatively low.

The key method for the preparation of powders of magnetic hexagonal ferrites with a grain size of more than 1 µm includes heating of a mixture of the starting compounds at temperature above 1000 °C (so-called ceramic method). An attempt has been made to use this method for the synthesis of barium ferrite nanoparticles [103]. The initial components (barium carbonate and iron oxide) were ground for 48 h in a ball mill and the resulting powder was mixed for 1 h at a temperature somewhat below 1000 °C. This gave rather large particles (200 nm and greater) with a broad size distribution. Similar results have been obtained in the mechanochemical synthesis of barium ferrite [104].

Nanocrystalline ferrites are often prepared by the coprecipitation method. The MnFe$_2$O$_4$ spinel nanoparticles with a diameter of 40 nm are formed upon the addition of an aqueous solution of stoichiometeric amounts of Mn^{2+} and Fe^{3+} chlorides to a vigorously stirred solution of alkali

[105]. The MgFe$_2$O$_4$ (6–18 nm) nanoparticles were obtained in a similar way. In contrast, the SrFe$_{12}$O$_{19}$ nanoparticles (30–80 nm) were synthesized by coprecipitation of Sr and Fe citrates followed by annealing of the resulting precipitate [106]. Coprecipitation upon decomposition of a mixture of Fe(CO)$_5$ and Ba(O$_2$C$_7$H$_{15}$)$_2$ under ultrasonic treatment has been successfully used for the synthesis of barium ferrite nanoparticles (∼50 nm) [107].

Methods for the preparation of ferrite nanoparticles of different compositions in solutions at moderate temperatures have been developed. First, worth mentioning is the sol–gel method resulting in highly dispersed powders with required purity and homogeneity. Low annealing temperatures allow one to control crystallization and to obtain single-domain magnetic ferrite nanoparticles with narrow size distributions and to easily dope the resulting particles with metal ions. This procedure was used to obtain Co- and Ti-doped barium ferrite nanoparticles (smaller than 100 nm) and, Zn-, Ti-, and Ir-doped strontium ferrite particles with a similar size [108].

Smaller nanoparticles (15–25 nm) of cobalt ferrite were obtained in a hydrogel containing lecithin as the major component. Judging by the good magnetic characteristics, these particles possessed a substantial degree of crystallinity without any annealing [109]. The sol–gel method was successfully used to synthesize a Co ferrite nanowire 40 nm in diameter with a length of up to a micrometer [110]. This wire can also be obtained within carbon nanotubes [111]. For the synthesis of ferrite nanoparticles, oil-in-water type micelles [112] and reverse (water-in-oil) micelles [113] are also widely used.

The homogeneity of metal ion distribution in final products can be enhanced, and the required stoichiometry can be attained by using presynthesis of heterometallic complexes of various composition. The thermal decomposition and annealing of the presynthesized [GdFe(OPr′)$_6$(HOPr′)]$_2$ complex give GdFeO$_4$ nanoparticles (∼60 nm) [114]. It is also pertinent to consider the procedure for the synthesis of cobalt ferrite CoFe$_2$O$_4$ nanoparticles, in which the first stage includes the preparation of the Fe–Co heterometallic particles and the second stage, their oxidation to CoFe$_2$O$_4$ [115]. Another route to analogous particle implies the use the heterometallic (η^5-C$_5$H$_5$)CoFe$_2$(CO)$_9$ cluster as the starting compound. The cobalt ferrite nanoparticles were also prepared by the microemulsion method from a mixture of Co and Fe dodecylsulfates treated with an aqueous solution of methylamine [116].

FeO (Wustite)
Cubic Fe^{2+} oxide is antiferromagnetic ($T_c = 185$ K) in the bulk state. Joint milling of Fe and Fe$_2$O$_3$ powders taken in a definite ratio give nanoparticles (5–10 nm) consisting of FeO and Fe [117]. On heating these particles at temperatures of 250–400 °C, the metastable FeO phase disproportionates to Fe$_3$O$_4$ and Fe, while above 550 °C it is again converted into nanocrystalline FeO [118].

FeOOH

The oxyhydroxides, nominally FeOOH, include goethite, lepidocrocite, akaganeite, and several other polymorphs. They often contain excess water. Ferrihydrite $Fe_5HO_8 \cdot 4(H_2O)$ is typically considered a metastable iron oxide that can act as a precursor to the more stable iron oxides such as goethite and hematite [119]. Oxyhydroxides are normally obtained by precipitation from an aqueous solution. The particle size is controlled by initial iron concentration, organic additives, pH, and temperature.

α-FeOOH (Goethite)

Among the known oxide hydroxides $Fe_2O_3 \cdot H_2O$, the orthorhombic α-FeOOH (goethite) is antiferromagnetic in the bulk state and has $T_c = 393$ K [120]. Synthetic goethite nanoparticles are typically acicular and are often aggregated into bundles or rafts of oriented crystallites. β-FeOOH (akagenite) is paramagnetic at 300 K [121]. Akaganeite always has a significant surface area and some amount of excess water, which increases tremendously with the decreasing particle size. Recent studies of nanoakaganite show that at very high surface areas, where the particle size becomes comparable to a few unit cells, akaganeite may contain goethite-like structural features possibly related to the collapse of exposed tunnels.

γ-FeOOH (lipidocrokite) is paramagnetic at 300 K and δ-FeOOH (ferroxyhite) is ferromagnetic [122]. Although the bulk α-FeOOH is antiferromagnetic, in the form of nanoparticles it has a nonzero magnetic moment due to the incomplete compensation of the magnetic moments of the sublattices. Goethite nanoparticles have been studied by Mössbauer spectroscopy [123]. As a rule, α-FeOOH is present in iron nanoparticles as an admixture phase. Ferrihydrite is widespread but the nature of its extensive disorder is still controversial Because of chemical and structural variability in FeOOH containing nanoparticles, it is also critical to determine their chemical composition, including water content, surface area, and particle size.

Co oxides

Cubic cobalt oxide is antiferromagnetic and has $T_N = 291$ K. Cobalt monoxide has played an important role in the discovery of the "exchange shift" of the hysteresis curve, first found for samples consisting of oxidized Co nanoparticles [124]. Data on the dependence of T_N on the particle size were obtained in a study of CoO nanoparticles dispersed in a LiF matrix [125]. The particles obtained by vacuum deposition contained a small metal core, according to powder X-ray diffraction. As the particle size decreased from 3 to 2 nm, T_N decreased from 170 to 55 K. Apparently, the presence of an oxide layer on cobalt nanoparticles can markedly increase the coercive force. For example, the coercive forces (at 5 K) of monodisperse 6 and 13 nm oxidized Co particles obtained by plasma gas condensation in an installation for the investigation of molecular beams were $\sim$5 and 2.4 kG, respectively [126].

Unfortunately, the blocking temperature for 6 nm nanoparticles was lower than room temperature (~200 K); therefore, under normal conditions, their coercive force was equal to zero.

Co_3O_4

The Co_3O_4 nanoparticles (cubic spinel) with sizes of 15–19 nm dispersed in an amorphous silicon matrix exhibited ferrimagnetic properties at temperatures below 33 K (for bulk samples, $T_N = 30$ K) [127]. A method for controlled synthesis of Co_3O_4 cubic nanocrystallites (10–100 nm) has been developed [128].

NiO

Bulk crystals of NiO are antiferromagnetic, the Néel temperature being 523 K, but when the nanoparticles sizes are of the order of a few nanometers, they become superparamagnetic or superantiferromagnetic [129]. NiO possess not only magnetic but also electrical properties. The conductivity increases by 6–8 orders of magnitude in nanosized NiO as compared to that of bulk crystals, something that is attributed to the high density of defects [130]. It has been pointed out that electrodes composed of NiO nanoparticles exhibit a higher capacity and better cyclability than the ordinary ceramic material [131].

1.3.6
Final Remarks

We discussed above "free" nanoparticles as powders or suspensions in liquid media. In practice, magnetic nanoparticles are normally used as films (2D systems) or compact materials (3D systems). The compacting of magnetic nanoparticles even those having a protective coating on the surface often results in the loss of or substantial change in their unique physical (magnetic) characteristics. If the nanosized magnetic particles are retained after compaction, the materials based on them can serve as excellent initial components for the preparation of permanent magnets. A highly promising method of stabilization is the introduction of nanoparticles in different types of matrices. An optimal material should be a nonmagnetic dielectric matrix with single-domain magnetic nanoparticles with a narrow size distribution regularly arranged in the matrix. Various organic polymers are mostly used as these matrices. Encapsulation of magnetic nanoparticles makes them stable against oxidation, corrosion and spontaneous aggregation, which allows them to retain the single-domain structure. The magnetic particles coated by a protective shell or introduced in matrix can find application as the information recording media, for example, as magnetic toners in xerography, magnetic ink, contrasting agents for magnetic resonance images, ferrofluids and so on. The appropriate material has been given adequate consideration in the subsequent chapters.

References

1. P. Moriarty, *Rep. Prog. Phys.*, **2001**, 64, 297.
2. A.I. Gusev, A.A. Rampel, *Nanokristallicheskie Materialy (Nanocrystalline Materials)*, Moscow: Fizmatlit, **2001**.
3. (a) A.P. Alivisatos, P.F. Barbara, A.W. Castleman, J. Chang, D.A. Dixon, M.L. Klein, G.L. McLendon, J.S. Miller, M.A. Ratner, P.J. Rossky, S.I. Stupp, M.E. Thompson, *Adv. Mater.*, **1998**, *10*, 1297; (b) T. Sugimoto, *Monodispersed Particles*, Elsevier, **2001**; (c) "Nanostructured Materials; Selected Synthesis, Methods, Properties and Applications," Eds., P. Knauth and J. Schoonman, Kluwer, Dordrecht, **2004**; (d) J.P. Wilcoxon, B.L. Abrams, *Chem. Soc. Rev.*, **2006**, *35*, 1162; O. Masala, R. Sesadri, *Ann. Rev. Mater. Res.*, **2004**, *34*, 41.
4. (a) J.-T. Lue, *J. Phys. Chem. Solids.* **2001**, *62*, 1599; (b) J. Jortner, C.N.R. Rao, *Pure Appl. Chem.*, **2002**, *74*, 1491; (c) N.L. Rosi, C.A. Mirkin, *Chem. Rev.*, **2005**, *105*, 1547.
5. (a) V.E. Fertman, *Magnetic Fluids Guide-Book: Properties and Application*, Hemisphere, New York, **1990**; (b) B.M. Berkovsky, V.F. Medvedev, M.S. Krakov, *Magnetic Fluides: Engineering Applications*, Oxford University Press, Oxford, **1993**.
6. J.L. Kirschvink, A. Kirschvink-Kobayashi, B.J. Woodford, *Proc. Natl. Acad. Sci*, **1992**, *89*, 7683.
7. D.G. Rancourt, *Rev. Mineral. Geochem.*, **2001**, *44*, 217.
8. (a) S.P. Gubin, Yu.A. Koksharov, G.B. Khomutov, G.Yu. Yurkov, *Russian Chem. Rev.*, **2005**, *74*, 489; (b) An-Hui-Lu, E.L. Salabas, F. Schuth, *Angew. Chem. Int. Ed.*, **2007**, *46*, 1222; (c) S.P. Gubin, Yu.A. Koksharov, *Neorg. Mater.*, **2002**, *38*, 1287.
9. R. Ferrando, J. Jellinek, R.L. Johnston, *Chem. Rev,*, **2008**, *108*, 845.
10. S.P. Gubin, *Ros. Khim. Zh.*, **2000**, 44(6), 23.
11. (a) W.J Zhang, *Nanopart. Res.*, **2003**, *5*, 323; (b) D.L. Huber, *Small* **2005**, *1*, 482.
12. S. Gangopadhyay, G.C. Hadjipanayis, B. Dale, C.M. Sorensen, K.J. Klabunde, V. Papaefthymiou, A. Kostikas, *Phys. Rev.*, **1992**, *B 45*, 9778.
13. J.F. Loffler, J.P. Meier, B. Doudin, J.-P. Ansermet, W. Wagner, *Phys. Rev.*, **1998**, *B 57*, 2915.
14. W.A. de Heer, P. Milani, A. Chatelain, *Phys. Rev. Lett.*, **1990**, *65*, 488.
15. L. Guo, Q.J. Huang, X.Y. Li, S.H. Yang. *Phys. Chem. Chem. Phys.* **2001**, *3*, 1661.
16. K.S. Suslick, C. Seok-burn, A.A. Cichowlas, M.W. Grinstaff, *Nature* **1996**, *353*, 414.
17. S.M. Ponder, J.G. Darab, J. Bucher, D. Caulder, I. Craig, L. Davis, N. Edelstein, W. Lukens, H. Nitsche, L. Rao, D.K. Shuh, T.E. Mallouk, *Chem. Mater.*, **2001**, *13*, 479.
18. A.S. Dehlinger, J.F. Pierson, A. Roman, P.H. Bauer, *Surf. Coat. Technol.*, **2003**, *174*, 331.
19. L. Del Bianco, A. Hernando, E. Bonetti, E. Navarro, *Phys. Rev.*, **1997**, *B 56*, 8894.
20. U. Gonser, H.G. Wagner, *Hyperfine Interact.*, **1985**, *24–26*, 769.
21. N. Saegusa, M. Kusunoki, *Jpn. J. Appl. Phys.*, **1990**, *29*, 876.
22. T. Majima, T. Ishii, Y. Matsumoto, M. Takami, *J. Am. Chem. Soc.*, **1989**, *111*, 2417.
23. K. Haneda, Z.X. Zhou, A.H. Morrish, T. Majima, T. Miyahara, *Phys. Rev.*, **1992**, *B46*, 13832.
24. J. van Wonterghem, S. Morup, S.W. Charles, S. Wells, J. Villadsen, *Phys. Rev. Lett.*, **1985**, *55*, 410.
25. M.W. Grinstaff, M.B. Salamon, K.S. Suslick, *Phys. Rev.*, **1993**, *B48*, 269.
26. D.P. Dinega, M.G. Bawendi, *Angew. Chem. Int. Ed. Engl.* **1999**, *38*, 1788.
27. O. Kitakami et al.., *Phys. Rev.*, **1997**, *B 56*, 849.
28. M. Pileni, *Appl. Surf. Sci.*, **2001**, *171*, 1.

29. M.P. Pileni, *Langmuir*, **1997**, *13*, 3266.
30. L. Yiping, G.C. Hadjipanayis, V. Papaefthymiou, A. Kostikas, A. Simopoulos, C.M. Sorensen, K.J. Klabunde, *J. Magn. Magn. Mater.*, **1996**, *164*, 357.
31. D.L. Peng, K. Sumiyama, T.J. Konno, T. Hihara, and S. Yamamuro, *Phys. Rev.*, **1999**, *B 60*, 2093.
32. L. Bi, S. Li, Y. Zhang, D. Youvei, *J. Magn. Magn. Mater.* **2004**, *277*, 363.
33. O. Cıntora-Gonzalez, C. Estournes, M. Richard Pionet, J.L. Guille, *Mater. Sci. Eng., C* **2001**, *15*, 179.
34. W.N. Wang, I. Yoshifimi, I. Wuled-Lengorro, K. Okuyama, *Mater Sci. Eng., B* **2004**, *111*, 69.
35. A. Gavirin, C.L. Chen, *J. Appl. Phys.*, **1993**, *73*, 6949.
36. S. Doppiu, V. Langlais, J. Sort, S. Surinach, M.D. Baro', Y. Zhang, G. Hadjinapayis, J. Nogue's, *Chem. Mater.*, **2004**, *16*, 5664.
37. C. Estourne's, T. Lutz, J. Happich, T. Quaranta, P. Wissler, J.L. Guille, *J. Magn. Magn. Mater.*, **1997**, *173*, 83.
38. D. de Caro, J.S. Bradley, *Langmuir*, **1997**, *13*, 3067.
39. D.-H. Chen, S.-H. Wu, *Chem. Mater.*, **2000**, *12*, 1354.
40. K.L. Tsai, J. Dye., *Chem. Mater.*, **1993**, *5*, 540.
41. A.M. Tishin, Yu.I. Spichkin, *The Magnetocaloric Effect and Its Applications*, Institute of Physics: Bristol, Philadelphia, **2003**.
42. S. Thongchant, Y. Hasegawa, Y. Wada, S. Yanagia, *Chem. Lett.*, **2001**, *30*, 1274.
43. S. Thongchant, Y. Hasegawa, Y. Wada, S. Yanagida, *Chem. Lett.*, **2003**, *32*, 706.
44. J.A. Nelson, L.H. Bennet, M.J. Wagner, *J. Am. Chem. Soc.*, **2002**, *124*, 2979.
45. D. Johnson, P. Perera, M.J. O'Shea, *J. Appl. Phys.*, *79*, 5299.
46. M.J. O'Shea, P. Perera, *J. Appl. Phys.*, **1999**, *85*, 4322.
47. W.A. de Heer, P. Milani, A. Chatelain, *Phys. Rev. Lett.* **1990**, *65*, 488.
48. X.Q. Zhao, Y. Liang, Z.Q. Hu, B.X. Liu, *J. Appl. Phys.*, **1996**, *80*, 5857.
49. M.D. Bentzon, J. van Wonterghem, S. Morup, A. Tholen, C.J.W. Koch, *Philos. Mag.*, **1989**, *B 60*, 169.
50. C. Prados, M. Multigner, A. Hernando, J.C. Sanchez, A. Fernandez, C.F. Conde, A. Conde, *J. Appl. Phys.*, **1999**, *85*, 6118.
51. S. Sako, K. Ohshima, M. Sakai, S. Bandow, *Surf. Rev. Lett.*, **1996**, *3*, 109.
52. M. Respaud, J.M. Broto, H. Rakoto, A.R. Fert, L. Thomas, B. Barbara, M. Verelst, E. Snoeck, P. Lecante, A. Mosset, J. Osuna, T. Ould Ely, C. Amiens, B. Chaudret, *Phys. Rev.*, **1998**, *B 57*, 2925.
53. F. Bodker, S. Morup, S.W. Charles, S. Linderoth, *J. Magn. Magn. Mater.*, **1999**, *196–197*, 18.
54. H. Bonneman, W. Brijoux, R. Brinkmann, N. Matoussevitch, N. Waldoefner, N. Palina, H. Modrow, *Inorg. Chim. Acta*, **2003**, *350*, 617.
55. S. Gangopadhyay, G.C. Hadjipanayis, B. Dale, C.M. Sorensen, K.J. Klabunde, V. Papaefthymiou, A. Kostikas, *Phys. Rev.*, **1992**, *B 45*, 9778.
56. L. Del Bianco, A. Hernando, M. Multigner, C. Prados, J.C. Sanchez-Lopez, A. Fernandez, C.F. Conde, A. Conde, *J. Appl. Phys.*, **1998**, *84*, 2189.
57. H.Y. Bai, J.L. Luo, D. Jin, J.R. Sun, *J. Appl. Phys.*, **1996**, *79*, 361.
58. T. Sourmail, *Prog. Mater. Sci.*, **2005**, *50*, 816.
59. X.G. Li, T. Murai, T. Saito, S. Takahashi, *J. Magn. Magn. Mater.*, **1998**, *190*, 277.
60. B.L. Cushing, V.L. Kolesnichenko, C.J. O'Connor, *Chem. Rev.*, **2004**, *104*, 3893.
61. X.G. Li, A. Chiba, S. Takahashi, *J. Magn. Magn. Mater.*, **1997**, *170*, 339.
62. A.M. Afanas'ev, I.P. Suzdalev, M.Ya. Gen, V.I. Gol'danskii, V.P. Korneev, E.A. Manykin, *Zh. Eksp. Teor. Fiz.*, **1970**, *58*, 115.

63. B.K. Rao, S.R. de Debiaggi, P. Jena, *Phys. Rev.*, **2001**, *B 64*, 418.
64. S. Sun, H. Zeng, *J. Am. Chem. Soc.*, **2002**, *124*, 8204.
65. S. Sun, E.E. Fullerton, D. Weller, C.B. Murray, *IEEE Trans. Magn.*, **2001**, *37*, 1239.
66. T. Hyeon, *Chem. Commun.*, **2003** 927.
67. M. Chen, J.P. Liu, S. Sun, *J. Am. Chem. Soc.*, **2004**, *126*, 8394.
68. K. Elkins, D. Li, N. Poudyal, V. Nandwana, Z. Jin, K. Chen, J.P. Liu, *J. Phys. D: Appl. Phys.*, **2005**, *38*, 2306.
69. H. Kodama, S. Momose, T. Sugimoto, T. Uzumaki, A. Tanaka, *IEEE Trans. Mag.*, **2005**, *41*, 665.
70. B. Stahl, J. Ellrich, R. Theissmann, M. Ghafari, S. Bhattacharya, H. Hahn, N.S. Gajbhiye, D. Kramer, R.N. Viswanath, J. Weissmuller, H. Gleiter, *Phys. Rev. B* **2003**, *67*, 14422.
71. R.C. O'Handley, *Modern Magnetic Materials: Principle and Applications*, Wiley-Interscience: New York 434, **2000**.
72. S. Sun, *Adv. Mater.* **2006**, *18*, 393.
73. H. Zeng, J. Li, J.P. Liu, Z.L. Wang, S. Sun, *Nature*, **2002**, *420*, 395.
74. B. Warne, O.I. Kasyutich, E.L. Mayes, J.A.L. Wiggins, K.K.W. Wong, *IEEE Trans. Magn*, **2000**, *36*, 3009.
75. D. Weller, M.F. Doerner, *Annu. Rev. Mater. Sci.*, **2000**, *30*, 611.
76. S. Sun, C.B. Murray, D. Weller, L. Folks, A. Moser, *Science* **2000**, *287*, 1989.
77. E.V. Shevchenko, D.V. Talapin, H. Schnablegger, A. Kornowski, O. Festin, P. Svedlindh, M. Haase, H. Weller, *J. Am. Chem. Soc.*, **2003**, *125*, 9090.
78. J.-I. Park, J. Cheon, *J. Am. Chem. Soc.*, **2001**, *123*, 5743.
79. N.S. Sobal, U. Ebels, H. Mohwald, M. Giersig, *J. Phys. Chem.*, **2003**, *B 107*, 7351.
80. R.M. Cornell, U. Schwertmann, *The Iron Oxides: Structure, Properties, Reactions, Occurrences and Uses*, 2nd ed.; Wiley-VCH: Weinheim, **2003**.
81. P. Tartaj, M.P. Morales, S. Veintemillas-Verdaguer, T. Gonzalez-Carren, C.J. Serna, *J. Magn. Magn. Mater.* **2005**, *290–291*, 28.
82. C. Feldmann, H.-O. Jungk, *Angew. Chem., Int. Ed.* **2001**, *40*, 359.
83. R. Vijaya Kumar, Yu. Koltypin, Y.S. Cohen, Y. Cohen, D. Aurbach, O. Palchik, I. Felner, A. Gedanken, *J. Mater. Chem.* **2000**, *10*, 1125.
84. Z. Li, H. Chen, H. Bao, M. Gao, *Chem. Mater.* **2004**, *16*, 1391.
85. D.R. Lovley, *Microbiol. Rev.* **1991**, *55*, 259.
86. U. Schwertmann, E. Murad, *Clays Clay Miner.*, **1983**, *31*, 277.
87. M.F. Hansen, C.B. Koch, S. Morup, *Phys. Rev.*, **2000**, *B 62*, 1124.
88. L. Zhang, G.C. Papaefthymiou, J.Y. Ying, *J. Appl. Phys.*, **1997**, *81*, 6892.
89. Y.Y. Fu, R.M. Wang, J. Xu, J. Chen, Y. Yan, A.V. Narlikar, H. Zhang, *Chem. Phys. Lett.*, **2003**, *379*, 373.
90. T. Hyeon, S.S. Lee, J. Park, Y. Chung, H.B. Na, *J. Am. Chem. Soc.*, **2001**, *123*, 12798.
91. J. Tang, M. Myers, K.A. Bosnick, L.E. Brus, *J. Phys. Chem.*, **2003**, *B 107*, 7501.
92. J. Rockenberger, E.C. Scher, A.P. Alivisatos, *J. Am. Chem. Soc.*, **1999**, *121*, 11595.
93. R. Janot, D. Guerard, *J. Alloys Compd.*, **2002**, *333*, 302.
94. (a) Z. Jing, S. Wu, *Mater. Lett.*, **2004**, *58*, 3637; (b) X.G. Wen, S.H. Wang, Y. Ding, Z.L. Wang, S.H. Yang, *J. Phys.Chem. B*, **2005**, *109*, 215.
95. T. Fried, G. Shemer, G. Markovich, *Adv. Mater.*, **2001**, *13*, 1158.
96. I. Nedkov, T. Merodiiska, S. Kolev, K. Krezhov, D. Niarchos, E. Moraitakis, Y. Kusano, J. Takada, *Monatsh. Chem.*, **2002**, *133*, 823.
97. S. Sun, H. Zeng, *J. Am. Chem. Soc.*, **2002**, *124*, 8204.
98. X. Wang, J. Zhuang, O. Peng, Y. Li, *Nature*, **2005**, *437*, 121.
99. Y. Hou, J. Yu, S. Gao, *J. Mater. Chem.*, **2003**, *13*, 1983.
100. Yu.F. Krupyanskii, I.P. Suzdalev, *Zh. Eksp. Teor. Fiz.*, **1974**, *67*, 736.
101. R.N. Panda, N.S. Gajbhiye, G. Balaji, *J. Alloys Compd.*, **2001**, *326*, 50.

102. W.W. Yu, X. Peng, *Angew. Chem. Int. Edn.*, **2002**, *41*, 2368.
103. G. Benito, M.P. Morales, J. Requena, V. Raposo, M. Vazquez, J.S. Moya, *J. Magn. Magn. Mater*, **2001**, *234*, 65.
104. J. Ding, T. Tsuzuki, P.G. McCormick, *J. Magn. Magn. Mater.*, **1998**, *177–181*, 931.
105. Z.J. Zhang, Z.L. Wang, B.C. Chakoumakos, J.S. Yin, *J. Am. Chem. Soc.*, **1998**, *120*, 1800.
106. A. Vijayalakshimi, N.S. Gajbhiye, *J. Appl. Phys.*, **1998**, *83*, 400.
107. K.V.P.M. Shafi, A. Gedanken, *Nanostruct. Mater.*, **1999**, *12*, 29.
108. G. Mendoza-Suarez, J.C. Corral-Huacuz, M.E. Contreras-Garcia, H. Juarez-Medina, *J. Magn. Magn. Mater.*, **2001**, *234*, 73.
109. S. Li, V.T. John, S.H. Rachakonda, G.C. Irvin, G.L. McPherson, C.J. O'Connor, *J. Appl. Phys.*, **1999**, *85*, 5178.
110. G. Ji, S. Tang, B. Xu, B. Gu, Y. Du, *Chem. Phys. Lett.*, **2003**, *379*, 484.
111. C. Pham-Huu, N. Keller, C. Estournes, G. Ehret, J.M. Greneche, M.J. Ledoux, *Phys. Chem. Chem. Phys.*, **2003**, *5*, 3716.
112. C. Liu, A.J. Rondinone, Z.J. Zhang, *Pure Appl. Chem.*, **2000**, *72*, 37.
113. C.J. O'Connor, Y.S.L. Buisson, S. Li, S. Banerjee, R. Premchandran, T. Baumgartner, V.T. John, G.L. McPherson, J.A. Akkara, D.L. Kaplan, *J. Appl. Phys.*, **1997**, *81*, 4741.
114. S. Mathur, H. Shen, N. Lecerf, A. Kjekshus, H. Fjellvag, G.F. Goya, *Adv. Mater.*, **2002**, *14*, 1405.
115. T. Hyeon, Y. Chung, J. Park, S.S. Lee, Y.W Kim, B.H. Park, *J. Phys. Chem.*, **2002**, *B 106*, 6831.
116. N. Moumen, M.P. Pileni, *Chem. Mater.*, **1996**, *8*, 1128.
117. J. Ding, W.F. Miao, E. Pirault, R. Street, P.G. McCormick, *J Alloys Compd.*, **1998**, *161*, 199.
118. (a) L. Minervini, R.W. Grimes, *J. Phys. Chem. Solids*, **1999**, *60*, 235; (b) K. Tokumitsu, T. Nasu, *Scr. Metall.*, **2001**, *44*, 1421.
119. U. Schwertmann, J. Friedl, H. Stanjek, D.G. Schulze, *Clay Miner.* **2000**, *35*, 613.
120. C.J.W. Koch, M.B. Madsen, S. Morup, *Hyperfine Interact.*, **1986**, *28*, 549.
121. S. Morup, T.M. Meaz, C.B. Koch, H.C.B. Hansen, *Z. Phys.*, **1997**, *D 40*, 167.
122. T. Meaz, C.B. Koch, S. Morup, in *Proceedings of the Conference ICAME-95, Bologna*, **1996**, *50*, 525.
123. M.B. Madsen, S. Morup, *Hyperfine Interact.*, **1988**, *42*, 1059.
124. M. Kiwi, *J. Magn. Magn. Mater.*, **2001**, *234*, 584.
125. S. Sako, K. Ohshima, M. Sakai, S. Bandow, *Surf. Rev. Lett.*, **1996**, *3*, 109.
126. D.L. Peng, K. Sumiyama, T. Hihara, S. Yamamuro, T.J. Konno, *Phys. Rev.*, **2000**, *B 61*, 3103.
127. M. Sato, S. Kohiki, Y. Hayakawa, Y. Sonda, T. Babasaki, H. Deguchi, M. Mitome, *J. Appl. Phys.*, **2000**, *88*, 2771.
128. J. Feng, H.C. Zeng, *Chem. Mater.*, **2003**, *15*, 2829.
129. R.H. Kodama, *J. Magn. Magn. Mater.*, **2000**, *221*, 32.
130. V. Biju, A.M. Khadar, *J. Mater. Sci.*, **2003**, *38*, 4005.
131. Z. Fei-bao, Z. Ying-ke, L. Hu-liu, *Mater. Chem. Phys.*, **2004**, *83*, 60.

2
Synthesis of Nanoparticulate Magnetic Materials

Vladimir L. Kolesnichenko

2.1
What Makes Synthesis of Inorganic Nanoparticles Different from Bulk Materials?

Most magnetic materials used in today's technology are either metals or metal oxides. From a chemist's point of view, their preparation in bulk form is a simple task, however a bit more challenging aspect relates to phase purity, crystal structure and morphology which are responsible for better performance of these functional materials. The reduction of the crystal dimensionality to the nanometer scale brings a new degree of complexity to their synthesis. The area of science dealing with the development of magnetic nanoparticles, originates from solid-state inorganic chemistry and physics on one hand, and from the colloid and surface chemistry on the other.

Ferrofluids, the colloidal dispersions of magnetic iron oxides in hydrocarbon oil, represent probably the first application of magnetic materials in the new form. Their preparation was based on mechanical grinding of the bulk oxide in the presence of surfactant oleic acid and a hydrocarbon solvent. By that time, an alternative approach which originated from the discoveries of colloidal chemistry was introduced, according to which the single metal ion precursors were used in chemical reactions leading to their condensation and subsequent precipitation. This technique was later called "bottom-up," to indicate its fundamental difference from mechanical grinding, and was applied mostly to metal oxides, sulfides, and noble metals. As the theory of magnetism and the discovery of quantum confinement effect ignited a new wave of interest to nanocrystalline inorganic materials, the third approach to synthesis, based on the vapor condensation, was developed.

The simplicity of the grinding method may be considered an advantage, but it can be used only (a) for metal oxides because most metals are malleable and (b) for those areas of application, where particle morphology and phase purity are unimportant. The vapor phase condensation methods are superior in terms of phase purity; they also have an advantage when multilayer composites are to be prepared, but they are not competitive in the scaled preparations and when uniform particle morphology is necessary.

Magnetic Nanoparticles. Sergey P. Gubin
© 2009 WILEY-VCH Verlag GmbH & Co. KGaA, Weinheim
ISBN: 978-3-527-40790-3

Many areas of science and technology require nanoscale materials with narrow particle size distribution and with uniform particle shape. This can be achieved only by chemical synthesis in solution methodology. The key to success in the synthesis of uniform nanocrystals lies in the control over the kinetics of all steps of crystallization, beginning with nucleation and growth, and ending with coarsening and agglomeration.

The substances of interest for nanotechnology have long-range network of ionic, covalent or metallic bonds in their crystals. Since reactions of their synthesis and their crystallization take place simultaneously, the control over the kinetics of the chemical reaction ensures control of the rate of crystallization.

Temperature and concentration are not the only parameters used for controlling the rate of the reaction; the type of precursors and the mechanism of the reaction also play a major role. Many precursors in traditional solution precipitation reactions for the preparation of metals and metal oxides are ionic compounds. Solution reactions involving ions are usually fast, which complicates control over kinetics and morphology of the products. A great variety of ligands gives rise to metal complexes with different molecular and electronic structures, and therefore different stability constants, reactivity, and kinetics. Furthermore, using covalent metal-containing compounds (organometallic, etc.) which react and/or decompose only at elevated temperatures yielding metals or metal oxides, new chemical reactions with different mechanisms and different kinetics can be introduced.

Choosing correct precursors and maintaining accurate reagent ratios helps to control the elemental composition and the stoichiometry of the products, however it is not as easy to control their crystal structure. For example, the same metal or metal oxide can crystallize into different crystal structures, and moreover, sometimes the meta-stable kinetically favored products form instead of the regular thermodynamically more stable ones. There is evidence that crystal structure of metal and metal-oxide nanoparticles depends on the composition of the reaction solutions and the conditions of synthesis.

What all synthetic approaches have in common is that in order to stabilize the nanocrystals against agglomeration, special attention must be paid to the state of their surface. This problem can be addressed in solution rather than in the solid state, and it can be regarded as a problem of colloidal and surface chemistry. The immediate environment of each atom in the crystal lattice is well determined for the atoms located inside of the crystal; however, the atoms at the crystal face lack some of their neighbors. The presence of coordinatively unsaturated atoms on the surface causes its increased chemical reactivity. The "dangling bonds" turn into chemical bonds after the reactive molecules available from the medium become attached, or otherwise agglomeration takes place. Agglomeration of the nanocrystals due to their high surface energy is a very common and difficult problem to overcome. It is usually addressed by adding a capping ligand to the reaction solution, by doing synthesis in the surfactant micelles, or by providing conditions for the adsorption of the solvated ions, which causes the electrostatic repulsion of the charged surfaces (electric double layers).

All of the previously mentioned aspects of chemical synthesis in solution are usually addressed simultaneously. This makes the synthesis of nanocrystalline inorganic materials far more complex than someone might think by relating it to the preparation of the bulk metals and metal oxides. The purpose of this chapter is to illuminate this fascinating area of chemical synthesis (Figures 2.1 and 2.2). As an additional reading, recent reviews covering these and other topics with somewhat different accents can be recommended [1–15].

Figure 2.1 The reactor setup which allows injection of air-sensitive reagent solution.

Figure 2.2 The reactor setup which allows simultaneous injection of two reagent solutions. This method helps maintaining the same reagent ratio during mixing.

2.2
Synthesis of Magnetic Metal Nanoparticles

In principle, synthetic approaches developed for preparation of magnetic metal nanoparticles are not different from the approaches used for other metals.

Keeping in mind that the reader would benefit from seeing the subject in a broader scope, we included in this chapter the most important methods used for the synthesis of all metals. This can be easily justified by known efforts to synthesize nanoparticles of magnetic alloys containing nonmagnetic metals.

We offer classification of metal nanoparticles syntheses based on different mechanisms; namely, we discuss reduction methods separately from thermal decomposition of single precursor methods. We have to admit however that it is not possible to draw a sharp borderline between both types because many reduction reactions occur via formation of the intermediate complexes, which then decompose more or less rapidly. Our judgment on which category should a particular example belong to, will be based on criteria whether the reducing agent was actually used, or the metal-containing precursor was capable for decomposition yielding metal, under conditions of the experiment.

Production of metal nanoparticles with uniform size and shape is possible under conditions of accurate control over kinetics of their nucleation, growth, and coarsening. The best control is achieved when all the three steps are separated in time; namely, the nucleation must be finished by the time when the growth begins. It is easier to get to this level of control with homogeneous solution systems than for heterogeneous reactions, whose kinetics is largely controlled by the rate of diffusion of the reducing agent.

Selection of the solvent suitable for production of colloidal or nanocrystalline powdered metal is determined by the nature of metal, nature of precursors, peculiarities of surface composition, and colloidal properties. Variation of solvents most commonly used for synthesis of the nanoscale metals, resembles the list of solvents used in inorganic and organic synthesis, beginning from water and liquid ammonia, going through polar aprotic solvents, less polar alcohols, ethers, and ending with hydrocarbons. The same metal can be prepared in different solvents; this is especially true for metals which are not easily oxidized and do not interact with protic (water and alcohols) and other solvents. As a reference for determining metal activity, the redox potentials should be used with caution because these numbers vary greatly with changed concentrations, pH, complexing strength or solvent. Literature analysis indicates that the margin of applicability of water as a solvent, lies near iron; this means that metals less active than iron, can be easily prepared in aqueous (and any organic solvent) systems.

2.2.1
Reduction of Metal Salts in Solution

2.2.1.1 Electron Transfer Reduction

This section outlines the reducing agents which do not undergo structural rearrangement during a redox process (Table 2.1). This makes the apparent mechanism relatively simple and potentially reversible. Obviously this applies to metal salt versus metal as components of the half-cell; however, this

2 Synthesis of Nanoparticulate Magnetic Materials

Table 2.1 Reducing agents acting as electron donors.

Reducing agent	Precursor	Product nanoparticles	Conditions	References
Cathode	Pd, Pd(II)	Pd (1.4–4.8 nm)	MeCN/THF, N(oc)$_4$Br, room temperature	[16]
H$_2$	Fe[N(SiMe$_3$)$_2$]$_2$	Fe (7 nm cubes)	Mesitylene, 3 bars H$_2$ + RNH$_2$ + RCO$_2$H, 150 °C	[42]
Electride	AuCl$_3$	Au (6 nm)	Me$_2$O, −50 °C, K$^+$ (15-crown-5)$_2$e$^-$	[25]
Alkalide	DyCl$_3$	Dy 8–16 nm (1000 °C)	THF, −50 °C, K$^+$ (15-crown-5)$_2$Na$^-$	[27]
Na$^+$ C$_{10}$H$_8^-$	FeCl$_2$	Fe (5 nm)	NMPO, 20–200 °C	[21], [33]

question is more complex applied to reducing agent which is a multiatomic ion or molecule. The comprehensive answer would be provided by cyclic voltammetry studies, which are not available in most cases discussed here.

The electrochemical reduction methods should be mentioned first; they are well-developed and used for refining, electroplating, and/or bulk manufacturing of all metals. The challenge with this method, which is applied to production of colloidal metals and nanoparticles, is due to the great tendency of the reduced metal to deposit on the electrode surface. This problem was successfully solved for a large number of transition metals [16–18] by using electrolysis with sacrificial anode. Metal ions formed in solution, migrate to cathode, become reduced and form clusters stabilized by solvent or surfactant. Tetraalkylammonium salts appeared to be especially effective as inhibitors of precipitation of metals on the cathode surface and as colloid stabilizers. The particle size can be controlled in this method by regulating the current density: smaller particles form at a higher current density.

Anode: $M_{bulk} \longrightarrow M^{n+} + n\,e^-$

Cathode: $M^{n+} + n\,e^- + \text{surfactant/solvent} \longrightarrow M_{cluster}$

The best examples of the chemical reduction by direct electron transfer mechanism employ electropositive metals, low-valent metal cations, solvated electrons, and related reducing agents. Bulk metals as reducing agents work well for metal refining, especially for noble metals, which precipitate in a bulk form. In order to obtain metals in colloidal or nanocrystalline powder form, it is however advantageous to use a dissolved reducing agent; therefore solutions of sodium in ammonia or amines would be preferred over bulk metal or its amalgam [19]. Transmetallation reactions by using a colloid of active metal as a reducing agent for another easily reducible metal, offer sometimes favorable pseudohomogeneous conditions for coating the nanoparticles and protecting them from oxidation [20–23].

Low-valent metal salts are not commonly used as reducing agents probably because of difficulty with maintaining the desired ionic strength in solution and therefore colloid stabilization. The most practical example in a sense of availability and handling is a Sn(IV)/Sn(II) pair which in aqueous solution has $E° = 0.151$ V; the following pairs would be more efficient, however, due to their lower reduction potentials: Cr(III)/Cr(II) pair with $E° = -0.408$ V, and Ti(IV)/Ti(III) pair with $E° = -0.04$ V.

Solutions of alkaline metals in ammonia represent "clean," powerful, and rapidly acting reducing agents, which are readily available. The related solubilized reducing agents include electrides and alkalides containing a cryptand or crown ether-encapsulated Li^+, Na^+ or K^+ cation counterbalanced with solvated electron (electron trapped in a cavity of the cryptand or crown ether) or with sodide (Na^-) or potasside (K^-) anions. [24–28]. These salts are conveniently prepared in ether-type solvents at low temperature by a reaction of 1 or 2 equivalents of a complexing agent per equivalent of metal for alkalide and electride, respectively. These solutions represent most powerful reducing agents that are capable for reduction of even the group 3 metal cations.

Solutions of lithium or sodium and aromatic amines (pyridine, bipyridyls, and o-phenanthroline), hydrocarbons with extended π-system (naphthalene, biphenyl), or other compounds with conjugated π-system (benzophenone) in ethers or polar aprotic solvents, contain anion radicals derived from these aromatic compounds. These anion radicals have different reducing power, which is presumably determined by the energy splitting between HOMO and LUMO in their parent aromatic molecules. Usually this splitting is smaller in more extended delocalized systems (as in condensed aromatics). The mechanism of the reduction reactions with these anion radicals can be different for main group metals or early transition metals and for the middle and late transition metals, involving either electron transfer, or formation of more or less stable intermediate low-valent metal complexes (see Section 2.2.2).

The anion radicals as reducing agents are usually prepared in ether-type solvents (diethyl ether, tetrahydrofuran, glyme, diglyme, etc.) [29–32]. This brings a known limitation to their applicability because metal salts are usually insoluble in these low-polar solvents. It has been found recently [33] that sodium naphthalide can be prepared and also used in N-methyl pyrrolidone (NMPO). Without naphthalene, sodium slowly reacts with this solvent and decomposes it; sodium naphthalide however appears to be stable at room temperature for at least several hours. NMPO is one of the "supersolvents" whose high dielectric constant $\varepsilon = 32.6$ and capability to solvate ions (Donor Number $= 27.3$) enable it to dissolve simple metal salts and to promote their dissociation. The instantaneous chemical reactions take place when sodium naphthalide solution is added to solutions of anhydrous Fe(II), Co(II), and Ni(II) chlorides at room temperature, as their color changed to dark brown. However, three metals behave differently when their solutions are heated at

190–200 °C: Fe appears as a colloid with uniform 5 nm nanoparticles; Co agglomerates easily; and Ni metal does not form under the same conditions, but Ni_3C precipitates instead. The difference between iron and cobalt is most likely attributed to the difference in the efficiency of solvation of their coordinatively unsaturated surface atoms. Explanation for different behavior of nickel would involve its intermediate arene complex whose structure and reactivity differ from the corresponding Fe and Co complexes. None of these intermediates have been isolated, however as an evidence for the validity of this hypothesis, it was found that iron colloid produced by this method can be used as a reducing agent for $NiCl_2$ solution, and the colloidal nickel is produced.

Hydrogen is used for preparation of nanoparticles and giant clusters of easily reducible metals, such as Pd, Pt, Ag, and Au [34–41]. The reactions can be performed either in aqueous or in nonaqueous solutions and the limitation of the applicability of this reducing agent to other metals is determined mainly by its reducing strength. An interesting example of using hydrogen for preparation of such an active metal as iron was reported by Chaudret et al. [42]. It was accomplished by reduction of iron(II) amide:

$$Fe[N(SiMe_3)_2]_2 + H_2 \longrightarrow Fe + 2HN(SiMe_3)_2.$$

The technique employed here relates to hydrogenation of metal alkene and arene complexes in a solution of nonpolar solvents (Section 2.2.2.2). The extended-time (48 h) hydrogen reduction of iron amide precursor was performed in a solution of hydrocarbon mesitylene at the temperature of 150 °C and at a presence of hexadecylamine and long-chain carboxylic acid. Variation of the length of the chain of amine and acid ligands did not influence the morphology of the product, and cube-shaped *bcc* particles with 7 nm edges formed. The particles assemble in superlattices with their crystallographic axes aligned and with the interparticle distance (1.6–2 nm) shorter than the ligand (oleic acid) chain length (~2.2 nm). This can be explained by the presence of the original amido ligands or hexamethyldisilazane as spacers between nanocrystals.

2.2.1.2 Reduction via Intermediate Complexes

A category of reducing agents discussed here contain multiatomic ions or molecules, which can act as good ligands (Table 2.2). The mechanism of the redox reactions considered here suggests the formation of the intermediate complexes, which decompose more or less rapidly under conditions of the experiment. Major restructuring of the reducing agent during this complex decomposition, makes these redox reactions irreversible. Again as in the previous category, it is difficult to determine sometimes whether the intermediate complex actually forms.

Complex hydrides (represented mainly by B and Al) are the best known examples of reducing agents, which act mostly via formation of the

Table 2.2 Coordinating reducing agents.

Representative reducing agent	Precursor	Product	Conditions	References
$NaBH_4$	$CoCl_2 \cdot 6H_2O$	Co 4.4 or 5.7 nm	Micelles with DDAB surfactant	[45]
$LiBHMe_3$	$CoCl_2$	ε-Co (2–11 nm)	Injection @ 200 °C, HOol, PR_3 in Oc_2O	[48]
N_2H_4	$Co(Oac)_2$	Co (20 nm)	Aqueous solution	[59]
H_3NOHCl	$PtCl_6^{2-}$	Au@Pt, Au@Pd	Aqueous solution	[52]
AlR_3	$Ni(acac)_2$	NiAl (colloid → solid)	Toluene, 80 °C, H_2	[54]
Citric acid	$AuCl_4^-$	Au (variable size)	Aqueous solution	[51], [52]
Alcohols	$Co(Oac)_2$	hcp-Co (6–8 nm)	Ph_2O, HOol, TOP, diol, 250 °C	[49]
Dimethyl-formamide	$AgNO_3$	Ag (6–20 nm)	DMF solvent, $Me_3Si(CH_2)_3NH_2$ stabilizer	[60]
Tartaric acid	Ag, Au, Pd, Pt salts	Ag, Au, Pd, Pt	Water solvent, variable pH, and stabilizer	[61]
Ascorbic acid	$AgNO_3$	Ag (15–26 nm)	Na sulfonate polymer MW 8000 stabilizer	[62]

intermediate hydrido complexes. The intermediate products of the reduction were isolated and structurally characterized [43, 44] as Co(I) derivatives with π-acceptor ligands according to the reactions:

$$[Co(terpy)Cl_2] + NaBH_4 \longrightarrow [Co(terpy)BH_4]$$

$$Co(H_2O)_6Cl_2 + P(ch)_3 + NaBH_4 \longrightarrow \{CoH(BH_4)[P(ch)_3]_2\}.$$

The most commonly used are $NaBH_4$, NR_4BH_4, $LiAlH_4$, $LiBHMe_3$ called "superhydride", and NR_4BHMe_3. Sodium tetrahydroborate (frequently called sodium borohydride) is very practical in terms of easy handling and low cost. It is fairly stable in dry air and soluble in water yielding solutions which are surprisingly stable. They cannot be stored, however freshly prepared at room temperature and in the absence of acid, they evolve hydrogen very slowly. To the contrast, attempts to dissolve $NaBH_4$ in methanol result in its rapid decomposition. This reducing agent found its wide application for preparation of metal nanoparticles due to its convenience, however metallic products obtained from aqueous solutions, are usually contaminated with boron (as an element and as a metal boride) [45, 46]. The $NaBH_4$ reduction of aryldiazonium tetrachloroaurate in the heterogeneous water/toluene system resulted in the formation of the toluene-soluble gold nanoparticles stabilized by covalently bound (Au–C) aryl groups [47].

$$(ArN_2)^+ AuCl_4^- + NaBH_4 \longrightarrow Au_n Ar_m + N_2.$$

As the authors stated, such bonding provides an enhanced stability to the nanoparticles.

Lithium trimethyl hydroborate acts as a more clean reducing agent and it is easily soluble in aprotic organic solvents and therefore it can be used under a broader range of conditions [48, 49]. According to Murray et al. the reduction of cobalt and nickel chlorides or acetates with $LiBEt_3H$ is performed in a high-boiling solvent (dioctyl or diphenyl ether) by injection of the precursors in the preheated solvent. This method involved using of two types of the capping ligands: strongly bound (carboxylic acid) and weakly bound (trialkyl phosphine and phosphine oxide). Changing their ratio as well as their chain length, allowed tuning of the nanocrystal sizes with tight size distributions ($\sigma = 7 - 10\%$).

Tetraalkylammonium salts also have advantage of being applicable in nonaqueous systems. The following reaction performed in THF solution at room temperature was used for the preparation of tetraoctylammonium chloride stabilized 2.5 nm cobalt nanoparticles [50].

$$CoCl_2 + 2(NR_4)BEt_3H \longrightarrow Co_{coll} + 2NR_4Cl + 2BEt_3 + H_2.$$

The structure of the obtained particles was studied by X-ray absorption spectroscopy, and it was determined that the nanoparticle core is surrounded by chloride ions and the external coating consists of tetraalkylammonium cations. As the length of the alkyl chain in NR_4^+ is varied from butyl to octyl, the interatomic distances Co–Cl and Co to N, change too. In the case of the longer chains, the Co to N distance increases for steric reason, and therefore positive centers of the coating move away from the surface of the metal core. The electrostatic attraction of NR_4^+ to Cl^- causes Co–Cl distance to become longer as well, than with the shorter-chain alkyl groups. The optimal size of the NR_4^+ cation is when R = octyl; the larger cations pull chloride ion from the surface too far, thus causing destabilizing effect. The smaller cations also fail to provide stabilizing effect probably for steric reason.

The reducing power of tetrahydroborate ion is not as strong as for most reagents from the previous category, however it was (and it is) successfully employed for the preparation of many metals, even as active as iron. Its reduction potentials in aqueous solutions are -1.24 and -0.481 V, at basic and acidic pH, respectively:

$$H_2BO_3^- + 5H_2O + 8e^- \longrightarrow BH_4^- + 8OH^-$$

$$B(OH)_3 + 7H^+ + 8e^- \longrightarrow BH_4^- + 3H_2O.$$

Hydrazine is a well-known reducing agent for metal cations having relatively high reduction potentials, due to its suitable aqueous reduction potentials in acidic (-0.23 V) and basic (-1.15 V) solutions:

$$N_2 + 5H^+ + 4e^- \longrightarrow N_2H_5^+$$
$$N_2 + 4H_2O + 4e^- \longrightarrow N_2H_4 + 4OH^-.$$

Compared to hydroborates, it acts slower and at a higher temperature. In addition to noble metals, it is used for reduction of Ag(I), Cu(II), Ni(II), and Co(II) ions into free metals in the form of nanoparticles. Evidently, these reactions begin from formation of complexes with hydrazine acting as a ligand. The extensive coordination chemistry of hydrazine is an evidence for this.

Alcohols and hydroxyacids in aqueous solutions do not exhibit reducing property significantly and the values of their reduction potentials are unknown. They associate with metal ions however; alcohols produce solvates or alkoxides, and hydroxy-carboxylic acids (such as citric, tartaric, etc.) produce chelated complexes. These associates are thermally unstable, especially with stronger oxidizing ions (Au(III), Pt(II)), and decompose easily in aqueous solutions, yielding colloidal metals. Reduction of gold(III) with citrate is a classical example of a homogeneous solution colloid synthesis developed and published in 1951 by Turkevich et al. [51]. The adducts with less oxidizing metal ions such as Co(II) and Ni(II) would decompose yielding metals at higher temperature in nonaqueous solutions. 1,2-diols appear to be more efficient as reducing agents, solvents and as colloid stabilizing media than regular alcohols because they (a) have higher boiling points and therefore offer more harsh reaction conditions necessary for preparation of more active metals; (b) form stronger associates with metal ions due to chelating; (c) have higher dielectric constants and therefore dissolve a wider variety of precursors; (d) higher dielectric constant and chelating ability assist with the colloid stabilization because the surface of nanoparticles is more efficiently solvated and because this promotes the electric double-layer formation around the nanoparticles; and (e) stronger association with metal ions permits a better control over the crystal growth and therefore improves the quality of the products in terms of particle uniformity and a reduced degree of agglomeration. Ruthenium nanoparticles were prepared by reduction of $RuCl_3$ in various polyols; the polyols served as the reducing agents and the solvents, and the acetate ion served as a stabilizing agent [53]. The particle size in the range of 1–6 nm was restricted by choice of reduction temperature; smaller particles formed at a higher temperature. Under each set of conditions, the product formed with very narrow size distribution. The monodisperse hcp-Co nanoparticles were synthesized by reduction of cobalt oleate (generated in situ from cobalt acetate and oleic acid) in solution of phenyl or octyl ether with 1,2-dodecanediol, at a presence of trioctylphosphine [49]. The reducing agent was injected in the preheated to 250°C solution of precursors, and this temperature is maintained for 15–20 min. Size of the particles was tuned by tailoring the concentration and composition of stabilizers: increasing the concentration of TOP and oleic acid reduced particle size to

3–6 nm; using tributylphosphine instead of TOP increases the particle size to 10–13 nm.

Aluminum alkyls are capable for the reduction of transition metal salts, and for stabilization of metallic colloids in nonaqueous solutions:

$$MX_n + n\, AlR_3 \longrightarrow M + n\, R_2AlX + 1/2n\, R_2,$$

where M = metals of groups 6–11 and X = acetylacetonate. These reactions are presumably promoted by the high affinity of Al to oxygen. Nanoparticles of metal remain surrounded by organoaluminum species enabling the control of their growth [1, 54–58]. Slow oxidation and derivatization of these core–shell species permit the tailoring of their solubility and colloidal stability in different media. The techniques involving organoaluminum reagents have recently been extended by the same authors to include the metal carbonyl decomposition route (Section 2.2.2.1).

2.2.2
Thermal Decomposition Reactions

Some reactions discussed in Section 2.2.1.1 are likely to go through the formation of the intermediate complexes, which then decompose more or less rapidly yielding metals. The rate of decomposition of these complexes varies and one may succeed with their isolation and characterization. Such complexes were isolated and structurally characterized for some metals, in the studies related to coordination and organometallic chemistry, but not to nanotechnology. Aromatic amines usually act in them as two-electron σ-donor with π-back bonding [63], and aromatic hydrocarbons form sandwich-type complexes [64]. Although synthesis of the nanoparticles proceeding through formation and decomposition of such intermediates qualify for the present section, we will discuss only methods where such potential intermediates are actually used as the reagents. This approach helped to shape this section as dealing only with metal carbonyls and alkene/arene complexes.

2.2.2.1 Decomposition of Metal Carbonyls

Metal carbonyls contain a zero-valent metal and therefore do not require any reducing agent assisting formation of free metal. They decompose at moderate temperatures yielding pure metal and carbon monoxide, and this reaction is widely used for metal coating and other processes. This unique property of metal carbonyls appeared to be the basis for newly developed methods of preparation of metal nanoparticles, especially for the synthesis of iron and cobalt metals and some ferromagnetic alloys. In order to maintain the control over metal crystal growth, the decomposition is performed in solutions of solvents with low polarity (metal carbonyls are nonpolar) and with high boiling point, at high temperature.

Control over the formation of nuclei and their subsequent growth into nanocrystals of metal requires assistance of the reagent specifically interacting with "hot" metal atoms and the medium. We will focus at three representative types here, namely polymers with nucleophilic functional groups, capping ligands, and organoaluminum agents. In general, polymers affect crystal growth of any inorganic substance, as they physically wrap growing crystals, slow down their further growth, and inhibit their agglomeration. Polymers capable of chemical interaction with metal centers influence all these steps more efficiently. Comparative studies of the influence of styrene and copolymers of styrene with N-vinylpyrrolidone or with γ-vinylpyridine showed that both copolymers acted as the nucleation centers (at their donor functional groups), catalyzed the decomposition and influenced kinetics of the crystal growth. Changing the degree of functionality and changing the reagent concentration allows control over final nanoparticle size [65]. The high-temperature thermolysis (200–260 °C) of $Fe(CO)_5$ in molten noncoordinating polymers revealed strong relationship between structure of polymer and nanoparticles [66, 67]. High-density polyethylene represents the extreme example of polar group-free polymer. The EXAFS and Mössbauer studies on iron-containing nanoparticles dispersed in polyethylene matrix indicated the direct contact between surface metal atoms and carbon, which is probably attributed to partial cleavage of the polymeric chains. Similar result was reported by the same authors for cobalt but not for copper nanoparticles obtained by thermolysis of their carboxylates dispersed in polyethylene matrix [68, 69].

Capping ligands containing functional groups with variable donor properties and variable size of their side substituent appeared to be in focus of many research efforts. Capping ligands further facilitated control over kinetics of nucleation, growth. and coarsening of metal nanoparticles, inhibit their agglomeration and influence the nanopowder solubility. Specific capping ligand-to-metal interaction sometimes leads to the formation of the unusual crystalline forms of this metal. A new crystalline phase of the metallic cobalt (ε-Co) related to the beta phase of manganese appeared as a kinetically controlled product of the thermal decomposition of $Co_2(CO)_8$ in hot toluene in the presence of trioctylphosphine oxide (TOPO); the same reaction performed in the absence of TOPO resulted in the formation of a pure *fcc* Co phase [70]. This new crystalline form of cobalt is metastable with respect to the *fcc* Co phase. The multiply twinned *fcc* Co nanoparticles formed by higher temperature "burst nucleation" followed by slower growth process [49]. Synthesis performed by injection of $Co_2(CO)_8$ solution into preheated phenyl ether solution of oleic acid and tributylphosphine, resulted in the formation of monodisperse 7–10 nm spherical nanoparticles of cobalt.

Simultaneous presence of two or three capping ligands that differently interact with metal permits a selective control over kinetics of crystal growth in different crystallographic directions. A study of the decomposition of $Co_2(CO)_8$ under a wide range of conditions revealed the stepwise character of the crystal growth [71, 72]. Rapid injection of cobalt carbonyl solution into a hot o-dichlorobenzene solution containing both the labile ligand TOPO and the stronger ligand oleic acid, followed by quenching of the reaction, resulted in the formation of different products, depending on composition, time, and temperature. The initial product is *hcp*-Co with a nanodisk shape with dimensions that are tunable from 4 × 25 to 4 × 75 nm. At a fixed concentration of oleic acid, the length of the nanodisks was directly proportional to the concentration of TOPO. As time progressed, the nanodisks dissolved and spherical 8 nm nanocrystals of ε-Co with a tight size distribution appeared within a few minutes as a final product. When the decomposition of $Co_2(CO)_8$ was performed in the presence of the long-chain aliphatic amines instead of, or in addition to, TOPO, the lifetime of the nanodisks in solution increased, which facilitated their preparation. Varying the reaction time and [capping ligand]/[precursor] ratio resulted in the synthesis of *hcp*-Co nanodisks with sizes ranging from 2 × 4 nm to 4 × 90 nm, although the narrowest size distribution was obtained for medium-sized (4 × 35 nm) disks.

Trialkylaluminum reagents appeared to be efficient stabilizing agents in the metal carbonyl decomposition method. The obtained 10 nm Co nanoparticles had a narrow size distribution (±1.1 nm) and they were surrounded by organoaluminum species. Slow air oxidation and peptization with suitable surfactants lead to air-stable magnetic fluids [73]. The XANES studies have proved the formation of zero-valent magnetic cobalt particles.

2.2.2.2 Decomposition of Metal Alkene and Arene Complexes

As it was mentioned in Section 2.2.1.1, reduction of metal salts with anion radicals derived from aromatic hydrocarbons goes likely through the formation of the intermediate π-bonded complexes. In this section, we will discuss examples of using these types of complexes as precursors in metal nanoparticles synthesis. As in the case of metal carbonyls, metal alkene and arene complexes contain a low-valent metal, so the assistance of the reducing agent is not necessary. Hydrogen gas is usually used, however that causes hydrogenation of alkene to alkane, which does not coordinate to metal.

The most commonly used ligands include 1,5-cyclooctadiene (COD), 1,3,5-cyclooctatriene (COT), and dibenzylidene acetone (DBA), π-allyl ligands, such as cyclooctenyl anion ($C_8H_{13}^-$) and cyclopentadienyl anion (Cp^-). This method was extensively used for the synthesis of Co, Ni, Ru, Pd, Pt, CoPt, CoRu, CoRh, and RuPt nanoparticles and Co and Ni nanorods [74–85] and trigonal particles [86]. Solution syntheses were typically performed in the presence of H_2 or CO and the stabilizing agent polyvinylpyrrolidone (PVP) at room or slightly elevated temperature, yielding 1–2 nm particles with clean surfaces. Magnetic nanoparticles obtained by this method exhibit magnetic properties analogous to those of clusters prepared in high vacuum.

The cyclopentadienyl complex of indium has been used for the synthesis of 15 nm uniform particles of In metal protected with PVP or other capping ligands [87, 88]. Performing this reaction in the presence of long-chain amines yielded the nanowires of In and In_3Sn alloy with high aspect ratios [89]. Similarly, colloidal 5 nm Cu has been obtained from the organometallic complex $CpCu(^tBuNC)$ in the presence of CO and PVP (3.5 nm) or polydimethylphenylene oxide (5 nm) in solutions of CH_2Cl_2 or anisole at room temperature [90].

2.2.3
Combination Methods Used for Synthesis of Alloy Nanoparticles

This section outlines the chemistry related exclusively to magnetic bimetallic nanoparticles containing iron or cobalt and platinum, and their core/shell structures. Since chemical properties of platinum compounds are very different from properties of iron and cobalt compounds, it is difficult to find similar precursors or similar reactions occurring with the same rate for both metals. It is necessary however to maintain similar kinetics for both metals, to ensure uniformity of composition of all nanoparticles internally and in a bulk. For this reason, the reaction systems presented here contain metal precursors of a different type, and the reactions occurring in these systems are of different types either.

Thermal decomposition of $Fe(CO)_5$ combined with the diol reduction of $Pt(acac)_2$ in the same pot resulted in the formation of intermetallic FePt nanoparticles [91]. Their composition was controllable by adjusting the molar ratio of Fe and Pt precursors. Particle size could be tuned by growing seed particles followed by addition of new portion of the reagents to enlarge them in the range of 3 to 10 nm.

The reaction between $Co_2(CO)_8$ and $Pt(hfacac)_2$ (hfacac = hexafluoroacetylacetonate) is interesting in a sense that cobalt carbonyl acted as a reducing agent for platinum salt. It resulted in the formation of bimetallic $CoPt_3$ and CoPt nanoparticles [20]. This reaction was performed under reflux in a toluene solution containing oleic acid as a stabilizing agent:

$$Co_2(CO)_8 + Pt(hfacac)_2 \longrightarrow CoPt + Co(hfacac)_2 + 8CO.$$

In another system, $Pt(hfacac)_2$ was reduced by presynthesized Co nanoparticles. Partially sacrificed cobalt particles appeared to be coated with platinum, and this layer provided a sufficient stabilization against oxidation.

$$2Co + Pt(hfacac)_2 \longrightarrow Co/Pt + Co(hfacac)_2.$$

The resulting 6 nm Co@Pt nanoparticles were coated with dodecyl isocyanide capping ligands, which were air stable and dispersible in nonpolar solvents.

2.3
Synthesis of Magnetic Metal Oxide Nanoparticles

Synthetic methods used for the preparation of ferrimagnetic metal oxides are essentially the same as general methods for preparation of all metal oxides. For this reason while keeping our primary focus at magnetic metal oxides, we will cover the subject in a broader scope so that a better sense of this topic can be attained. Two types of ferrimagnetic oxides of commercial interest are inversed: spinel-structured iron oxides and ferrites, and hexagonal ferrites. Compared to the synthesis of metallic nanoparticles, preparation of metal oxides presents one degree of relief but at least two additional challenges. Most of magnetic metal oxides are not air sensitive, which facilitates their processing. One has to keep in mind, however, that highly demanded magnetite (Fe_3O_4) oxidizes by oxygen into maghemite (γ-Fe_2O_3). The challenges are the uniformity of composition for ternary oxides and the type and the quality of their crystal structure. Magnetic properties appear to be very sensitive to composition, crystallinity, and even the morphology of metal-oxide nanoparticles. For this reason, a chemist working on the development of synthetic strategy meets a lot of challenges associated not only with doing the right chemistry, but also with its optimization.

Most of the reported syntheses for the nanoparticles of metal oxides belong to one of the following reaction types, (a) hydrolysis, (b) oxidation, and (c) thermal decomposition of the oxygen-rich precursors. There is no straight answer to the question of what reaction type is better, however; it is clear that conditions for performing the reaction of choice must assure control over all steps of the crystal growth. Similarly to syntheses of other nanoparticles, the most beneficial methods addressing this problem are based on solution technique.

Traditional solid-state syntheses will not be covered here since they cannot be used for nanoparticle preparation. Hexaferrites fall under this category because they form typically at temperatures much higher than any conventional solvent can tolerate.

The area of single-molecule magnets will not be covered here, however the review paper focusing at iron(III) oxo clusters, which relate to iron oxide nanoparticles, is recommended for further reading [92].

2.3.1
Reactions of Hydrolysis

2.3.1.1 Hydrolysis in Aqueous Solutions
All transition metal ions undergo hydrolysis in aqueous solutions:

$$M(H_2O)_6^{2+} + H_2O \rightleftharpoons M(OH)(H_2O)_5^+ + H_3O^+. \quad (2.1)$$

How far this equilibrium goes to the right can be determined from the corresponding metal hydroxo complex formation constant. In a practical sense, in order to initiate condensation of hydroxo complexes into the polynuclear clusters, the acid formed as a byproduct of hydrolysis, must be neutralized. These clusters act as seeds for further condensation until they grow large enough for precipitation. Most of transition metals precipitate from aqueous solutions as hydrated oxides. Subsequent calcination helps to convert them into the crystalline oxides, however agglomeration is unavoidable. Further details on condensation mechanisms and formation of iron oxides can be found in the review papers [93, 94].

Precipitation of thermodynamically favorable phases helps to obtain a better-crystallized anhydrous oxide even from diluted aqueous solutions. The best example is the formation of spinel-structured ternary oxides by coprecipitation of M(III) with M(II) present in reaction solution in 2:1 molar ratio. Fe_3O_4, for example, has been prepared by a simple coprecipitation of ($Fe^{2+} + 2Fe^{3+}$) with NaOH at temperatures above 70 °C [95]; 5–25 nm particles of $MnFe_2O_4$ were similarly prepared from aqueous Mn^{2+} and Fe^{2+} at temperatures up to 100 °C [96]. Variation in size and shape of the nanoparticles was observed under conditions of strict control of acidity and ionic strength in aqueous solutions containing no complexing agents [97, 98]. These variables influence the electrostatic surface charge density of the nanoparticles, the interfacial tension, and consequently their surface energy.

Nanocrystalline $NiFe_2O_4$, $CuFe_2O_4$ and $ZnFe_2O_4$ were synthesized by rapid pouring of aqueous solutions of metal salts into preheated to 100 °C 1 M NaOH solution under vigorous stirring [99]. The obtained solids were used for the preparation of electric double-layer stabilized aqueous colloids. Nearly monodisperse 8.5 nm particles of Fe_3O_4 were obtained by using a different solution mixing mode [100]. Aqueous solution containing $Fe^{2+} + 2Fe^{3+}$ was added dropwise into 1.5 M NaOH solution under vigorous stirring at ambient temperature.

The products obtained by the above methods did not have to be annealed at a higher temperature, which minimized their agglomeration. This is especially useful for the preparation of their colloids, which is done by careful acidification of the precipitated oxides causing the adsorption of H^+ at the particle surfaces. Excessive positive charge at the surface of the nanoparticles provides the electrostatic repulsion and stabilizes the colloid [99–101].

A systematic study on precipitation of $CoFe_2O_4$ resulted in determining the influence of reaction temperature, reactant concentration, and reactant addition rate on the size of the products [102]. In each case, aqueous solutions of Fe^{3+} and Co^{2+} were precipitated with dilute NaOH. The results indicated that increasing the temperature from 70 to 98 °C increased the average particle size from 14 to 18 nm. Increasing the NaOH concentration from 0.73 to 1.13 M increased the particle size from 16 to 19 nm, and slowing the NaOH addition rate appeared to broaden the particle size distribution.

Small nanocrystals of magnetite (hydrodynamic radius ∼5.7 nm) were obtained by continuous mixing of 0.1M metal ion solution (Fe(II)/Fe(III) in 1:2 ratio) with 1.0 M NH_3(aq) at room temperature followed by passing the resulted solution through a heated at 75 °C bath for 1 min [103]. In the subsequent steps, the nanoparticles were stabilized with hydrophilic polymer, chelating aminoalcohol, tris(hydroxymethyl) aminomethane, or tetramethylammonium hydroxide.

Aiming the preparation of the MRI contrast agent precursor, magnetite nanoparticles were precipitated using tetramethylammonium hydroxide or its mixtures with aqueous ammonia [104]. The inner and the outer-sphere relaxivities appeared to be dependent on the particle size, which varied as the conditions of precipitation were changed. The same authors used this method for extensive nanoparticle surface chemistry studies [105]. The small-molecule ligands used in this study included carboxylic, sulfonic, phosphonic, and phosphoric acid functional groups. The strongest binding was determined for bifunctional molecules due to a cooperative effect.

Precipitation from basic aqueous solutions was applied for the preparation of magnetite particles, which in subsequent steps were coated with polymers (starch or methoxypolyethylene glycol) and oleate ligands [106].

The rate of reaction and crystal growth can be elegantly controlled by using the reagent, which slowly hydrolyses in solution yielding a base. Pr^{3+}-doped CeO_2 in the form of monodispersed 13 nm particles was obtained by heating aqueous solutions of $Ce(NO_3)_3$, $PrCl_3$ and hexamethylenetetramine at 100 °C [107]. Urotropin in this case acted as a precursor of slowly generated base and as a stabilizer.

$$(CH_2)_6N_4 + 6H_2O \rightleftharpoons 6CH_2O + 4NH_3.$$

The experimental setup for aqueous precipitation reactions is relatively simple, which makes this method attractive. It is common, however, that the nanoparticles form with rather broad size distribution. This is attributed to very high rate of precipitation and a very low solubility of metal oxides in water. These problems are partially solved by strict control of the reaction conditions, particularly the rate of mixing of the reagent solutions.

2.3.1.2 Hydrolysis in Nonaqueous Solutions

Reactions of hydrolysis suggest using water as a reagent but do they suggest using it as a solvent as well? It is convenient to assign water its dual role, especially because of its exceptional solvent properties. However, water is a chemically active solvent and its involvement in every step of all chemical reactions is unavoidable. It is not only that hydrolysis in aqueous solution is fast because water is present in excess, but also because of its participation in the acid–base equilibrium in the solution and in the coordination sphere of the metal ion. Due to high charge density of the transition metal ions, the hydrogens in the coordinated water appear to be more positively charged than in free water, and therefore more acidic. This is illustrated by the reaction (2.1) and it gives a driving force for further hydrolysis. The just-formed hydroxo ligands have great tendency for bridging and the binuclear complexes form in solution, as it is shown in reaction (2.2). Furthermore, bridging hydroxo ligands are likely to be more acidic than terminal.

$$2M(OH)(H_2O)_5^{2+} \rightleftharpoons [(H_2O)_4M(\mu\text{-}OH)_2M(H_2O)_4]^{4+} + 2H_2O \quad (2.2)$$

$$2M(OH)(H_2O)_5^{2+} \rightleftharpoons [(H_2O)_5M(\mu\text{-}O)M(H_2O)_5]^{4+} + H_2O \quad (2.3)$$

$$[(H_2O)_5M(\mu\text{-}O)M(H_2O)_5]^{4+} + H_2O \rightleftharpoons [(H_2O)_5M(\mu\text{-}O)M(OH)(H_2O)_4]^{3+} + H_3O^+.$$

Further condensation would occur until all (sterically accessible) coordinated water molecules and/or until all proton acceptors in solution are used up. In aqueous solution, condensation processes will (almost) stop only after the aggregates grow large and precipitate out from the solution.

Using the nonaqueous solvent instead of water helps with controlling the condensation. If the solvent of choice has good donor properties, it can complete the coordination sphere of metal ions and thus restrict condensation; if the solvent of choice is a poor donor, a capping ligand would be needed in order to passivate the surface of the nanocrystals. In either case, an optimal balance between complete metal complexation and the desired degree of condensation (so that the clusters grow) can be achieved by tuning the reagent ratios, concentrations, temperature, and time. Therefore, replacing water with nonaqueous solvents, adds a certain degree of freedom to control the crystallization of metal oxides.

Case One: Coordinating Solvent as a Stabilizing Agent
Alcohols are good alternatives for water as coordinating solvents possessing relatively high dielectric constant and high donor number. Polyols are even better than regular alcohols because they are more polar ($\varepsilon = 41$ for ethylene glycol and 32 for diethylene glycol) and they form stronger associates with metal ions. This is especially true for 1,2-diols, 1,2,3-triols, and diethylene glycol which form chelates with metal ions. Cobalt ferrite in the form of 5.3 nm crystalline particles was obtained by hydrolysis of Co(II) and Fe(III) acetates in propylene glycol solution upon heating at 160 °C [108]. A series of nanoscale binary and ternary metal oxides were obtained by hydrolysis of metal acetates and alkoxides in diethylene glycol solutions at 180 °C [109, 110].

The 4–7 nm nanoparticles of $MnFe_2O_4$, $FeFe_2O_4$, $CoFe_2O_4$, $NiFe_2O_4$, and $ZnFe_2O_4$ have been prepared by hydrolysis of chelated alkoxide complexes of corresponding metals in solution of parent alcohol diethylene glycol (DEG) [111, 112]. Synthesis is accomplished by the sequence of the following reactions (see Scheme 2.1): (a) metal ion chelated alcohol complexes formation; (b) ligand deprotonation – alkoxide complexes formation; (c) high-temperature

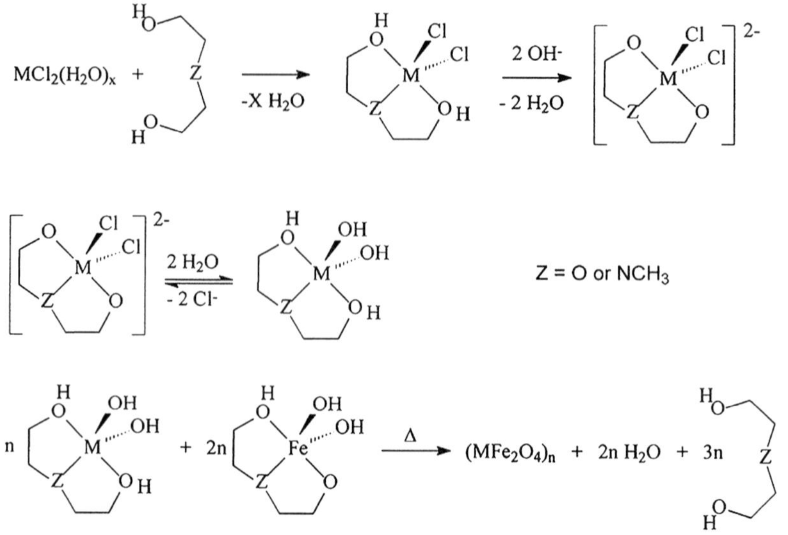

2 Synthesis of Nanoparticulate Magnetic Materials

Scheme 2.1 Scheme of hydrolysis of chelated alkoxide complexes of metals in solution of parent alcohol diethylene glycol [111, 112].

hydrolysis – crystal nucleation and growth. All these reactions run according to a well-determined stoichiometry, and therefore generate the final product with almost a quantitative yield. The suggested sequence of steps is illustrated in Scheme 2.1. The rate of this reaction is highly dependent on temperature, ranging from indefinitely low at a room temperature to the instantaneous at 200–220 °C. This property is very useful, since it assures a fine control over the nucleation and the growth of the nanocrystals. Since no formation of iron oxides takes place at room temperature, all reagents can be premixed, and therefore any nonhomogeneity of the product, due to reactions happening faster than the rate of mixing, is eliminated. Additionally, these conditions help achieving of excellent ordering in the crystal lattice of the metal oxide, which is the quality responsible for the magnetic properties. Most of the obtained nanoparticles are single crystalline.

Bigger nanocrystals of magnetite were obtained when the complexing strength of the medium was increased by replacement of DEG with structurally similar N-methyl diethanolamine. The results indicate that the reaction is slower in a more complexing medium, and that the Ostwald ripening (mass transfer) is presumably taking place, which makes it different from the original DEG-based system.

The function of the diethylene glycol is not only to be a solvent and a complexing/chelating agent, but also to be a stabilizing agent. As a result, even in hot solution during the synthesis, the nanocrystals remain in colloidal form and do not agglomerate. The agglomeration of the nanocrystals due to their high surface energy is a very common and difficult problem to overcome.

It is usually addressed by adding a capping ligand in the reaction solution, by doing synthesis in surfactant micelles, or by providing the conditions for the adsorption of the solvated ions (electric double-layers formation), which would cause the electrostatic repulsion of the charged surfaces. In this method, diethylene glycol seems to act as a capping ligand, and a favorable medium for electric double-layer stabilization, so as a consequence the agglomeration is barely noticeable.

Once precipitated, the nanocrystalline powders still contain the adsorbed diethylene glycol, and if kept under the DEG solvent they remain nonagglomerated for an indefinitely long time. The additional advantage of this method is that the "hot" inorganic chemistry, occurring when crystal cores are synthesized, does not interfere with the surface chemistry that can be done later under ambient conditions for the preparation of biocompatible nanocomposites.

Polar aprotic solvent 2-pyrrolidone was used as a reaction medium for preparation of magnetite nanoparticles by hydrolysis and partial reduction of Fe(III) [113]. Refluxing the solutions of the only precursor $FeCl_3 6H_2O$ (normal boiling point of 2-pyrrolidone is 245 °C) resulted in the formation of the colloids containing Fe_3O_4 nanoparticles with sizes strongly dependent on reaction time, 4, 12, or 60 nm for 1, 10, and 24 h. The evidence is presented that the hydrolysis reaction is promoted by trapping the byproduct HCl with amine, and the reduction of Fe(III) to Fe(II) is done with CO; both the amine and CO are *in situ* generated from 2-pyrrolidone. Remarkably, the solvent also played a role of the colloid stabilizing medium; the precipitation of nanopowders had to be initiated by adding the 1:3 methanol/diethyl ether mixtures. Smaller nanoparticles were easily dispersible in either alkaline or acidic aqueous solutions.

Case Two: Capping Ligand as a Stabilizing Agent
The 6–12 nm monodisperse particles of $BaTiO_3$ (ferroelectric material) have been prepared by hydrolysis of a mixed-metal alkoxide precursor, $BaTi(O_2C(CH_3)_6CH_3)[OCH(CH_3)_2]_5$ with aqueous H_2O_2 in diphenyl ether solution with oleic acid used as a stabilizer [114]. Condensation occured as the solution was heated at 100 °C, but the particle growth was constrained by the presence of the oleic acid stabilizer. This method resulted in crystalline particles that did not require calcination.

The 5 nm particles of γ-Fe_2O_3 have been prepared by refluxing the solution of $FeCl_3 6H_2O$, sodium acetate, *n*-octylamine, and water in 1,2-propanediol [115]. Octylamine acted as a base and a stabilizing agent in this reaction which produced a toluene-soluble nanopowder. Controlled hydrolysis of titanium tetraisopropoxide in oleic acid as a solvent and a surfactant produced TiO_2 with different morphologies [116]. Fast hydrolysis induced by injecting aqueous amine solution at 80 °C resulted in the formation of a variable aspect ratio nanorods; slow hydrolysis by water generated *in situ* from esterification reaction between oleic acid and ethylene glycol, resulted in nearly spherical nanocrystals.

2.3.2
Oxidation Reactions

A commonly used precursor for the preparation of iron oxide nanoparticles is iron pentacarbonyl. Air oxidation of decane solutions of $Fe(CO)_5$ containing a capping agent 11-undecenoic acid, dodecyl sulfonic acid, or octyl phosphonic acid was performed at low temperature (273 K) under high-intensity ultrasonication [117]. The nanoparticles were found to be amorphous but superparamagnetic with magnetic properties strongly influenced by the type of a capping ligand attached to their surface.

Highly crystalline monodispersed maghemite (γ-Fe_2O_3) with variable sizes was obtained by high-temperature solution oxidation reactions [118]. According to one method, iron pentacarbonyl was decomposed (at 100 °C) in a solution of octyl ether containing oleic acid to yield an iron oleate complex, which was then decomposed at 300 °C to yield a colloid of monodisperse amorphous iron metal. In a subsequent step, these nanoparticles were oxidized by trimethylamine oxide at high temperature. As the initial molar ratio of $Fe(CO)_5$ to oleic acid was changed from 1:1 to 1:2 and 1:3, the particle size of the product changed from 4 to 7 to 11 nm. Further increase of the size up to 16 nm was achieved by adding more iron oleate complex to the presynthesized 11 nm Fe_2O_3 colloid followed by aging the solution at 300 °C. According to another method, iron pentacarbonyl was oxidized directly by using $(CH_3)_3NO$ in octyl ether solution. Injection of $Fe(CO)_5$ into a preheated to 100 °C solution of $(CH_3)_3NO$ and lauric acid as a stabilizing agent, resulted in exothermic reaction yielding dark-red solution, which turned black upon heating it at 300 °C. This method allowed the preparation of 13 nm uniform γ-Fe_2O_3 nanocrystals dispersible in hydrocarbon solvent. Magnetic measurements performed on these materials revealed that they are all superparamagnetic with blocking temperatures 25, 185, and 290 K for 4, 13 and 16 nm particles, respectively.

Single-crystalline iron oxide nanoparticles were obtained by direct air oxidation of iron pentacarbonyl in a hot (180 °C) solution containing dodecylamine or trioctylphosphine oxide capping ligand [119]. Trioctylphosphine oxide as a strongly binding ligand suppressed the equilibrium between the growing crystals and the solution, necessary for Ostwald ripening, resulting in the formation of only small (~6 nm) monodisperse nanocrystals. Dodecylamine is a relatively weakly binding ligand, which can reversibly coordinate to the surface of the growing nanocrystals, and therefore permit Ostwald ripening. The size of the nanocrystals produced with this capping ligand was strongly influenced by the molar ratio of $Fe(CO)_5$ to ligand, as well as by the reaction time. As the ratio was 1:1 and the time 9 h, a mixture of diamond, sphere, and triangle-shaped crystals of similar size (~12 nm) was obtained. Simple precipitation method allowed the separation of spherical component. When the molar ratio of the precursor to ligand was changed to 1 : 10, the crystal size greatly increased, being a mixture of 10 and 40 nm after 9 h of heating. The hexagon-shaped nanocrystals grow to about 50 nm and become more monodispersed after the reaction time of 16 h; remaining smaller crystals are rare and some observed in a reduced to <3 nm size. The 12 nm nanocrystals are superparamagnetic at room temperature, but the 50 nm crystals are ferromagnetic with $H_c = 52$ Oe at 300 K. The details of the crystal structure of these products are discussed in this chapter.

2.3.3
Thermal Decomposition of Metal Complexes with O-Donor Ligands

Many research reports concerning synthesis of the nanoparticular metal oxides by thermolysis of the oxygen-rich precursors represent detailed systematic studies aiming a high level of control over the particle size and shape. The recent paper by Peng et al. is offering an alternative explanation of the mechanisms occurring in the noninjection-type systems [120].

A high-temperature reaction of iron(III) acetylacetonate with 1,2-hexadecanediol, oleic acid, and oleylamine in high-boiling ether solution allowed the preparation of Fe_3O_4 nanoparticles with well-controlled size and size distribution [121].

$$Fe(acac)_3 + C_{14}H_{29}CH(OH)CH_2OH + C_{17}H_{33}CO_2H$$
$$+ C_{18}H_{35}NH_2 \longrightarrow \text{nano-}Fe_3O_4(L)_n.$$

If a refluxing phenyl ether (b.p. 259 °C) is used, the 4 nm particles are formed; refluxing in a higher boiling benzyl ether (b.p. 298 °C) allowed the preparation of 6 nm particles. The key to success in preparation of monodisperse particles is in preheating the reaction solutions at 200 °C and exposing them to this condition for some time before bringing the mixture to a reflux. If the temperature is raised continuosly from ambient to refluxing, the product has broad size distribution – from 4 to 15 nm. If the refluxing time was minimized to 5 min instead of a regular 30 min, the nonmagnetic precipitate with X-ray diffraction pattern similar to FeO was isolated. Aiming the preparation of larger magnetite particles, the seed-mediated process was used for this system, bringing the final particle size to 20 nm. Similar reaction employing $Fe(acac)_3$ in combination with Co(II) or Mn(II) acetylacetonate resulted in the formation of manganese and cobalt ferrites also with a variable size and a narrow size distribution.

The size-controlled synthesis of indium oxide nanoparticles was accomplished by thermal decomposition of $In(acac)_3$ in oleylamine at 250 °C [122].

Multiple additions of the precursor permitted to expand range of sizes from 4 to 8 nm. Weak size dependence of photoluminescence was demonstrated.

Thermal decomposition of iron(III) acetylacetonate dissolved in 2-pyrrolidone resulted in the formation of 5 nm particles of magnetite [123]. Remarkably, 2-pyrrolidinone acted as a polar solvent and as a complexing agent controlling the crystal growth and protecting the crystals from agglomeration.

$$Fe(acac)_3 + \text{2-pyrrolidinone} \longrightarrow nano\text{-}Fe_3O_4(L)_n.$$

A seed-mediated process applied to this system helped to obtain particles as large as 11 nm. The products of these reactions are soluble in water and 2-pyrrolidone. Aiming the preparation of stable biocompatible magnetite nanoparticles suitable for MRI application, the same authors further developed this method by using hydrophilic polymers, methoxypolyethylene glycol monocarboxylic acid, and polyethyleneglycol dicarboxylic acid [124, 125]. The obtained nanoparticles were characterized by TEM, electron diffraction, IR spectroscopy, magnetic measurements, and used for *in vitro* MRI experiments, after they were conjugated with a cancer-targeting antibody.

Pyrolysis of iron(III) or iron(II) long-chain carboxylates dissolved in long-chain hydrocarbons at a presence of variable quantity of corresponding carboxylic acid resulted in the formation of Fe_3O_4 nanocrystals coated with the parent carboxylic acid [126]. The rate of decomposition was higher for a shorter-chain (10–14 carbon atoms) acids resulting in a greater variation of particle size as reaction progressed before size defocusing stage. The longer chain oleic and stearic acids showed more steady behavior, which facilitated tuning the conditions for preparation of particles with different sizes. The reactions were carried out in air-free conditions at 300 °C with the iron carboxylate concentration kept constant, but with a variable concentration of acid. The progress of the reactions was monitored by TEM. At the beginning (10 min of heating) very small particles were detected. After this, the quasicube-shaped crystals formed which gradually changed to the dot-shaped crystals of approximately the same volume. The cubes were more stable when high concentration of long-chain acid was used. After several tens of minutes, spherical particles started to show a sign of broadening of sizes probably as a result of Ostwald ripening. Control over the size of the nanocrystals was achieved by varying the concentration of the fatty acid when the concentration of iron salt was fixed. The higher the concentration of acid, the larger the particle size was obtained, 0.1 eq. excess of oleic acid produced 8 nm particles, and 3 eq. excess of it produced 30 nm particles. Introducing the long-chain primary amines or alcohols in the reaction systems, catalyzed the reactions and caused the formation of the smaller nanocrystals of 3–4 nm. This method was extended to other metal oxides, such as Cr_2O_3, MnO, Co_3O_4, and NiO.

$$Fe(O_2CR)_n + HO_2CR \longrightarrow nano\text{-}Fe_3O_4(L)_n$$

Thermal decomposition of iron(III) oleate was studied as a function of temperature and concentration of oleic acid; fine separation of the nucleation and growth allowed the preparation of monodisperse magnetite particles [127]. Variation of the particle size was achieved by using the solvents with different boiling points; 5, 9, 12, 16, and 22 nm particles were obtained in 1-hexadecene, octyl ether, 1-octadecene, 1-eicosene, and trioctylamine with the boiling points 274, 287, 317, 330, and 365 °C, respectively. The authors extended this easily scalable method to other metal oxides such as MnO, FeO, and CoO.

Thermal decomposition of Fe(III), Cu(II), and Mn(II) cupferonates in trioctylamine/octylamine solutions resulted in preparation of the γ-Fe_2O_3, MnO, and Cu_2O nanoparticles [128]. A solution of the precursor was injected into preheated to 300 °C

$$Fe[O(NO)NC_6H_5]_3 + N(C_8H_{17})_3 \longrightarrow nano\text{-}Fe_3O_4(L)_n$$

trioctylamine followed by aging the solution at 225 °C for 30 min. The maghemite obtained by this method had the particle size of 6.7 nm; a smaller 4 nm particles were obtained at a lower injection temperature and lower precursor concentration. Injection of the additional precursor resulted in increasing the size by 1 nm; multiple injections resulted in the formation of 10 nm particles. The crystallinity of these particles studied by TEM imaging revealed tetragonal γ-Fe_2O_3 with an ordered superlattice of cation vacancies.

Solution pyrolysis of the anhydrous zinc acetate [129] and manganese acetate [130] resulted in the formation of ZnO and MnO nanoparticles. Reactions were carried out in presence of long-chain amines and phosphonic acid (ZnO) or oleic acid (MnO). Changing conditions of the synthesis allowed tuning of the final particle size.

2.4
Technology of the Preparation of Magnetic Nanoparticles

This section considers the technological approaches to protection and stabilization of the nanoparticles used during their synthesis and during further steps of processing. The issues related to assembling, functionalizing, and making composites for specific areas of application, are beyond of its scope.

The primary focus of the previous sections was on the solution chemical reactions leading to the formation of the nanocrystalline metals and metal oxides. All of these reactions more or less successfully permitted the control over the morphology of the nanocrystals. This is accomplished by either controlling the kinetics of nucleation and the growth of the nanocrystals, or by physical restriction to the crystal growth. Syntheses permitting control of the rate of crystallization are usually performed in homogeneous solutions and using a stabilizing agent; the other methods based on physical restriction of the crystal growth are based on heterogeneous techniques, such as microemulsions (reverse micelles), Langmuir–Blodgett films, etc. Inhibiting the nanocrystals agglomeration is usually addressed at the same time with their formation in both techniques. This classification is not exclusive, however, because some of the methods, for example formation of the aqueous colloids stabilized by electric double layers can qualify for both groups. Formation and dynamics of the charged and the other dispersion systems are described in details in the book [131].

2.4.1
Stabilizing Agents in Homogeneous Solution Techniques

Since metals and metal oxides are insoluble in any solvents used for their synthesis, controlling the rate of their crystal growth is achieved by controlling the rate of reaction of their formation. This was possible when a complexing agent was used, whose function was to coordinate to metal ions or atoms which were just formed in solution as the reaction intermediates, and to metal atoms on the surface of the growing crystal. These complexing agents are classical ligands from coordination chemistry, modified by attaching a bulky substituent acting as a spacer surrounding the nanoparticle, and as an interface adjusting the affinity of the nanocrystal to the medium. For biological application, this substituent also plays the role of a linker for attaching biomolecules, drugs, etc. Depending on the structure of the substituent, capping ligands, dendrimers or coordinating polymers can be distinguished. Capping ligands are discrete molecules containing a chain or brunched-chain substituent; dendrimers are highly branched structures formed by condensation of well-defined number of the monomer molecules [132–138]; coordinating polymers are macromolecules with the donor groups repeatedly attached to them along the chain. Examples of lipophilic capping ligands are long-chain carboxylic and phosphonic acids, amines, phosphines, thiols, phosphine oxides; the hydrophilic capping ligands contain oxygen-rich substituents such as (oligo)ethylene glycol chains or carbon chains terminated with $-OH$, $-COOH$, $-NH_2$, $-SO_3^-$, and other polar groups. The commonly used coordinating polymers are of the polyether-type (dextran, polyethylene glycol), polyamine (polyethyleneamine), or carbon-chain polymers with side substituents the same as listed for capping ligands. Other examples of commonly used coordinating C-chain polymers are polyvinyl pyridine and polyvinyl pyrrolidone. The aspects of coordination chemistry at the nanoparticle surface are covered in more details in the recent review [12].

The criteria for discussing the cons and pros of the mentioned types of stabilizing agents are their efficiency for morphology control, and colloid stabilizing effect and solubility. Capping ligands are more convenient for formulation of the reaction solutions and for tuning their coordination strength, and therefore the rates of crystallization. Coordination polymers are more efficient as stabilizing agents because cooperative effect of their multiple spot attachment to the nanocrystal surface makes binding of the macromolecule stronger than binding of a capping ligand. In order to stabilize and solubilize larger nanoparticles, larger spacer (capping ligand) is required. This issue is easily addressed with capping ligands, as length and branching of their substituents is varied. Macromolecules of polymers are frequently too long; the nanocomposite with even small nanoparticle appears bulky, and the diamagnetic contribution of the macromolecule suppresses its useful magnetic properties.

Tetraalkylammonium (usually R = octyl) chlorides and bromides and aluminum alkyls should be mentioned as separate groups of stabilizing agents. The halide ion from ammonium salt is coordinated to the surface metal atoms thus providing a negative charge to the surface of the nanoparticles. Ammonium cations form a positively charged layer around the nanoparticles and enhance the stabilizing effect due to electrostatic forces and steric effect of four alkyl substituents. Aluminum alkyls are presumably bound to the nanocrystal surface in the form $XAlR_2$ through bridging atom X = O or Cl and provide steric effect responsible for the stability against agglomeration and solubility in nonpolar solvents.

2.4.2
Heterogeneous Solution Techniques

The most common media for heterogeneous solution techniques are the microemulsions and Langmuir–Blodgett films. The amphiphilic surfactant containing long-chain (typically C_8 to C_{18}) substituent attached to an ionic group such as $-SO_3^-$ or $-NR_3^+$ forms spherical micelles in a dilute hydrocarbon solution with hydrocarbon tails facing outward and ionic groups facing inward. The most commonly used surfactants are cetyltrimethylammonium bromide (CTAB), sodium dodecylsulfate (SDS), and sodium bis(2-ethylhexyl)sulfosuccinate (AOT). The micelles absorb appreciable amount of liquid water which localizes in their polar cores. The micelles swell and their size can be controlled by adjusting the water to surfactant ratio ω_0.

$$\omega_0 = \frac{[H_2O]}{[S]} = \frac{n_w}{n_s}$$

where n_s and n_w are the moles of surfactant and water per micelle, respectively. Radius of the micelles changes linearly as a function of concentration of water. Thus formed "water pools" have nanometer scale size and can be used as "nanoreactors" for synthesis of the nanocrystals whose size would be comparable to the size of micelles. Microemulsions are highly dynamic; the micelles continuously collide, coalesce and split, and therefore exchange their content. As two similar micellar solutions, one containing metal salt and another one containing the precipitating agent are mixed, rapid intermicellar exchange results in crystallization of the product inside the micelles, which produces a colloidal dispersion. In subsequent steps, surface chemistry can be performed, followed by isolation of the nanoparticles as powders free from byproducts, from the original surfactant and solvents.

Most of syntheses for metal nanoparticles, if performed by microemulsion technique, employ BH_4^- or $N_2H_5^+$ as reducing agents (Section 2.2.1.2). Most of syntheses for metal-oxide nanoparticles, if performed by microemulsion techniques, employ NaOH, NH_4OH or NMe_4OH as precipitating agent (Section 2.3.1.1). More details on the micellar synthesis can be found in the reviews [9, 13].

The solution of precursors deposited on the interface between another nonmixable liquid and a gas, adopts the structure of Langmuir monolayer. The reaction of synthesis carried out in such a thin film takes place in two-dimensional conditions since the diffusion of the precursor molecules occurs only in its plain. This method allows the preparation of disk-shaped nanoparticles with high aspect ratio. Introducing the surfactant or stabilizing agent in the reaction solution permits control over the rate of diffusion and therefore the morphology of the products. The UV-initiated decomposition of $Fe(CO)_5$ deposited on water surface as the solution in chloroform containing also stearic acid, resulted in the formation of plate-shaped iron-containing nanostructures [139, 140]. Gold nanoparticles were obtained by the same method but using $Au(PPh_3)Cl$ as a precursor.

According to an alternative method, the ionic metal precursor is dissolved in aqueous subphase and a spontaneous reduction is carried out by Langmuir monolayer containing a reducing and a stabilizing agent. Reduction of $AuCl_4^-$ with monolayer of hexadecylaniline, which also acts as a capping agent, results in the formation of highly oriented, flat gold sheets and ribbons bound to the monolayer [141].

An interesting application of Langmuir–Blodgett technique aimed the surface derivatization of the nanoparticles. The nanosized "Janus" Au particles whose different sides were coated with either hydrophilic or hydrophobic capping ligands, were prepared by ligand exchange reactions in a Langmuir monolayer [142]. The presynthesized hydrophobic alkanethiolate-passivated gold particles reacted with hydrophilic thiol derivatives injected into the water subphase. The ligand exchange reactions were limited only to the side of the particles facing the aqueous phase. The unique amphiphilic character of the Janus particles is presented by their solubility in water and the formation of micelle-like aggregates.

2.5 Conclusions

Metals and metal oxides are used as the primary magnetic materials. It might seem surprising that the number of different synthesis methods developed for such simple chemicals in the nanocrystalline form, far exceeds the number of synthesis methods known for more complex inorganic and organic compounds. The nanoparticles however have a complicated composition and structure due to their extensive surface chemistry.

The target product of any area of developmental research is expected to exhibit a specific set of functional properties. While the higher saturation magnetization is beneficial for any area of application of the magnetic nanoparticles, the requirements to the other properties are variable depending on specific use. Composition and properties of the organic coating are particularly important for biomedical application. This coating must diminish the interparticle interaction and stabilize aqueous colloids at physiological conditions (pH, temperature). The issues of toxicity, biodistribution, and of specific biorecognition of the nanocomposite are addressed by surface and conjugation chemistry and are subjects of a very active research targeting biomedical application. It is evident that this area will receive even more attention in the nearest future.

One of the areas of materials science targets the development of high-density information recording media and microelectronic devices whose function is greatly affected by the particle size, shape, crystal structure, and surface composition of the magnetic nanoparticles. For this reason, many research efforts aim synthesis of the nanoparticles with controlled size and size distribution. The accomplishments in this area are very impressive; however, we are still far away from being able to manage the composition of the surface as well.

Preparation of the nanoparticles that are different from spherical or pseudospherical is one of the very intriguing focus points for many researchers. Shape anisotropy dramatically influences their electronic structure which opens a window to new useful properties. From the fundamental point of view, such studies help to better understand the quantum confinement effect in the light of relationship of geometrical and electronic structure. These studies will also explain the mechanism of the crystal growth and, therefore, will make further developmental efforts more rational. Advancements in the preparation of anisotropic metal nanoparticles are relatively faint, but the accomplishments in preparation of anisotropic metal oxide particles are even weaker.

Preparation of the core–shell and onion-type composite particles meets the issue of compatibility of crystal lattices for different inorganic components. For example, many areas of application of the magnetic nanoparticles rely almost exclusively on iron oxides; however, iron and some magnetic metals alloys would be preferred due to their superior magnetic susceptibility. Protection of metals from oxidation at the nanoscale level represents one of the major challenges of this area.

References

1. Bönnemann, H.; Richards, R.M. *Eur. J. Inorg. Chem.* **2001**, 2455.
2. Lu, A.-H.; Salabas, E.L.; Schüth, F. *Angew. Chem. Int. Ed.* **2007**, *46*, 1222.
3. Roucoux, A.; Schulz, J.; Patin, H. *Chem. Rev.*, **2002**, *102*, 3757.
4. Schmid, G. *J. Chem. Soc. Dalton Trans.* **1998**, 1077.
5. Ritchie, I.M.; Bailey, S.; Woods, R. *Adv. Colloid Interface Sci.*, **1999**, 183.
6. Daniel, M.-C.; Astruc, D. *Chem. Rev.*, **2004**, *104*, 293.
7. McHenry, M.E.; Laughlin, D.E. *Acta Mater.* **2000**, 223.
8. Burda, C.; Chen, X.; Narayanan, R.; El-Sayed, M.A. *Chem. Rev.*, **2005**, *105*, 1025.
9. Lisiecki, I. *J. Phys. Chem. B*, **2005**, *109*, 12231.
10. Gubin, S.P.; Koksharov, Yu.A. *Inorg. Mater.*, **2002**, *38*, 1287.
11. Gubin, S.P.; Koksharov, Yu.A.; Khomutov, G.B.; Yurkov, G.Yu. *Russ. Chem. Rev.*, **2005**, *74*, 489.
12. Gubin, S.P.; Kataeva, N.A. *Russ. J. Coord. Chem.*, **2006**, *32*, 849.
13. Pileni, M.P. *Nature Mater.*, **2003**, *2*, 145–150.
14. Cushing, B.; Kolesnichenko, V.; O'Connor, C.J. *Chem. Rev.*, **2004**, *104*, 3893–3946.
15. Schmid, G. *Nanoparticles*, Wiley-VCH, Weinheim, **2004**.
16. Reetz, M.T.; Helbig, W. *J. Am. Chem. Soc.* **1994**, *116*, 7401–7402.
17. Reetz, M.T.; Lohmer, G. *Chem. Commun.*, **1996**, 1921.
18. Reetz, M.T.; Helbig, W.; Quaiser, S.A.; Stimming, U.; Beuer, N.; Vogel, R. *Science*, **1995**, *267*, 367–369.
19. Leslie-Pelecky, D.L.; Bonder, M; Martin, T; Kirkpatrick, E.M.; Yi. Liu, Zhang, X.Q.; Kim, S.H.; Rieke, R.D. *Chem. Mater.*, **1998**, *10*(#11) 3732–3736.
20. Park, J.-I.; Cheon, J. *J. Am. Chem. Soc.* **2001**, *123*, 5743.
21. Ban, Z.; Barnakov, Yu.A.; Li, F.; Golub, V.; O'Connor, C.J. *J. Mater. Chem.*, **2005**, *15*, 4660–4662.
22. Lu, Z.; Prouty, M.D.; Guo, Z.; Golub, V.; Kumar, C.S.S.R.; Lvov, Y. *Langmuir*, **2005**, *21*, 2042–2050.
23. Shon, Y.-S.; Dawson, G.B.; Porter, M.; Murray, R.W. *Langmuir*, **2002**, *18*, 3880–3885.
24. Tsai, K.-L.; Dye, J.L. *J. Am. Chem. Soc.* **1991**, *113*, 1650–1652.
25. Tsai, E.H.; Dye, J.L. *Chem. Mater.* **1993**, *5*, 540–546.
26. Nelson, J.A.; Wagner, M.J. *Chem. Mater.*, **2002**, *14*(#4) 1639–1642.
27. Nelson, J.A.; Bennett, L.H.; Wagner, M.J. *J. Mater. Chem.*, **2003**, *13*(#4) 857–860.
28. Nelson, J.A.; Bennett, L.H.; Wagner, M.J. *J. Am. Chem. Soc.*, **2002**, *124*(#12) 2979–2983.
29. Leslie-Pelecky, D.L.; Zhang, X.Q; Kim, S.H.; Bonder, M.; Rieke, R.D. *Chem. Mater.*, **1998**, *10*(#1) 164–171.
30. Kirkpatrick, E.M.; Leslie-Pelecky, D.L.; Kim, S.H.; Rieke, R.D. *J. Appl. Phys.*, **1999**, *85*(2B) 5375–5377.
31. Baldwin, R.K.; Pettigrew, K.A.; Ratai, E.; Augustine, M.P.; Kauzlarich, S.M. *Chem. Commun.*, **2002**, *17*, 1822–1823.
32. Hope-Weeks, L.J. *Chem. Commun.*, **2003**, *24*, 2980–2981.
33. Kolesnichenko, V.L.; Vu, V.; Goloverda, G. *Ukrainian Chem. J.*, **2005**, *71*, 65–70.
34. Vargaftik, M.N.; Zagorodnikov, V.P.; Stolyarov, I.P.; Moiseev, I.I.; Likholobov, V.A.; Kochubey, D.I.; Chuvilin, A.L.; Zaikovsky, V.I.; Zamaraev, K.I.; Timofeeva, G.I. *J. Chem. Soc., Chem. Commun.*, **1985**, *14*, 937–939.
35. Vargaftik, M.N.; Zagorodnikov, V.P.; Stolyarov, I.P.; Moiseev, I.I.; Kochubey, D.I.; Likholobov, V.A.; Chuvilin, A.L.; Zamaraev, K.I. *J. Mol. Catal.*, **1989**, *53*, 315–348.
36. Vargaftik, M.N.; Moiseev, I.I.; Kochubei, D.I.; Zamaraev, K.I. *Faraday Discuss.* **1992**, *92*, 13–29.
37. Volkov, V.V.; Tendeloo, G.Van; Vargaftik, M.N.; Moiseev, I.I. *J. Cryst. Growth*, **1993**, *132*, 359–363.

38. Poulin, J.C.; Kagan, H.B.; Vargaftik, M.N.; Stolarov, I.P.; Moiseev, I.I. *J. Mol. Catal. A*, **1995**, *95*, 109–113.
39. Volkov, V.V.; Van Tendeloo, G.; Tsirkov, G.A.; Cherkashina, N.V.; Vargaftik, M.N.; Moiseev, I.I.; Novotortsev, V.M.; Kvit, A.V.; Chuvilin, A.L. *J. Cryst. Growth*, **1996**, *163*, 377–387.
40. Volokitin, Y.; Sinzig, J.; de Jongh, L.J.; Schmid, G.; Vargaftik, M.N.; Moiseev, I.I. *Nature*, **1996**, *384*, 621–623.
41. Schmid, G.; Morun, B.; Malm, J.O. *Angew. Chem., Int. Ed.* **1989**, *28*, 778–780.
42. Dumestre, F.; Chaudret, B.; Amiens, C.; Renaud, P.; Fejes, P. *Science*, **2004**, *303*, 821–823.
43. Nakajima, M.; Saito, T.; Kobayashi, A.; Sasaki, Y. *J. Chem. Soc. Dalton Trans.* **1977**, 385.
44. Corey, E.J.; Cooper, N.J.; Canning, W.M.; Lipscomb, W.N.; Koetzle, T.F. *Inorg. Chem.*, **1982**, *21*, 192–199.
45. Lin, X.M.; Sorensen, C.M.; Klabunde, K.J.; Hadjipanayis, G.C. *Langmuir* **1998**, *14*, 7140–7146.
46. Petit, C.; Taleb, A.; Pileni, M.P. *J. Phys. Chem. B*, **1999**, *103*, 1805–1810.
47. Mirkhalaf, F.; Paprotny, J.; Schiffrin, D.J. *J. Am. Chem. Soc.* **2006**, *128*, 7400–7401.
48. Sun, S.; Murray, C.B. *J. Appl. Phys.* **1999**, *85*, 4325–4330.
49. Murray, C.B.; Sun, S.; Doyle, H.; Betley, T. *Mater. Res. Soc. Bull.* **2001**, 985.
50. Modrow, H.; Bucher, S.; Hormes, J.; Brinkmann, R.; Bönnemann, H. *J. Phys. Chem. B*, **2003**, *107*, 3684–3689.
51. Turkevich, J.; Stevenson, P.S.; Hillier, J. *Discuss. Faraday Soc.* **1951**, *11*, 58.
52. Schmid, Gunter; Lehnert, Andreas; Malm, Jan Olle; Bovin, Jan Olov. *Angew. Chem., Int. Ed. Engl.*, **1991**, *30*, 874–6.
53. Viau, G.; Brayner, R.; Poul, L.; Chakroune, N.; Lacaze, E.; Fievet-Vincent, F.; Fievet, F. *Chem. Mater.* **2003**, *15*, 486–494.
54. Bönnemann, H.; Brijoux, W.; Hofstadt, H.-W.; Ould-Ely, T.; Schmidt, W.; Waßmuth, B.; Weidenthaler, C. *Angew. Chem., Int. Ed.* **2002**, *41*, 599.
55. Angermund, K.; Bühl, M.; Dinjus, E.; Endruschat, U.; Gassner, F.; Haubold, H.-G.; Hormes, J.; Köhl, G.; Mauschick, F.T.; Modrow, H.; Mörtel, R.; Mynott, R.; Tesche, B.; Vad, T.; Waldöfner, N.; Bönnemann, H. *Angew. Chem., Int. Ed.* **2002**, *41*, 4041.
56. Bönnemann, H.; Waldöfner, N.; Haubold, H.-G.; Vad, T. *Chem. Mater.* **2002**, *14*, 1115.
57. Haubold, H.-G.; Vad, T.; Waldöfner, N.; Bönnemann, H. *J. Appl. Crystallogr.* **2003**, *36*, 617.
58. Angermund, K.; Bühl, M.; Endruschat, U.; Mauschick, F.T.; Mörtel, R.; Mynott, R.; Tesche, B.; Waldöfner, N.; Bönnemann, H.; Köhl, G.; Modrow, H.; Hormes, J.; Dinjus, E.; Gassner, F.; Haubold, H.-G.; Vad, T.; Kaupp, M. *J. Phys. Chem. B* **2003**, *107*, 7507.
59. Gui, Z.; Fan, R.; Mo, W.; Chen, X.; Yang, L.; Hu, Y. *Mater. Res. Bull.* **2003**, *38*, 169–176.
60. Pastoriza-Santos, I.; Liz-Marzán, L.M. *Langmuir* **1999**, *15*, 948.
61. Tan, Y.; Dai, X.; Li, Y.; Zhu, D. *J. Mater. Chem.* **2003**, *13*, 1069.
62. Sondi, I.; Goia, D.V.; Matijevic, E. *J. Colloid Interface Sci.* **2003**, *260*, 75.
63. Groshens, T.G.; Henne, B.; Bartak, D.; Klabunde, K.J. *Inorg. Chem.*, **1981**, *20*, 3629.
64. Kundig, E.Peter; Jeger, Patrick; Bernardinelli, Gerald. *Inorgan. Chim. Acta* **2004**, *357*, 1909–1919.
65. Smith, T.W.; Wychick, D. *J. Phys. Chem.* **1980**, *84*, 1621–1629.
66. Yurkov, G.Yu.; Fionov, A.S.; Koksharov, Yu.A.; Kolesov, V.V.; Gubin, S.P. *Inorgan. Mater.* **2007**, *43*, 834–844.
67. Gubin, S.P. *Colloids Surf., A*, **2002**, *202*, 155.
68. Gubin, S.P.; Spichkin, Yu.I.; Koksharov, Yu.A.; Yurkov, G.Yu.; Kozinkin, A.V.; Nedoseikina, T.I.; Korobov, M.S.; Tishin, A.M. *J. Magn. Magn. Mater.* **2003**, *265*, 234.

69. Yurkov, G.Yu.; Kozinkin, A.V.; Nedoseikina, T.I.; Shuvaev, A.T.; Vlasenko, V.G.; Gubin, S.P.; Kosobudskii, I.D. *Inorgan. Mater.*, **2001**, *37*, 997.
70. Dinega, D.; Bawendi, M.G. *Angew. Chem., Int. Ed.* **1999**, *38*, 1788.
71. Puntes, V.F.; Krishnan, K.M.; Alivisatos, A.P. *Science* **2001**, *291*, 2115.
72. Puntes, V.F.; Zanchet, D.; Erdonmez, C.K.; Alivisatos, A.P. *J. Am. Chem. Soc.* **2002**, *124*, 12874.
73. Bönnemann, H.; Brijoux, W.; Brinkmann, R.; Matoussevitch, N.; Waldöfner, N.; Palina, N.; Modrow, H. *Inorg. Chim. Acta*, **2003**, *350*, 617.
74. Osuna, J.; de Caro, D.; Amiens, C.; Chaudret, B.; Snoeck, E.; Respaud, M.; Broto, J.-M.; Fert, A. *J. Phys. Chem.* **1996**, *100*, 14571.
75. Zitoun, D.; Amiens, C.; Chaudret, B.; Fromen, M.C.; Lecante, P.; Casanove, M.J.; Respaud, M. *J. Phys. Chem. B* **2003**, *107*, 6997.
76. Pellegatta, J.-L.; Blandy, C.; Choukroun, R.; Lorber, C.; Chaudret, B.; Lecante, P.; Snoeck, E. *New J. Chem.* **2003**, *17*, 1528.
77. Pelzer, K.; Vidoni, O.; Philippot, K.; Chaudret, B.; Colliere, V. *Adv. Funct. Mater.* **2003**, *13*, 118.
78. Gomez, M.; Philippot, K.; Colliere, V.; Lecante, P.; Muller, G.; Chaudret, B. *New J. Chem.* **2003**, *27*, 114.
79. Ely, T.O.; Amiens, C.; Chaudret, B. *Chem. Mater.* **1999**, *11*, 526.
80. Verelst, M.; Ely, T.O.; Amiens, C.; Snoeck, E.; Lecante, P.; Mosset, A.; Respaud, M.; Broto, J.M.; Chaudret, B. *Chem. Mater.* **1999**, *11*, 2702.
81. Ely, T.O.; Pan, C.; Amiens, C.; Chaudret, B.; Dassenoy, F.; Lecante, P.; Casanove, M.-J.; Mosset, A.; Respaud, M.; Broto, J.M. *J. Phys. Chem. B* **2000**, *104*, 695.
82. Pan, C.; Pelzer, K.; Philippot, K.; Chaudret, B.; Dassenoy, F.; Lecante, P.; Casanove, M.-J. *J. Am. Chem. Soc.* **2001**, *123*, 7584.
83. Cordente, N.; Respaud, M.; Senocq, F.; Casanove, M.-J.; Amiens, C.; Chaudret, B. *Nano Lett.* **2001**, *1*, 565.
84. Dumestre, F.; Chaudret, B.; Amiens, C.; Fromen, M.-C.; Casanove, M.-J.; Renaud, P.; Zurcher, P. *Angew. Chem., Int. Ed.* **2002**, *41*, 4286.
85. Dumestre, F.; Chaudret, B.; Amiens, C.; Respaud, M.; Fejes, P.; Renaud, P.; Zurcher, P. *Angew. Chem., Int. Ed.* **2003**, *42*, 5213–5216.
86. Bradley, J.S., Tesche, B.; Busser, W.; Maase, M.; Reetz, M. *J. Am. Chem. Soc.* **2000**, *122*, 4631–4636.
87. Soulantica, K.; Maisonnat, A.; Fromen, M.-C.; Casanove, M.-J.; Lecante, P.; Chaudret, B. *Angew. Chem., Int. Ed.* **2001**, *40*, 448.
88. Soulantica, K.; Erades, L.; Sauvan, M.; Senocq, F.; Maisonnat, A.; Chaudret, B. *Adv. Funct. Mater.* **2003**, *13*, 553.
89. Soulantica, K.; Maisonnat, A.; Senocq, F.; Fromen, M.-C.; Casanove, M.-J.; Chaudret, B. *Angew. Chem., Int. Ed.* **2001**, *40*, 2984.
90. de Caro, D.; Agelou, V.; Duteil, A.; Chaudret, B.; Mazel, R.; Roucau, C.; Bradley, J.S. *New J. Chem.* **1995**, *19*, 1265.
91. Sun, S.; Murray, C.B.; Weller, D.; Folks, L.; Moser, A. *Science* **2000**, *287*, 1989.
92. Gatteschi, D.; Sessoli, R.; Cornia, A. *Chem. Commun.* **2000**, 725.
93. Jolivet, J.-P.; Chanéac, C.; Tronc, E. *Chem. Commun.* **2004**, 481.
94. Jolivet, J.-P.; Tronc, E.; Chanéac, C. *C.R. Geosci.* **2006**, *338*, 488.
95. Kuo, P.C.; Tsai, T.S. *J. Appl. Phys.* **1989**, *65*, 4349.
96. Tang, Z.X.; Sorensen, C.M.; Klabunde, K.J.; Hadjipanayis, G.C. *J. Colloid Interface Sci.* **1991**, *146*, 38–52.
97. Vayssières, L.; Chanéac, C.; Tronc, E.; Jolivet, J.-P. *J. Colloid Interface Sci.* **1998**, *205*, 205.
98. Jolivet, J.-P.; Froidefond, C.; Pottier, A.; Chanéac, C.; Cassaignon, S.; Tronc, E.; Euzen, P. *J. Mater. Chem.* **2004**, *14*, 3281.

99. Sousa, M.H.; Tourinho, F.A.; Depeyrot, J.; da Silva, G.J.; Lara, M.C. *J. Phys. Chem.* **2001**, *105*, 1168.
100. Kang, Y.S.; Risbud, S.; Rabolt, J.F.; Stroeve, P. *Chem. Mater.* **1996**, *8*, 2209.
101. Garcell, L.; Morales, M.P.; Andres-Vrergés, M.; Tartaj, P.; Serna, C.J. *J. Colloid Interface Sci.* **1998**, *205*, 470–475.
102. Chinnasamy, C.N.; Jeyadevan, B.; Perales-Perez, O.; Shinoda, K.; Tohji, K.; Kasuya, A. *IEEE Trans. Magn.* **2002**, *38*, 2640–2642.
103. Zaitsev, V.S.; Filimonov, D.S.; Presnyakov, I.A.; Gambino, R.J.; Chu, B. *J. Colloid Interface Sci.* **1999**, *212*, 49–57.
104. Babes, L.; Denizot, B.; Tanguy, G.; Le Jeune, J.J.; Jallet, P. *J. Colloid Interface Sci.* **1999**, *212*, 474–482.
105. Portet, D.; Denizot, B.; Rump, E.; Lejeune, J.J.; Jallet, P. *J. Colloid Interface Sci.* **2001**, *238*, 37–42.
106. Kim, D.K.; Mikhaylova, M.; Zhang, Y.; Muhammed, M. *Chem. Mater.* **2003**, *15*, 1617–1627.
107. Rojas, T.C.; Ocana, M. *Scripta Mater.* **2002**, *46*, 655–660.
108. Ammar, S.; Helfen, A.; Jouini, N.; Fiévet, F.; Rosenman, I.; Villain, F.; Molinié, P.; Danot, M. *J. Mater. Chem.* **2001**, *11*, 186.
109. Feldmann, C. *Adv. Mater.*, **2001**, *13*, 1301.
110. Feldmann, C.; Jungk, H.-O. *Angew. Chem. Int. Ed.* **2001**, *40*, 359.
111. Caruntu, D; Remond, Y; Chou, N.H.; Jun, M.-J.; Caruntu, G.; He, J.; Goloverda, G.; O'Connor, C.; Kolesnichenko, V. *Inorg. Chem.*, **2002**, *41*, 6137.
112. Caruntu, D.; Caruntu, G.; Yuxi, C.; O'Connor, C.; Goloverda, G.; Kolesnichenko, V. *Chem. Mater.*, **2004**, *16*, 5527.
113. Li, Z.; Sun, Q.; Gao, M. *Angew. Chem. Int. Ed.* **2005**, *44*, 123–126.
114. O'Brien, S.; Brus, L.; Murray, C.B. *J. Am. Chem. Soc.* **2001**, *123*, 12085–12086.
115. Rajamathi, M.; Ghosh, M.; Seshadri, R. *Chem. Commun.*, **2002** 1152.
116. Cozzoli, P.D., Kornowski, A.; Weller, H. *J. Am. Chem. Soc.* **2003**, *125*, 14539.
117. Shafi, K.; Ulman, A.; Yan, X.; Yang, N.-L.; Estournès, White, H.; Rafailovich, M. *Langmuir*, **2001**, *17*, 5093–5097.
118. Hyeon, T.; Lee, S.S.; Park, J.; Chung, Y.; Na, H.B. *J. Am. Chem. Soc.* **2001**, *123*, 12798–12801.
119. Cheon, J.; Kang, N.-J.; Lee, S.-M.; Lee, J.-H.; Yoon, J.-H.; Oh, S.J. *J. Am. Chem. Soc.* **2004**, *126*, 1950–1951.
120. Chen, Y.; Johnson, E.; Peng, X. *J. Am. Chem. Soc.* **2007**, *129*, 10937–10947.
121. Sun, S.; Zeng, H.; Robinson, D.B.; Raoux, S.; Rice, P.M.; Wang, S.X.; Li, G. *J. Am. Chem. Soc.* **2004**, *126*, 273–279.
122. Seo, W.S.; Jo, H.H.; Lee, K; Park, J.T. *Adv. Mater.* **2003**, *15*, 795–797.
123. Li, Z.; Chen, H.; Bao, H.; Gao, M. *Chem. Mater.* **2004**, *16*, 1391–1393.
124. Li, Z.; Wei, L.; Gao, M.; Lei, H. *Adv. Mater.* **2005**, *17*, 1001–1005.
125. Hu, F.; Wei, L.; Zhou, Z.; Ran, Y.; Li, Z.; Gao, M. *Adv. Mater.* **2006**, *18*, 2553–2556.
126. Jana, N.R.; Chen, Y.; Peng, X. *Chem. Mater.* **2004**, *16*, 3931–3935.
127. Park, J.; An, K.; Hwang, Y.; Park, J.G.; Noh, H.J.; Kim, J.Y.; Park, J.H.; Hwang, N.M.; Hyeon, T. *Nature Mater.*, **2004**, *3*, 891–895.
128. Rockenberger, J.; Scher, E.; Alivisatos, A.P. *J. Am. Chem. Soc.* **1999**, *121*, 1595–1596.
129. Cozzoli, P.D.; Curri, M.L.; Agostiano, A.; Leo, G.; Lomascolo, M. *J. Phys. Chem. B* **2003**, *107*, 4756–4762.
130. Yin, M.; O'Brien, S. *J. Am. Chem. Soc.* **2003**, *125*, 10180–10181.
131. Pashley, R.M.; Karaman, M.E. *Applied Colloid and Surface Chemistry*, John Wiley & Sons, Chichester, **2004**.
132. Badetti, E.; Caminade, A.-M.; Majoral, J.-P.; Moreno-Manas, M.; Sebastian, R.M. *Langmuir* **2008**, *24*, 2090–2101.
133. Shifrina, Z.B.; Rajadurai, M.S.; Firsova, N.V.; Bronstein, L.M.; Huang, X.; Rusanov, A.L.;

Muellen, K. *Macromolecules* **2005**, *38*, 9920–9932.
134. Wang, C.; Zhu, G.; Li, J.; Cai, X.; Wei, Y.; Zhang, D.; Qiu, S. *Chem. Eur. J.* **2005**, *11*, 4975–4982.
135. Xie, H.; Gu, Y.; Ploehn, H.J. *Nanotechnology* **2005**, *16*, 492–501.
136. Richardson, D.D.; Ely, S.R.; McMurdo, M.J.; Van Patten, P.G. *Mater. Res. Soc. Symp. Proc.* **2002**, *726*, 343–348.
137. Esumi, K.; Suzuki, A.; Yamahira, A.; Torigoe, K. *Langmuir*, **2000**, *16*, 2604–2608.
138. Zhao, M.; Crooks, R.M. *Angew. Chem. Int. Ed.* **1999**, *38*, 364–366.
139. Khomutov, G.B.; Bykov, I.V.; Gainutdinov, R.V.; Gubin, S.P.; Obydenov, A.Yu.; Polyakov, S.N.; Tolstikhina, A.L. *Colloids Surf., A* **2002**, *198–200*, 347–358.
140. Khomutov, G.B.; Kislov, V.V.; Gainutdinov, R.V.; Gubin, S.P.; Obydenov, A.Yu.; Pavlov, S.A.; Sergeev-Cherenkov, A.N.; Soldatov, E.S.; Tolstikhina, A.L.; Trifonov, A.S. *Surf. Sci.*, **2003**, *532–535*, 287–293.
141. Swami, A.; Kumar, A.; Selvakannan, P.R.; Mandal, S.; Pasricha, R.; Sastry, M. *Chem. Mater.*, **2003**, *15*, 17–19.
142. Pradhan, S.; Xu, L.-P.; Chen, S. *Adv. Funct. Mater.*, **2007**, *17*, 2385–2392.

3
Magnetic Metallopolymer Nanocomposites: Preparation and Properties

Gulzhian I. Dzhardimalieva, Anatolii D. Pomogailo, Aleksander S. Rozenberg and Marcin Leonowicz

3.1
Introduction

The various physical and mechanical parameters of metallopolymeric nanocomposites can be considerably improved with retention or sometimes even with the increase in the conventional magnetic characteristics (saturation magnetization, coercive force, etc.) that show considerable promise for various applications. Such substances are very perspective as high-density information media, materials for permanent magnets, magnetic cooling systems, or fabrication of magnetic sensors. They can also be used for medical and biological applications, e.g., for magnetic resonance tomography, magnetic targeted drug delivery systems, magnetic resonance imaging (MRI) contrast enhancement, color imaging, and cell sorting [1, 2].

Ferromagnetic metal particles can be formed in a polymer matrix (ferroplastics or ferroelastics) by practically all methods for synthesis of metal nanoparticles in polymers. The manufacturing method can be subdivided into three large groups: physical, chemical, and physicochemical. The division is very conventional and is based mostly on the method of nanoparticle formation and the character of their interactions with the matrix. The first group of methods includes the procedures seemingly devoid of any chemical interactions between the dispersed phase and the dispersion medium. Composites of this type are rare. More popular chemical methods of manufacturing nanocomposites are based on interactions between components anticipated or detected. The reduction reactions and precursors are well known. Nevertheless, the new approaches have been elaborated from practical demands and many details of the chemical reduction are still unclear. The third group of methods are those when metal or their oxide nanoparticles are formed using high-energy processes of atomic metal evaporation, including low-temperature discharge plasma, radiolysis, and photolysis in the presence of monomer and polymer. Recently, such methods have become widely used in vacuum metallization

Magnetic Nanoparticles. Sergey P. Gubin
© 2009 WILEY-VCH Verlag GmbH & Co. KGaA, Weinheim
ISBN: 978-3-527-40790-3

3.2
The General Methods of Synthesis and Characterization of Magnetic Nanoparticles in a Polymer Matrix

3.2.1
Magnetic Nanoparticles in Inorganic Matrices

Many ferromagnetic polymeric nanocomposites are already synthesized with nanosize transition metal particles [3, 4], ferrites [5], and nanosize magnetic particles of Fe [6], Co [7, 8], Ni [9], γ-Fe_2O_3 [10–13], and Fe_3O_4 [11]. An ensemble of cobalt nanoparticles, isolated into a polymer matrix, has the same magnetic properties as in molecular beams. But the average magnetic moment per atom arises at 13–20% caused by the decrease in the atomic coordination number, or an increase in the interatomic distances [14, 15]. Magnetic nanocomposites were also obtained by introducing metal-containing nanoparticles into dielectric zeolite [16], intercalated graphitic [17], and aluminosilicate [18] matrices, and into the surface silica layer [19]. For example, using regulated thermodestruction of $Co_2(CO)_8$ into NaY zeolite voids (H_2, 470 K), nonoxidized cobalt particles (0.6–1 nm in diameter) can be formed [20] and into the molecular sieves MCM-41 (an impregnation by an aqueous solution of $CoCl_2$, calcinations at 1020 K in an O_2 current), the disappearance of the antiferromagnetic phase transition at 33 K of bulk Co_3O_4 is observed in the amorphous silicate matrix (Figure 3.1), containing 6 mol.% of Co_3O_4 (with mean particle sizes of 1.6 nm) [21, 22].

3.2.2
Magnetic Nanoparticles in Polymer Matrices

The magnetic properties of *in situ* polymerizable materials are better than those produced by the standard solid-state reaction method [23] (Figure 3.2).

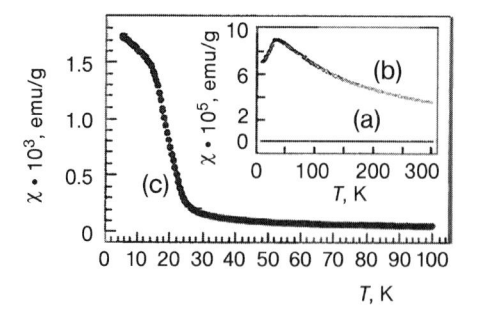

Figure 3.1 Temperature dependent DC magnetic susceptibilities in $H = 1000$ G of the molecular sieve (a), the Co_3O_4 powders (b), and the diluted system of Co_3O_4 nanocrystals (c).

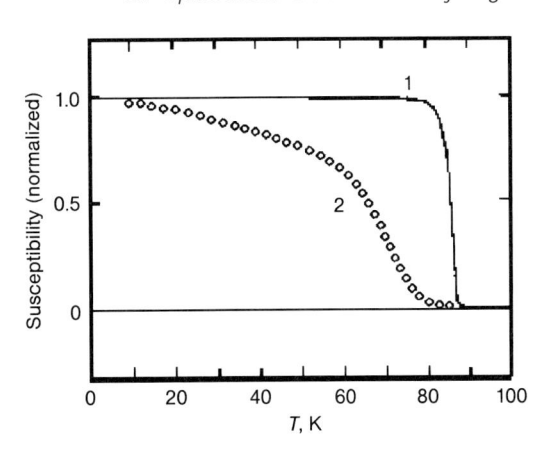

Figure 3.2 Real parts of the complex magnetic susceptibility for $Bi_2Sr_2Ca_{0.8}Y_{0.2}Cu_2O_{8.2}$ prepared by the *in situ* polymerizable complex method (1) and by the conventional solid-state reaction method (2).

An original method was proposed for the production of a uniform material with a mean particle size of <15 nm [24]. It is based on the use of atom transfer radical polymerization of styrene and modified magnetic $MnFe_2O_4$ (9 nm) as chemically attached macroinitiators. For this purpose, a reverse micelle microemulsion procedure can also be used [25]. Received core–shell nanoparticles show a core of 9.3 ± 1.5 nm with a 3.4 ± 0.8 nm polystyrene shell. Polymer particles without a magnetic core have not been observed. In such materials, a decrease in coercivity was observed, which is consistent with the reduction of magnetic surface anisotropy upon polymer coating.

It is interesting to form polymeric matrices including Fe_3O_4 and γ-Fe_2O_3 nanoparticles (20 nm) by the exchange reactions of Fe-containing salts with some perfluorated ion-exchange membranes (such as Nafion® 117, Du Pont) with a following alkaline hydrolysis of exchange product [26]. The sonochemical decomposition of $Fe(CO)_5$ was carried out in the presence of different surfactants, including polymeric surfactants, such as poly(vinylamine). It allows us to form coated Fe_2O_3 nanoparticles 5–16 nm in diameter [27]. In another method [28], Co^{2+} were adsorbed on an ion-exchange resin and later reduced by an excess of $NaBH_4$ that allows one to form (Co–B) alloys nanocomposites, having a very wide distribution over the sizes (3–30 nm). These cobalt–boron alloys are intensively studied owing to their superior properties (such as high strength, high magnetic permeability combined with high resistivity and large resistance to corrosion). One other similar method is based on the reduction of Co salts in a PVP (or THF) solution [14, 29]. A five times greater coercive force of (Co in PS)-nanomaterial can be achieved by low-temperature (below 720 K) annealing [10]. Cobalt carbonyl ($Co_2(CO)_8$) was introduced into a styrene copolymer matrix from solutions (in DMF or in isopropanol) and thermally treated at 470 K. The content of the spherical Co nanoparticles with a mean diameter of 2.6 nm increases to 8 wt.%, but this fact and the nature of the precursor had no considerable effect on the FMR (ferromagnetic resonance) spectrum [30]. The magnetic behavior of the Ni–LDPE and Co–LDPE systems depends on the synthesis conditions [31]. In the former system at high temperatures (>370 K) the Ni clusters irreversibly interact with the matrix; in the latter (Co–LDPE) the temperature rise to 470 K changes the thermomagnetic curve radically (as a result of coarsening of the Co particles).

The magnetic hysteresis of the samples obtained in a crazing mode of PP filled by nickel shows that the coercive force does not depend on the magnetic field direction (the parallel and perpendicular directions to the stretching axis) and equals 2.5 kA m^{-1}, though there is a considerable difference between the J_r and J_s values [32]. The magnetic behavior was studied for systems obtained by the reduction of metal salts in various polymeric media, such as a swollen interpolymeric PPA (polyacrylic acid) complex with urea formaldehyde oligomer [33], P4VPy and styrene copolymer with 4-vinylpyridine [34]. In these cases, the composites synthesized under a reduction of $FeCl_3$ (coordinated with the pyridine groups) by hydrazine and their following oxidation, contain spherical Fe_3O_4 nanoparticles (20–200 nm) with H_c values up to 31.5 kA m^{-1}. Nanocomposites with spherical γ-Fe_2O_3 nanoparticles (5 nm in diameter), obtained by alkaline-hydrolytic "development" of $FeCl_3$ macrocomplexes in block copolymers, also have similar characteristics [35]. Magnetic polymer composites were synthesized by thermodestruction of arenecyclopentadienyl–ferrum complexes [36] or by a thermodestruction of Co and Fe carbonyls in PVF matrices [37]. The Fe, Fe_2O_3, and Co nanoparticles embedded in the polyethylene matrix have been produced via thermal decomposition of iron pentacarbonyl $Fe(CO)_5$, iron(III) acetate $Fe(CH_3COO)_3$, and cobalt(II) formate $Co(HCOO)_2$, respectively, in the polyethylene–oil solution melt [38]. Also, core–shell Co-containing nanoparticles fixed on the surface of ultradispersed polytetrafluoroethylene (UPTFE) microgranules (which are a few hundred nm in size) have been produced via thermal decomposition of cobalt(II) acetate $Co(CH_3COO)_2$ in mineral oil containing UPTFE [39]. Metal-containing polymer matrices with magnetic properties can be synthesized by hydrogen reduction of cobalt-organic precursors [40] and Ni carbonyl in epoxy films in various hardening modes [41], as well as by tetrahydroborate reduction of nickel compounds into stretched and crazed PP matrices (with Ni crystallite size <40 nm) [32]. In some other methods polymeric composites with monodispersed Co nanoparticles (with diameter 1 nm) were obtained by metal spraying [40] or by introducing magnetoactive particles into various gels, such as silicone [43], polyacrylamide [44], polyethyleneoxide [45], etc.

There are approaches allowing the forming of some polymer-mediated self-organized magnetic nanoparticles with control of the assembly dimensions and thickness. For example, one of these methods relates to magnetic materials based on binary FePt [46] or CoPt [47] nanoparticles, formed by $Pt(AcAc)_2$, $FeCl_2$ reduction. Recently, a method of $Fe_{50}Pt_{50}$ nanoparticle production using PVP (poly-N-(vinyl-2-pyrrolidone) or PEI (polyethyleneimine) was described [48]. Such multilayer self-assembled particles were localized on a PEI-modified silicon oxide surface. The formation rate of the annealed thin film depends on the assembly thickness and increases with the annealing time and temperature (the optimum is achieved during 30 min at 770–870 K). Such layer-by-layer assembled nanoparticles have great potential for practical applications.

3.2.3
Preparation of Magnetic Polymer Nanocomposites in Magnetic Fields

Another very interesting method is the formation of polymer-immobilized particles in statu nascendi in magnetic fields when ferroplastics are first magnetically oriented in the superimposed fields. For example, neodium ferrite can be obtained in fields with magnetic induction of 3×10^{-4} to 3×10^{-3} and forms the final ferroplastics after the following mechanochemical treatment, such as thermal pressing [44, 49–51], Brigman anvil treatment [52], or a combination of different actions [53]. Ordered submicron ferromagnetic Co–B arrays (about 30 nm, aggregated up to 450–600 nm in size) were obtained [54] by $NaBH_4$ reduction, using the orientation in the magnetic field of the soluble macrocomplexes (Co^{2+} with starch). The driving of the magnetic field (1 T) and the mediation of soluble starch play a critical role in the formation of well-defined patterns. The prepared chains with composition $Co_{23}B_{10}$ exhibit strong ferromagnetic signals in the range 100–420 K. It is supposed that such a convenient strategy should be helpful in the production of ordered arrays of other magnetic materials. This process can be inspired by magnetotactic bacteria, which provide excellent templates to direct the synthesis of advanced materials. Such methods in principle allow the production of nanocomposites with maximal magnetic characteristics, especially if the anisotropic form nanoparticles are treated, such as, for example, magnetically hard barium and strontium ferrites, having a hexagonal crystalline lattice with monoaxial magnetic anisotropy (the single axis of easy magnetization in the monocrystal is directed perpendicular to the hexagonal basal plane). The orientation by linearly or circularly moving magnetic fields can be used practically for improvement of the electrophysical or magnetic characteristics of the composites and for the production of magnetic matrices with oriented chain-like structures that can be used in varnish and film technologies for information recording systems. The coating formation in the moving magnetic fields allows one to prevent the filler particle deposition and to concentrate them either nearly on the film surface or uniformly through the film volume.

3.2.4
Peculiarities of Magnetic Behavior of Metallic Nanoparticles in Polymer Matrix

A very perspective method is connected with the electrochemical formation of modified thin (3–7 nm) magnetic coatings on the self-assembling

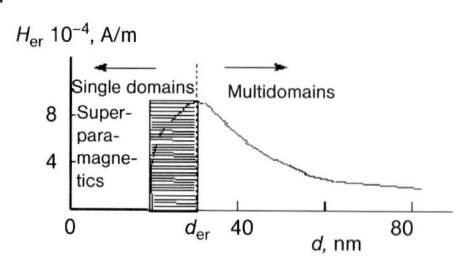

Figure 3.3 The dependence of the coercive force H_c on the ferrous nanoparticle diameter d.

Langmuir–Blodgett films [55]. The charged amphiphilic monolayer can be used as a template to organize inorganic materials by adsorption of ionic species along the interface. A special technique for studying the unusual magnetic properties of so-formed objects (on the base of $Cu_3[Fe(CN)_6]_2 \cdot 12H_2O$) was demonstrated in [56]. The results show that those multilayers present a spin ordering below 25 K. The coercive force of powder ferromagnetic materials depends on the particle sizes and forms, but the dimensional dependence of H_c is very peculiar. In systems with very big multidomain particles, the coercive force increases with the decrease in the particle size (Figure 3.3) [51].

It achieves the maximum value $H_{c,max}$ when the mean particle size reaches one domain dimension. For spherical particles the critical diameter (when the coercive force value equals $H_{c,max}$) can be calculated from the following equation [57]:

$$d_{cr} = (1.9/J_s)[10cI/(\mu_0 a N_p)]1/2,$$

where c is a constant depending on the lattice crystalline structure; I is the so-called exchange constant; a is the lattice constant; $N_p = -H_p/J$ is a demagnetization coefficient (for a demagnetization field H_p, it is directed opposite to the external field). For monodomain Fe and Co particles the d_{cr} value is ~20 nm, and for Ni it is ~60 nm at $H_{c,max}$ $10^3 - 10^4$ A m^{-1}.

When the monodomain particle size becomes less than critical (every such particle can be considered as a ferromagnetic particle, but with variable magnetization) a drop in the H_c value is observed. The total magnetization (arising under an external magnetic field H sufficient for saturation) relaxes to zero after removing the external field according to the law:

$$J = J_s \exp(-t/\tau),$$

where $\tau = \tau_0 \exp[E_M/(k_B T)]$ is the relaxation time; τ_0 is a constant (in seconds); $E_M = KV$ is the potential barrier (J); K is the constant of uniaxial anisotropy (J m^{-3}); and V is the particle volume.

Simultaneously small particles are similar to a paramagnetic substance with a large paramagnetic moment and can transfer to the superparamagnetic state. This process is a magnetic-phase transition of the second order, characterized by continuous change of the first derivatives of the thermodynamic potential. The properties of such supermagnetic systems can be characterized by the values of τ_0 and K. Manifestation of supermagnetism is often indirect evidence for the ferromagnetic phase low content compared with the filler total content. For example, in the Ni-crazed PP system [32], there is a large number of Ni superparamagnetic nanoparticles (<10 nm) in the samples, whose magnetization (at $H = 75$ kA m^{-1}) is considerably higher than that for block nickel, which leads to an apparent decrease in nickel mass in the whole. Superparamagnetic particles were found in many metal-lopolymeric nanocomposites, for example, in the following systems: Nafion® 117–Fe$_3$O$_4$ [26]; poly(dimethylphenylenoxide)-α-Fe (spherical particles with diameter $d \leq 2.5$ nm, $K_{\text{eff}} = 1.83 \times 10^6$ J m^{-3} at 100 K) [58]; in block copolymers γ-Fe$_2$O$_3$ ($d \approx 5$ nm, $K = 1.58 \times 10^5$ J m^{-3}, $\tau_0 = 4.2 \times 10^{-12}$ s) [59]. Possible ways to increase the relative magnetic permeability were studied by scanning a polymer composite with various dispersed hybrid fillers [60]. Such compositions were obtained on the base of composites PE–amorphous metal-locomposite (69% Co, 12% B, 4% Si, 4% Ni, 2% Mo), PE–nickel–zinc ferrite, PE–metallocomposite ferromagnetic (80% Ni, 5% Mo, 0.5% Mg, 0.15% Si, 0.1% C) with maximum density and they are of great interest for the fabrication of permanent metallopolymeric magnets with optimal technical parameters.

Another group of magnetic nanocomposites must be especially noted. We are dealing with the problem of the electromagnetic wave–matter interactions in the microwave range that is created mainly by the need to improve the stealthiness of aircraft, missiles, and ships, against detection by radar receivers. The best way to produce such materials is to form magnetic nanocomposites to cover the metallic surface with microwave radar absorption (0.1–18 GHz range, in general, in X-band, 8.2–12.4 GHz and P-band, 12.4–18 GHz) [61]. A ceramic microwave absorbent can be obtained by three main methods [18]: (a) mixing of micronic or even submicron metal powders within a liquid matrix precursor; (b) preparation of a porous host matrix impregnated by a concentrated solution of transition metal nitrates salts; (c) mixing of alkoxides with an aqueous solution of metal ions. The last method was used [62] for polymeric nanocomposite fabrication from complex ferrite Ni$_{0.5}$Zn$_{0.4}$Cu$_{0.1}$Fe$_2$O$_3$ or CoFe$_2$O$_3$ by adding conducting materials (5 wt.% of graphite powder). This complex ferrite was obtained as a copreciptant at various pH values after drying and silanization by 3-aminopropyltrimetoxysilane or 3-methacryloxypropyltrimethoxysilane until a three-dimensional polysiloxane network was formed. Finally, the silanized Ni$_{0.5}$Zn$_{0.4}$Cu$_{0.1}$Fe$_2$O$_3$ or CoFe$_2$O$_3$ was dispersed in MMA and its radical suspension polymerization was carried out. The permittivity is affected by the sample thickness (4–10 nm) and nanofiller concentration in the polymeric matrix, as well as by the thermal treatment conditions. The optimal variant presents nanocomposites (with 20 wt.% of complex ferrite, thickness $\sim$10 nm) that show strong microwave absorption at 15.2 GHz with magnitude of reflection loss of -25.4 dB. Polymer composites, containing conducting fillers or ferroelectric particles, are effective microwave adsorption materials important to suppress microwaves reflected from metal structures (in both civil and stealth defense system for military platforms). These multicomponent materials can be effective absorbers of electromagnetic waves due to the large imaginary part of the complex dielectric permittivity responsible for dissipation of electromagnetic energy and due to their low bulk conductivity. Fibrous magnetic materials can also be produced using the technology of joint spraying in a gaseous current (the melt-blowing technique) of polymer (LDPE) melts and a highly dispersed ($\sim$1000 nm) filler of strontium or barium ferrite [63]. Such composites can be used for gas and liquid filtration, oil product sorption, industrial and domestic waste, etc. In recent years, there has been increased interest in so-called magnetic fluids, being colloidal solutions with very strong magnetic properties. Polymers such as oligoorganosiloxanes can be used as carriers capable of working under extreme conditions (in a wide temperature range, in vacuum, in aggressive or biologically active media). Recent investigations of composites with conducting and ferroelectric particles showed that they really possess some specific characteristic properties of interest. Such materials have also been proposed for use as sound absorbers and damping materials. In this case, the elastic waves interact with ferroelectric particles and elastic energy is transformed into electrical energy that is dissipated in carbon black particle chains [64]. Another alternative way in the problem of magnetic conductivity in the metal-containing polymeric systems is the creation of ferromagnetic element-organic compounds [65], and multispinous paramagnetic molecules on the base, for example, of polymetallorganosiloxanes [66].

Thus, matrix-stabilized magnetic nanoparticles can be prepared by different methods to make various forms such as polycrystalline, aggregates, thin and thick films, and single crystals. A common problem in many cases is the agglomeration of particles during the precipitation or removal of the solvent. Relatively simple and reproducable approaches for the synthesis of crysrtalline nanoparticles of controllable size are of great fundamental and technological interest. Preparation of high-quality nanocrystals is important for investigation of their size-determined properties. The synthesis of high-quality transition metal nanocrystals requires often high-temperature furnace processing because these particles do not crystallize well at room temperature. The processability and applications of polymer-bonded magnet are also limited because of their poor heat resistance properties. Therefore, there is a need to develop polymer-bonded magnets with enhanced magnetic properties for high-temperature and agressive environments.

Below we consider in detail one of the ways of preparing magnetic nanomaterials from a choice and type of precursors as well as kinetic peculiarities of synthesis of nanocomposites and properties of the products obtained. This approach and experimental techniques used are rather demonstrative and allow one to observe potential and limitations of directed synthesis of magnetic nanomaterials.

3.3 Magnetic Metal Nanoparticles in Stabilizing the Polymer Matrix *In Situ* via Thermal Transformations of Metal-Containing Monomers

The search and investigation of the self-regulated systems, in which both the synthesis of polymeric matrix and generation and growth of nanoparticles proceed simultaneously, can be useful for the solution of nanoparticle stabilization problems. For synthesis of matrix-stabilized nanoparticles the thermal transformations of corresponding metal-containing monomers in the solid phase are of great interest. The main advantage of the synthesis of composites by thermal transformation of metal-containing precursors is in the possibility of formation of nanocomposites with a relatively high concentration of metal phase. Another advantage deals with technological simplicity and the ease to control the processes and the properties of material obtained.

Thermal transformations go through three main macrostages with different temperatures [67–75]: (1) Initial monomer dehydration (desolvation) at 303–473 K; (2) the stage of solid-state homo-and copolymerization of the dehydrated monomer (at 473–573 K); (3) the produced polymer decarboxylation to a metal-containing phase and oxygen-free polymer matrix at $T_{ex} > 523$ K with an intense gas emission. The analysis of the thermal transformation of the gaseous products and the composition of the solid product (decarboxylated polymer, including metal or its oxide) allows determining a general scheme of metal acrylate thermal transformations:

Initiation

$$M(CH_2=CHCOO)_n \longrightarrow M(CH_2=CHCOO)_{n-1} + CH_2=CHCOO^\bullet (R^I).$$

The formed R^I initiates the polymerization to produce the linear or networked polymers.

Polymerization

$$R^I + s[CH_2=CHCOO]\,M^{n+}{}_{1/n} \longrightarrow R^I - [-CH_2-CH_{COOM1/n}-]_s - R^I.$$

With temperature the metal-containing fragments of formed polymers decompose to produce a metal (or its oxide) and CO_2.

Decarboxylation

$$R^I - [-CH_2-CH_{COOM1/n}-]_s - R^I \longrightarrow CH_2=CH-CH$$
$$=CH-[-CH_2-CH=CH-CH_2-]_{s/2} -CH$$
$$=CH-CH=CH_2 + 2(s+1)CO_2 + (s+2)M.$$

The polymers resulted in the decarboxylation reaction can be additionally thermopolymerized to form the net structure with conjugated multiple bonds.

In general, the composition of solid-phase products of thermolysis can be represented as a sum of the C–H–O-fragment fractions:

$$MO_z(CH_2CHCOO)_{p-x}(CHCHCOO)_{q-y}(CH_2CH)_x(CHCH)_\mathcal{H}.$$

The metal oxides can be formed by the following oxidation reactions:

$$M + \lambda_1 CO_2 = MO_z + (\lambda_1 - z)CO_2 + zCO,$$

$$M + \lambda_2 H_2O = MO_z + (\lambda_2 - z)H_2O + zH_2.$$

One of the main transformations is the origin of the acrylic $CH_2=CHCOO$ radical in the primary decomposition act that initiates metal-containing monomer polymerization with a consequent decarboxylation of metal-containing units. The process temperature has an impact on the product yield and their composition. At high temperatures, the decarboxylation is accompanied by practically complete removal of the oxygen-containing units from the polymer matrix.

3.3.1
The Kinetics of Thermolysis of Metal-Containing Monomers

The thermal transformations of a series of unsaturated metal carboxylates [67–75] (transition metal acrylates: $Cu(OCOCH=CH_2)_2$ ($CuAcr_2$), $Co(OCOCH=CH_2)_2 \cdot H_2O$ ($CoAcr_2$), $Ni(OCOCH=CH_2)_2 \cdot H_2O$ ($NiAcr_2$); cluster acrylate: $[Fe_3O(OCOCH=CH_2)_6 \cdot 3H_2O]OH$ ($FeAcr_3$); monomer cocrystallites: $CoAcr_2$ or $NiAcr_2$–$FeAcr_3$ with a Fe:Co atomic ratio of 1:0.8 and 2:1 and Fe:Ni = 2:1; cobalt and iron maleates: $Co[OOCCH=CHCOO] \cdot 2H_2O$, $Fe_3O(OH)[OOCCH=CHCOOH]_6 \cdot 3H_2O$ maleates) and acrylamide complexes of metal nitrates (CoAAm, FeAAm) have been studied. Thermolysis of monomers under study is accompanied by the gas evolution and the mass loss due to the dehydration followed by thermal transformation of dehydrated specimens. The process stages proceed sequentially in different temperature ranges.

3.3.1.1 Dehydration
At low thermolysis temperatures ($T_{term} < 473$ K), the dehydration of monomer crystal hydrates occurs. From the data of thermal analysis (DTA, TG, DTG), the temperature ranges of dehydration of the monomers have been found to be 353–487 ($FeAcr_3$), 413–453 ($CoAcr_2$), 373–473 ($NiAcr_2$, see Refs. [76–78]), 393–433 (CoMal), and 373–433 K (FeMal). It was shown [68] that under isothermal conditions at $\langle T_{term} \rangle = 303$–433 K dehydration of $CoAcr_2$ was reversible. This was also confirmed by the results of IR spectroscopy. The dehydration resulted in a disappearance of absorption bands [79] attributed to the modes of crystal water.

In a number of cases one can distinguish two temperature ranges of vaporization. Thus, in the dehydration of CoAcr$_2$ under isothermal conditions at 303–348 K vaporization is limited by evaporation of the liquid phase, $P_{H_2O}(T_{term}) = 1.7 \times 10^7 \exp[-9200/RT)]$ KPa, and the heat of vaporization, $\Delta H_{st}(H_2O) = 38.5$ kJ mol^{-1}, which is close to that of pure H_2O, $\Delta H_{evap}(H_2O) = 43.9$ kJ mol^{-1} [80]. The vaporization at $T_{term} > 348$ K is determined by the dependence of $P_{H_2O}(T_{term}) = 2.7 \times 10^4 \exp[-4800/RT)]$ kPa and is limited by the dehydration of CoAcr$_2$, $\Delta H_{evap}(H_2O) = 20.1$ kJ mol^{-1}. It should be mentioned that the regime of dehydration of crystal hydrates can affect the reactivity and subsequent thermal transformations of dehydrated products [81, 82].

3.3.1.2 Polymerization

The increase in temperature up to $\langle T_{term} \rangle = 473$–573 K leads to the solid-phase polymerization of dehydrated monomer. In this temperature range, under conditions of both thermal analysis and self-generated atmosphere (SGA), the thermal transformations of the monomers are accompanied by slight mass loss ($\ll 10$ wt.%) and a slight gas evolution. In the case of acrylates and maleates, in addition to CO_2 the gaseous products contain vapors of CH_2=CHC(O)OH and HOC(O)CH=CHC(O)OH, which condense on the "cool" (T_{room}) wall of the reactor vessel. This is confirmed by the results of IR- and mass-spectroscopic studies. According to the data of TA [73–78, 83], representative temperatures (T_{polym}) at which polymerization proceeds are as follows: ~543 K (CoAcr$_2$), ~563 K (NiAcr$_2$), ~510 K (CuAcr$_2$), ~518 K (FeAcr$_3$), 488–518 K (CoMal), ~518 K (FeMal), 364 K (CoAAm). The EXAFS study of the solid-phase product at the initial stage of thermolysis ($t_{term} = 5$ min at 643 K) also indicates the rearrangement of ligand environment of FeMal during its dehydration and polymerization [74, 75]. As a result, Fe atoms coordinate O atoms of both carboxylate groups of maleate anion.

Temperature scanning by the optical spectroscopy technique reveals the main sequences of thermal transformations of the CoAAm complex. According to the thermometric analysis, the $T = f(t)$ curve (Figure 3.4) has several quasistationary portions at 320, 341, 356, 407, 453, and 582 K, which are consistent with the basic stages in thermal transformations of the cobalt nitrate acrylamide complex.

3.3.1.3 Kinetics of Decarboxylation

At $T_{term} > 523$ K (for CuAcr$_2$ $T_{term} > 453$ K) the intensive gas evolution of thermally polymerized samples is observed. The kinetic peculiarities of this process were studied under isothermal SGA-conditions for CuAcr$_2$ ($\langle T_{term} \rangle =$ 363–513 K), CoAcr$_2$ (623–663 K), NiAcr$_2$ (573–633 K), FeAcr$_3$ (473–643 K), FeCoAcr (613–633 K), Fe$_2$CoAcr (613–633 K), Fe$_2$NiAcr (603–643 K), CoMal (613–643 K), FeMal (573–643 K), CoAAm (463–503 K). The rate of gas

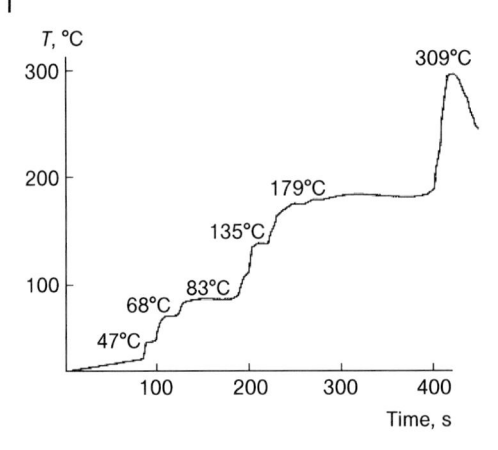

Figure 3.4 Thermogram of Co(NO3)$_2 \cdot$ (AAm)$_4 \cdot$ 2H$_2$O. The quartz ampule with $d = 20$ mm and $l = 30$ mm; a chromel–alumel thermocouple.

evolution, $W = d\eta/dt$, decreases monotonically with the degree of conversion, $\eta = \Delta\alpha_{\Sigma,t}/\Delta\alpha_{\Sigma,f}$, where $\Delta\alpha_{\Sigma,t} = \alpha_{\Sigma,t} - \alpha_{\Sigma,0}$, $\Delta\alpha_{\Sigma,f} = \alpha_{\Sigma,f} - \alpha_{\Sigma,0}$, $\alpha_{\Sigma,f}$, $\alpha_{\Sigma,t}$, and $\alpha_{\Sigma,0}$ are the final, current (corresponding to time t) and the initial number of moles of gaseous products released per mole of the starting substance at T_{room}, respectively. The kinetics of gas evolution $\eta(\tau)$ in the general case (up to $\eta \leq 0.95$) is satisfactorily approximated by the equation for two parallel reactions:

$$\eta(\tau) = \eta_{1f}[1 - \exp(-k_1\tau)] + (1 - \eta_{1f})[1 - \exp(-k_2\tau)], \qquad (3.1)$$

where $\tau = t - t_0$ (t_0 is the time of heating); $\eta_{1f} = \eta(\tau)|_{k_2t \to 0, k_1t \to \infty}$, k_1, k_2 are the effective rate constants. The k_1, k_2, η_{1f}, and $\Delta\alpha_{\Sigma,f}$ parameters depend on T_{term} as follows:

$$\eta_{1f}, \Delta\alpha_{\Sigma,f} = A\exp[-E_{a,\text{eff}}/(RT_{\text{term}})],$$

$$k_{\text{eff}} = k_{0,\text{eff}}\exp[-E_{a,\text{eff}}/(RT_{\text{term}})],$$

where A, $k_{0,\text{eff}}$ are the preexponential coefficients (s^{-1}), $E_{a,\text{eff}}$ is the activation energy (kJ mol^{-1}).

The initial rate of gas evolution $W_{\tau=0} = W_0$ will be

$$W_0 = \eta_{1f}k_1 + (1 - \eta_{1f})k_2. \qquad (3.2)$$

The kinetics of gas evolution in thermolysis of NiAcr$_2$, FeCoAcr, Fe$_2$CoAcr, Fe$_2$NiAcr, and FeMal$_3$ is described by Eqs. (3.1) and (3.2).

At $k_2 \approx 0$, $\eta_{1f} \to 1$ and one can write

$$\eta(\tau) \approx 1 - \exp(-k_1\tau), \quad W_0 \approx k_1. \qquad (3.3)$$

These kinetic equations were used to describe the thermolysis of CoAcr$_2$, and CoMal.

When $\tau \ll 1/k_2$, $k_1 \gg k_2$, and one obtains

$$\eta(\tau) \approx \eta_{1f}[1 - \exp(-k_1\tau)] + (1 - \eta_{1f})k_2\tau, \quad W_0 \approx \eta_{1f}k_1. \qquad (3.4)$$

Equation (3.4) describes the kinetics of gas evolution of $CuAcr_2$. The kinetic parameters of thermal transformation of compounds under study are shown in Table 3.1.

The change of m_0/V does not affect the rate of thermolysis. It should be noted that two gas evolution regions are observed in the decomposition of $FeAcr_3$: low-temperature region ($\langle T_{term} \rangle = 473 - 573$ K) and high-temperature region ($\langle T_{term} \rangle = 603 - 643$ K) (Figure 3.5). Here the rate of gas evolution is well approximated by Eq. (3.3) but with different values of k and $\Delta\alpha_{\Sigma,f}$ (see Table 3.1). Most probably, the differences of kinetic parameters in low- and high-temperature regions for thermolysis of $FeAcr_3$ as well as the variation of η_{1f} values at constant values of other kinetic parameters ($\Delta\alpha_{\Sigma,f}$, k_1, k_2) in the case of FeCoAcr, Fe_2CoAcr, and Fe_2NiAcr are determined by the occurrence of two parallel processes of gas evolution.

According to the W_0 values, the reactivity of metal acrylates in thermolysis decreases as follows: Cu ≥ Fe > Co > Ni. The values of effective activation energies of gas evolution for $CuAcr_2$ ($E_{a,eff} = 202.7$ kJ mol^{-1}) and $NiAcr_2$ ($E_{a,eff} = 246.6$ kJ mol^{-1}) thermolysis under SGA-conditions are close to the calculated one $E_{a,eff}$ for thermolysis at the TA-regime [76]: 211.1 and 244.1 kJ mol^{-1}. At the same time, for $CoAcr_2$ in the SGA-regime, the value of $E_{a,eff}$ (238.3 kJ mol^{-1}) of thermolysis is higher than that of thermolysis under TA-conditions ($E_{a,eff} = 206.1$ kJ mol^{-1}). It is of interest that the difference in rate constants of gas evolution observed at thermolysis of cobalt acrylate and maleate. Activation parameters of the rate of gas evolution for FeMal are close to those for $FeAcr_3$, FeCoAcr, Fe_2CoAcr, and Fe_2NiAcr in the same region of T_{term}.

Qualitative phase analysis performed by time-resolution X-ray diffractometry during frontal polymerization of CoAAm showed (Figure 3.6) that below 323 K, no changes in the diffraction pattern characteristic of the starting crystalline monomer take place. Above this temperature, weak reflections appear which indicate nucleation of a new phase. Aging at a constant temperature (343 K) brings about an increase in the intensity of reflections corresponding to this phase, and within several minutes, a number of additional peaks attributed to the other high-temperature phase emerge on the X-ray pattern. At 363 K, these transformations stop and a pure high-temperature phase preceding polymerization appears. The final X-ray pattern corresponds to the amorphous polymer; however, it exhibits a well-pronounced reflection with an interplanar distance of 8.6 Å. This is in keeping with the corresponding reflection corresponding to the monomer and suggests a certain crystallinity of the resulting product. This observation agrees with the presence of at least short order in the polymerizing system [84].

Table 3.1 Kinetic parameters of gas evolution at thermolysis of transition metal acrylates and maleates.

Sample	$\eta_{1f}, \Delta\alpha_{\Sigma,f}$	$\eta_{1f}, \Delta\alpha_{\Sigma,f} = A\exp[-E_{a,eff}/(RT_{term})]$		k	$k_{eff} = k_{0,eff}\exp[-E_{a,eff}/(RT_{term})]$		$W_0 = k_{0,eff}\exp[-E_{a,eff}/(RT_{term})]$	
		A	$E_{a,eff}$ (kJ/mol)		$k_{0,eff}$ (s^{-1})	$E_{a,eff}$ (kJ/mol)	$k_{0,eff}$ (s^{-1})	$E_{a,eff}$ (kJ/mol)
CuAcr$_2$	η_{1f},	1.8×10^4	48.1	k_1	9.5×10^{11}	154.7	9.5×10^{11}	202.7
	$\Delta\alpha_{\Sigma,f}$	3.6	12.5	k_2	9.2×10^{11}	163.0		
CoAcr$_2$	η_{1f},	1.0	0	k_1	3.0×10^{14}	238.3	3.0×10^{14}	238.3
	$\Delta\alpha_{\Sigma,f}$	1.55	0	k_2		0		
FeAcr$_3$	η_{1f},	1.0	0	k_1	4.2×10^{21}	246.5	4.2×10^{21}	246.5
(473–573 K)	$\Delta\alpha_{\Sigma,f}$	1.6×10^2	25.5	k_2		0		
(573–643 K)	η_{1f},	1.0	0	k_1	1.3×10^6	127.5	1.3×10^6	127.5
	$\Delta\alpha_{\Sigma,f}$	1.7×10^2	26.3	k_2		0		
NiAcr$_2$	η_{1f},	2.6	4.6	k_1	1.7×10^{17}	242.4	4.4×10^{17}	247.0
	$\Delta\alpha_{\Sigma,f}$ <593 K	1.4×10^{11}	125.4					
	$\Delta\alpha_{\Sigma,f}$ >633 K	1.2	10.5	k_2	7.5×10^8 0	156.7		
FeCoAcr	$\eta_{1f} = 0.45$ (663 K)–0.65(613 K)			k_1	2.3×10^{12}	207.0	1.3×10^{12}	207.0
	$\Delta\alpha_{\Sigma,f}$	5.25×10^2	31.3	k_2	6.0×10^6	138.0		
Fe$_2$CoAcr	$\eta_{1f} = 0.35$ (663 K)–0.50(613 K)			k_1	2.6×10^{12}	205.0	1.1×10^{12}	205.0
	$\Delta\alpha_{\Sigma,f}$	1.5×10^2	25.1	k_2	6.6×10^5	125.5		
Fe$_2$NiAcr	η_{1f},	4.4×10^7	75.0	k_1	6.1×10^6	129.5	2.7×10^{14}	205.0
	$\Delta\alpha_{\Sigma,f}$	6.5×10^2	25.5	k_2	0.6×10^2	79.4		
CoMal	η_{1f},	1.0	0	k_1	1.1×10^6	125.4	1.1×10^6	125.4
	$\Delta\alpha_{\Sigma,f}$	1.3×10^2	23.4	k_2	0			
FeMal$_3$	η_{1f},	0.59×10^2	23.4	k_1	3.3×10^7	133.8	1.9×10^9	157.2
	$\Delta\alpha_{\Sigma,f} = 4.78(573$ K$)$–$7.40(643$ K$)$			k_2	1.0×10^7	110.8		

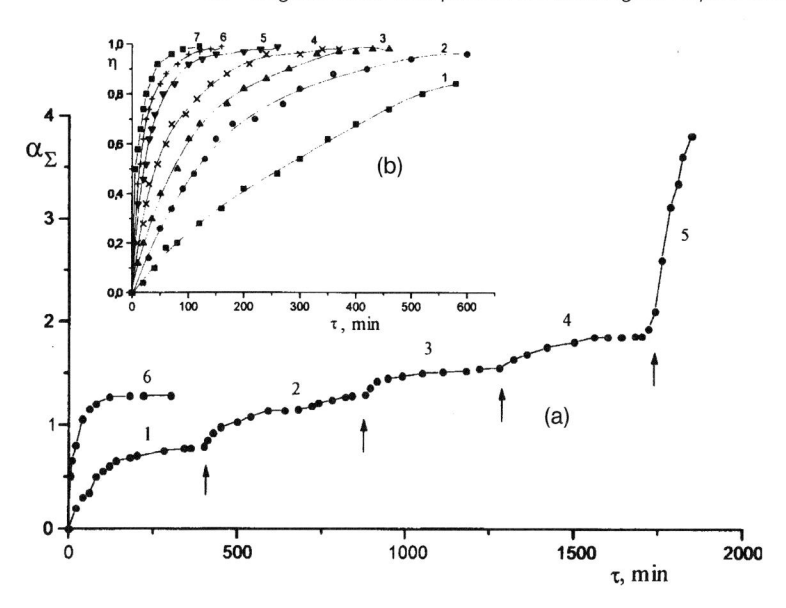

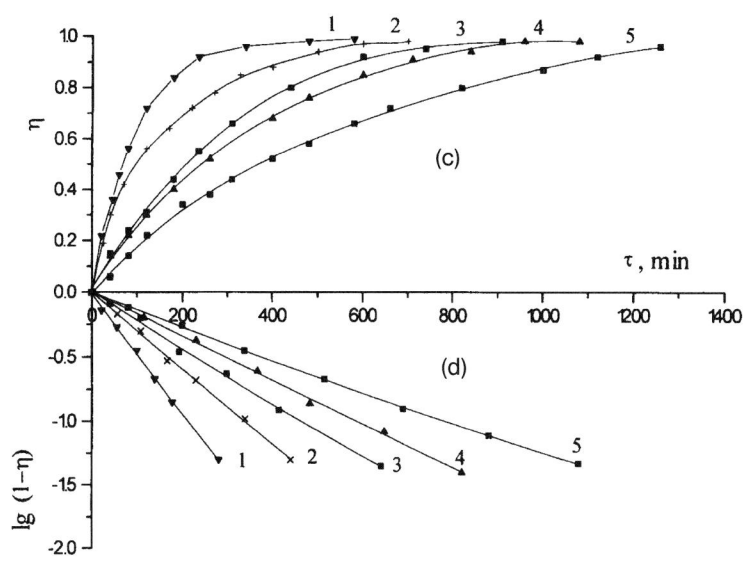

Figure 3.5 (a) The kinetics of gas evolution from FeAcr$_3$ at T_{exp} (°C): 1, 215; 2, 250; 3, 275; 4, 300; 5, 350; 6, 240. The moment of T_{exp} increasing is shown by pointer. (b) Dependence η(T) on T_{exp} (°C): 1, 200; 2, 205; 3, 210; 4, 215; 5, 220; 6, 230; 7, 240 ($m_o/v = 6.7 \times 10^{-3}$ g cm^3 where m_o is start mass of sample). (c) The dependence η(t) on $T_{exp} = 300 - 370\,°C$: 1, 370; 2, 360; 3, 350; 4, 340; 5, 330 °C. (d) The semilogarithmic anamorphous of the dependence η(t).

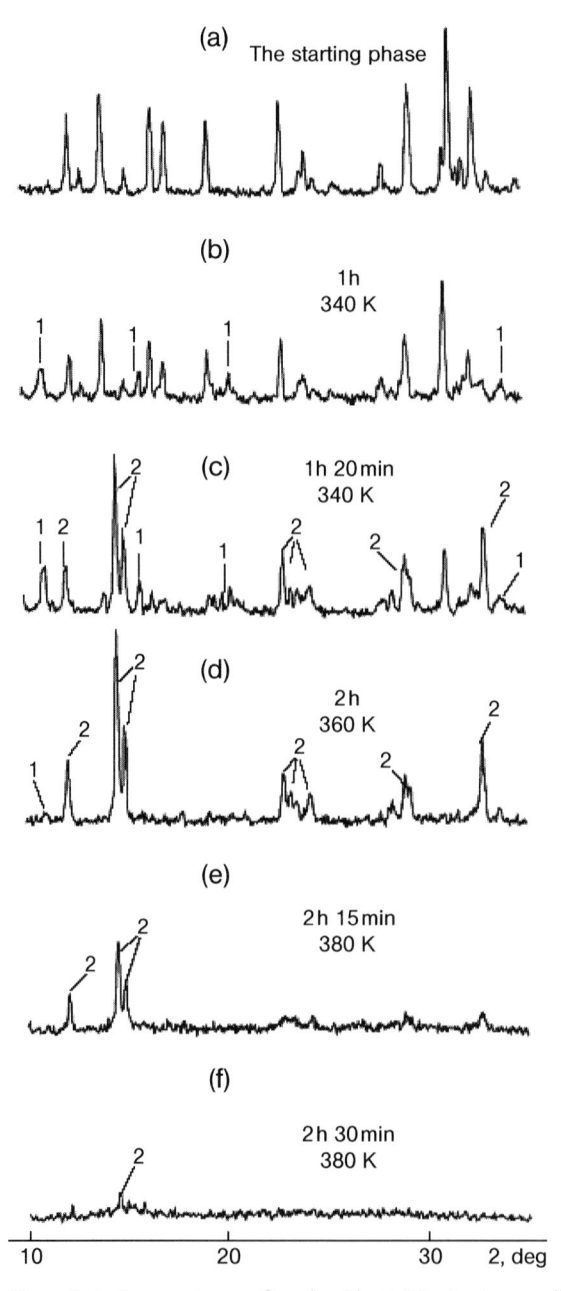

Figure 3.6 X-ray patterns of acrylamide Co(II) nitrate complex samples: (a) the starting monomer, (b, c) the appearance of monocrystal hydrate 1, (d) anhydrous phase 2, (e) a mixture of phase 2 and the polymer product, and (f) the polymer product.

Given this, each stage characterizes a certain step during crystal structure disintegration. Actually, the experiments lend support to the view that the thermal transformation of the system under consideration proceeds in stages: the starting monomer initially loses one water molecule to give rise to monocrystal hydrate, elimination of the second water molecule promotes formation of an anhydrous phase, and finally, polymerization takes place. Special studies demonstrated that dehydration of the monomer at 343 K within 12 h is accompanied by weight losses corresponding to elimination of two water molecules. The X-ray diffraction pattern of this sample is analogous to that of the anhydrous phase. It should be noticed that dehydration is a reversible process: according to X-ray diffraction analysis, the anhydrous phase converts via one step into crystal hydrate containing two water molecules.

3.3.2
The Topography and Structure of Magnetic Metallopolymer Nanocomposites

During thermolysis the sample-specific surface $S_{sp,f}$ was found to increase in the case of $CuAcr_2$, $CoAcr_2$, and $NiAcr_2$. The major changes occur at the earlier stage of conversion, and at the end, the $S_{sp,f}$ value exceeds that of $S_{sp,0}$ by two to three times (Table 3.2). In the case of $CuAcr_2$, the behavior of $S_{sp,f}(T_{term})$ is peculiar. First the $S_{sp,f}$ value increases with T_{term} up to 493 K, and then it decreases. A decrease in $S_{sp,f}$ at $T_{term} > 493$ K is apparently determined by caking of particles. However, in the thermal transformation of $FeAcr_3$, $FeCoAcr$, Fe_2CoAcr, Fe_2NiAcr, $CoMal$, and $FeMal$, the values of $S_{sp,f}$ do not substantially change.

Table 3.2 Dispersity of starting metal carboxylate samples and their thermolysis products.

Sample	$S_{0,sp}$ (m²/g)	$S_{f,sp}$ (m²/g)	$L_{OM,av}$ (μm)
$CuAcr_2$	14.7	48.0 (463 K)–53.8 (473 K)–43.8 (503 K)	5–50
$CoAcr_2$	20.2	24.1 (623 K)–42.1 (663 K)	100–150
$FeAcr_3$	15.0	15.0	1–5
$NiAcr_2$	16.0	55.0–60.5	60–100
$FeCoAcr$	9.0	13.6	5–10
Fe_2CoAcr	8.1	11.3	10–15
Fe_2NiAcr	8.5	13.5	100–200
$CoMal$	30.0	30.0	5–70
$FeMal$	24.0	26.0	30–50

The analysis of the data on specific surface and topography of the solid phase both during and at the end of thermolysis leads to a supposition that metal carboxylates studied have common properties.

(i) At a low degree of gas evolution (during the samples heating) a remarkable loss of the particle transparency is observed. The surface of particles becomes rough due to desolvation processes and dehydration, in particular. Such surface structure is typical especially for $CuAcr_2$.

(ii) At a low degree of conversion, the crystalline particles, in particular CoMal (fraction 2), lose their ability to rotate the polarized plane of transmitted light. This indicates that the sample became amorphous and it is, apparently, associated with high rates of dehydration and polymerization, which occur before the major gas evolution.

(iii) In the case of $CoAcr_2$, $NiAcr_2$, and $CuAcr_2$, the transformation process is accompanied by material dispersion so that the sizes of the formed particles appear to be within the 1 – 10 µm range, and hence, the $S_{sp,f}$ value increases. For $FeAcr_3$, $FeCoAcr$, Fe_2CoAcr, Fe_2NiAcr, CoMal, and FeMal, no changes in both the particle dimensions and the $S_{sp,f}$ values are observed at the end of gas evolution. During the gas evolution (at mass loss of 15–30%), the initially small particles of $FeAcr_3$, $FeCoAcr$, and Fe_2CoAcr enlarge their size due to the formation of agglomerates consisting of porous, fragile glass-like plates with an average size of 20–100 µm. The agglomerates can be mechanically separated into irregularly shaped small particles of 1–3 µm size.

(iv) During transformation, the sample transparency diminishes to become opaque by the end of the process. This indicates that the process proceeds homogeneously in the whole sample volume. The portion of these particles constitutes 40–60% of the total volume. Together with the general loss of the particle transparency, the transformation process is also developed in the macrodefect regions in the form of small opaque zones. The appearance of a dark film on the surface of the particles in the course of conversion points to the occurrence of surface reactions. The appearance of a dislocation-decorated network means that the reaction regions are localized on growth defects [9].

(v) At the later stage of thermolysis, small opaque particles (<1 µm) are formed. In some cases ($FeAcr_3$, $NiAcr_2$, $FeCoAcr$, Fe_2CoAcr) fractal-like structures such as a chain of agglomerates with an average length of 50–70 µm are observed. They consist of five to seven agglomerates of small particles.

The results of the electron microscopic study of the products of metal-containing monomer transformations show that $FeAcr_3$, CoMal, Fe_2NiAcr,

Table 3.3 Average size of spherical clusters in decarboxylated matrix.

Sample	FeAcr$_3$	FeCoAcr	Fe$_2$CoAcr	Fe$_2$NiAcr	CoMal
$d_{EM,av}$ (nm)	7–9	5–6	5–6	6–8	3–4

FeCoAcr, and Fe$_2$CoAcr have similar morphology. It is characterized by practically spherical electron-dense clusters having a narrow size distribution and allotted in a low electron-dense matrix. The clusters are either single particles or agglomerates of 3–10 particles. The average diameters ($d_{EM,av}$) of clusters are listed in Table 3.3. The nanosized particles are uniformly distributed in the matrix at an average distance of 8–10 nm. In the CoMal sample along with nano-sized spherical clusters, the formation of relatively large aggregates such as cubic crystals of 10–20 nm size is observed.

The distinct crystalline reflections, which appeared in the pattern, matched well with the characteristic lines of metallic cobalt (Figure 3.7) for the product of CoAAm thermolysis. The broad spectra suggested a nanocrystalline structure. In fact, the crystallites are very fine and their size depends on the thermolysis temperature. The Scherrer assessment gave the mean crystallite size 7 and 20 nm, for the samples thermolyzed at temperatures 873 and 1073 K, respectively. The latter temperature resulted also in the transformation of the single-phase cobalt in the product formed at 873 K (Figure 3.7(a)) into two forms differing by the lattice constants: (i) cubic – $a_1 = 3.54470$ Å, (ii) cubic $-a_2 = 3.61265$ Å, represented by double peaks on the X-ray pattern. The second contained some amount of dissolved carbon (CoC$_x$).

The TEM images (Figure 3.8) proved that the crystallites obtained at the thermolysis temperature of 873 K had a uniform size close to 7 nm, which well corresponds to the value obtained from the X-ray pattern. Increasing

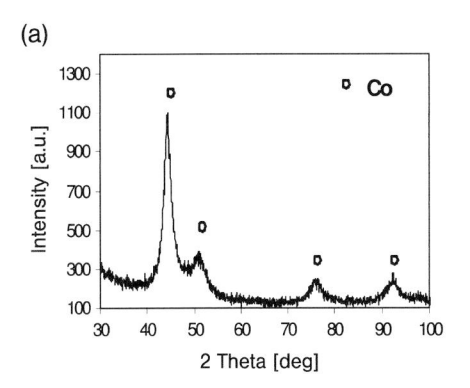

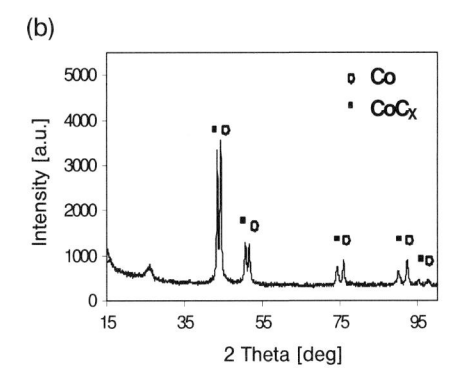

Figure 3.7 X-ray patterns of the CoAAm complex thermolyzed at temperatures (a) 873 K and (b) 1073 K.

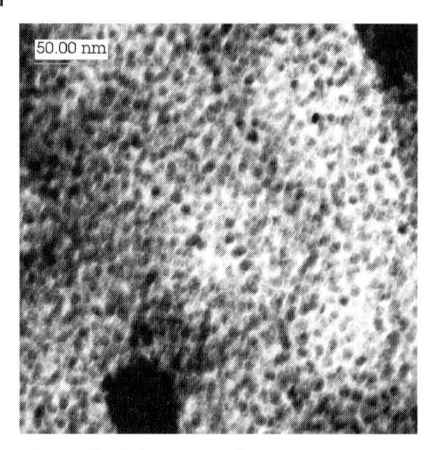

Figure 3.8 TEM microstructures of nanocomposite obtained by thermolysis of frontally polymerized CoAAm at temperature 873 K.

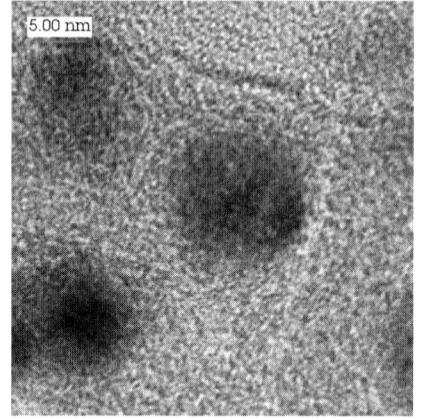

the thermolysis temperature resulted in coarsening of the microstructure and agglomeration of the nanoparticles.

The composition of the metal-containing phase in the solid-phase products depends on T_{term} at the end of gas evolution. The fraction of the metal-oxide phase increases with T_{term} due to acceleration of the oxidation reactions.

3.3.3
The Magnetic Properties of the Metallopolymer Nanocomposites

The formation of magnetic particles proceeds during decarboxylation of metal-containing carboxylate groups at the final stages of thermal transformation. This is confirmed by the dynamics of magnetic properties (Figure 3.9): a drastic increase in χ_σ, σ_s, H_c values is associated with formation of the $CoFe_2O_4$ phase during the thermolysis of Co–Fe acrylate cocrystallites.

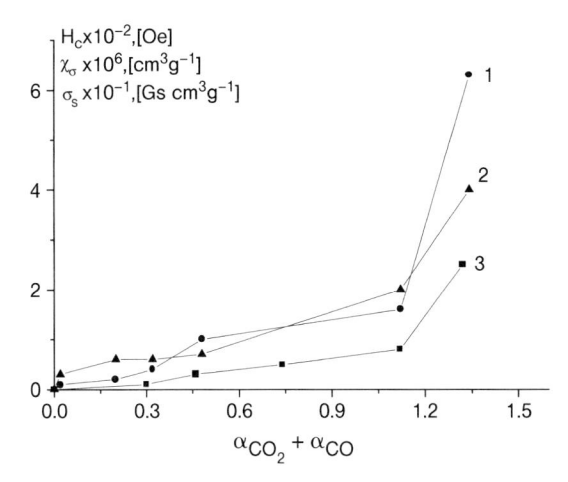

Figure 3.9 The dependences of $H_C(1)$, $\sigma_S(2)$, and χ_σ (3) on $\alpha(CO_2) + \alpha(CO)$ for FeCoAcr ($T_{exp} = 573$ K).

Room temperature Mössbauer spectra for the starting iron oxoacrylate and specimens annealed at different temperatures are shown in Figure 3.10. The Mössbauer parameters of these spectra are collected in Table 3.4.

The spectrum analysis of the starting material shows the existence of trivalent low-spin iron (Figure 3.10(a)). In the course of annealing at 493 K, a partial reduction of Fe(III) to divalent high-spin iron (Fe^{2+}) occurs. In Figure 3.10(b), one can see an additional doublet showing large quadruple splitting, resulting from Fe^{2+}. The increase in the annealing temperature to 643 K brings about a decrease in the Fe^{2+} content from 66.5% down to 49.6%; and a new phase, magnetite, which spectral contributions is about 7%, is formed. This fact is evidenced by the values of hyperfine fields, 49 and 45.3 T, in the A and B sites, respectively (Figure 3.10(c)). Further annealing, at 663 K, causes a further increase in the magnetite content up to about 35%, leading also to the complete disappearance of the phase containing Fe^{2+} (Figure 3.10(d)). One can also notice that in distinction to the bulk magnetite, the relative intensities of Mössbauer lines corresponding to the B site (Fe^{3+}, Fe^{2+}) are much smaller than those at the A site (Fe^{3+}) that points to the nonstoichiometric structure of the magnetite.

We assume that this is an effect of partial only filling of vacancies of Fe ions in B location, which leads to the reduced number of the $Fe^{2+} - Fe^{3+}$ pairs responsible for jumps of electrons. The idea of the nonstoichiometry of the magnetite is further supported by a low value of the hyperfine fields, especially in the B site (48.6 T (A), 44.3 T (B)), in relation to those reported in the literature values for the stoichiometric magnetite (48.6 T (A), 46.0 T (B)) [85].

The values of hyperfine fields representing Fe_3O_4 in tetrahedral and octahedral sites are 51.2 T and 46.6 T, respectively, for the acrylate annealed at 663 K. The hyperfine field representing the unidentified phase is 37 T.

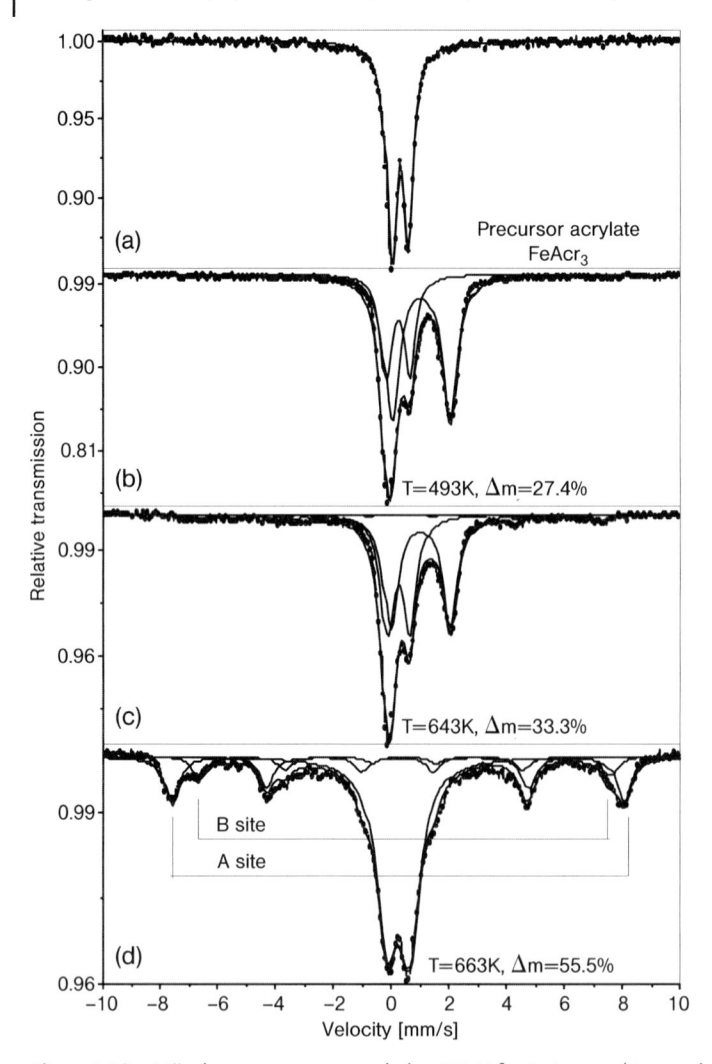

Figure 3.10 Mössbauer spectra recorded at 295 K for FeAcr$_3$ and its products of thermolysis.

A monotonic decrease in the content of the phase containing Fe(III), from 65.47% to 13%, at 300 K and 80 K, respectively, accompanied by the increasing amount of magnetite suggests the existence of superparamagnetic particles of magnetite. These results may also arise from the wide distribution of magnetite particle size and their partial agglomeration. Apparently the room temperature doublet, resulting from small, superparamagnetic crystallites of the magnetite phase, is included in the doublet representing Fe(III).

For the product of iron acrylate obtained at 663 K, the hysteresis loops, recorded at temperatures 5 K and 300 K, are shown in Figure 3.11. The loops

Table 3.4 Parameters of Mössbauer spectra at 295 K.

Sample/ annealing conditions	δ (IS) (mm/s)	QS (mm/s)	Hhf (T)	Iron forms	Relative percentage of iron (%)
Precursor FeAcr$_3$	0.184	0.617	0	Fe(III)	100
$T = 493$ K	0.945	2.128	0	Fe^{2+}	66.5
	0.110	0.878	0	Fe(III)	33.5
$T = 643$ K	0.927	2.165	0	Fe^{2+}	49.6
	0.131	0.831	0	Fe(III)	43.5
	0.086	0	49.0	Fe^{3+} (A)	1.4
	0.186	0	45.3	Fe^{3+}, Fe^{2+} (B)	5.5
$T = 663$ K	0.117	0.778	0	Fe(III)	65.4
	0.106	0	48.6	Fe^{3+} (A)	24.2
	0.321	0	44.3	Fe^{3+}, Fe^{2+} (B)	10.4

exhibit a large high-field susceptibility, and the magnetization does not saturate in the field applied. The coercivity at 5 K is ∼73 mT and reduces to 8 mT at room temperature [86].

The hysteresis loops for the product of CoAAm thermolysis were recorded at temperatures 50, 100, and 250 K. The loops are closed and symmetrical vs. the origin of the coordinate system. The shape of the loops evidences the ferromagnetic character of the material. The coercivity depends on temperature

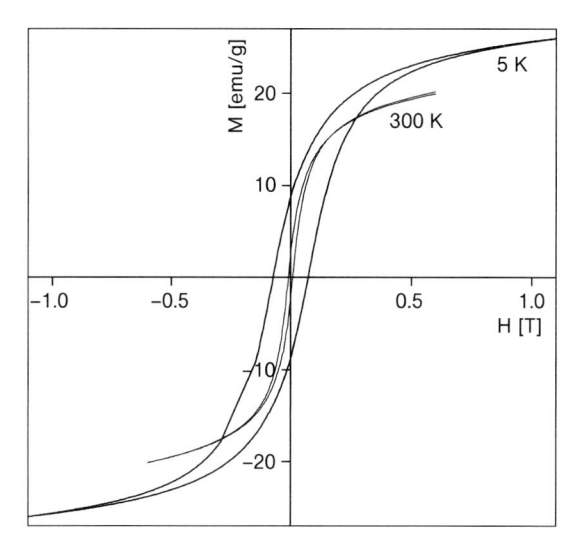

Figure 3.11 Hysteresis loops for the FeAcr$_3$ annealed at 663 K, recorded at 5 and 300 K.

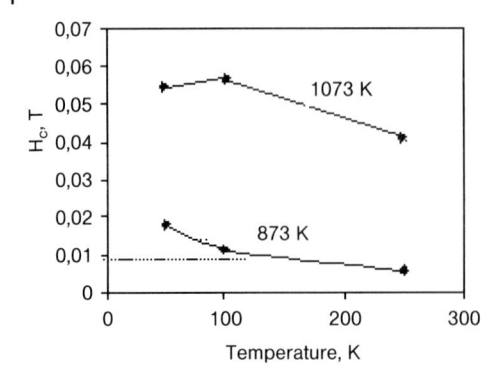

Figure 3.12 Temperature dependence of the coercivity for the nanocomposites obtained by thermolysis of frontally polymerized CoAAm at temperatures 873 K and 1073 K.

and generally decreases with increasing temperature (Figure 3.12). At room temperature the material thermolyzed at 873 K showed the coercivity 0.01 T.

The zero-field-cooled (ZFC) and field-cooled (FC) magnetization versus temperature curves for thermolysis product of FeAAm were taken in the fields 0.002 T and 0.6 T [87]. The curve recorded in the lower field shows the FC and ZFC magnetizations decreasing and increasing, respectively. The lines, however, do not meet each other at room temperature. This reveals a rather broad distribution of crystallite sizes with domination of the large crystals. The FC vs. temperature curve, recorded in the field 0.6 T, shows, similar to the dependence of the TRM vs. temperature, a deflection at about 5 K, which supports the idea of the irreversible magnetization processes occurring at low temperatures, which may result from canted surface spins, which do not achieve the full alignment with the external field at the lowest temperatures. The freezing processes observed can be (i) of a spin-glass type and appear in the partly disordered surface layers of particles due to their topological disorder or (ii) of a cluster-glass character due to a freezing of randomly oriented particles forming agglomerates and coupled by magnetic interactions.

3.4
Conclusion

Thus, both the science and practice of polymeric metal-containing magnetic nanocomposites are now intensively developing and this trend is kept going for the new possibilities of synthesis, new methods of composite forming, and the great practical interest. The final scientific aim of the synthesis of such nanostructured materials and studies is modeling of their structures for a correct interpretation of the experimental facts. On the other hand, the modeling and deeper understanding of the formation and properties of such nanomaterials will open new horizons both for their design and for practical application. Ferromagnetic nanoparticles stabilized by a polymer matrix were obtained during solid-phase polymerization of metal-containing monomers followed by controlled thermolysis of the products obtained. By varying the synthesis conditions, it is possible to obtain nanoparticles of metals, their oxides, and of their carbides with the required size, shape and structure, and homogeneity of size distribution through the matrix. The static magnetic properties of the sample are dominated by the interparticle magnetic interactions and glass-type freezing effects.

Acknowledgments

The authors thank Professor S.P. Gubin for the fruitful discussions. Financial supports from Russian Foundation for Basic Researches (Projects 07-03-00113 and 07-03-91582), and INTAS (grant N 05-1000008-7834) are gratefully acknowledged.

References

1. S.P. Gubin, I.D. Kosobudskii. *Russ. Chem. Rev.*, 52, 1350 (**1983**).
2. S.P. Gubin, Y.A. Koksharov, G.B. Khomutov, G.Y. Yurkov. *Uspekhi Khimii*, 74, 539–574 (**2005**).
3. El-Awansi, N. Kinawy, M. Emitwally. *J. Mater. Sci.*, 24, 2497 (**1989**).
4. H.S. Gokturk, T.J. Fiske, D.M. Kaylon. *J. Appl. Polym. Sci.*, 50, 1891 (**1993**).
5. J. Yacubowicz, M. Narkis. *Polym. Eng. Sci.*, 30, 46 (**1990**).
6. D.E. Nikles, J.L. Cain, A.P. Chacko, R.I. Webb. *IEEE Trans. Magn.*, 30, 4068 (**1995**).
7. D.C. Edwards. *J. Mater. Sci.*, 25, 4175 (**1990**).
8. X.-S. Yi, G. Wu, Y. Pan. *Polym. Int.*, 44, 117 (**1997**).
9. C. Laurent, D. Mauri, E. Kay, S.S.P. Parkin. *J. Appl. Phys.*, 65, 2017 (**1989**).
10. D.L. Leslie-Pelesky, X.Q. Zhang, R.D. Ricke. *J. Appl. Phys.*, 79, 5312 (**1996**).
11. R.H. Marchessault, S. Ricard, P. Rioux. *Carbohyd. Res.*, 224, 133 (**1992**).
12. R.F. Ziolo, E.P. Gianneis, B. Weinstein, M.P. O'Horo, B.N. Ganguly, V. Mehrotra, M.W. Russell, D.R. Huffman. *Science*, 257, 219 (**1992**).
13. R.F. Ziolo, E.P. Gianneis, R.D. Shull. *Nanostruct. Mater.*, 3, 85 (**1993**).
14. M. Respaud, J.M. Broto, H. Racoto, A.R. Fert, L. Thomas, B. Barbara, M. Verelest, E. Snoek, P. Lecante, A. Mosset, J. Osuna, T.O. Ely, C. Amiens, B. Chaudret. *Phys. Rev. B.*, 57, 2925 (**1998**).
15. M. Respaud, J.M. Broto, H. Racoto, J.C. Ousset, J. Osuna, T.O. Ely, C. Amiens, B. Chaudret, S. Askenary. *Physica. B.*, 247, 532 (**1998**).
16. F.J. Lozaro, J.L. Garcia, V. Schunemann, A.X. Trautwein. *IEEE Trans. Magn.*, 29, 2652 (**1993**).
17. J.J. Host, V.P. Dravid. *Mater. Res. Soc. Symp. Proc.*, 457, 225 (**1997**).
18. Ph. Colomban, V. Vendange. *Mater. Res. Soc. Symp. Proc.*, 457, 451 (**1997**).
19. V.M. Smirnov, N.P. Bobrysheva, V.B. Aleskovskii. *Dokl. Chem. Phys.*, 356, 492 (**1997**).
20. Z. Zhang, Y.D. Zhang, W.A. Hines, J.I. Badnick, W.M.H. Sachtler. *J. Am. Chem. Soc.*, 114, 4843 (**1992**).
21. Y. Yayakawa, S. Kohiki, M. Sato, T. Babasaki, H. Deguchi, A. Hidaka, H. Shimooka, S. Takahashi. *Physica E.*, 9, 250 (**2001**).
22. C.T. Kresge, M.E. Leonowics, W.J. Roth, J.C. Vartuli, J.S. Beck. *Nature*, 359, 710 (**1992**).
23. H. Mazaki, M. Kakihana, H. Yasuoka. *Jpn. J. Appl. Phys.*, 30, 38 (**1991**).

24. C.R. Vestal, Z.J. Zhang. *J. Am. Chem. Soc.*, *124*, 14312 (**2002**).
25. C. Liu, B. Zou, A.J. Rondinone, Z.J. Zhang. *J. Phys. Chem. B*, *104*, 1141 (**2000**).
26. L. Raymond, J.-F. Revol, D.H. Ryan, R.H. Marchessault. *J. Appl. Polym. Sci.*, *59*, 1073 (**1996**).
27. R.V.P.M. Shafi, A. Ulman, X. Yan, N.-L. Yang, C. Estournes, H. White, M. Rafailovich. *Langmuir*, *17*, 5093 (**2001**).
28. T. Ji, H. Shi, J. Zhao, Y. Zhao. *J. Magn. Magn. Mater.*, *212*, 189 (**2000**).
29. J. Ramos, A. Millan, F. Palacio. *Polymer*, *41*, 8461 (**2000**).
30. S.N. Sidorov, L.M. Bronstein, V.A. Davankov, M.P. Tsyurupa, S.P. Solodovnikov, P.M. Valetsky, E.A. Wilder, R.J. Spontak. *Chem. Mater.*, *11*, 3210 (**1999**).
31. R.F. Ziolo, E.P. Gianneis, B. Weinstein, M.P. O'Horo, B.N. Ganguly, V. Mehrotra, M.W. Russell, D.R. Huffman. *Science*, *257*, 219 (**1992**).
32. N.I. Nikonorova, S.V. Stakhanova, I.A. Chmutin, E.S. Trofimchuk, P.A. Chernavskii, A.L. Volynskii, A.T. Ponomarenko, N.F. Bakeev. *Polym. Sci., B 34*, 487 (**1998**).
33. I.M. Papisov, Yu.S. Yablokov, A.I. Prokofev, A.A. Litmanovich. *Polym. Sci. A*, *35*, 515 (**1993**); *36*, 352 (**1994**).
34. L. Chen, W.-J. Yang, C.-Z. Yang. *J. Mater. Sci.*, *32*, 3571 (**1997**).
35. H.B. Sohn, R.E. Cohen. *Chem. Mater.*, *9*, 264 (**1997**).
36. A.I. Aleksandrova, A.I. Prokofev, V.N. Lebedev, E.B. Balagurov, N.N. Bubnov, I.Yu. Metlenkova, S.P. Solodovnikov, A.N. Ozerin. *Russ. Chem. Bull.*, *44*, 2355 (**1995**).
37. R. Tannenbaum, C.I. Flenniken, E.P. Goldberg. *J. Polym. Sci. B. Polym. Phys.* *28*, 2421 (**1990**).
38. G.Yu. Yurkov, A.S. Fionov, Yu.A. Koksharov, V.V. Kolesov, S.P. Gubin. *Inorg. Mater.*, *43*, 834–844 (**2007**).
39. G.Yu. Yurkov, D.A. Baranov, A.V. Kozinkin, Yu.A. Koksharov, T.I. Nedoseikina, O.V. Shvachko, S.A. Moksin, S.P. Gubin. *Inorg. Mater.*, *42*, 1012–1019 (**2006**).
40. J. Osuna, D. Caro, C. Amiens, B. Chaudret, E. Snoeck, M. Respaund, J.-M. Broto. A. Fert. *J. Phys. Chem.*, *100*, 14571 (**1966**).
41. M.V. Shamurina, V.I. Roldugin, T.D. Pryamova, T.D. Vysotskii. *Colloid. J.*, *56*, 450 (**1994**); *57*, 580 (**1995**); *61*, 473 (**1999**).
42. J.K. Vassilion, V. Mehrotra, M.W. Russell, E.P. Gianneis. *Mater. Res. Soc. Symp. Proc.*, *206*, 561 (**1991**).
43. T. Shiga, A. Okadda, T. Kurauchi. *J. Appl. Polym. Sci.*, *58*, 787 (**1995**).
44. T. Ktapeinski, A. Galeski, M. Kruszewski. *J. Appl. Polym. Sci.*, *58*, 1007 (**1995**).
45. J. Twomey, S.H. Chew, T.N. Blanton, A. Schmid, K.L. Marshall. *J. Polym. Sci. B Polym. Phys.*, *32*, 1687 (**1994**).
46. S. Sun, C.B. Murray, D. Weller, L. Folks, A. Moser. *Science*, *287*, 1989 (**2000**).
47. J.H. Park, J. Cheon. *J. Am. Chem. Soc.*, *123*, 5473 (**2001**).
48. S. Sun, S. Anders, H.F. Haman, J.-U. Thiele, J.E.E. Baglin, T. Thomson, E.E. Fullerton, C.B. Murray, B.D. Terris. *J. Am. Chem. Soc.*, *124*, 2884 (**2002**).
49. M. Fujiwara, T. Matsushita, K. Yamagichi, T. Fueno. *Synth. Met.*, *41–43*, 3267 (**1991**).
50. M. Fujiwara, W. Mori, K. Yamagichi. *Mol. Cryst. Liq. Cryst.*, *274*, 175 (**1995**).
51. D.L. Lesli-Pelecky, R.D. Rieke. *Chem. Mater.*, *8*, 1771 (**1996**).
52. A.G. Golubkov, N.P. Evrukov. *Plast. Massy.*, (3), 22 (**1998**).
53. Y. Osada. *Adv. Mater.*, *3*, 107 (**1991**).
54. X. Cao, Y. Xie, F. Yu, Z. Yao, L. Li. *J. Mater. Chem.*, *13*, 893 (**2003**).
55. M.R. Bryce, M.C. Petty. *Nature*, *374*, 771 (**1995**).
56. C. Lafuente, C. Mingotaud, P. Delhaes. *Chem. Phys. Lett.*, *302*, 523 (**1999**).
57. D.D. Mishin. Magnetic materials. *Manual for Colleges* (High School, Moscow, **1991**).
58. D. de Caro, T.O. Ely, A. Mari, B. Chaudret, E. Snoeck, M. Respaund, J.-M. Broto, A. Fert. *Chem. Mater.*, *8*, 1987 (**1996**).

59. B. Sohn, R.E. Cohen, G.C. Papaefthymiou. *J. Magn. Magn. Mater.*, *182*, 216 (**1998**).
60. T.J. Fiske, H. Gokturk, D.M. Kaylon. *J. Appl. Polym. Sci.*, *65*, 1371 (**1997**).
61. S. Praveen, V.K. Babbar, A. Razdan, R.K. Puri, T.K. Goel. *J. Appl. Phys.*, *89*, 4362 (**2000**).
62. D.-K. Kim, M.S. Toprak, M. Mikhaylova, Y.-S. Jo, S.J. Savage, T.T sakalakos, M. Muhammed. *Solid State Phenom.*, *99–100*, 165–168 (**2004**).
63. V. Goldade, L. Pinchuk, A. Kravtsov. Polymers friendly for the environment. *7th Eur. Polym. Federation Symp. Polym. Mater.* (**1998**).
64. K. Uchino. *MRS Bull.*, *18*, 42 (**1993**).
65. A.L. Buchachenko. *Russ. Chem. Rev.*, *59*, 529 (**1990**).
66. M.M. Levitskii, A.L. Buchachenko. *Russ. Chem. Bull.*, *46*, 1367 (**1997**).
67. E.I. Alexandrova, G.I. Dzhardimalieva, A.S. Rozenberg, A.D. Pomogailo. *Russ. Chem. Bull.*, *42*, 254–258 (Engl. Trans.) (**1993**).
68. E.I. Alexandrova, G.I. Dzhardimalieva, A.S. Rozenberg, A.D. Pomogailo. *Russ. Chem. Bull.*, *42*, 259–264 (Engl. Trans.) (**1993**).
69. A.S. Rozenberg, G.I. Dzhardimalieva, A.D. Pomogailo, *Polym. Adv. Technol.*, *9*, 527–535 (**1998**).
70. A.D. Pomogailo, G.I. Dzhardimalieva, A.S. Rosenberg. *Acta. Physica. Polonica.*, *102*, 135–145 (**2002**).
71. A.S. Rozenberg, E.I. Alexandrova, G.I. Dzhardimalieva, A.N. Titkov, A.D. Pomogailo. *Russ. Chem. Bull.*, *42*, 1666–1672 (Engl. Trans.) (**1993**).
72. A.S. Rozenberg, E.I. Alexandrova, G.I. Dzhardimalieva, N.V. Kir'ykov, P.E. Chizhov, V.I. Petinov, A.D. Pomogailo. *Russ. Chem. Bull.*, *44*, 858–866 (Engl. Trans.) (**1995**).
73. A.S. Rozenberg, E.I. Alexandrova, N.P. Ivleva, G.I. Dzhardimalieva, Raevskii, O.I. Kolesova, I.E. Uflyand, A.D. Pomogailo. *Russ. Chem. Bull.*, *47*, 259–264 (Engl. Trans.) (**1998**).
74. A.T. Shuvaev, A.S. Rozenberg, G.I. Dzhardimalieva, N.P. Ivleva, V.G. Vlasenko, T.I. Nedoseikina, T.A. Lyubeznova, I.E. Uflyand, A.D. Pomogailo. *Russ. Chem. Bull.*, *47*, 1460–1465 (Engl. Trans.) (**1998**).
75. A.D. Pomogailo, V.G. Vlasenko, A.T. Shuvaev, A.S. Rozenberg, G.I. Dzhardiamlieva. *Colloid., J.* *64*, 472–477 (Engl. Transl.) (**2002**).
76. Z. Wojtczak, A. Granowski. *J. Therm. Anal.*, *26*, 233–239 (**1983**).
77. I. Skupinska, H. Wilezura, H. Bonink. *J. Therm. Anal.* *31*, 1017 (**1986**).
78. Z. Wojtczak. *J. Therm. Anal.*, *36*, 2357–2362 (**1990**).
79. K. Nakamoto. *Infrared and Raman Spectra of Inorganic and Coordination Compounds* (John Wiley & Sons, New York, **1987**).
80. A.R. Lidin, L.L. Andreeva, V.A. Molochko. *Handbook of Inorganic Chemistry. The Constants of Inorganic Compounds* (Khimiya, Moscow, **1987**).
81. E.I. Alexandrova, A.V. Raevskii, A.S. Rozenberg, A.N. Titkov. *Chem. Phys.*, *8*, 1630–1639 (**1989**).
82. E.I. Alexandrova, A.V. Raevskii, A.N. Titkov, A.S. Rozenberg. *Chem. Phys.*, *9*, 1244–1249 (Engl. Trans.) (**1990**).
83. A.S. Rozenberg, A.V. Raevskii, E.I. Aleksandrova, O.I. Kolesova, G.I. Dzhardimalieva, A.D. Pomogailo. *Russ. Chem. Bull., Int. Ed.*, *50*(5), 901–906 (**2001**).
84. *Physical Encyclopedic Dictionary* (Sovetskaya Éntsiklopediya, Moscow, **1984**) [in Russian].
85. T.K. McNab, R.A. Fox, A.J.F. Boyle. *J. Appl. Phys.*, *39*, 5703–5711 (**1968**).
86. M. Leonowicz, M. Ławecka, A. Olawska-Waniewska, G. Dzhardimalieva, A.S. Rozenberg, A.D. Pomogailo. *Macromolecular Symposia*, *204*, 257–265 (**2003**).
87. E. Sówka, M. Leonowicz, A.D. Pomogailo, G.I. Dzhardimalieva, J. KaYmierczak, A. Olawska-Waniewska, M. Kopcewicz, *J. Magn. Magn. Mater.*, *316*(2) E749–E752 (**2007**).

4
Magnetic Nanocomposites Based on the Metal-Containing (Fe, Co, Ni) Nanoparticles Inside the Polyethylene Matrix

Gleb Yu Yurkov, Sergey P. Gubin, and Evgeny A. Ovchenkov

4.1
Introduction

The high reactivity of nanoparticles and their tendency toward spontaneous compaction accompanied by deterioration of basic physical (magnetic) properties make stabilization a major challenge in fabrication of materials based on metal-containing magnetic nanoparticles [1].

There are two ways for producing nanoparticle materials: the first one is to compact nanoparticles, and the second one is to introduce nanoparticles into isolating materials whose production is well known. Among the large number of polymers currently used as the matrices for magnetic nanoparticle stabilization, polyethylene occupies a special place. This semicrystalline carbon-chain polymer has a rather "loose" structure and a large number of nanoscaled cavities in it. We have been experimenting with this material since the last 30 years and learned how to introduce nanoparticles of different compositions into it and how to distribute them rather uniformly in the volume of the polymer. The possibility of a material combining the properties of a polymer and metal and the control over the properties of such a material via compositional changes have long been discussed [2]. The study of metal-containing nanoparticles in polymer matrices is stimulated by the increasing interest in this issue in many areas of chemistry, physics, and materials science, especially magnetic materials. Research efforts are most commonly (but not nearly always) focused on metallic (more precisely, metal-containing) nanoparticles – spheroidal particles of diameter $d = 1-10$ nm ($N = 10$ to 10^3 atoms), with $N_{surface}/N_{bulk} \geq 1$, surrounded by low-Z atoms [3].

Iron and cobalt, are the basic metals for numerous magnetic compounds used in practice. Moreover, these bulk metals are characterized by high ferromagnetic transition temperatures (1043 K, 1408 K, and 650 K, respectively [4]) and rather high magnetic anisotropy (4.72×10^4 J/m^3, 4.53×10^5 J/m^3, and 5.7×10^3 J/m^3, respectively [5]). The physical parameters of the nanoparticles differ from those of the corresponding bulk materials. Most experiments show

---

Magnetic Nanoparticles. Sergey P. Gubin
© 2009 WILEY-VCH Verlag GmbH & Co. KGaA, Weinheim
ISBN: 978-3-527-40790-3

that the magnetic anisotropy is larger (by a factor of $10-10^2$) for nanoparticle samples than for bulk materials (see, for example, [6, 7]).

The possibility of using polymers containing no functional groups or heteroatoms as matrices for stabilization of magnetic nanoparticles seems quite attractive, as these polymers (e.g., polyethylene and polypropylene) are good dielectrics; they are stable, inexpensive, and easily processable into articles of any shape. The advantages of magnetic materials obtained using these polymers are low specific gravity, high homogeneity of nanoparticle distribution and, which is the most important, an isolated arrangement of the nanoparticles at distances much greater than their dimensions. However, the lack of solubility (or very poor solubility) of most polymers of this type in organic solvents appreciably restricts the possibilities of introduction and homogeneous distribution of magnetic nanoparticles with a narrow size distribution in these polymers. The well-developed method for forming nanoparticles (2–8 nm) in a solution (melt) of polymers in mineral oil is more promising. This method is versatile as regards both the composition of the resulting magnetic nanoparticles (Fe, Co, Ni, Fe–Co, Fe–Mo, Fe–Pt, Fe–Nd, Fe–Sm, Fe_2O_3, Fe_3O_4, ferrites) and the type of polymer (low- and high-density polyethylene, polypropylene, polyamides). The content of nanoparticles in the sample can reach 60 mass%. The synthesis of magnetic nanomaterials as washers (25 × 5 mm) and thick (3 mm) films possessing good magnetic characteristics has been reported.

4.2
Experimental Details

4.2.1
Synthesis

According to current views [8, 9], molten polymers consist of separate spherulites, in which platelets or lamellae persist in a slightly distorted, twisted form, as stable structural components of the polymer. Without a specially engineered method, one cannot synthesize material consists of structurally uniform nanoparticles distributed inside nanocomposites with optimized electrical, magnetic, optical, and mechanical properties. A key issue is how to introduce nanoparticles into natural nanopores in the real polymer.

We used high-temperature solutions of the low-density polyethylene (LDPE) and other polymers in a hydrocarbon oil. It is believed that, above 140 °C, such systems contain lamellae of polyethylene separated from each other by more flowing hydrocarbon molecules [10]. Above 250 °C, polyethylene–oil system is transparent, homogeneous, and low-viscosity liquids. It is reasonable to assume that, under such conditions, both the amorphous and crystalline parts of the dissolved polymer are accessible to appropriate reagents.

Metal-containing nanoparticles in the polyethylene volume were prepared by a method based on the standard "cluspol" technique [11–14]. Thermolysis of metal-containing precursors (MR_n; M = Fe, Co, Ni; R = CO, HCOO, CH_3COO, etc.), such as inorganic and organic salts or organometallic compounds, is widely used to produce metal and oxide nanoparticles, both free-standing [11] and incorporated in nanocomposites, including polymer-containing materials [7, 12–22].

Below, the procedure for introducing metal-containing nanoparticles into a polyethylene matrix is illustrated in detail by an example introduction of γ-Fe_2O_3 nanoparticles [23].

The flow rate of argon was adjusted so as to rapidly remove the ligands and solvent from the reactor. An appropriate amount of an iron(III) acetate solution was added to molten polyethylene with vigorous stirring. In this method, the precursor was decomposed in such a way that each drop of the solution entered a heated hydrophobic medium, and all subsequent transformations (solvent evaporation, thermal decomposition, and nucleation) occurred within a drop – nanoreactor. It is believed that the decomposition of metal-containing compounds typically involves two steps: molecular-scale dispersion, atomization or reduction and subsequent condensation of metal atoms into nanoparticles. The polymeric nanocomposites produced were separated from oil via rinsing with organic solvents. An enhanced variant of this method, which involves the usage of a metal reactor instead of the glass one, allows us to produce 10 kg of nanocomposite per week.

There is also great interest in bi- and trimetallic nanoparticles and alloys. The corresponding preparation techniques are similar to those for homometallic nanoparticles. In this work, such materials were prepared via thermal decomposition of appropriate compounds in high-temperature solutions of polymers. In order to produce heterometallic nanoparticles, compounds with the general formula $R_nMM'X_q$ were used, where R is an organic radical; M = Fe, Co, Mn, Cr, or other transition metals; M' is lanthanide, Cu or Ag; and X is an unstable ligand. We obtained lanthanide-containing clusters with the general formula $[M(CO)_nL]Sm$, where M = Fe or Co; Fe–Pt clusters were used to produce Fe–Pt nanoparticles in a polyethylene matrix. Attempts were also made to simultaneously decompose two metal-containing compounds. Barium (strontium) and iron acetates decomposed in a polymer–oil high-temperature solution according to the schemes

$$Ba(CH_3COO)_2 \longrightarrow BaO + (CH_3)_2CO + CO_2;$$

$$Sr(CH_3COO)_2 \longrightarrow SrO + (CH_3)_2CO + CO_2;$$

$$Fe(CH_3COO)_3 \longrightarrow Fe_2O_3 + CO + CO_2 + H_2O.$$

Varying the relative amounts of the starting reagents, we obtained $BaFe_2O_4$ and $BaFe_{12}O_{19}$ ferrite nanoparticles by the above reactions.

Fe–Co nanoparticles in a polyethylene matrix were produced by the thermal decomposition of iron pentacarbonyl with $CoCl_2$. This mixture of compounds transformed into Fe–Co nanoparticles at 280–290 °C. Another way to produce Fe–Co nanoparticles is through the codecomposition of cobalt and iron formates or acetates.

Fe–Sm nanoparticles were prepared via thermal destruction of $[FeSm(CO)_n]_x$ clusters at 280 °C according to the scheme $[FeSm(CO)_n] \rightarrow Fe + Sm + CO$. In order to produce Fe–Pt nanoparticles, $Fe(CO)_5$ and H_2PtCl_6 were reacted at 260 °C in a polymer–oil solution. The nanomaterials prepared are listed in Table 4.1.

Table 4.1 The list of synthesizing of nanomaterials.

	Matrix: high-pressure polyethylene			
Precursor (MCC)	Size of particles (nm)[a]	Composition of nanoparticles[b]	Weight concentration (%)	Reference
$Fe(CO)_5$	2–9	a-Fe, Fe_2O_3, Fe_xC_y, Fe_3O_4	0.2 to 50	[24–27]
$Fe(CH_3COO)_3$ + oxidant[c]	2–7	γ-Fe_2O_3	1 to 24.7	[24–29]
$Fe(CH_3COO)_3$	5–19	Fe_3O_4	1 to 10	
$Fe_2(C_2O_4)_3$	9–23	Fe_2O_3, Fe_3O_4	2 to 20	
FeC_2O_4	8–25	Fe_2O_3, FeO, Fe_3O_4	20	
$Fe(HCOO)_3$	4–60	Fe, FeO, Fe_3O_4, Fe_2O_3	1 to 62.7	[30, 31]
$Co(HCOO)_2$	5–30	Co	2 to 40	[32]
$Co(HCOO)_2$ + oxidant[a]	5–19	CoO	5; 10; 15	
$Co(CH_3COO)_2$ + oxidant[a]	4–20	CoO	10	
$Co(CH_3COO)_2$	4–23	Co, CoO	5, 10	
$Fe(CO)_5$ (excess) + $CoCl_2$	3–10	Fe–Co	5, 10	
$Fe(CH_3COO)_3$ + $Co(CH_3COO)_2$	6–20	Fe–Co (different ratios)	Up to 20	
$Fe(HCOO)_3$ + $Co(HCOO)_2$	5–20	Fe–Co (different ratios)	Up to 50	[24]
$Fe(CO)_5$ + $Co_2(CO)_8$	3–10	Fe–Co	5, 10	[24]
$Fe(CO)_5$ + H_2PtCl_6	5–13	Fe_2O_3-Pt, Fe–Pt	5	[24]
$Ni(CH_3COO)_2$	5–19	NiO	5	[24]
$Ni(CH_3COO)_2$ + $Fe(CH_3COO)_3$	2–5	$NiFe_2O_4$	5 to 30	
$Ba(CH_3COO)_2$	5–21	BaO	5, 10	

Table 4.1 (continued).

Precursor (MCC)	Size of particles (nm)[a]	Composition of nanoparticles[b]	Weight concentration (%)	Reference
Matrix: high-pressure polyethylene				
$Ba(CH_3COO)_2 +$ $Fe(CH_3COO)_3$	6–22	$BaFe_2O_4$, $BaFe_{12}O_{19}$	2 to 40	[24]
$Ba(CH_3COO)_2 +$ $Fe_2(C_2O_4)_3$ or FeC_2O_4	4–22	$BaFe_2O_4$, $BaFe_{12}O_{19}$	5.51 to 50	
$Fe(CH_3COO)_3 +$ $Sr(CH_3COO)_2$	4–19	$SrFe_2O_4$	5, 10	[24]
$Mn(CH_3COO)_2 +$ Oxidant[a]	3–12	Mn_2O_3, $Mn_{12}O_{12}$	2, 10	
$Cr(CO)_6$	5–14	Cr	5	
$Cr(CH_3COO)_2$	4–20	CrO, Cr_2O_3	5	
Matrix: low-pressure polyethylene				
$Fe(CO)_5$	2–15	a-Fe, Fe_2O_3, Fe_xC_y	5, 10	[27]
Matrix: polypropylene				
$Fe(CO)_5$	5–30	a-Fe, Fe_2O_3, Fe_xC_y	0.2 to 33	[27]
Matrix: polyamide				
$Fe(CO)_5$	5–20	a-Fe, Fe_2O_3, Fe_xC_y, Fe_xN_y	2 to 49	[24]

[a] The sizes of the nanoparticles may vary from experiment to experiment in the specified limits; the nanoparticles' size distribution in a single experiment did not exceed 15%.
[b] Varying experiment conditions allows producing individual samples with the following compositions.
[c] O_2, H_2O_2, or $KMnO_4$ were used as the oxidants.

The above examples demonstrate that the proposed approach allows one to produce powder polymers accommodating metal-containing nanoparticles. The composition and concentration of nanoparticles can be varied over wide ranges. It is possible to impart any form to the obtained powders by well-known methods (Figure 4.1).

Several parameters of the obtained magnetic nanocomposites which are important for further utilization of the latter have been determined.

In metal–polymer nanocomposites, there is a strong interaction (at the level of chemical bonding) between nanoparticles and polymer chains. The adhesion between a metal and a polymer molecule depends on the nature of the molecular forces involved, the structural perfection of the surface layer of the nanoparticles, internal stress, and electric charges [33–35]. This requires a more rigid control over the parameters of further processing steps in comparison with the parent polymers.

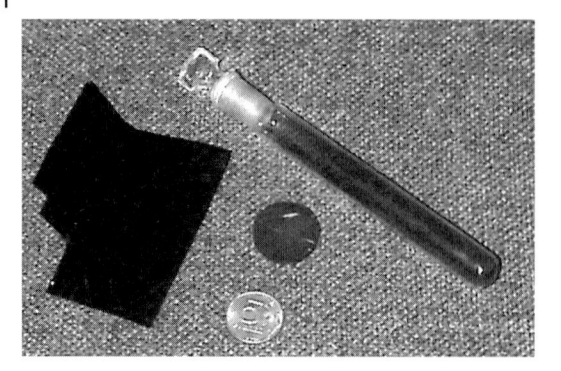

Figure 4.1 Fe + LDPE nanocomposite materials in various forms (powder in a tube, compacted powder in the form of a sheet and a washer). Russian "5 rubles" coin is shown for comparison.

Table 4.2 Effect of the Fe content on the decomposition onset temperature of polyethylene–Fe nanocomposites.

Content of Fe (wt.%)	0.0	0.5	1.0	5.0	10.0	15.0	20.0	30.0
Decomposition temperature (°C)	320	335	355	385	395	400	410	415

The introduction of a metal enhances the thermal stability of the polymer. Results on the thermal decomposition of polyethylene–matrix composites are summarized in Table 4.2.

These data demonstrate that the thermal stability of the materials increases rather sharply at Fe contents of up to 5 wt% and varies more gradually at higher Fe contents.

Thus, the fabricated magnetic nanocomposites demonstrate increased thermal stability which is significantly higher than the thermal stability of the initial polymer.

At the same time, it is well known that the introduction of metal powders 5–20 μm in size was reported to reduce the thermal stability of classical metallopolymers [31].

4.2.2
Composition and Structure of Magnetic Nanometallopolymers

In order to determine the composition and structure of nanoparticles and investigate their interaction with matrices, a variety of physical characterization techniques were used (TEM, HRTEM, SEM, HRSEM, Mössbauer spectroscopy, X-ray RED, X-ray emission, SAXS, EXAFS). Figure 4.2 shows a TEM micrograph of the dispersed nanocomposite powder

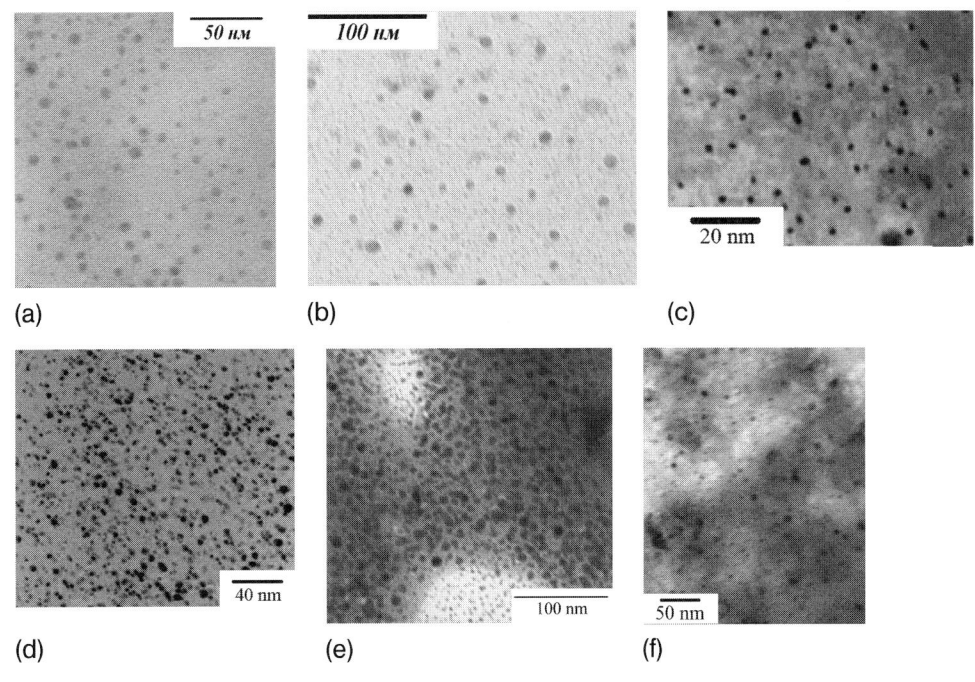

Figure 4.2 TEM image of nanocomposites on the LDPE-based and (a) Fe-containing nanoparticles ($d = 5.1$ nm); (b) Fe_2O_3 nanoparticles (4.6 nm); (c) Co nanoparticles ($d = 3.9$ nm); (d) Ni nanoparticles ($d = 4.2$ nm); (e) Pt@Fe_2O_3 nanoparticles ($d = 4.9$ nm); and (f) Co@Fe_2O_3 nanoparticles ($d = 5.8$ nm). Transmission electron microscopy was performed using JEM-100B working at 80 kV. Powder of nanocomposites (PE + 5 wt.% metal-containing nanoparticles) has been dispersed by ultrasound in alcohol; a drop of the resulted suspension was placed on a copper grid.

containing nanoparticles. The particles are morphologically mainly spherical and their sizes are distributed in the range from 3.0 to 9.0 nm.

4.3
Magnetic Properties of Metal-Containing Nanoparticles

4.3.1
Iron Containing Nanoparticles

As was shown above, the method developed allows obtaining significant amounts of polymer powder in an easy way. This powder containing magnetic nanoparticles of different composition can be used for the fabrication of products of arbitrary forms. It was important in future investigations to study possible changes of magnetic characteristics under the conditions of the exploitation of these materials. For investigating magnetic properties of the samples prepared, the effect of the hot pressing on magnetic properties of the compositions produced, the effect of the temperature on magnetic properties of the materials comprised of LDPE and Fe-containing nanoparticles, and the effect of the environment on the magnetic properties of the samples during their aging were studied.

Work has focused on the study of Fe-containing composite nanomaterials. These samples were synthesized by means of the thermodestruction of iron carbonyl. According to the TEM data, the average particle size is 5.1 nm. The composition of the particles was determined by means of Mössbauer spectroscopy, XRD, and EXAFS. These researches have shown that the particles have a complex structure and consist of several components such as α-Fe, γ-Fe_2O_3, and Fe_2C_5.

As was highlighted before, the properties of the obtained nanocomposites strongly depend on many parameters of preparation stages. For example, in the case of Fe nanocomposites, it is very difficult to control oxidation in the final stages of preparation – extraction and drying. This often leads to a noticeable difference in the magnetic properties of the samples obtained under the same conditions. Under these circumstances, the most reliable information can be obtained from the study of the sample properties before and after different sample treatments. A vibrating sample magnetometer (model PARC-155) with helium cryostat was used for all the magnetic measurements reported here. Sensitivity at the magnetic moment was better than 5×10^{-5} emu. The value of a typical signal induced by the samples was about 2×10^{-3} emu.

A set of measurements was carried out at temperature intervals from 4.2 to 291 K and in magnetic fields of up to 5 kOe. The first set of experiments was devoted to the analysis of the change of magnetic properties of a powder sample exposed to open air. For this purpose the sample with 5% of Fe nanoparticles in PE was measured immediately after ex traction in Ar atmosphere and then measurements were repeated after period of 1-day, 5-day, and 2-week expositions at air. The corresponding demagnetization curves are shown in Figure 4.3. To clarify the picture, the curves are not signed but arrows show the direction of evolution with time.

One of the peculiarities of these curves is a kink in the zero field. The similar kink often appears for the Fe nanoparticles prepared in PE, and as will be discussed later, can be explained by the interaction of a ferromagnetic metal core with an antiferromagnetic shell.

Table 4.3 lists the values of coercive force and remnant magnetization for these samples. Both the coercive force and the remnant magnetization decrease near 7% during 2 weeks. Gradual decomposition of the metal core is apparently reflected. This process is relatively slow since changes in magnetic properties are moderate. On the other hand, it continues for several days without any manifestation of slowing down. The second set of experiments was devoted to the investigation of the plate form of this material. For this purpose the powder after short exposition at air was compacted to the plate.

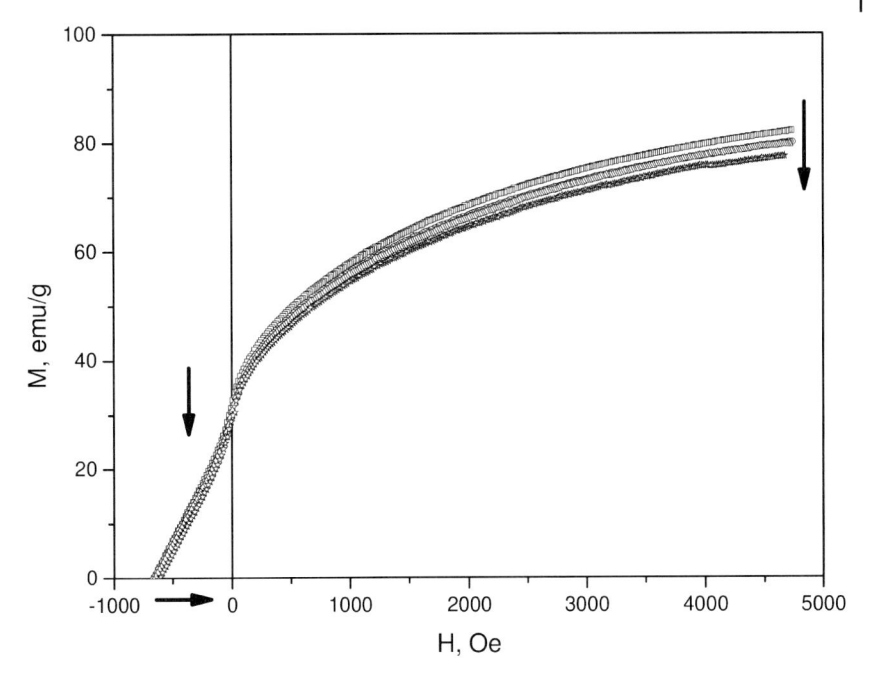

Figure 4.3 Demagnetization curves for 5 wt.% Fe nanoparticles in polyethylene at room temperature measured after extraction from Ar atmosphere and after exposition to open air for 1 day, 5 days, and 2 weeks. Arrows show evolution of curves with time.

Table 4.3 The coercive force H_C and remnant magnetization M_R for 5% Fe nanoparticles in polyethylene after different expositions at air.

Exposition time	H_C (Oe)	M_R (emu/g)
Starting sample	675	31.2
1 day	670	30.6
5 days	660	29.7
2 weeks	635	28.9

The full hysteresis loops for the original powder and the compacted sample are shown in Figure 4.4. The values of coercive force and remnant magnetization for the compacted sample (540 Oe and 25 emu/g, respectively) are markedly smaller than they were for the starting powder. The kink on the curve at zero fields became more pronounced. All these changes reflect noticeable chemical modification of material during forming processes.

Further investigations reveal good stability of the plate form of material. We did not find any changes in the magnetic properties of the plate sample

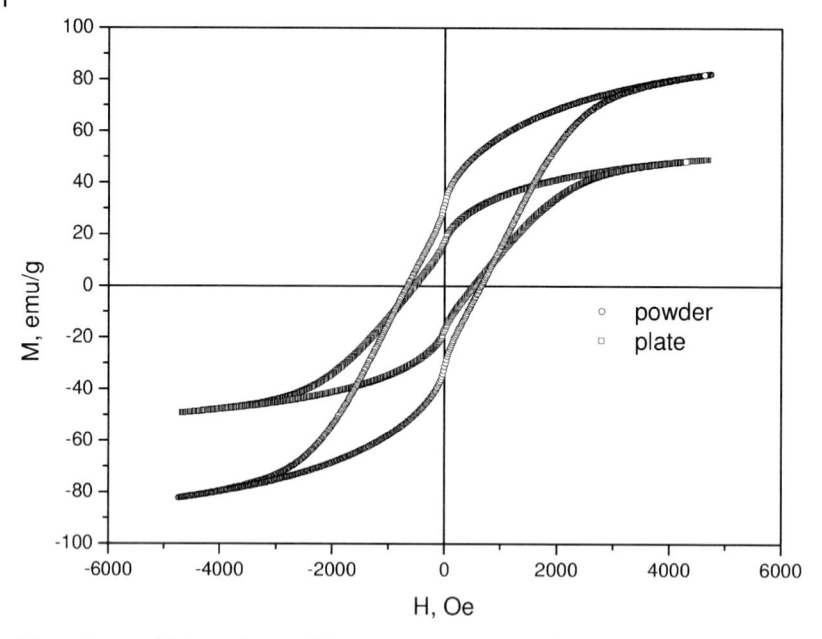

Figure 4.4 Field dependence of the magnetization M(H) for 5 wt.% Fe nanoparticles in polyethylene at room temperature measured for an original powder and a compacted samples.

kept at room temperature during 6 months. The time relaxation of remnant magnetization for this sample was studied. The sample was magnetized and then the values of magnetization were measured in the zero field after 1 h, 1 day, and 1 week (see Table 4.4).

Besides, the sample demonstrates good stability even to the moderate heating. The demagnetization curves of the sample measured at elevated temperatures are shown in Figure 4.5. Heating to 100 °C leads to a marked decrease of coercive force and remnant magnetization. After cooling down to room temperature, the magnetic properties of the sample are exactly the same as before heating. So heating of the compacted sample at open air up to 100 °C did not lead to the sample degradation.

Table 4.4 Relaxation of magnetization for the plate sample of 5% Fe nanoparticles in polyethylene.

Time of demagnetization	M_R (a.u.)
1 h	1
1 day	0.97
1 week	0.96

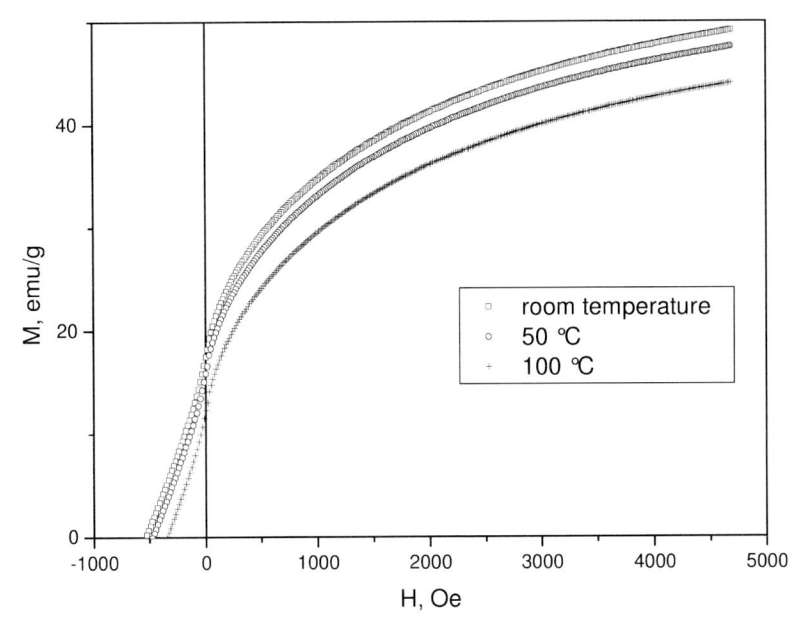

Figure 4.5 Demagnetization curves of the plate sample of 5% Fe nanoparticles in polyethylene measured at room temperature (open squares), at 50 °C (open circles), and at 100 °C (crosses).

The final set of experiments was devoted to the investigation of the properties of the samples heat-treated at high temperatures. The texture of the sample by placing it in the magnetic field after heat treatment was studied. For this purpose the samples were heat-treated in the air atmosphere at the temperatures of 195, 215, 240, 260, and 290 °C. Two pieces of the sample were taken at each temperature. Both pieces were kept at a given temperature for about 5–10 min, and then one of them was placed into a specially made magnetic system with the value of magnetic field equal to 7 kOe.

For the samples cooled in the magnetic field we measured the magnetization curves along and transverse to the direction of texturing magnetic field. There was no difference found between these curves and the magnetization curves of the samples cooled without an external magnetic field at all the temperatures of the heat treatment. It reflects a strong chemical bonding of metallic particles with the surrounding polymer and nearly isotropic magnetic properties of the particles.

The magnetization curves of the samples heat-treated at different temperatures are shown in Figure 4.6. The values of the remnant magnetization, the magnetization in the field of 4.5 kOe, and the coercive force at room temperature and at 100 °C are collected in Table 4.5.

It can be clearly seen that the heat treatment leads to a progressive decrease of the coercive force of the samples. The fastest coercive force decrease is

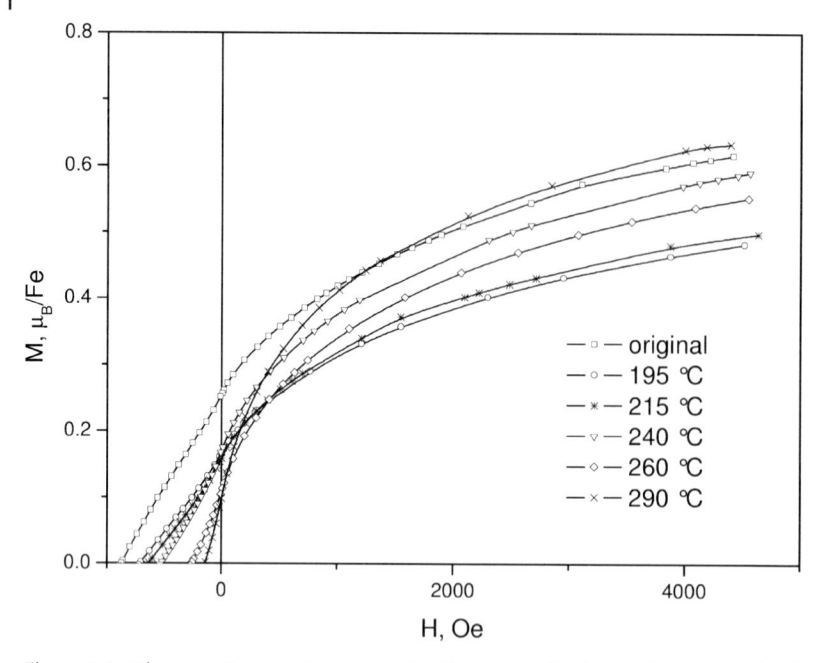

Figure 4.6 The room temperature magnetization curves for the starting material and samples after heat treatment.

Table 4.5 The values of the remnant magnetization, M_R, magnetization in the field of 4.5 kOe, $M_{H=4.5}$, the coercive force, H_C, at room temperature and at 100 °C for different temperatures of heat treatment of samples obtained in experiment.

	H_C	M_R	$M_{H=4.5}$	H_C (Oe) (100 °C)
	(Oe)	(μ_B)	(μ_B)	
Original	880	0.25	0.62	500
195 °C	720	0.16	0.48	
215 °C	570	0.14	0.50	
240 °C	490	0.15	0.56	
260 °C	270	0.10	0.55	230
290 °C	170	0.08	0.64	

observed in the temperature range of 240–260 °C. The coercive force decrease may be explained as a consequence of particles' oxidation. The temperature range of 240–260 °C of the fastest oxidation apparently corresponds to the polyethylene melting temperature.

The behavior of the specific magnetization is more complex and unusual. Whereas the remnant magnetization consistently decreases with the increase of the heat treatment temperature reflecting the decrease of the sample coercive force, the magnetization in the field of 4.5 kOe markedly decreases for low temperatures of the heat treatment and then increases with the increase of the temperature of the heat treatment. The oxidation of the sample decreases the volume of metal core and must result in a decrease of the sample magnetization. It is evidently the reason for the decrease observed at low temperatures of the heat treatment. The increase of magnetization with the increase of the temperature of the heat treatment was, in our opinion, due to the presence of Fe precursors in the sample. Therefore, during high temperature treatment the reaction of chemical decomposition is continued and leads to the formation of new particles. This idea is supported by data obtained on samples with Co nanoparticles. As will be described later, the heat treatment of these samples at 200–300 °C does not lead to any changes in coercive force. At the same time, the samples often show an increase in specific magnetization especially those which demonstrated low specific magnetization after preparation.

The obtained results explain the discrepancy of the properties of the samples obtained in different experiments. On the one hand, the increase of the synthesis temperature must lead to a more complete decomposition of the precursor but, on the other hand, it leads to oxidation. The temperature of the syntheses near 240 °C chosen experimentally is on the border of the fast sample oxidation and does not correspond to completing the particles' formation. This temperature most likely corresponds to the unstable region.

The influence of the heat treatment time on the sample properties was investigated. The magnetization curves for the samples heat-treated at the temperature of 280 °C for different periods are shown in Figure 4.6. Evolution of curves is similar to that observed in the case of the heat treatment temperature increase. The coercive force consequently decreases and magnetization shows a local increase followed by a further decrease. This behavior does not contradict supposition about the background reasons of changes, since both the oxidation and degree of reaction completeness are proportional to the duration of heat treatment.

The inset in Figure 4.7 shows the full hysteresis loop for the sample heat-treated for 6 h. The shape of the curve is a conventional double-shifted hysteresis loop observed for a system composed of interacting ferromagnetic and antiferromagnetic phases. The system obviously consists of a ferromagnetic metal core and an antiferromagnetic shell. The best candidate for the antiferromagnetic phase is an oxidized form of Fe. At early stages oxidation manifests itself by a kink on the hysteresis loop in the zero field, which is a consequence of superposition of the double-shifted hysteresis loop with an ordinary hysteresis loop. The conclusion concerning oxidation of Fe nanoparticles is confirmed by Mössbauer investigation of the same samples.

Figure 4.8 shows the zero-field cooled (ZFC) and field-cooled (FC) dependences of the original (not heat-treated) sample containing 5 wt.% Fe in a polyethylene matrix and of the sample heat-treated during 6 h at 280 °C. The

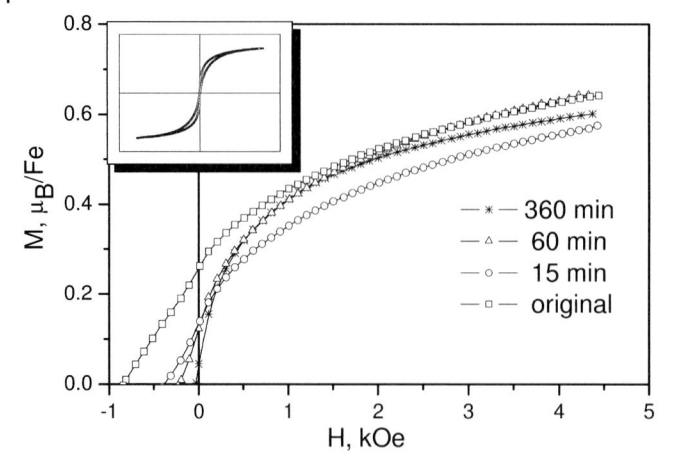

Figure 4.7 The room temperature magnetization curves for the original material and heat-treated samples at 280 °C for 15, 60, and 360 min. The inset shows the full hysteresis loop for the letter case.

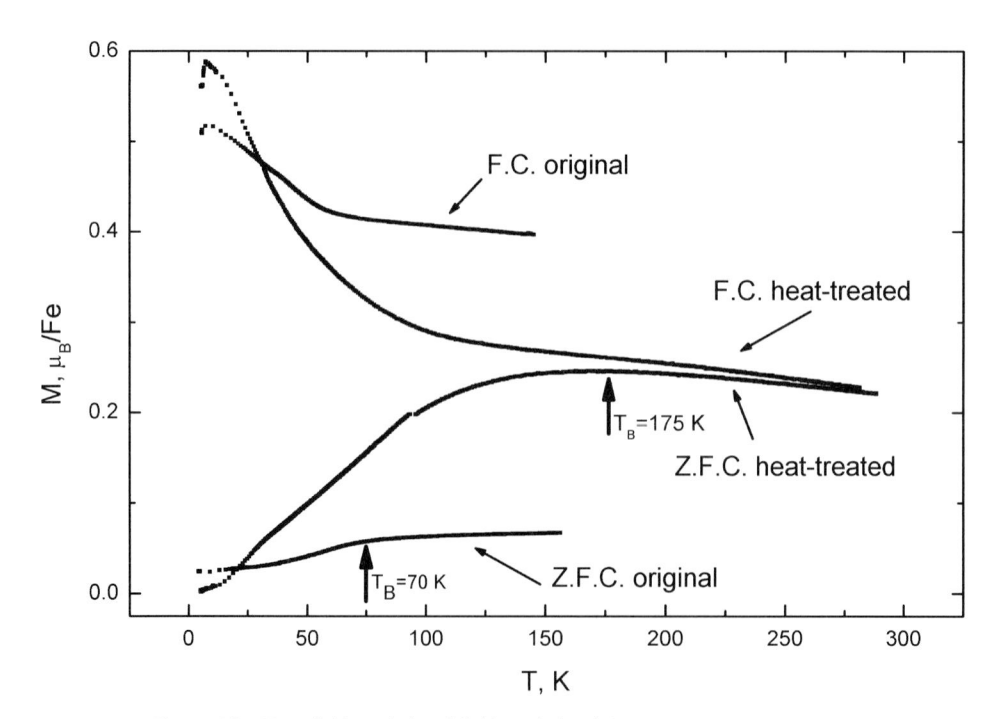

Figure 4.8 Zero-field-cooled and field-cooled (4 kOe) $M(T)$ dependences for original and heat-treated (6 h at 280 °C) samples containing 5 wt.% Fe in polyethylene matrix, measured in the field of 300 Oe.

curves for the heat-treated sample allow determining the blocking temperature that has the value about 175 K. The anomaly at the temperatures lower than 70 K on the curves for the original sample reflects the presence of an oxidized shell in the original sample.

4.3.2
Iron Oxide Nanoparticles

These samples were synthesized by means of the thermodestruction of iron(III) acetate. According to the TEM data, the average particle size is 4.6 nm. The composition of the particles determined by means of Mössbauer spectroscopy, XRD, and EXAFS is γ-Fe_2O_3. A more detailed description of these samples may be found in [28]. The purpose of the study of dc magnetization was to directly observe the blocking process and to determine such main magnetic parameters as saturation magnetization, coercive force, remnant magnetization, and the rate of magnetization decay. The magnetic properties of γ-Fe_2O_3 nanoparticles embedded in the LDPE matrix were investigated at different temperatures.

The field dependence of magnetization $M(H)$ was measured at room temperature for a set of samples with the following concentrations of Fe_2O_3: 1, 3, 5, 30, and 50 wt.% (samples F1, F3, F5, F30, and F50, respectively). The results for samples F5 and F50 are presented in Figures 4.8 and 4.10. Magnetic

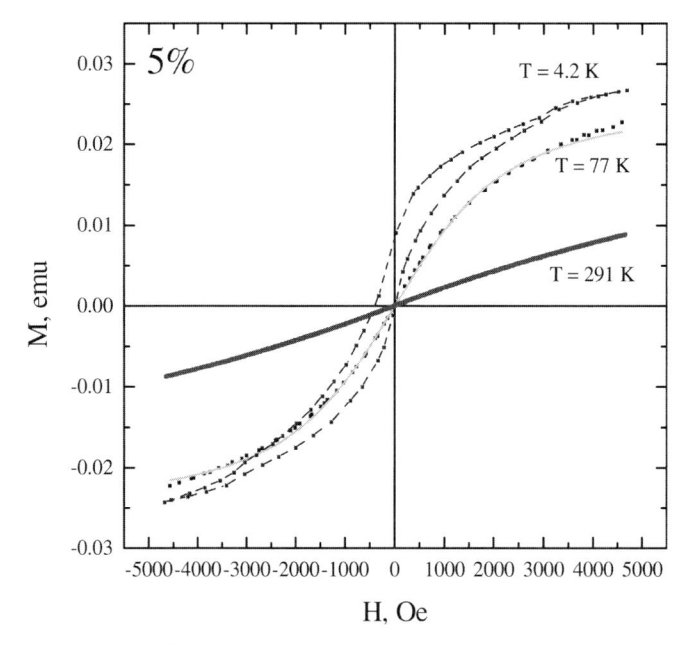

Figure 4.9 Field dependence of the magnetization at different temperatures for the sample with 5 wt.% of Fe_2O_3.

hysteresis was observed for all the samples except F5. The values of remnant magnetization and coercive force were found to have the largest values for the F30 sample. In the case of the F5 sample the nonlinear magnetization curve obeys rather well the Langevin function $L(x) = \text{cth}(x) - 1/x$, where $x = \mu H/k_B T$, H is the external field, T is the temperature, μ is the magnetic moment of one particle, and k_B is the Boltzmann constant.

The appearance of hysteresis on room temperature $M(H)$ curves can be the result of the transition either to the stable state [36, 37] or to the collective magnetic ordering state (like the spin-glass state). The latter is possible if there are enough strong interparticle interactions in the system. The collective ordering seems to be inappropriate because the absence of hysteresis for the F5 sample is not in accordance with the observation of hysteresis for the F1 and F3 samples, where the distances between particles are larger and, therefore, interactions are reduced. At room temperature, the saturation of the magnetization was practically achieved for only two samples: F50 and F30.

For three of the samples (F5, F30, F50) the field dependence of magnetization was measured at $T = 77$ K. The results for samples F5 and F50 at this temperature are also presented, in Figures 4.9 and 4.10. For the F5 sample hysteresis was not observed. The magnetization curve also obeys the Langevin function quite well at room temperature. For the other two samples an increase of coercive force and remanent magnetization with the decrease of

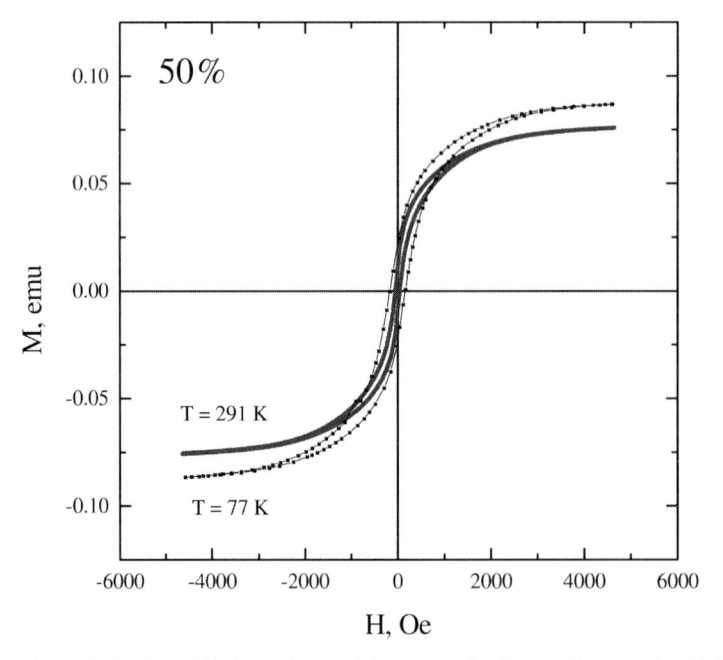

Figure 4.10 The field dependence of the magnetization for the sample with 50 wt.% of Fe_2O_3.

temperature was noted. An increase in saturation magnetization was observed for these samples with the decrease of temperature. One of the possible explanations of this phenomenon is the presence of two types of nanoparticles with different sizes and, consequently, different blocking temperatures in the sample. One part of the particles possesses a T_b of higher than room temperature. These particles make a main contribution to magnetization at near room temperature. Other particles have blocking temperatures lower than the boiling point of liquid nitrogen and remain superparamagnetic in the temperature interval of 77–291 K. Their contribution to the magnetic moment of the sample is completely reversible (without hysteresis) and increases with the decrease of temperature, roughly as $1/T$. In all the experiments the total signal was measured.

In particular, ZFC and FC magnetization curves do not coincide below T_b and magnetic hysteresis appears in the M versus. the H loop. Figure 4.11 represents ZFC and FC magnetization for sample F5. Based on the data about the magnetization of the measured samples, the value of T_b was determined as the temperature of the maximum on the ZFC curve. For the F5 sample,

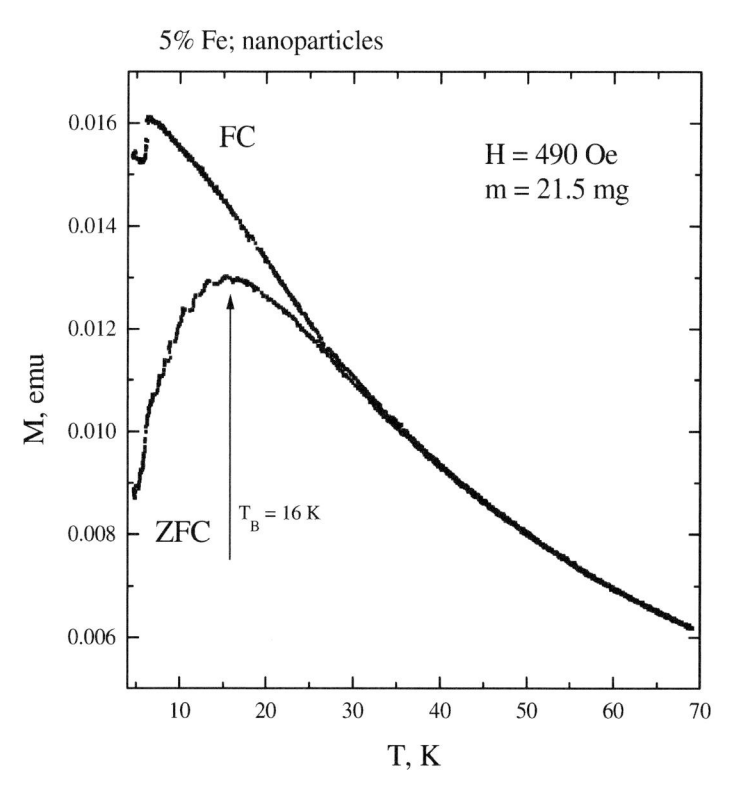

Figure 4.11 Zero-field cooling and field-cooling magnetization for the sample with 5 wt.% of Fe_2O_3.

T_b was equal to 16 K and magnetic hysteresis was observed at $T = 4.2$ K ($H_C = 490$ Oe) (see Figure 4.9). Analogous measurements were made for the F30 sample and $T_b = 77$ K was obtained for this case.

According to the theoretical studies [37], the remanent magnetization decays toward a thermodynamically equilibrium state below T_b. The characteristic time of relaxation strongly depends on temperature. In the vicinity of the blocking point, the relaxation time becomes comparable with the time in the experiment when this parameter can be measured.

The measurement of the remanent magnetization decay was carried out for sample F50 at room temperature. During a 10-h period the value of remanent magnetization decreases by several percent. However, the time dependence of a magnetic moment is neither exponential nor logarithmic. This means that the mechanism of magnetization decay cannot be explained in the framework of the known simple models. The possible reason for such a complicated behavior may be the dispersion in particle sizes or the interaction between particles and the magnetic field remained after switching off the magnetic field.

4.3.3
Cobalt Nanoparticles

These samples were synthesized by means of the thermodestruction of Co formate. According to the TEM data, the average size of the particles is 3.9 nm. The composition of the particles determined by means of XRD and EXAFS is metallic Co. The structure of the particles was determined by means of EXAFS. A more detailed description of these samples is available in [32]. The magnetization curves for Co nanoparticles in polyethylene at different temperatures are shown in Figure 4.12.

The main peculiarity of magnetic behavior of Co samples is near constant value of the coercive force in the investigated temperature range. While for Fe samples the value of H_C is rapidly increased at temperatures lower than 77 K, for Co samples the values of H_C at 4.2 and 77 K are almost equal. If field-cooled $M(H)$ dependences for Co and Fe nanoparticles are compared (Figure 4.13), it is clear seen that in the case of Co sample there is no bend on the curve at 75 K. The bend on the $M(H)$ curve and the increase of coercive force at lower temperatures for Fe samples show the presence of different oxidized phases in the particle shell. In the case of Co particles, in the PE matrix the oxidation and role of the particle shell in magnetic properties are not important. This supposition is supported by data on the magnetic properties of the heat-treated sample that will be discussed below.

In contrast to Fe samples, heating of Co samples up to 100 °C does not lead to a noticeable decrease of H_C. Since the value of H_C for a small particle is determined by the product of the particle volume and anisotropy energy, this behavior must be a consequence of much higher anisotropy energy in the case of Co particles. The lower value of anisotropy energy in the case of Fe particles,

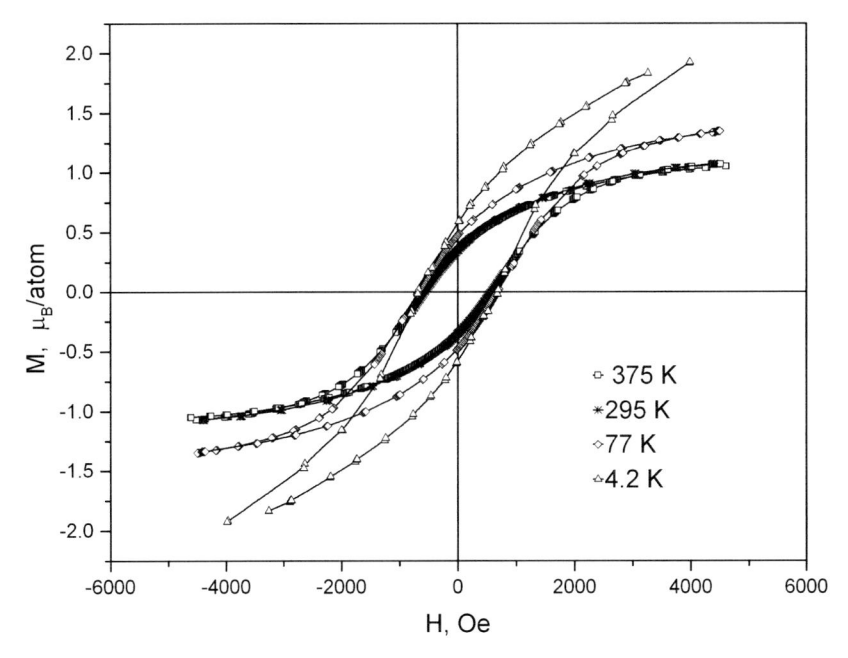

Figure 4.12 The magnetization curve for Co nanoparticles at different temperatures.

in particular, may be a consequence of low magnetic transition temperature as it is often observed for Fe-containing intermetallic compounds.

Another feature of curves in Figure 4.12 is the noticeable increase of the saturation magnetization at low temperatures. A similar increase in this temperature range was observed for Fe samples as well.

Table 4.6 lists the values of the coercive force, remnant magnetization, and magnetization in the field of 4.5 kOe for the original Co samples and Co samples heat-treated at 280 °C.

Heat treatment of Co-containing samples in contrast to the Fe ones does not lead to progressive oxidation of the particles and the decrease of coercive force. Moreover, heat treatments in the case of Co-containing samples lead

Table 4.6 The values of the remnant magnetization, M_R, magnetization in the field of 4.5 kOe, $M_{H=4.5}$, and the coercive force, H_C, at room temperatures of Co samples heat-treated at 280 °C for 2 h and original Co samples.

	H_C	M_R	$M_{H=4.5}$
	Oe	μ_B	μ_B
Original	590	0.35	1.05
Heat treated	590	0.62	1.96

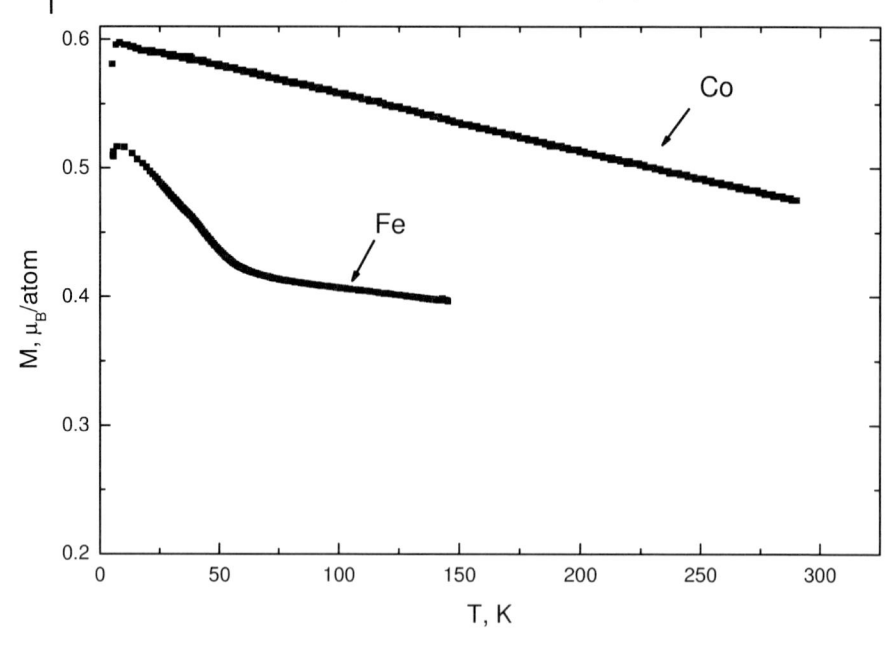

Figure 4.13 The field-cooled temperature dependences of magnetization for Co and Fe nanoparticles.

to an increase of specific magnetization that is explained by the presence of chemical precursors in the as-prepared material.

From the point of view of the preparation method the increase of preparation temperature could not only lead to a more complete particle formation but it may also lead to the formation of too large particles. The possible decision is a two-stage process. On the first stage, using low-viscosity melt/solution of polymer in oil allows obtaining a uniform distribution of Co in the polymeric matrix. On the second agglomeration stage (heat treatment at higher temperatures), nanoparticles are formed from small Co particles. On this stage, the high viscosity of polymer prevents the formation of too large particles.

4.3.4
Co@Fe$_2$O$_3$ Particles

These samples were synthesized via the thermal destruction of cobalt and iron carbonyls at once. According to the TEM data, the average particle size is 5.8 nm. The composition of the particles determined by means of ^{57}Fe Mössbauer spectroscopy is γ-Fe$_2$O$_3$, XRD, and EXAFS analysis was show Co and γ-Fe$_2$O$_3$ phases.

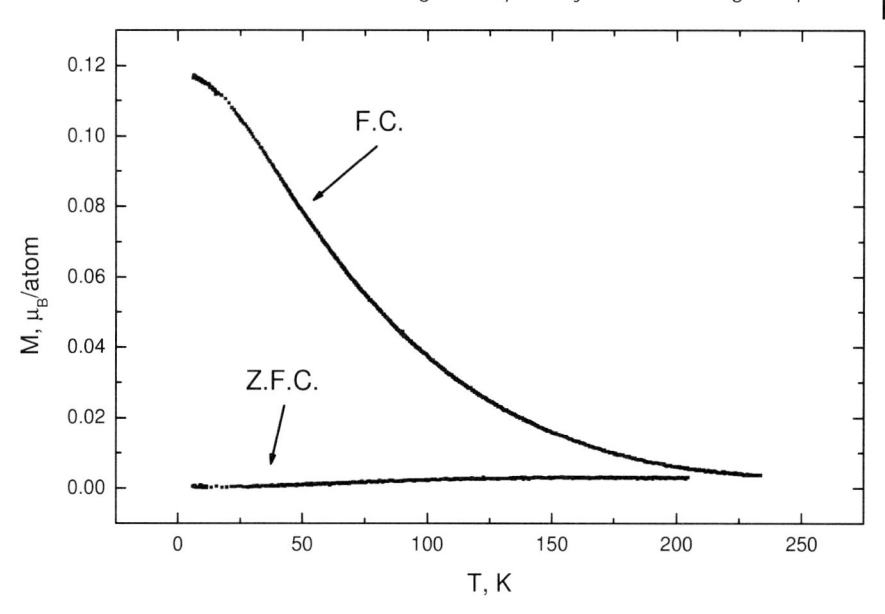

Figure 4.14 Zero-field-cooled and field-cooled $M(T)$ dependences for the Co@Fe$_2$O$_3$ (Co:Fe=1:5) sample.

The magnetic properties of the Co@Fe$_2$O$_3$ composition with estimated chemical were investigated. The magnetic properties of this material are unexpectedly different from those observed for Co and Fe samples. Figure 4.14 shows ZFC and FC temperature dependences of magnetization for this sample. The blocking temperature for this sample determined as a point where ZFC and FC curves start to diverge is 200 K.

It is well known that in the case of intermetallics on-site anisotropy of Co and Fe has different signs. For that reason in Co–Fe bulk alloys due to mutual compensation the anisotropy energy can be very low. On the other hand, Co–Fe alloys demonstrate largest known magnetic moment per atom that attracts technological interest to these materials. The observed low blocking temperature of nanoparticles reflects low intrinsic anisotropy energy of the obtained Co–Fe composition.

Figure 4.15 presents a magnetization versus field plot measured at 4.2 K, 77 K, and room temperature. Fitting the room temperature $M(H)$ curve with the Langevin function gives the value of magnetic moment per Fe or Co atom near 0.5 μ_B. It allows supposing that a sufficiently large number of the precursors successfully decomposed during the sample preparation stage.

Magnetization processes at 4.2 and 77 K in fields up to 4500 Oe are not re-entrant. In particular, the curve measured at 77 K is lower than the curve measured at room temperature because the measurements were made after cooling in the zero field. The measurements at 4.2 K were made after cooling

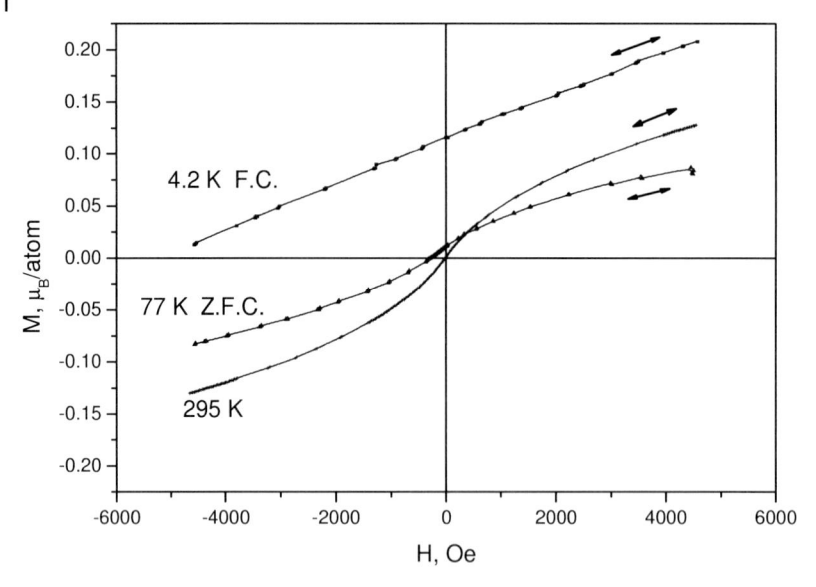

Figure 4.15 The $M(H)$ curves for CoFe$_5$ nanoparticles measured at 4.2 K, 77 K, and room temperature.

in the field 4500 Oe. Both curves are shifted from zero point and show low amplitude of changes at field sweep between −4.5 and 4.5 kOe.

This behavior can be explained by supposing that for this sample the ferromagnetic metal core of nanoparticles strongly exchange-biased by the antiferromagnetic shell. It can occur for example because of strong oxidation of the Fe shell in Co–Fe particles.

4.4
FMR Investigations of Nanocomposites

The FMR experiments were conducted on the basis of the computerized X-band EPR spectrometers "Varian E-4" and "Varian E-109 with the help of nitrogen and helium flow cryostat systems. We used powder-like samples as well as compacted samples that were pressed from the prepared powder at a pressure of about 200 atm. No marked difference was found in the magnetic properties of the samples of these two types in the FMR measurements.

These samples were synthesized by means of the thermal decomposition of nickel acetate. According to the TEM data, the average particles size is 4.2 nm. The XRD results reveal the presence of nickel oxide and metallic nickel phases in the nanoparticles. The structure of the particles was determined by means of EXAFS.

Figure 4.16 shows the selected FMR spectra at moderate and high temperatures for the Ni-containing samples. The room temperature spectrum consists

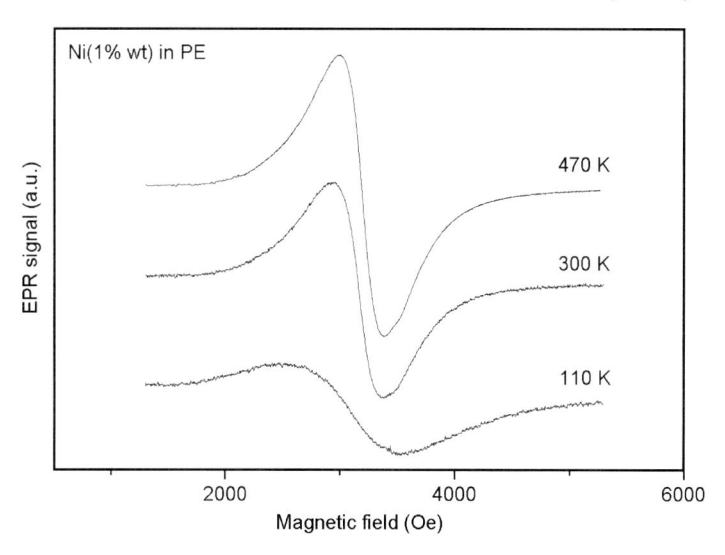

Figure 4.16 Spectra of microwave absorption of Ni-containing samples at different temperatures.

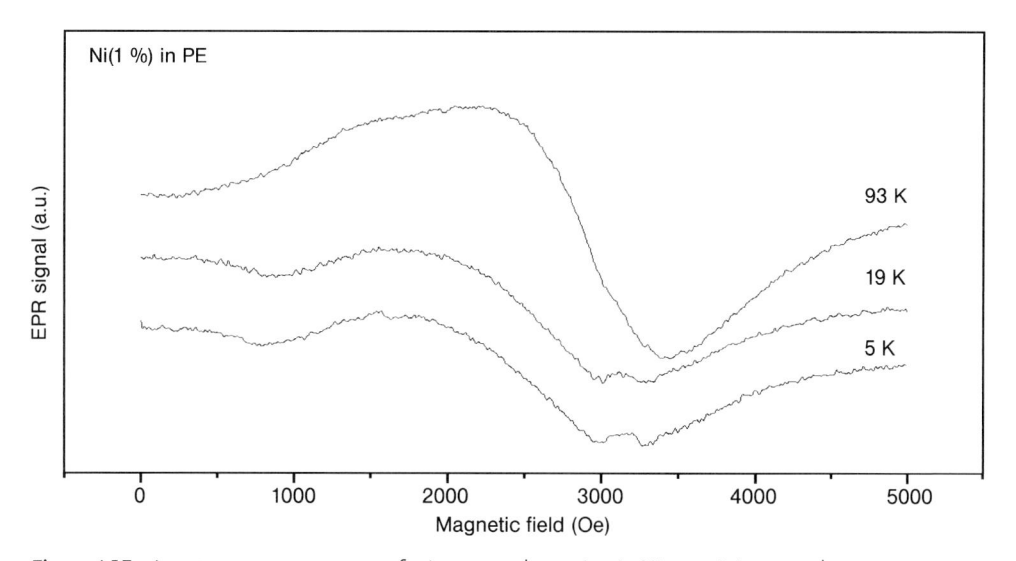

Figure 4.17 Low-temperature spectra of microwave absorption in Ni-containing samples.

of a single symmetric line with $g \approx 2$ and $\Delta H = 500$ Oe. As the temperature increases, the line slightly narrows (down to 380 Oe at 470 K). Sample cooling results in broadening of the line (up to 1000 Oe at 110 K) without marked lineshape transformation. Below approximately 100 K, the spectra become more complicated (see Figure 4.17), with pronounced low field broad components.

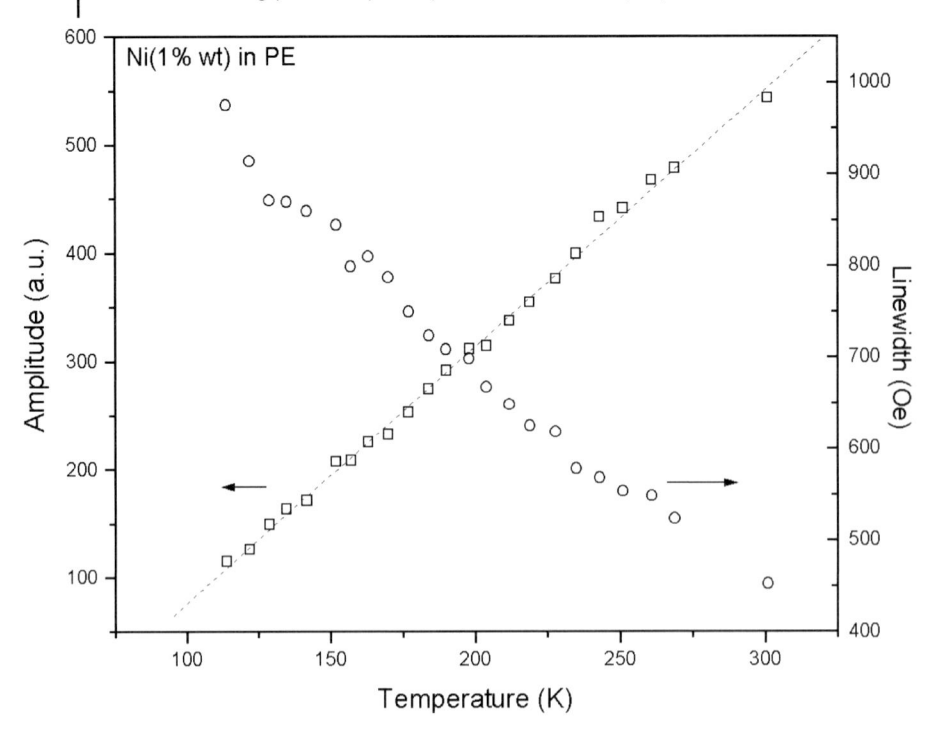

Figure 4.18 Thermal behavior of the spectra of microwave absorption of Ni-based samples.

At very low temperatures, the resonance field shifts to lower values (the g-factor increases, correspondingly); however, the dramatic line change that is typical of any magnetic transition does not take place. It should be stressed that only negligible microwave absorption hysteresis was observed at helium temperatures. Such behavior resembles the antiferromagnetic ordering in bulk systems [26].

Figure 4.18 shows, for Ni-based samples, the temperature dependences of the FMR parameters: linewidth ΔH and signal amplitude A. The amplitude is a linear function of temperature, whereas the thermal behavior of ΔH is more complicated. In general, both FMR characteristics demonstrate the temperature dependence that is typical of nanoparticle. Figure 4.19 presents the FMR line intensity I, calculated as $I = A^* \Delta H^2$, versus temperature. The curve $I(T)$ has a broad maximum, between 150 and 250 K.

Figure 4.20 demonstrates the FMR spectra for Fe-based samples at selected temperatures. As with the Ni-based samples, the room temperature spectrum contains a single line with $g \approx 2$, which, however, is much narrower ($\Delta H = 80$ Oe). As the temperature decreases the spectra broaden and become asymmetric. This tendency is especially marked below approximately 50 K (Figure 4.21). The hysteresis of microwave absorption at low magnetic fields is insignificant. Figure 4.22 shows the thermal variation of ΔH and A for

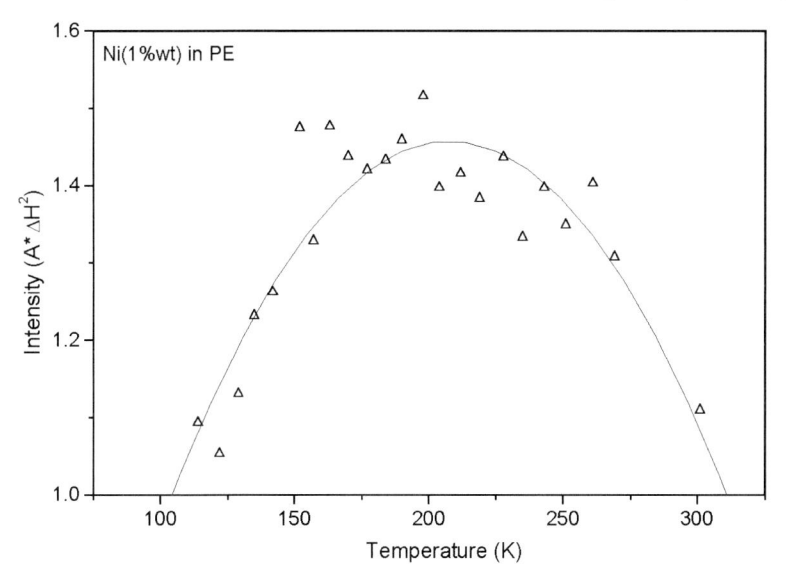

Figure 4.19 FMR line intensity versus temperature.

Fe-based nanoparticles. It is interesting that some anomalies are observed near 200 K for both $\Delta H(T)$ and for $A(T)$ curves.

The FMR data for homometallic Fe-, Co-, and Ni-based nanoparticles in LDPE with 1% wt content of metal were obtained. Independently of the metal, for all the nanoparticles that were studied the spectrum behavior

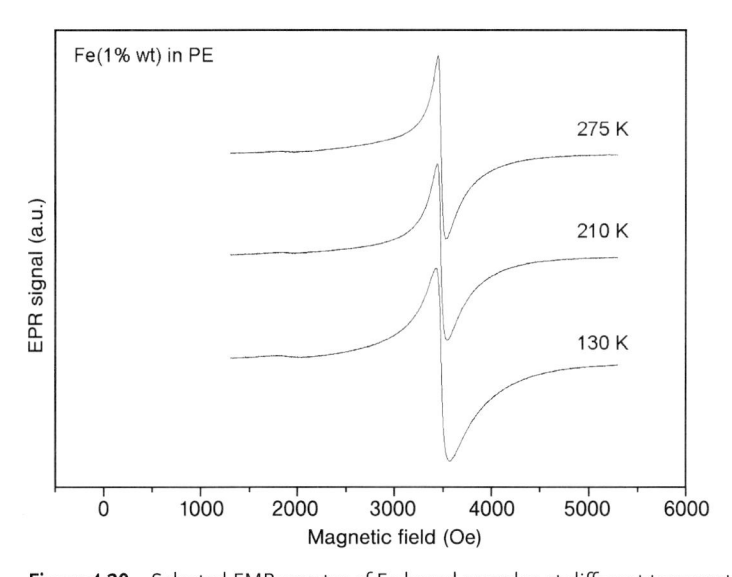

Figure 4.20 Selected FMR spectra of Fe-based samples at different temperatures.

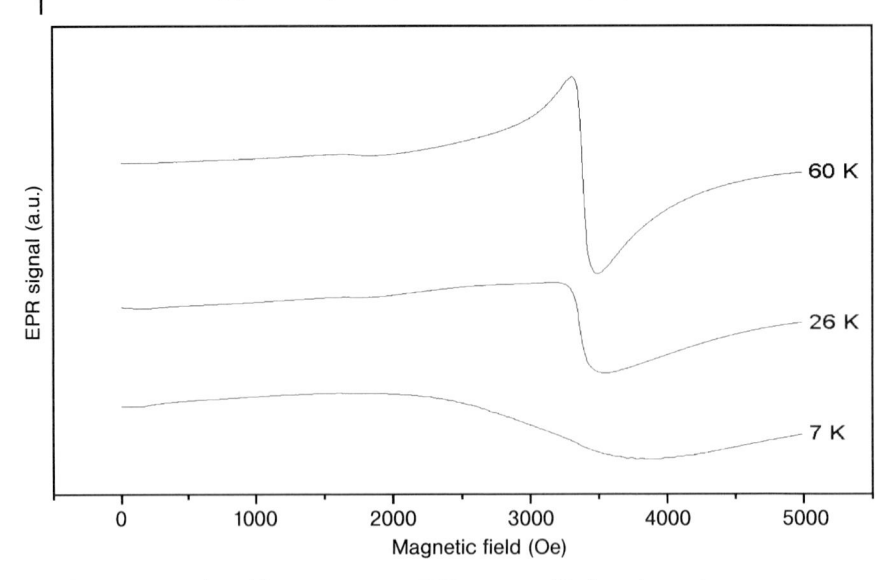

Figure 4.21 Selected low-temperature FMR spectra of Fe-based samples.

is superparamagnetic at moderate temperatures. Namely, the linewidth increases with sample cooling, whereas the signal amplitude decreases (Figures 4.18 and 4.22). No marked hysteresis was detected, which indicates the absence of the transition to blocking state at the temperature region that was investigated. This fact is also proved by the dc magnetic measurements.

The anomalies of FMR parameters near 200 K (Figure 4.22) in Fe-based samples can be due to the antiferromagnetic (or spin canting) structure of the nanoparticles or, at least, of their surface. Indeed, antiferromagnetic FeO has a Neel temperature of 198 K [10]. Surface oxidation can also be responsible for the maximum FMR intensity at 200 K for Ni-based nanoparticles (Figure 4.19). In Ref. [23], a maximum blocking temperature of 200 K was observed for NiO particles with a mean size of 5.3 nm. The anomaly in the thermal behavior of the FMR spectra of Co-based nanoparticles will also result from particle oxidation. It is well known [24, 25] that the oxidation of nanoparticles results in the distortion of the magnetic structure (at least near the particle surface) and in a reduction of the total particle magnetic moment.

4.5
Conclusions

The homometallic Fe-, Ni-, and Co-based nanoparticles embedded in a polyethylene matrix was studied. Magnetic nanoparticles stabilized by a polyethylene matrix were obtained by thermolysis of the metal-containing

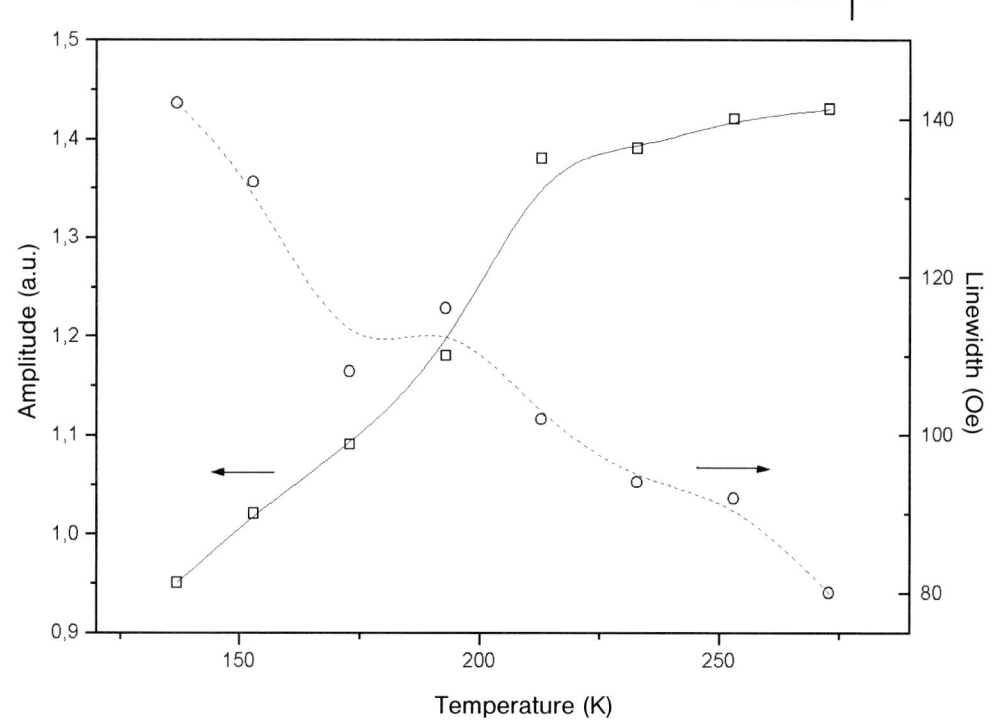

Figure 4.22 Thermal behavior of the spectra of microwave absorption on Fe-based samples.

precursors. By varying the synthesis conditions, it is possible to obtain nanoparticles of metals and their oxides with the required size, structure, and homogeneous size distribution in the volume of the matrix.

Since most metal oxides are antiferromagnetic, it is necessary to prevent particle oxidation which will reduce significantly the saturation magnetization and the coercive force of the magnetic material.

The FMR spectra of Ni- and Co-based samples demonstrate properties which indicate sufficient particle oxidation. Such oxidation was also found in magnetization experiments at high temperatures (near 100 °C).

The process of heat treatment of the sample containing 5 wt.% of Fe and Co in a polyethylene matrix at the temperatures of up to 280 °C was studied. The changes in nanoparticles' content, especially the degree of Fe oxidation during the process, were tracked, but Co was not oxidized during this treatment. It was shown that at temperatures up to 100 °C not any noticeable changes in nanoparticle composition or oxidation took place.

Thus, a novel magnetic nanomaterial with unusual magnetic properties has been produced.

Acknowledgments

The authors thank Dr. Yu. A. Koksharov for data of FMR experiments. This work was partially financially supported by the Russian Foundation for Basic Research (grant nos. 07-03-00885_a, 07-08-00523-a, 08-03-00681_a, 08-08-90250-Uzb_a), INTAS-05-1000008-7834, ISTC nos. 3457 Program of Federal Agency of Science and Innovations No.02.513.12.0042.

References

1. Pomogailo, A.D., Rozenberg, A.S., and Uflyand, I.E., *Nanochastitsy metallov v polimerakh (Metal Nanoparticles in Polymers)*, Khimiya, Moscow, (**2000**).
2. Edelstein, A.S. and Cammarata, R.C. (Eds), *Nanomaterials: Synthesis, Properties and Applications*, Institute of Physics Publishing, Bristol and Philadelphia, (**1998**), 596 pp.
3. Gubin, S.P. (**2000**) *Russian Chemical Journal*, Vol. XLIV, No. 6, p. 23.
4. Sharma, V.K. and Waldner, F. (**1977**) *JAP*, Vol. 48, p. 4298.
5. Chikazumi, S. (**1987**) *Physics of Ferromagnetism. Magnetic Properties and Applications*, Mir, Moscow.
6. Hempel, K.A., and Roos, W. (**1981**) *IEEE Transactions on Magnetics*, Vol. MAG17, No. 6, p. 2642.
7. Koksharov, Yu.A., Blyumenfel'd, L.A., Tikhonov, A.N., and Sherle, A.I.,Zh. (**1999**) *Fiz. Khim.*, Vol. 73, No. 10, p. 1862.
8. Volmer, M. (**1983**) *Zur Kinetik der Phasenbildung und der Elektrodenreaktionen*, Akademie, Leipzig.
9. Fedotov, V.D., and Abdrashitov, I.A. (**1979**) *High-Molecular Compounds*, Vol. 21A, No. 10, p. 2275.
10. Marton, L., and Marton, C. (**1980**) *Methods of Experimental Physics*, Vead. Press, New York.
11. Gubin, S.P., and Kosobudskii, I.D. (**1983**), *Usp. Khim.*, Vol. 52, p. 1350.
12. Kozinkin, A.V., Sever, O.V., Gubin, S.P., et al., (**1994**) *Neorg. Mater.*, Vol. 30, No. 5, p. 678 [*Inorg. Mater.* (Engl. Transl.), Vol. 30, No. 5, p. 634].
13. Kozinkin, A.V., Vlasenko, V.G., Gubin, S.P., et al., (**1996**) *Neorganicheskie Materialy*, Vol. 32, No. 4, p. 422 [*Inorg. Mater.* (Engl. Transl.), Vol. 32, No. 4, p. 376].
14. Gubin, S.P., Kozinkin, A.V., Afanasiev, M.I., Popova, N.A., Sever, O.V., Shuvaev, A.T., Tsirlin, A.M. (**1999**) *Neorg. Mater*, Vol. 35, No. 2, p. 237 [*Inorg. Mater.* (Engl. Transl.), Vol. 35, No. 2, p. 180].
15. S.V. Vonsovskii (Ed.) (**1966**) *Ferromagnetic Resonance*, Pergamon, Oxford.
16. de Biasi, R., and Devezas, T.C. (**1978**) *JAP*, Vol. 49, p. 2466.
17. Gates, B.C., Guczi, L., and Knosinger, H. (**1986**) *Metal Cluster in Catalysis*, Elsevier, Amsterdam.
18. Kodas, T.T., and Hampden-Smith, M.J. (**1994**) *The Chemistry of Metal CVD*, Wiley-VCH, Weinheim.
19. Hanipden-Smith, J., and Kodas, T.T. (**1995**) *Chem. Vap. Deposition*, Vol. 1, p. 8.
20. Gates, B.C., Guczi, L., and Knosinger, H. (**1986**) *Metal Cluster in Catalysis*, Elsevier, Amsterdam.
21. Kodas, T.T., and Hampden-Smith, M.J. (**1994**) *The Chemistry of Metal CVD*, Wiley-VCH, Weinheim.
22. Hanipden-Smith, J., and Kodas, T.T. (**1995**) *Chemical Vapor Deposition*, Vol. 1, p. 8.
23. Yurkov, G.Yu., Gubin, S.P., Pankratov, D.A., Koksharov, Yu.A., Kozinkin, A.V., Spichkin, Yu.I., Nedoseikina, T.I., Pirog, I.V., and Vlasenko, V.G. (**2002**) *Neorganicheskie Materialy*, Vol. 38, No. 2, p. 186, [*Inorg. Mater.* (Engl. Transl.), Vol. 38, No. 2, p. 137].
24. Gubin, S.P., Spichkin, Yu.I., Yurkov, G.Yu., and Tishin, A.M. (**2002**)

Russian Journal of Inorganic Chemistry, Vol. 47, Supp. l. p. 32.

25. Gudoshnikov, S., Liubimov, B., Matveets, L., Ranchinski, M., Usov, N., Gubin, S., Yurkov, G., Snigirev, O., and Volkov I. (**2003**) *Journal of Magnetism and Magnetic Materials*, Vol. 258–259, p. 54.

26. Koksharov, Yu.A., Gubin, S.P., Kosobudsky, I.D., Yurkov, G.Yu., Pankratov, D.A., Ponomarenko, L.A., Mikheev, M.G., Beltran, M., Khodorkovsky, Y., and Tishin, A.M. (**2001**) *Journal of Physical Review B*, Vol. 63, No. 17, p. 12407.

27. Rostovshikova, T.N., Kiseleva, O.I., Yurkov, G.Yu., Gubin, S.P., Pankratov, D.A., Perfiliev, Yu.D., Smirnov, V.V., Chernavsky, P.A., and Pankina, G.V. (**2001**) *Vestnik Moskovskogo Universiteta, Seria 2 Khimiya*, Vol. 42, No. 5, p. 419.

28. Yurkov, G.Yu., Gubin, S.P., Pankratov, D.A., Koksharov, Yu.A., Kozinkin, A.V., Spichkin, Yu.I., Nedoseikina, T.I., Pirog, I.V., and Vlasenko, V.G. (**2002**), *Neorganicheskie Materialy*, Vol. 38, No. 2, p. 186, [*Inorg. Mater.* (Engl. Transl.), Vol. 38, No. 2, p. 137].

29. Yurkov, G.Yu., Kozinkin, A.V., Nedoseikina, T.I., Shuvaev, A.T., Vlasenko, V.G., Gubin, S.P., and Kosobudsky, I.D. (**2001**) *Neorganicheskie Materialy*, Vol. 37, No. 10, pp. 1175 [*Inorg. Mater.* (Engl. Transl.), Vol. 37, No. 10, p. 997.].

30. Rebinder, P.A. (**1978**) *The Elected Works: The Superficial Phenomena in Disperse Environments*. G.I. Fuks (Ed.)), Nauka, Moscow.

31. Kats, G.S., and Milevsky, D.V. (**1981**) *Loading for Composite Materials*. (Reference-book). Khimia, Moscow.

32. Gubin, S.P., Spichkin, Yu.I Koksharov, Yu.A., Yurkov, G.Yu., Kozinkin, A.V., Nedoseikina, T.A., Vlasenko, V.G., Korobov and Tishin, A.M. (**2003**) Magnetic and structural properties of Co nanoparticles in polymeric matrix. *Journal of Magnetism and Magnetic Materials*, Vol. 265, No. 2, pp. 234–242.

33. Deriagin, B.V. (**1986**) *The Theory of Stability Colloids and Thin Films. Surface Forces*, Nauka, Moscow.

34. Berlin, A.A., and Basin, V.E. (**1974**) *Bases of Adhesion of Polymers*. Khimia, Moscow.

35. Vakula, V.A., and Pritykin, L.M. (**1984**) *Physical Chemistry of Adhesion of Polymers*. Khimia, Moscow.

36. Liu, X., Wang, J., Gan, L.M., Ng, S.C., and Ding, J. (**1998**) *Journal of Magnetism and Magnetic Materials*, Vol. 184, p. 344.

37. Jacobs, I.S., and Bean, C.P. (**1963**) *Magnetism* (G.T. Rado and H. Suhl), Vol. 3, p. 271, Academic Press, New York.

38. Taylor, R.H. (**1975**) *Advances in Physics*, Vol. 24, p. 681.

5
Organized Ensembles of Magnetic Nanoparticles: Preparation, Structure, and Properties

Gennady B. Khomutov and Yury A. Koksharov

5.1
Introduction

Magnetic nanoparticles, nanocomposite materials, and organized magnetic nanoparticulate assemblies and superstructures are currently a subject of wide basic and applied research due to their interesting and practically important properties (see, for example [1–10]). Magnetic nanoparticles are an important class of functional materials, possessing unique magnetic properties due to their reduced size (below 100 nm) with potential for use in devices with reduced dimensions, in functional nanostructured composite materials and in various biomedical applications. Magnetic nanoparticles can be prepared and functionalized by a number of physical–chemical techniques including precipitation, redox reactions, hydrothermal synthesis, reverse micelles, polyol, sol–gel, thermolysis and high-temperature reduction of metal salts, photolysis, sonolysis, multisynthesis processing, electrochemical, and other techniques [9, 11–20]. Stable colloid dispersions of magnetic nanoparticles in organic or inorganic media (often called magnetic fluids or ferrofluids) have attracted much attention and interest of basic and applied researchers during last decades [21–24]. Colloid dispersions of magnetic nanoparticles can be prepared in a wide variety of liquid media: simple polar or nonpolar liquids (water, kerosene, etc.), various complex media composed of self-assemblies of surfactants, or polymers [25]. The colloid magnetic particles dispersed in a magnetic fluid are usually composed of iron, cobalt, iron oxides (magnetite, maghemite, etc.), and various ferrites. Ferrite nanoparticles are widely used because they are chemically stable and available without any organic coating and dispersed in water, or coated by surfactants or polymers and dispersed in oils [25]. Water-based ferrofluids from colloidal FePt nanoparticles were also reported [26].

Studies on nanoparticles were mainly concerned with preparation and characterization of quasiuniform spherical forms of various sizes. However, recently anisotropic nanoparticles have attracted many attention. Anisotropic magnetic nanoparticles and nanostructures are of particular interest because they are expected to exhibit interesting magnetic properties due to the shape anisotropy.

Organized ensembles of magnetic nanoparticles, nanoparticulate magnetic nanosystems, and nanostructured composites are advancing to the forefront of modern research of magnetism and magnetic materials. The behavior of arrays of nanoscale magnetic particles is interesting both from a fundamental point of view, and also for applications in magnetic recording media, magnetic cellular automata, or magnetoelectronic devices.

Appropriate assembling of nanoparticles generates new organized nanostructures with novel collective physical properties, which can be exploited for multipurpose applications in nanoelectronics, spintronics, ultrahigh density magnetic storage, magnetooptics, chemical catalysis, sensors, biomedical separation, diagnostics, therapy, etc. In particularly, development of spintronics, which exploits the phenomena that electrical current carriers have not only charge but also spin, requires new magnetic nanostructures with desired magnetic properties. It becomes increasingly important to develop efficient methods for preparing and assembling magnetic nano-objects at smaller and smaller length scales, from ultrathin films and multilayers to 1D chains and nanowires, and eventually to organized individual and patterned collective nanoparticulate structures [27].

Also, of substantial interest and importance are integrated and multicomponent structures, which are the basis for creation of new functional and polyfunctional nanomaterials. The integration and combining of nanocomponents of different nature into novel hybrid systems opens wide possibilities for design and creation of new nanosystems and nanomaterials with advanced or even novel unique properties and functional advantages important for real-world applications. In that aspect, the multicomponent heterostructured magnetic nanoparticles (alloys, core–shells, and binary superlattices) are prospective next-generation nanomaterials due to the combination of their magnetism and other properties resulting in potential multifunctionalities [3].

Many different approaches to magnetic ordered nanostructure synthesis and processing have been developed till now [28]. These strategies can be roughly categorized as either "top-down" or "bottom-up."

There are various top-bottom methods for fabricating arrays of magnetic nanostructures. These include, first of all, different lithography methods: electron beam lithography [29], focused ion beam irradiation etching [30, 31] and sputtering [32], X-ray interference lithography [33], UV lithography [34], laser interference lithography, and other physical methods including scanning-probe lithography, step growth methods, shadow masks, radiation damage, etc. (see, for example [35]).

Bottom-up processes are based on the synthetic and assembling chemical–physical approaches and self-assembling phenomenon.

Bottom-up and top-down approaches each have their advantages and disadvantages. Top-down approaches can be extremely effective at reproducibly defining nanostructure dimensions – lithography is the foundation of the microelectronics industry and sub-100 nm transistor gate lengths are defined by lithography in commercially available transistors. However, lithography faces fundamental limitations in defining features smaller than 10 nm in diameter. Top-down processing costs are also becoming prohibitively high with ever shrinking feature size.

Bottom-up routes to nanostructures, on the other hand, such as colloidal syntheses, are inexpensive and scalable. These methods have the potential to produce nanoparticles with characteristic dimensions less than 10 nm in large quantities with low cost. One significant challenge facing bottom-up processes, however, is that the nanostructures are "free-standing" and, therefore, must be assembled at specific positions on a substrate for device applications. This can be a significant technological hurdle. Nonetheless, there are applications, as in the biomedical fields, in which dispersed particles are in fact desired. The dispersibility of nanoparticles in various solvents and the ability to deposit them by spin-coating, inkjet printing, stamping, roll-to-roll processing, etc., can also be a processing advantage compared to the top-down processes by enabling low-temperature deposition on alternative substrates like plastics at quasiambient conditions. This capability could lead to new low-cost electronic and photonic technologies [28].

The available synthetic and assembling methods for the formation of organized magnetic nanoparticulate structures are based on two general approaches: first, the adsorption and deposition of presynthesized colloidal nanoparticles onto surfaces and interfaces or incorporation of such nanoparticles into organized matrices; second is connected with the synthesis and generation of nanoparticles and nanostructures directly on surfaces or into organized molecular or inorganic matrix.

The purpose of this chapter is to discuss known and novel experimental physical–chemical and colloidal nanofabrication methods for the preparation of organized ensembles of magnetic nanoparticles and nanostructured nanoparticulate systems and materials. Emphasis has been placed on various low-dimensional and ordered nanostructures. The characteristic properties of some organized ensembles of magnetic nanoparticles are discussed here. A number of examples of magnetic nanoparticulate structures are presented based mainly on the original author's results.

5.2
Two-Dimensional Systems: Layers and Nanofilms

Regular planar patterns and organized assemblies of magnetic nanoparticles at surfaces and interfaces or composite nanofilms are related to two-dimensional

(2D) nanostructures. Nanoparticulate layer structures on solid surfaces or freestanding nanofilms in gaseous or liquid phases are currently widely studied and are attractive for both fundamental research and potential applications in biomedical devices and as hyper high-density magnetic storage media with high perfomance. Ordered 2D arrays of magnetic nanoparticles are useful in studies of interparticle dipole–dipole interactions in a spatially well-defined system [36]. A number of methods including solvent evaporation and self-assembly, spin-coating, Langmuir–Blodgett (LB) technique, and layer-by-layer (LbL) assembly have been employed to fabricate those ordered nanoparticulate structures. Among the techniques for the deposition of organized nanofilms of magnetic nanoparticles on solid substrate surfaces, LbL and LB techniques are some of the most efficient and promising methods because they enable control of the composition and structure of every deposited monolayer, and allow formation of multilayer structures on different substrates.

Spin-coating technique was often used for the preparation of thin polymeric and composite films on solid substrates. In particular, special spin coating was employed for FePt magnetic nanoparticle layer deposition on 2.5 in. disk substrate [37]. Spin coating was used for deposition of Co nanoparticles obtained by thermolysis of $Co_2(CO_8)$ onto the surface of solid substrates [38] and for the formation of iron oxide hydrosol layer on the TiO_2 nanoparticulate film [39]. Ferromagnetic composite polymeric films containing homogeneously dispersed Fe and Ni nanoparticles with 30-nm diameter were deposited via spin coating onto glass and semiconductor substrates [40].

In general, spin-coating technique meets principal difficulties in control the deposited layer thickness and homogeneity down to monolayer of its components.

A solid substrate can be modified with multifunctional molecules that further replace the surfactant around nanoparticles, forming a monolayer assembly of the nanoparticles on the substrate surface. This self-assembly technique is known as molecule-mediated self-assembly [41]. When assembled via binding with–NH_2 groups, the 2D array of FePt nanoparticles on silica surface is extremely robust and can withstand annealing up to 800 °C without aggregation [42]. Self-assembled monolayers of mercaptopropionsulfonic acid on gold were used as a substrate for nucleation and growth of FeO(OH) nanocrystals [43].

Bilayer lipid (glyceryl monooleate) membranes were demonstrated to be a substrate with efficient binding of cationic magnetic Fe_3O_4 nanoparticles onto one or both the surfaces of the membranes with the formation of corresponding mono- and bilayers of magnetic nanoparticles [44].

In this section, we discuss the basic principles and embodiments of methods for the preparation of nanoparticulate layers and composite nanofilm materials. Complex and patterned nanoparticulate structures obtained by the use of LB and LbL technique will be discussed in Section 5.4.

5.2.1
Self-Assembled Nanostructures

Highly ordered magnetic nanostructures with advanced properties can be prepared by using self-assembly of colloidal magnetic nanoparticles [4, 45].

Monodisperse nanoparticles with insignificant size distributions can form close-packed arrays on a variety of solid substrates as the solvent from the particle dispersion is allowed to evaporate. Such a self-assembly process into a long-range-ordered structure may be comparable with crystallization, where elements rearrange themselves into a periodic crystal structure [42]. In a self-assembled nanoparticle superlattice (colloidal crystal), the nanoparticles act as the elements. In contrast to a crystal structure, where strong chemical bonds are usually present, nanoparticle superlattices often do not have such strong chemical interactions; instead, the particles are linked by weak hydrogen bond, van der Waals, and electric/magnetic dipole interactions. The dipolar interactions compete successfully with other magnetic interaction types in nanoparticle systems [46].

Solvent evaporation is very simple, versatile, and less time-consuming approach to form layers of nanoparticles on substrate surfaces from colloid suspension of nanoparticles [47–49]. Formation of ordered nanoparticulate structures by this method is possible in case of suspensions of quasimonodisperse nanoparticles. One of the characteristics of colloid monodisperse nanoparticles is the ability to form ordered close-packed arrays under evaporation of their suspension. The morphology of solution drop-deposited nanoparticle films is dependent on evaporation kinetics and particle interactions with the liquid–air interface. In low-dimensional packing structures, the hexagonal array with sixfold symmetry is by far the most common one, although fourfold symmetry is also possible depending upon the particle shape and surface properties. Potentially, using self-assembly one can produce planar nanoparticulate structures with exceptional long-range ordering [50].

Solvent evaporation method was used to prepare superlattices of monodisperse cobalt nanocrystals [51]. Co nanocrystals ranging in size from 2 to 11 nm were prepared by high temperature, solution phase reduction of cobalt chloride in the presence of stabilizing agents. Deposition of these uniform cobalt particles on solid substrates by evaporation of the carrier solvent resulted in the spontaneous self-assembly of 2D and 3D nanoparticulate magnetic superlattices (colloidal crystals) [51]. It was found that nanoparticles of various nature are able to self-organize in compact 2D structures (Ag and AgS nanoparticles) or form ribbons, dots, or labyrinths (cobalt and ferrites nanocrystals) [52]. Optical, magnetic, and transport properties of these structures were different from those of the isolated nanocrystals and of the bulk-phase materials. Highly ordered superlattices were formed via self-assembly of monodisperse 6-nm $Fe_{50}Pt_{50}$ nanoparticles from a hexane/octane (1 : 1) dispersion onto a SiO-coated substrate [53]. Thermal annealing converted the internal particle structure from chemically disordered fcc phase to ordered face-centered tetragonal phase and transformed the nanoparticle superlattices into chemically and mechanically robust ferromagnetic nanocrystal assemblies [53].

It was observed that trioctylphosphine-coated magnetic Co nanoparticles (6–8 nm in diameter) were stable to oxidation in pyridine dispersion and tended to form self-organized close-packed structures on substrate surface under drying of suspension droplet [54]. Air-stable, ligand-coated 8 nm Co nanoparticles with narrow size distribution self-assembled after deposition on a flat substrate and formed organized 2D superlattices [55, 56]. The magnetic properties of organized 2D nanoparticulate structures were different from nanoparticle dispersion in a solution where the saturation was not reached at applied field about 2 T even at 3 K and they were still superparamagnetic. When nanoparticles were organized in 2D superlattices, saturation was reached at lower applied field in comparison with colloid dispersion of those nanoparticles due to the increase of dipolar interaction in 2D ensembles of self-assembled Co nanoparticles [55, 56].

Application of magnetic field during the solvent evaporation in deposition process of magnetic nanoparticles resulted in the orientation of the easy magnetization axes of nanoparticles along the external field direction with the induction of collective magnetic properties of formed nanoparticulate structures revealed by substantial changes in the hysteresis loop [52]. Cobalt nanocrystals self-assembled into a hexagonal network [57] and in a face centered cubic (fcc) structures, were obtained and characterized in [58]. The resulted self-assembled nanoparticulate structures were dependent on the nature of ligand molecules used to stabilize the colloid nanoparticles [58].

Particularly regular, ordered 2D arrays of Co nanoparticles on Si/Si_3N_4 substrates (planar hexagonal ordered arrays with length and width both between 200 and 500 nm) were obtained with the application of external magnetic field perpendicular to the substrate surface during the deposition procedure [38]. Monodispersed defect-free ε-cobalt nanocrystals prepared by the rapid pyrolysis of cobalt carbonyl in an inert atmosphere and stabilized by a number of surfactants (oleic acid, phosphonic oxide, etc.) also formed 2D self-assemblies when evaporated at low rates in a controlled atmosphere with manifestation of collective behavior due to the collective dipolar interactions corresponding to a highly ordered nanoparticles system [59]. The collective magnetic behavior in 2D Co nanoparticle assemblies was studied by magnetic force microscopy and it was observed that magnetic structure and properties of Co nanoparticle layer were dependent on the density of magnetic nanoparticles per unit area [60].

Various ordered nanoparticulate structures and superlattices were obtained via self-assembly of monodisperse iron oxide nanoparticles with diameters of 3, 5, 10, 16, and 25 nm prepared by the thermal decomposition of iron carbonyl in octyl ether in the presence of oleic and stearic acids [61]. Hexagonal close-packed arrays of 3.5 nm diameter superparamagnetic $Fe_xCo_yPt_{100-x-y}$ nanoparticles with distorted fcc structure were prepared by hydrocarbon solvent evaporation on carbon-coated copper TEM grids [62]. The annealing

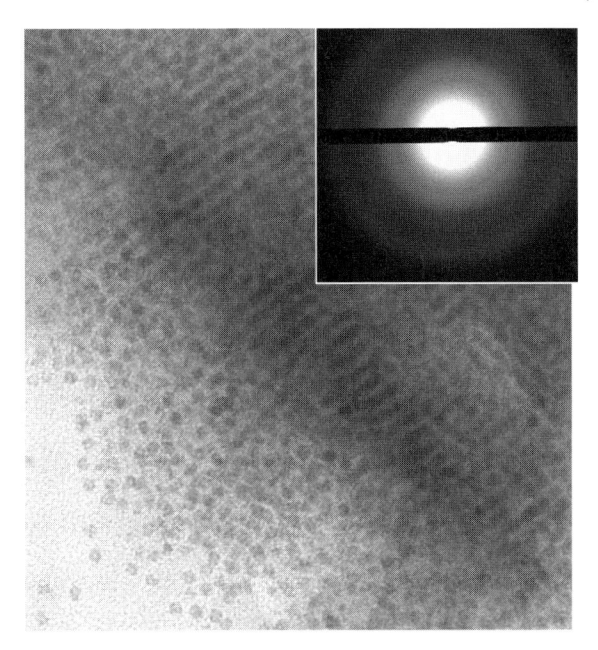

Figure 5.1 Transmission electron micrograph of ferritin. Samples were prepared by drying a droplet of aqueous ferritin solution onto TEM substrate (copper grid with carbon film). The insert shows typical selected-area electron diffraction picture obtained from the ferritin samples [285].

at temperatures ranging from 550 to 700 °C transformed the particles to the tetragonal (L10) phase with the increase in coercivity of the annealed films with increasing annealing temperature [62].

Figure 5.1 shows ordered structure formed by natural bioinorganic Fe-containing nanoparticles – ferritin molecules – via drying a drop of aqueous ferritin suspension on the substrate surface. Ferritin is known as a spherical protein bioinorganic complex composed of protein shell and inorganic iron-containing core in the form of hydrous ferric oxide. The inner inorganic core of the protein complex is usually 5–8 nm in diameter and is able to incorporate roughly 4500 iron atoms in the form of paracrystalline iron oxyhydroxide [63–65]. It follows from Figure 5.1 that ferritin inorganic cores are monodisperse with the mean diameter about 6 nm. The absence of marked reflections on the corresponding electron diffractogramm (the insert in Figure 5.1) points to the low crystallinity of ferritin samples in accordance with the literature data [63–65]. The identity of dimensions of ferritin molecules allows them to form ordered self-assembled nanostructures under the solvent evaporation similar to aforesaid monodisperse nanoparticles.

The substantial advantage of the solvent evaporation self-assembly technique is connected with its technical simplicity. However, this method works only with monodisperse nanoparticles and can only offer limited packing orders with almost no controls on the nanoparticles domain structure, packing density, and number of layers deposited.

5.2.2
Langmuir–Blodgett Technique

The experimental method widely used for operating with amphiphilic compounds and nanoparticles at the gas–liquid interfaces and for the formation of layers and nanoparticulate films on solid substrates is Langmuir–Blodgett (LB) technique. The principal advantages of the method are its relative simplicity, possibilities to operate at ambient conditions, and to deposit organized films on a wide variety of solid substrates.

Using LB technique, one can fabricate organized planar monolayer and multilayer films on solid substrate surfaces via formation of floating Langmuir monolayers at the gas–liquid interface and their subsequent transferring onto solid substrates surfaces via substrate dipping and lifting through the monolayer-covered liquid-phase surface. The technique was introduced by the pioneering works of Langmuir and Blodgett [66, 67] and since then it is used in numerous works related to studies of fundamental phenomena at the gas–liquid interface and in researches directed to the fabrication of new supramolecular functional assemblies and organized nanostructured and nanocomposite films [68–75]. Vast monolayer and multilayer structures with controlled composition and architecture composed of various amphiphilic compounds and/or inorganic nanocomponents can be formed and then deposited on solid substrate surfaces by this method.

LB technique was used as an instrument for the formation of planar magnetic nanostructures (LB films with magnetic properties) containing various magnetic components [76]. LB films containing planar arrays of Mn^{2+} cations were used as a model system for the investigation of 2D magnetism [77]. In our group, 2D magnets based on highly ordered Gd stearate LB films were first prepared via controlling of pH value and ligand composition of Gd^{3+} containing aqueous phase [78]. The data obtained by a number of techniques (EPR, SQUID, magnetization-induced second-harmonic generation) indicated the magnetic ordering for gadolinium stearate LB films [79–82].

Metal soup LB films were used as layered metal-organic precursors for the fabrication of ultrathin inorganic layers via thermal or plasma treatment [83]. Thus, iron oxide nanolayers (10–150 nm thick) were prepared by heating decomposition of multilayer Fe^{3+}-arachidate LB films [83].

LB technique was used in a number of research to form ordered 2D arrays of magnetic nanoparticles. Multilayer LB films of surfactant encapsulated molecular magnetic nanoparticles (cetyltrimethylammonium complexes with sodium hexametaphosphate stabilized nickel hexacyanoferrate nanoparticles) were formed successfully on ITO substrate [84]. Various polyoxometalates were organized in 2D multilayer structures via their binding from aqueous solution with cationic Langmuir monolayer of dimethyldioctadecylammonium and subsequent deposition of composite monolayer onto solid substrates by conventional LB technique [85]. Hydrophobic surfactant-coated magnetic maghemite nanoparticles were able to form stable Langmuir monolayers on the aqueous subphase surface [86]. The structure of deposited nanoparticulate layers was dependent on the nanoparticle size – small particles were organized in more compact aggregates than larger ones. It was observed using electron microscopy technique that the aggregate domains were rather compact for small particles (7.5 nm in diameter), while they were more digitated and behaved like an elastic gel for larger ones (15.5 nm in diameter) [86]. Organized multilayers of magnetic nanoparticles (10 nm magnetite Fe_3O_4, 100 nm γ-Fe_2O_3, 50 nm barium ferrite $BaFe_{12}O_{19}$) were deposited using LB technique onto the surface of fresh poly(ethylene terephthalate) and MgO substrates. The nanoparticulate fims were characterized by ordered lamellar structure and revealed magnetic anisotropy behavior [87]. Formation of complexes between colloidal γ-Fe_2O_3 nanoparticles and stearic acid Langmuir monolayer was studied in [88]. It was observed by FTIR that complexation between maghemite nanoparticles and stearic acid molecules occurred by electrostatic interactions between carboxylate ions of stearate and γ-Fe_2O_3 nanoparticle surface.

Organized 2D nanoparticulate structures composed of colloid cationic magnetite nanoparticles (about 8 nm diameter) sandwiched between polar headgroups of arachidate molecules were formed on oxidized silicon substrates using LB technique. Those structures were formed via binding of colloid magnetite nanoparticles from their aqueous solution with arachidic acid layers [89]. Hydrophobic 13 nm lauric-acid-stabilized Fe_3O_4 iron oxide nanoparticles (magnetite) dispersed in hexane were used to form nanoparticulate Langmuir monolayer on the surface of aqueous subphase, and the corrersponding multilayer nanoparticulate LB films were deposited successfully [90]. Compression of nanoparticulate Langmuir monolayer resulted in the formation of circular domain structures of magnetite nanoparticles and finally to a close-packed 2D array of magnetite nanoparticles [90]. Multilayer α-Fe_2O_3-stearate LB films were formed and order structure and size effect of the films were studied by FTIR spectroscopy and linear infrared dichroism [91].

LB technique also allowed preparation of well-defined 2D arrays of monodisperse magnetite and cobalt ferrite ($CoFe_2O_4$) nanoparticles described in [36]. Close-packed monolayers of colloidal cobalt ferrite nanocrystals synthesized using sol–gel-like method were prepared via LB technique [92]. It was discovered that colloidal γ-Fe_2O_3 nanoparticles with an average diameter of 8.3 nm can form complexes with water-insoluble fatty acids (in particular, arachidic acid) at the air–aqueous phase interface. The domain structures in Fe_2O_3 nanoparticles/arachidic acid complex Langmuir monolayers on the aqueous-phase surface were observed with morphologies dependent on the monolayer compression extent [93]. The γ-Fe_2O_3 nanoparticles and deposited nanoparticulate LB films showed superparamagnetic properties [93]. Nanocomposite magnetic LB films composed by poly(maleic monoester)/magnetite nanoparticle complexes were formed using LB technique [94]. The polymeric surfactant interacted with Fe_3O_4 nanoparticles by electrostatic attraction between carboxylate groups of poly(maleic monoester) and the positively charged Fe_3O_4 nanoparticles. Magnetic hysteresis loops were measured in the formed nanoparticulate LB films.

The features of structural organization of Langmuir monolayers and corresponding LB films of oleate stabilized magnetic $CoFe_2O_4$ nanocrystallites were studied in [95]. It was found that the increasing surface pressure resulted in a transition from a complex with well-separated aggregates of nanocrystallites to well-compressed, monoparticulate layers, and, ultimately, to multiparticulate layers. Multilayer nanoparticulate structures of FeO nanoparticle–stearate complexes were prepared by the LB technique [96]. By using this technique, ordered low-dimensional arrays of magnetic iron oxide nanoparticles on substrate surface were formed and characterized by AFM [97]. It was shown that stable Langmuir monolayers of stearate-α-Fe_2O_3 nanoparticle complexes can be formed via spreading stearic acid monolayer onto the surface of colloidal solution of α-Fe_2O_3 nanoparticles (2 and 7 nm dameter) [96]. Iron oxide nanoparticles were closely packed in the monolayer and well-ordered composite nanoparticulate multilayer LB films were formed on a number of different substrates by conventional LB deposition technique. Optical properties of LB films formed by stearate complexes with α-Fe_2O_3 nanoparticles (2 and 7 nm dameter) were investigated in [98]. The collected data indicated to the highly ordered layered structure of prepared LB films with hexagonal-like close packing of nanoparticles. Also, quantum-size effect in UV-vis spectra shift was observed in nanoparticulated α-Fe_2O_3 LB film [98].

The preparation of vast magnetic monolayer nanoparticulate film was demonstrated by the formation of Langmuir monolayers of the oleic-acid-coated Fe_3O_4 nanoparticles mixed with stearic acid molecules at the air/water interface following the deposition of monolayer onto the solid substrate surface [99]. LB technique was used successfully for the formation of layers of oleic-acid-stabilized 11–13 nm γ-Fe_2O_3 nanoparticles onto solid substrates such as silicon wafer or patterned poly(dimethylsiloxane) stamps. The latter were used for the formation of patterned nanoparticulate structures onto a silicon wafer via soft lithographic technique by imprinting (see Section 5.4.1) [100].

A number of works reported the use of LB technique to prepare ordered 2D arrays of FePt nanoparticles. The role of structure of amphiphilic fatty acid molecules on the properties of Langmuir monolayer containing FePt nanoparticles was investigated [101]. It was found that when number of CH_2 groups in the fatty acid molecules exceeded 14, it was possible to form regular 2D arrangements of the FePt nanoparticles.

Formation of patterned magnetic structures on silicon substrates based on nanoparticulate LB fims containing platinum-iron oxide core–shell nanoparticles with their subsequent transformation to FePt composite microstructures was reported in [102]. Nanocomposite films with displayed magnetic hysteresis were prepared from silica-encapsulated face-centered tetragonal (fct) FePt alloy nanoparticles via forming of ordered 2D arrays of presynthesized core–shell nanoparticles on substrate surface using LB technique followed by high-temperature heat treatment [103].

Monolayer of FePt (5 nm in diameter) nanoparticles synthesized via hydrazine hydrate reduction of $H_2PtCl_6 \cdot 6H_2O$ and $FeCl_2 \cdot 4H_2O$ in ethanol–water system was deposited by LB technique [104]. It was found that as-synthesized FePt nanoparticles were a kind of active electrochemical catalyst. The as-synthesized particles had disordered fcc structure and were transformed into ordered fct structure after annealing at temperatures above 400 °C. The coercivity of ordered fct FePt phase was about 2515 Oe [104].

Monolayer LB film of 4 nm FePt nanoparticles formed on a glass substrate surface was characterized by a smooth surface and close packing of magnetic nanoparticles. Corresponding multilayer nanoparticulate LB films were also characterized by ordered superlattice structure as was revealed by small-angle X-ray diffraction measurements [105]. Polymeric LB films formed by the layers of amphiphilic poly(alkylacrylamide)s were used to form organized assemblies of magnetic nanoparticles [106]. Iron oxide nanoparticles were coated by poly(N-alkylmethacrylamide)s with various alkyl chain lengths and used to form Langmuir monolayers and LB films with superparamagnetic properties [106].

Using LB technique, diluted magnetic semiconductor $(Cd_{1-x}Mn_xS)$ nanoparticles (about 3 nm diameter) incorporated into arachidate LB film were prepared [107]. These nanoparticles were synthesized through the reaction of arachidate LB films containing Cd and Mn ions with H_2S gas in an atmosphere of NH_3.

Magnetic and noble metal nanoparticles and nanostructures were synthesized directly in Langmuir monolayer by the method in which nanoparticles were fabricated via decomposition of an insoluble metalorganic precursor compound in Langmuir monolayer at the gas–water interface. In that method, reaction area is highly anisotropic quasi-2D structure and nanoparticle growth is an example of a 2D process where 2D diffusion of precursor molecules, active intermediates, complexes, nucleus and growing nanoparticles, surfactants, and additives occurs only in the plain of the monolayer [75]. Magnetic iron oxide nanoparticles were photochemically generated by the ultraviolet decomposition of a volatile precursor compound iron pentacarbonyl in a mixed Langmuir monolayer containing surfactant molecules [108, 109]. Such system represents an ultimately thin and anisotropic planar photochemical reaction system. It was found that the shape, size, and crystallinity of the resulting nanoparticles were dependent substantially on the monolayer composition and state during the growth process. It was demonstrated that the shape of synthesized iron oxide magnetic nanoparticles can be changed from 2D isotropic plate and ringlike to the field-aligned ellipsoidal and needle-like when external magnetic field parallel to the plane of particulate monolayer was applied during

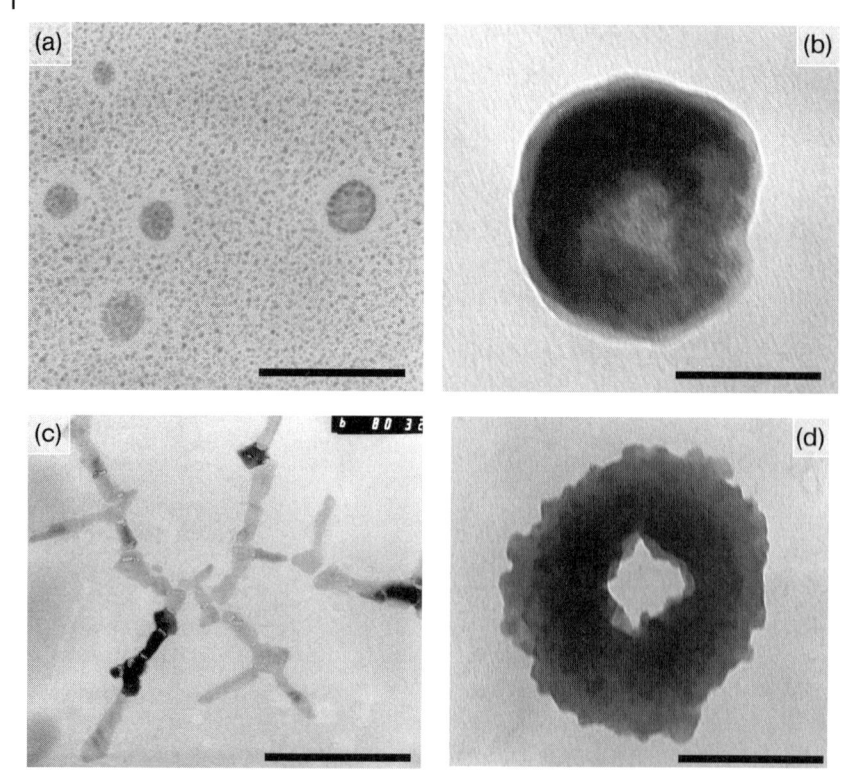

Figure 5.2 Transmission electron micrographs showing iron oxide nanoparticles and nanostructures synthesized photochemically at ambient conditions (21 °C) in Langmuir monolayer on the aqueous subphase surface (pH = 5.6) at low surface pressure values ($\pi \cong 0$) and deposited onto the copper grid with Formvar coating at $\pi = 25$ mN/m. Image (a): initial Fe(CO)$_5$/stearic acid (SA) ratio in monolayer was 10:1, UV exposure time 6 s. Images (b) and d): initial Fe(CO)$_5$/SA ratio was 10:1, UV exposure time 4 min. Image (c): self-organized nanostructure grown in a monolayer with initial Fe(CO)$_5$/SA ratio 5:1, UV exposure time 6 s, dark incubation of the monolayer after UV illumination 4 min. Bar size: (a) 120 nm; (b) 85 nm; (c) 300 nm; and (d) 155 nm [75].

the synthesis. The effects of self-organization of nanoparticles and formation of 2D nanostructures were also observed [110].

Figure 5.2 shows TEM micrographs of synthesized iron oxide nanoparticles and nanostructures with characteristic morphologies. Figure 5.2(a) presents the characteristic initial stages of the interfacial formation of large plate-like nanoparticles via planar aggregation of small (∼4 nm diameter) nanoparticles under short period of UV illumination (6 s) with immediate compression and deposition of the nanoparticulate monolayer. It is clearly seen from Figure.5.2(a) that every such planar aggregate is surrounded by an area with exhausted small nanoparticles. Different characteristic nanoparticle morphologies including disk (image 5.2(b)) and nanoring (image 5.2(d)) were observed in the experiments with 4 min UV illumination. Interesting type of magnetic nanostructure shape – nanoring – was obtained only without applied magnetic field and under conditions of permanent UV illumination. Ring shape can be a result of the 2D self-organization of growing nanoparticles in the plane of a monolayer due to the magnetic dipolar interactions and formation of coalesced ring aggregates with minimal magnetostatic free energy in accordance with theoretical considerations of formation of circular aggregates of magnetic colloids and its transformations under applied magnetic field [111] (see also Sections 5.4.2 and 5.6). Also, ramified iron oxide nanostructures were observed in a monolayer deposited after long dark incubation. The nanostructure was composed of quasiorthogonally organized nanorods with nm-sized gaps in some junctions (Figure 5.2(c)). Those nanostructures were observed in the mixed monolayers with surfactant (stearic acid), while individual nanoparticles and disordered aggregates were formed without surfactant under the same reaction conditions. The last observation implies the important role of surfactant molecules in the interfacial self-organization processes of iron-containing magnetic nanoparticles.

Figure 5.3 shows the characteristic AFM-tapping mode topographic images of nanoparticulate monolayer LB film with photochemically generated iron oxide nanoparticles deposited onto the surface of mica substrate by the vertical lifting method. Corresponding TEM images of such nanoparticles are presented on Figures. 5.2(a) and (b). Circular nanoparticles are clearly seen in Figure 5.3(a). Figure 5.3(b) demonstrates the AFM phase-contrast regime image corresponding to the image 5.3(a) revealing the difference in the material of circular objects and surrounding matrix, thus indicating grown inorganic nanoparticles. Figure 5.3(c) shows typical height cross-section profile of the image 5.3(a) and indicates an overall film roughness of ~ 1 nm with clearly observable nanoparticles of volcano-like morphology with obvious cavity in the central part of the nanoparticle. One can see from Figures 5.3(a) and (c) that grown nanoparticles are very flatten (height about 1 nm) with very high surface-to-volume ratio (diameter–height ratio ~ 100). The data presented in Figure 5.3 show that ultraflatten and even dented iron oxide nanoparticles with diameters 30–100 nm can be formed under conditions of permanent UV illumination in the gaseous phase of Langmuir monolayer. The presented data suggest that the role of anisotropic geometric factors, surfactant interactions with intermediates, nuclei, and nanoparticles, and monolayer state may be important for interfacial distribution and diffusion control of all components, and also for restriction of rotation and aggregation of nanoparticles. Interfacial synthesis approach allows to control effectively the nanoparticle growth rate via the distribution of precursor molecules in the plane of interfacial layer and its mixing with surfactants and additives along with varying the rate of precursor decomposition and generation of active intermediates and metal atoms. Thus, the variation in UV illumination conditions (and, particularly, flash regime) allows to change the morphology of nanoparticles grown photochemically

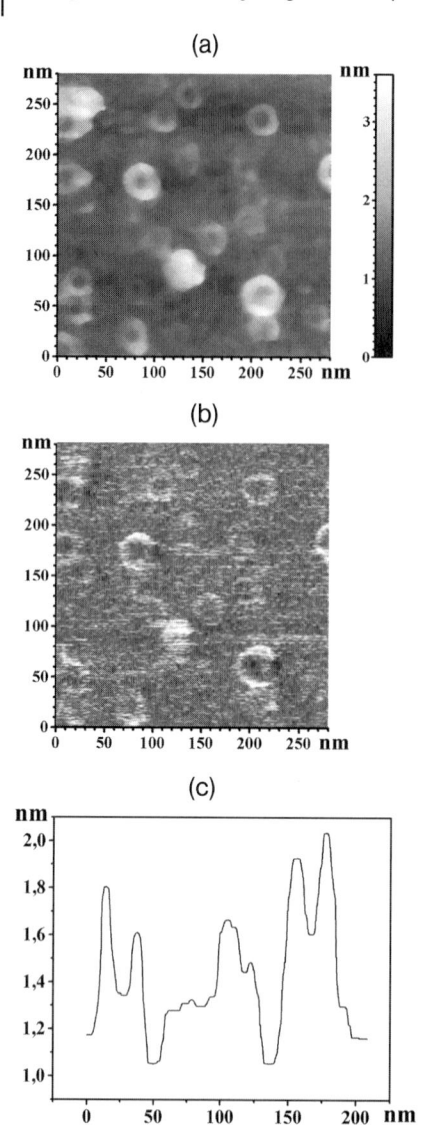

Figure 5.3 AFM tapping mode topographic images of iron-containing nanoparticles synthesized in uncompressed Langmuir monolayer ($\pi \cong 0$) at $T = 21\,°C$, and deposited onto the mica substrate at $\pi = 25$ mN/m using vertical substrate-lifting method. Image (a): nanorings interfacially formed with initial Fe(CO)$_5$/SA ratio in monolayer 10:1, UV exposure time 4 min, $T = 21\,°C$, subphase pH = 5.6. Image (b): phase-contrast image corresponding to the image 5.15(a). Image (c): typical height cross-section profile of the image (a) [74].

via iron pentacarbonyl decomposition due to kinetic factors and to obtain amorphous and crystalline nanoparticles and nanostructures [109].

5.2.3
Layer-by-Layer Assembly

The other general approach to fabrication of organized layered structures on surfaces is based on the formation of monolayer on a surface via adsorption of monolayer components from the adjacent bulk phase. Multilayer structures in that approach are formed via stepwise alternate adsorption procedure with sequential LbL assembly of di- or multicomponent multilayer films. This approach was first introduced by Iler [112] who exploited electrostatic interactions of charged colloids and obtained multilayer films composed by layers of positively charged alumina fibrils and negatively charged silica colloids. Later, the LbL alternate adsorption deposition technique was widely used to form various inorganic [113, 114], hybrid organic–inorganic [115–117], and organic films composed by oppositely charged polymers (latexes [118, 119], polyelectrolyte molecules [120, 121]), and by many other molecules [122, 123] on flat substrate surface, and on surfaces of particles [124–128]. The important advantage of the step-wise multilayer self-assembly via alternate adsorption of the structure-forming components is the principal possibility for controlling the composition, structure, and properties of individual layers within the film, thus giving rise to effective design of novel supramolecular and composite nanostructured thin films and coatings and allowing a variety of materials to be incorporated within the film structures. The other principal advantage of this method is the possibility to form layers on surfaces of substrates of arbitrary shape due to the adsorptional character of the LbL assembly technique. The technique is rapid, inexpensive, and experimentally very simple.

Monolayer and multilayer nanofilms with efficiently in-layer controlled composition incorporating nanocomponents of various nature (organic, polymeric, inorganic, biological, etc.) can be formed on solid substrate surfaces by this method. Due to the adsorption procedures used in LbL method, such nanofilms can be prepared not only on flat substrate surfaces but also on complex, nonplanar, and developed surfaces including colloidal particles.

Nanostructured films prepared by LbL assembly technique are now widely used for the development of functional systems for physical applications from photovoltaic devices and field-effect transistors to biomedical applications including drug delivery systems [129].

5.2.3.1 Supported Films

The LbL alternate adsorption deposition method was widely used during last decades by a large number of researchers and has proven to be very efficient in preparing a wide variety of organized layered structures and coatings on surfaces composed of various organic, inorganic, and biological components [130–132]. In particular, highly luminescent films of CdTe nanoparticles and polycations [133] and organic light-emitting diodes [134] were fabricated using LbL alternate adsorption deposition approach. LbL-assembled films were used as nanoreactors to generate inorganic nanoparticles [135].

One-dimensional structures as individual carbon nanotubes were used as templates for the fabrication of polyelectrolyte shell layers via LbL self-assembly method [136]. The LbL alternate adsorption deposition technique in combination with lithographic methods was used for the fabrication of patterned structures on solid surfaces [137, 138] and of polyelectrolyte multilayers deposited at liquid–liquid interfaces [139]. Electrostatic LbL assembly was used for the preparation of various composite films containing semiconductor [116, 140–143], metallic [144], nanoparticles, and biological components including chitosan [145], enzymes [146, 147], and DNA [148, 149].

A number of ultrathin magnetic nanocomposite films formed using LbL deposition technique on substrate surfaces were reported in literature. 2D nanofilm multilayer assemblies of 4 nm FePt nanoparticles with controlled thickness were formed onto a solid substrate surface (silicon oxide) via alternate sequential adsorption of polyethylenimine and presynthesized FePt nanoparticles prepared using high-temperature solution-phase synthesis [150–152]. Chemical analysis of the nanoparticulate assemblies revealed that more iron oxide was present in the thinner assemblies annealed at lower temperature or for shorter time. Thermal annealing induced the internal particle structure change from chemically disordered to ordered and transformed the nanoparticulate assembly from superparamagnetic to ferromagnetic [151]. A metallic magnetic core surrounded by a weakly magnetic or nonmagnetic shell resulted from annealing FePt nanoparticles in the PEI–FePt assembly [153]. As synthesized and deposited, the assemblies contain weakly magnetic material whose volume is equivalent to an outer 0.5-nm shell of the nanoparticles. During a 580 °C annealing, this material was incorporated into the magnetic domains of the nanoparticles. The fraction of nanoparticles transformed by annealing into ferromagnetic L10 FePt phase with structure essential for magnetic storage applications is found to vary with sample thickness. The samples thinner than four nanoparticle layers show no significant fraction of ferromagnetic (at 300 K) L10 structured nanoparticles under these annealing conditions, while in films comprised of four nanoparticle layers, less than half of the particles are ferromagnetic at 300 K [152]. Structural analysis using XRD showed that a minimum temperature of 450 °C was required to start the formation of the ordered ferromagnetic phase. Annealing for longer times and at higher temperatures not only led to higher coercivity and a larger fraction of ordered phase but also to the onset of some agglomeration of the nanoparticles [154].

The LbL-assembled films containing yttrium iron garnet nanoparticles were used to study the effect of magnetization-induced second-harmonic generation [155, 156].

Composite polymeric nanofilm structures with 2D arrays of iron oxide magnetic nanoparticles were fabricated using LbL assembly method and characterized by a number of techniques [157, 158]. It was found that the thickness of the polycation/iron oxide film and its refractive index increased with the increase of the layer number. The microwave irradiation affected the thickness of the polycation/iron oxide film and their refractive index.

Complex composite stratified assemblies of magnetite nanoparticles, poly(diallyldimethylammonium bromide), and exfoliated montmorillonite clay particles were prepared using LbL technique [159]. Distinct stratification of the Fe$_3$O$_4$/PDDA/clay films is obtained due to the sheetlike structure of the clay particles. This feature distinguished these assemblies from their polyelectrolyte–polyelectrolyte and polyelectrolyte–spherical particle analogs, where the layers of individual polyelectrolytes and particles were strongly interdigitated. Being adsorbed on PDDA strictly parallel to the substrate surface, montmorillonite produces a dense layer of overlapping alumosilicate sheets, which virtually flawlessly separates one magnetite layer from another. The difference in magnetic properties between assemblies of various architectures was attributed to the insulation effect of clay layers inserted between magnetic layers. The montmorillonite sheets disrupt the electron exchange interactions between the magnetite nanoparticles in adjacent layers, thereby limiting the magnetization reversal to 2D.

Iron oxide nanoparticles were synthesized via Fe oxidative hydrolysis in multilayer polyelectrolyte films formed by LbL assembly of polyions poly(diallyldimethylammonium)chloride and polystyrenesulfonate molecules on solid substrates [135].

It was found that microwave irradiation can be applied to control packing order of such self-assembled monolayers of magnetite nanoparticles with and without SiO$_2$ coating [160]. Microwave treatment of the poly(dimethyldiallylammonium chloride) layer prior to the nanoparticle adsorption results in substantial reduction of the surface roughness of the particulate films. This effect was attributed to the reduction of the length of partially desorbed segments of macromolecules protruding into the aqueous phase at a distance of about 70 nm as estimated by force–distance curves. Aggregation of nanoparticles on these segments was responsible for a relatively high degree of disorder in LbL self-assembled films. The reduced number of loose segments due to microwave irradiation improved 2D packing of adsorbed magnetic nanoparticles. For optimized conditions, the rms roughness, R, of magnetite self-assembled films can be as low as 1.5–3.5 nm. [160].

Monolayer and multilayer nanofilms comprising magnetic Fe$_3$O$_4$ nanoparticles and polyimide molecules have been fabricated on single crystal silicon and quartz substrates by LbL self-assembly method [161]. The assembling process involved the alternate dipping of a substrate into an aqueous solution of anionic polyimide precursor (polyamic acid salt), followed by dipping into an aqueous solution of polycation polydiallyldimethylammonium chloride coated magnetite nanoparticles. The structural characterization data suggest that well-ordered uniform monolayer and multilayer magnetic films have been formed on silicon and silica surfaces.

Multifunctional thin films having both electrical conductivity and ferrimagnetic properties were successfully fabricated by the LbL self-assembling method via successive adsorption of polypyrrole and ferrite nanoparticles from their aqueous solutions. The thin film consisting of six PPy and two ferrite nanoparticle layers had a conductivity of 0.18 S/cm and simultaneously showed a magnetic hysteresis [162].

Photoswitchable magnetic films consisting of azobenzene derivatives and iron oxide nanoparticles were prepared by using LbL alternate adsorption deposition technique. It was found that in these films photoinduced changes in the dipole moment due to the *cis–trans* photoisomerization affected the magnetization of iron oxide nanoparticles. As a result, the magnetic properties of these films were dependent on photoillumination at room temperature [163].

The LbL deposition method was used successfully for coating carbon nanotubes with magnetic iron oxide nanoparticles (diameter 6–10 nm). The particle-coated multiwall carbon nanotubes were superparamagnetic and can be aligned at room temperature on any substrate by deposition from an aqueous solution in an external magnetic field $H = 2$ kOe [164]. Also, magnetic nanoparticles shell on a thermoresponsive microgel core was formed successfully by LbL assembly method [165].

5.2.3.2 Nanofilm Capsules

Formation of insoluble nanofilm layers on the surface of colloidal particles via LbL alternate adsorption deposition technique, following decomposition of the colloidal core, allowed the preparation of polyelectrolyte and nanocomposite microcapsules [166], which may be useful for fabrication of colloidal nanoreactors and carriers [136, 167, 168].

A number of magnetic colloid capsule structures were reported in the literature. The general practically useful property of those capsules is the possibility for their directed spatial transport and localization by the external magnetic field. Hollow magnetic microspheres were obtained by calcinating the core–shell particles at elevated temperatures. Also composite hollow spheres were prepared by calcinating polystyrene latex particles coated with multilayers of silica and Fe_3O_4 nanoparticles [169].

Composite magnetic microcapsules with magnetite nanoparticles as a magnetic component were prepared via formation of composite nanoparticulate polyelectrolyte multilayer on the surface of different decomposable colloidal templates (melamine formaldehyde resin, glutaraldehyde fixed red blood cells, emulsion oil droplets, etc.) with subsequent dissolution of the colloidal core [170]. Such magnetic microcapsules were formed by incorporating presinthesized Fe_3O_4 colloid nanoparticles and also by selective synthesis of magnetite nanoparticles inside the polyelectrolyte capsules filled with polycations [168].

It was shown that permeability to macromolecules like FITC-labeled dextran of composite polyelectrolyte poly(sodium styrene sulfonate)/poly(allylamine hydrochloride) microcapsules prepared by LbL self-assembly technique and embedded with magnetic core–shell, gold-coated cobalt (Co@Au) nanoparticles (3 nm diameter) can be substantially increased by external alternating magnetic fields of 100–300 Hz and 1200 Oe applied to rotate the embedded Co@Au nanoparticles [171]. This method was considered for controlled drug delivery in biomedical applications; however, the required special equipment to create alternating magnetic field exactly in the area of targeted delivery makes the practical prospects of the method rather questionable.

Polyelectrolyte microcapsules formed by LbL assembly technique with magnetite Fe_3O_4 nanoparticles deposited onto the polyelectrolyte shell possessed increased protection efficiency against oxidation of encapsulated biocompounds (bovine serum albumin) by low-molecular-weight oxidizing agents (H_2O_2 dissolved in aqueous solution) [168].

The simultaneous encapsulation of both highly luminescent CdTe semiconductor nanocrystals and magnetic Fe_3O_4 nanoparticles in polymer microcapsules resulted in the formation of multifunctional capsules with luminescent properties and possibilities for external manipulation of the capsules by applied magnetic field [172]. Using locally created magnetic field gradient caused by a permanent magnet, the magnetic targeting and cellular uptake of polymeric nanocomposite microcapsules simultaneously functionalized with magnetic nanoparticles and luminescent CdTe nanocrystals was demonstrated [173].

Iron oxide nanoparticle/polymer microcapsules containing different number of layers of iron oxide (Fe_3O_4) nanoparticles have been prepared using the LbL alternating adsorption technique. The capsule-shell structure and permeability were sensitive to temperature treatment of microcapsules and increased dramatically after their microwave radiation treatment [174].

The main reason of the observed phenomena can be the local heating of nanoparticles and their surrounding polyelectrolyte shell matrix resulting in structural disturbing and destruction of the composite shell. Figure 5.4 shows characteristic structural transformations of composite polyelectrolyte microcapsules containing Fe_3O_4 nanoparticles (~10 nm diameter) caused by thermal and microwave treatment. These data are interesting for practical applications in medicine and bioengineering for remote control over the permeability of microcapsule shells using microwave irradiation. Thus, composite nanofilm magnetic microcapsules can be the base for development of novel containers and carriers for controlled transport and delivery drugs and other compounds.

5.2.3.3 Free-Standing and Free-Floating Films

Interesting novel nanofilm material – free-standing ultrathin nanocomposite film – composed of alternating layers of magnetite nanoparticles and poly(diallyldimethylammonium bromide) was obtained [175]. The film was initially assembled by LbL deposition technique on a cellulose acetate substrate, which was subsequently dissolved in acetone resulting in free-standing or free-floating composite nanofilm. Colloidal solution of negatively charged 8–10 nm magnetite nanoparticles was used for the film preparation. The developed assemble-and-dissolve principle imposes two major requirements on the substrate: (1) the organic solvents used in the lift-off step must not damage the film and (2) the substrate must be hydrophilic with a positively or

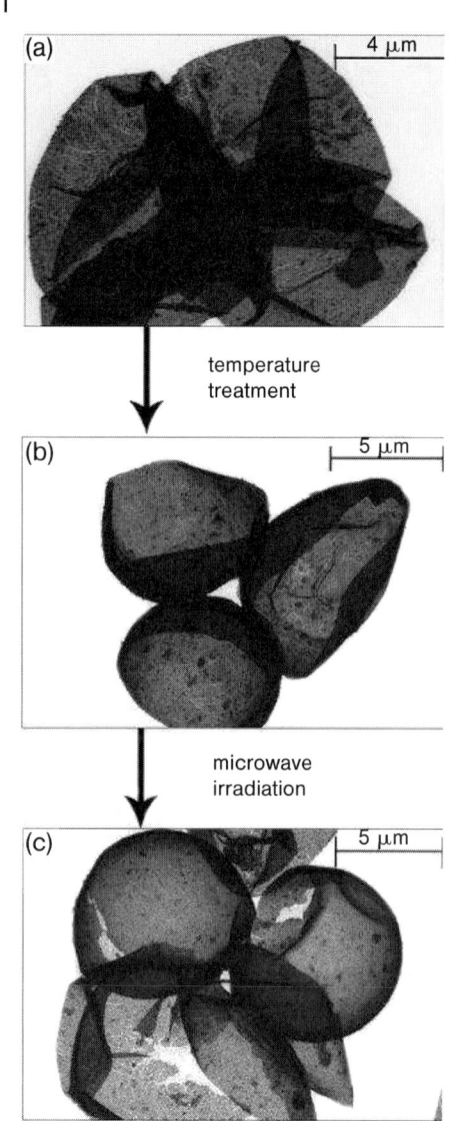

Figure 5.4 TEM images of microcapsules with oxide layers in the shell at different treatment and initial conditions; sample 2i (a) – initial capsule before microwave irradiation; sample 2t (b) – capsule after thermal treatment during 30 min at a temperature of 77 °C; sample 2tr (c) – capsule after microwave irradiation during 120 s at a power of 750 W [157].

negatively charged surface. From the suspended state, the film can be transferred onto any solid or porous support. The obtained nanofilm retained the magnetic properties of nanoparticles – it moved through the solution toward a permanent magnet placed near the side of the beaker. The procedure of these free-standing films preparation resembles the preparation of above-described nanocomposite polyelectrolyte microcapsules by dissolving micrometer scale core colloids. This technique can be extended to a variety of other compounds utilized in LbL film preparation, such as polymers, proteins, dyes, metal and semiconductor nanoparticles, vesicles, viruses, DNA, and others [175]. Free-standing films can consist of nanoparticles based on magnetic and nonmagnetic metals like Ag [176] and Au [177–180].

5.2.4
Bulk Phase Self-Assembled Sheetlike Nanofilms

Novel nanofilm magnetic materials were formed via an approach based on controlled processes of self-assembly and self-organization of colloid nanocomponents during the formation of their complexes with polyfunctional ligands in a bulk liquid phase in the absence of any surfaces or interfaces [181]. The approach opens possibilities for the preparation of novel highly organized nanofilm nanostructured materials and planar colloidal nanostructures representing the free-standing films composed of nanocomponents chemically bonded in the plain of the film. Using this approach, organic–inorganic ordered nanofilm material based on the complexes of polyamines and colloid magnetite Fe_3O_4 nanoparticles was formed and characterized for the first time.

Novel hybrid biomolecular nanofilm structures – complexes of natural polyamine (spermine) and magnetite nanoparticles – are presented in Figure 5.5. This figure shows characteristic TEM images of aggregate structures composed of spermine complexes with colloidal cationic magnetite nanoparticles (mean size ~10 nm) formed under various experimental conditions in the bulk aqueous phase. When suspension of magnetite nanoparticles was mixed with spermine solution under low ionic strength and pH values ~5, the formation of disordered nanoparticulate aggregates of various size and shape with loose and porous structure was observed (Figure 5.5(a)). The addition of spermine to suspension of magnetite nanoparticles at pH value ~4 resulted in the formation of highly organized free-floating planar sheetlike spermine/magnetite nanoparticles complexes with dense compact ultrastructure (Figures 5.5(b) and (c)). Those complexes belong to novel class of highly organized nanofilm nanostructured materials representing the free-floating self-assembled planar colloidal nanostructures (nanosheets) composed of chemically bonded nanocomponents [181]. The preliminary AFM analyses of obtained nanocomposite spermine/magnetite nanoparticulate sheetlike material indicate to the thickness of the nanofilm material corresponding to about two to four magnetite nanoparticle diameters (20–40 nm) with the thickness/length or width ratio about 1 : 1000 or less. The planar character of discovered structures and their quasistraight boundaries are result of the minimization of free energy of the system with competition and balancing of interparticle attracting interactions (van der Waals forces and chemical binding via polyfunctional ligands) and repulsive electrostatic interactions of cationic colloid nanoparticles with bound polyamine molecules. The organic ligand nature along with electrostatic and/or magnetic interactions are important factors of morphological control of inorganic nanostructures [46], and the

Figure 5.5 Transmission electron micrographs of nanostructures based on spermine complexes with colloidal cationic magnetite nanoparticles. Complexes were formed in a bulk aqueous phase. Image (a): disordered aggregates of magnetite nanoparticles formed in the presence of spermine in aqueous phase. Images (b) and (c): free-standing sheet-like nanofilm structure of spermine/magnetite nanoparticles complex formed under appropriate solution conditions. Image (b) bar size 500 nm, image (c) bar size 100 nm [181].

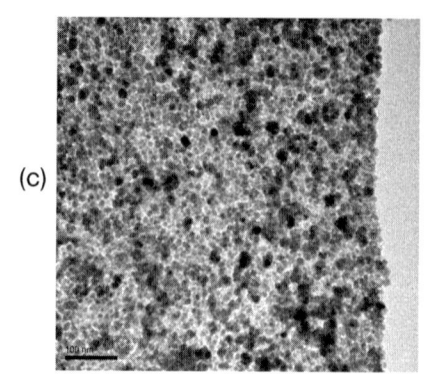

presented data illustrate this fact for iron oxide nanoparticles and polyamine ligand (spermine).

In this method, charged polyamines serve as "glue" ligands to organize magnetic colloidal nanoparticles in ultrathin, highly ordered free-floating films.

Initially the colloids (Fe_3O_4 nanoparticles) are electrostatically stabilized. The total free energy of any stable nanoparticle structure should be negative, and since the Coulomb energy of similar charged particles is positive for

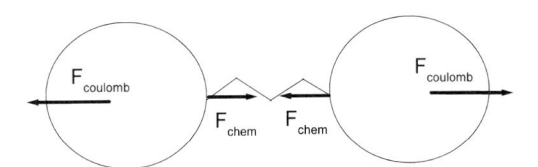

Figure 5.6 The mechanical analogy of chemical "gluing" of two equally charged nanoparticles by the polyamine molecule.

any configuration, an additional interaction with negative change of free energy of the system has to provide the system stability. Linear relatively short polyamine spermine molecule has four amine groups, which can bind to iron oxide nanoparticles. If the chemical binding of a nanoparticle with polyamine molecule is strong enough, then the negative change of chemical binding free energy can compensate and even overcome the contribution of positive electrostatic energy to the total free energy of the system. As a result, two nanoparticles are attached to each other via the polyamine "bridges." As each magnetite nanoparticle has a few binding sites to interact with an amine group, the nanoparticle can bind to several polyamine molecules and can decrease the free energy of the system even more. The "mechanical" analogy of such stable system of two nanoparticles is presented in Figure 5.6. The electrostatic repulsion between equally charged colloid particles is compensated by the "chemical" spring with fixed deformation energy.

Let Q be the average nanoparticle charge and a the average interparticle distance. Then the ratio of Coulomb energy $W_{\text{Coulomb}} = Q^2/a$ to chemical binding energy λ can be regarded as dimensionless system parameter. The total energy of N nanoparticles interconnected to each other by the polyamine "bridges" depends on the system size and geometry. We can imagine, for example, the following compact structures: 3D – sphere (Figure 5.7(a)), cube (Figure 5.7(b)), two-2D – square (Figure 5.7(c)), rectangle (Figure 5.7(d)), 1D – string (Figure 5.7(e)). We can also introduce some disorders like boundary irregularities (Figure 5.7(f)).

In our calculations of the energy of structures presented in Figures 5.7(a)–(d), we supposed that the particles are placed in points of cubic mesh with the period a, each particle has the same charge Q, and electrostatic interaction between particles is Coulomb (nonscreening). Each pair of neighbor particles has the chemical binding negative energy λ ($\lambda < 0$). In case of the structure in Figure 5.7(f), the irregularities of the boundaries were simulated by the random deviation from nonperturbed line. The step of the deviation was varied between 0 and $2a$.

It is evident that Coulomb interaction increases the structure total energy, while the chemical particle–polyamine binding decreases it. To maximize the negative change of "chemical" free energy, each particle should have as large coordination number as possible and, therefore, the nanostructure should be as compact as possible. To minimize the positive electrostatic energy of the system, the average inverse interparticle distance should be as

(a) (b) (c) (d) (e) (f)

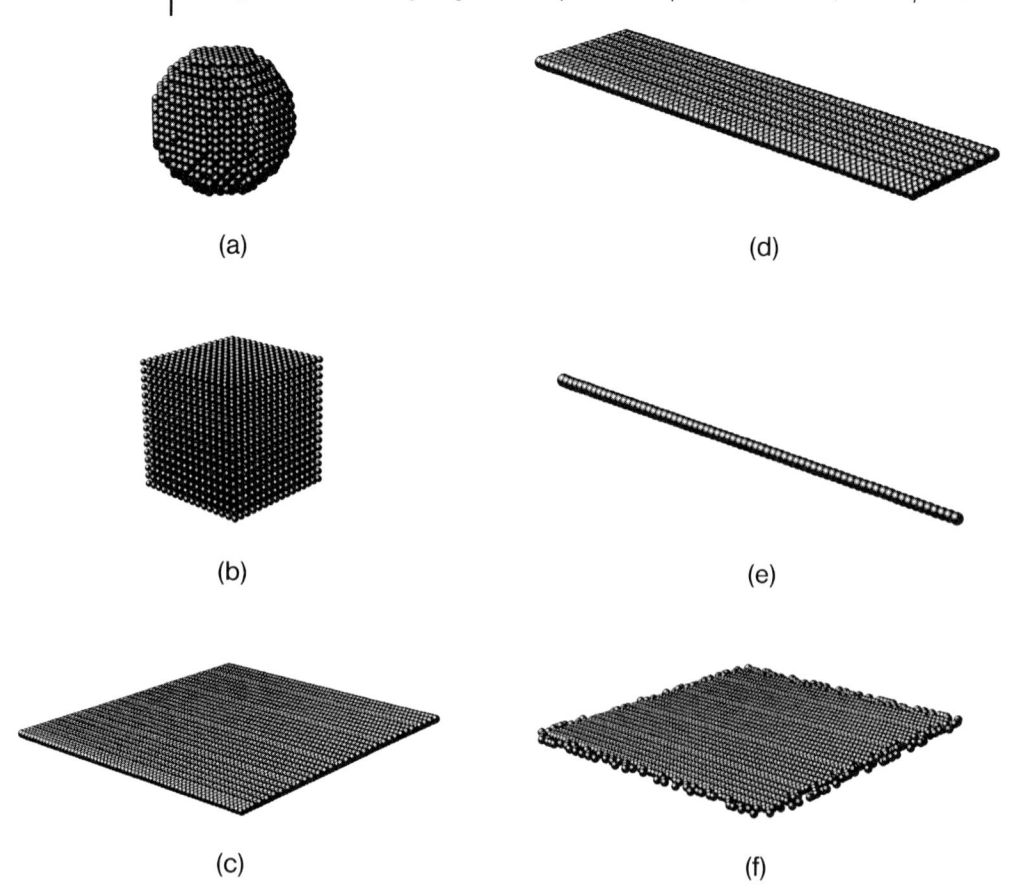

Figure 5.7 Nanoparticle structures with different geometries, considered in our calculations.

small as possible and, therefore, the nanostructure should be as extended as possible. There is a fundamental contradiction between these two tendencies showing possibilities for self-organization phenomena in chemically bonded electrostatically charged colloidal systems.

In case of the cubic arrangement of N nanoparticles, in the interior of the 3D bulk system each particle has the maximal coordination number $Z_3 = 6$, on the surface of the bulk system $Z_{3,\text{surf}} \leq 5$. The spherical system has the least surface and, therefore, the maximal negative chemical energy. However, the electrostatic energy of the spherical structure is not minimal within the 3D systems with fixed N.

Indeed, the negative change of chemical binding energy of a spherical particulate system (Figure 5.7(a)) with radius R can be calculated in the following way. Assuming that each particle takes the unite volume ($a = 1$), the system volume is equal to $4\pi R^3/3 = N$. In surface layer of the system, there are $4\pi R^2 = 4\pi \times (3 \times N/4\pi)^{2/3} \approx 4.9 \times N^{2/3}$ particles. Therefore, the chemical binding energy of the spherical system is equal to

$$W_{\text{sphére}} \approx (\lambda/2) \left\{ 6(N - 4.9 \times N^{2/3}) + 5 \cdot 4.9 \cdot N^{2/3} \right\}$$
$$\approx (\lambda/2) \left\{ 6N - 4.9 \cdot N^{2/3} \right\}. \tag{5.1}$$

If N particles form a cube (Figure 5.7(b)), then each of 12 cube edges contains $p = N^{1/3}$ particles, each of 6 cube faces contains p^2 particles, and 8 particles are in cube corners. The corresponding coordination numbers are as follows: $Z_{3,\text{face}} = 5$, $Z_{3,\text{edge}} = 4$, $Z_{3,\text{corner}} = 3$. The number of particles with $Z_3 = 6$ is equal to $p^3 - 6p^2 + 12p - 8$. The total chemical binding energy of the cubic system can be written as

$$W_{\text{cube}} = (\lambda/2)\{6(p^3 - 2(3p^2 - 6p + 4)) + 5(2(3p^2 - 6p + 4)$$
$$- 4(3p - 4)) + 4(4(3p - 4)) + 3 \cdot 8\} = (\lambda/2) \left\{6p^2(p-1)\right\}$$
$$= (\lambda/2) \left\{6N - 6N^{2/3}\right\}. \tag{5.2}$$

For a plane rectangular system (Figure 5.7(d)), each particle has the coordination number $Z_2 = 4$, except the cases when it is at the rectangle edge ($Z_{2,\text{edge}} = 3$) or at the corner ($Z_{2,\text{corner}} = 3$). If edge lengths are equal to N_1 and N_2, correspondingly, the rectangle interior contains $N_1 N_2 - \{2(N_1 + N_2) - 4\}$ particles, the edges contain $2N_1 + 2(N_2 - 2) = 2(N_1 + N_2) - 4$ particles and four particles are in the corners. The total chemical binding energy of the rectangular planar system is equal to

$$W_{\text{rectangle}} = (\lambda/2)\{4(N_1 N_2 - \{2(N_1 + N_2) - 4\})$$
$$+ 3(2(N_1 + N_2) - 8) + 2 \cdot 4\}$$
$$= (\lambda/2) \left\{4N - 2\sqrt{N}(\sqrt{n} + 1/\sqrt{n})\right\}, \tag{5.3}$$

where $n = N_2/N_1$, $N = N_1 N_2$.

Assuming $n = 1$ for the square system (Figure 5.7(c)), we get from Eq. (5.3).

$$W_{\text{square}} = 2\lambda\{N - \sqrt{N}\}. \tag{5.4}$$

In case of the ideal linear system (Figure 5.7e),

$$W_{\text{string}} = \lambda(N - 1). \tag{5.5}$$

By analogy, we also have for the plane disk:

$$W_{\text{disk}} \approx (\lambda/2) \left\{4(N - 0.56 \cdot N^{1/2}) + 3 \cdot 0.56 \cdot N^{1/2}\right\}$$
$$\approx (\lambda/2) \left\{4N - 0.56 \cdot N^{1/2}\right\}. \tag{5.6}$$

Figure 5.8 shows the chemical binding energy dependence (Eqs. (5.1), (5.2), (5.4), (5.5)) on the particle number for various particulate structures: cube, sphere, monolayer square, and disk as well as the string.

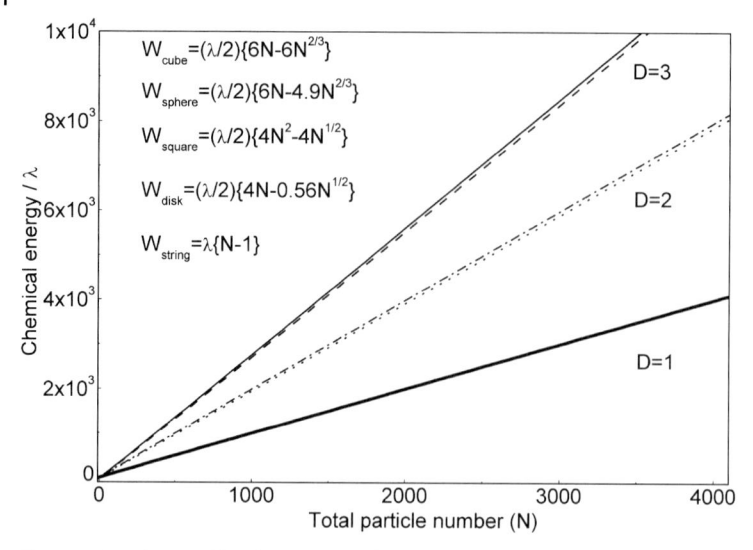

Figure 5.8 Chemical binding energy of the nanoparticle system versus particle number (Eqs. (5.1)–(5.6)) for different system geometry and dimensionality D. Thin solid line – the sphere, thick solid line – the string, dashed line – the cube, dot-dashed line – the disk, dot line – the square.

It is clear from Figure 5.8 that the chemical binding energy change is strongly dependent on the system dimensionality D. The free "chemical" energies of structures with the same D values (cube and sphere, disk and square) are very close, but decreasing D results in substantial decreasing of W_{chem}. Due to higher coordination numbers and in the absence of other interactions (e.g., electrostatic), compact 3D structures are energetically favorable (note that $\lambda < 0$) in comparison with planar and linear structures. Therefore, neutral or weakly charged nanoparticles bound together by polyamine molecules should aggregate in 3D structures. Our experiments confirmed this conclusion.

However, the situation can change significantly if the charge Q of particles is not negligible. Figure 5.9 presents the dependences of numerically calculated electrostatic energy of nanoparticulate structures with $D = 3$ (sphere and cube) and $D = 2$ (square) versus the particle number. The electrostatic energy of the string ($D = 1$) is much less than those presented in Figure 5.9.

It is obvious from Figure 5.9 that the Coulomb energy is significantly larger for 3D nanoparticulate structures in comparison with that of 2D and 1D structures. The least electrostatic energy (within systems with equal N) relates to 1D structures. Let us change the value $N_1/N_2 = n$ in Eq. (5.3) assuming that the total particle number is constant ($N = 4096$). As shown in Figure 5.10, the Coulomb energy sharply decreases with increasing of system anisotropy (the transformation from $D = 2$ to $D = 1$). The last point in Figure 5.10 corresponds to the linear structure ($N_1 = 1$; $N_2 = 4096$).

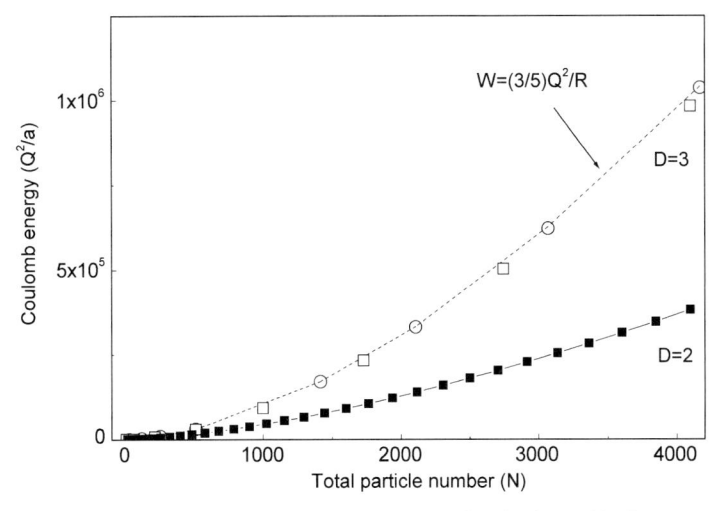

Figure 5.9 Electrostatic energy of the nanoparticle system versus particle number for different system geometry and dimensionality D. Dashed line shows the analytical result for the electrostatic energy of uniformly charged ball. It is assumed that the charge of individual particle and the distance between neighboring particles are unit ($Q = 1$, $a = 1$).

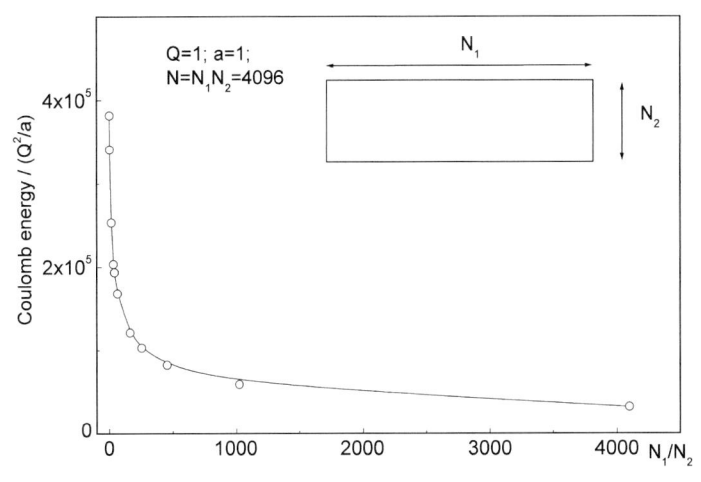

Figure 5.10 The electrostatic energy of rectangular nanoparticle system as a function of the anisotropy parameter N_1/N_2. Solid line is guide for eye. It is assumed that the charge of individual particle and the distance between neighboring particles are unit ($Q = 1$, $a = 1$). The total particle number $N = 4096$.

The total energy for nanoparticulate structures of different dimensionality and shapes is shown in Figure 5.11 as a function of the particle charge Q. The negative chemical binding energy of neighbor particles and the total number of particles in the system are assumed to be constant. When $Q = 0$, the most energetically favorable structures are 3D, (cube and sphere). Note that energy of cube and sphere are nearly the same (see Figures 5.8 and 5.9). The linear structure has the highest total energy in case of negligible charge. However, with increasing of the Q value, the energy of 3D structures dramatically grows in comparison with the linear structures. There is wide interval of the Q values in which the 2D nanoparticulate structures are the most energetically favorable (Figure 5.11).

We can describe the mechanism of discovered self-assembling and self-organization phenomena of cationic magnetite nanoparticles due to their binding with polyamine molecules in the following way. In one limiting case when the particles are weakly charged, they tend to form the 3D ensembles with significant (negative) chemical binding energy. The size of these aggregates is limited only by magnetite nanoparticles and polyamine concentrations.

In other limit case when the charge of particles is too large, the chemical binding energy of nanoparticles binding with polyamines cannot compensate the high electrostatic repulsive energy, and nanoparticles remain colloidally stabilized in suspension without forming any aggregates and compact

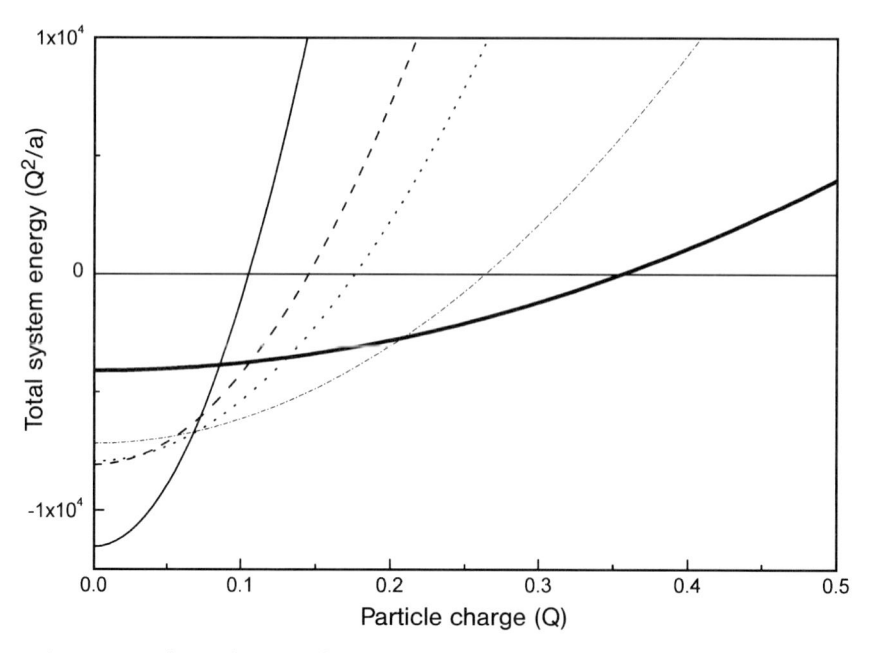

Figure 5.11 The total energy of nanoparticle systems as a function of the particle charge. Thin solid line – sphere, dashed line – planar square, dot line – rectangle with $n = 16$, dashed-dot line – rectangle with $n = 256$, thick line – linear structures (string). The distance between neighboring particles is unit ($Q = 1$, $a = 1$). The total particle number $N = 4096$. Chemical binding energy parameter $\lambda = -1$.

structures. In intermediate cases when the free "chemical" and electrostatic energies of the system nearly compensate each other, planar and linear structures are the most energetically favorable. The linear structures have the widest range of Q in which they are energetically stable (Figure 5.11). However, the linear structures are extremely unstable to various structural defects and fluctuations. We also found that the boundary defects of planar structures (Figure 5.7(f)) increase the system energy so that the ideal (defectless) system is energetically stable.

Thus, the energetic balance between electrostatic repulsive interactions of structure-forming colloid components of the nanofilm material and attractive forces caused by chemical bonding of components via bridge ligand molecules have substantial effect on the morphology of resulting self-assembled nanoparticulate structures. These findings show possibilities for efficient control of structural-functional characteristics of that novel type of nanofilm materials. The presence of magnetic iron oxide nanoparticles in these sheet-like complexes implies the possibilities for remote control of the behavior and spatial localization of such complexes by magnetic field and via magnetic interactions, which shows the perspective for their practical applications [181]. Biocompatible nanocomposite magnetic free-floating nanosheets can be useful for systems of controlled binding, separation, coating, covering, transport, and delivery in liquid media (particularly, for manipulations with drugs, bioactive compounds, and biocolloids) [181].

5.3
Anisotropic and Quasilinear (1D) Systems

Anisotropic quasi-1D systems based on nanoparticles include high aspect-ratio nanoparticles (nanowires, nanorods, and nanotubes) and linear arrays of nanoparticles. Such nanostructures have properties and structural features substantially different from quasispherical nanoparticles and 2D nanostructures, which makes them unique class of nanomaterials and important object for nanotechnological developments [182, 183]. Anisotropic magnetic nanoparticles and quasi-1D nanoparticulate structures are expected to exhibit interesting magnetic properties because of their shape anisotropy. In particular, 1D linear chains of nanoparticles exhibit unique transport properties and have been considered for applications such as anisotropically conducting materials, plasmon waveguides, single electron tunneling devices, magnetic logic, and magnetic quantum cellular automata [184, 185]. It was demonstrated that suspensions of nickel nanowires can behave as magneto-optical switches [186].

The shape control of magnetic nanoparticles was recognized as a very important issue in the nanoparticle synthesis. Anisotropic elongated magnetic nanoparticles are characterized by enhanced coercive force dependent on the shape and size of nanoparticles [187, 188]. There are number of approaches to form anisotropic magnetic nanoparticles and nanostructures based on chemical and physical interactions, and on the structural features of reaction systems. In particular, highly anisotropic prolate nanoparticles and quasi-1D nanoparticulate structures can be formed under applied external magnetic field. Definite specific ligands and synthesis in nanostructured and porous media can result in anisotropic nanoparticles. The use of appropriate templates allows preparing quasilinear nanoparticulate structures, magnetic nanowires and nanotubes.

5.3.1
Synthesis under Applied Magnetic Field

External magnetic field along with other factors can have an effect on the shape of synthesized magnetic nanoparticles. Anisotropic interactions of nanoparticle dipole magnetic moment with external magnetic field and dipole–dipole interactions of growing nanoparticles are the physical factors responsible for the generation of anisotropic magnetic nanoparticles and nanostructures under applied magnetic field [46]. Anisotropic acicular α-Fe nanoparticles were synthesized by reduction of a low concentration of Fe^{2+} ions with sodium borohydride in a tubular lecithin/cyclohexane/water reverse micelle solution under an applied magnetic field about 1200 Oe [189]. Also, acicular α-Fe nanoparticles 50–150 nm long and 5 nm wide were obtained using similar reaction procedure of Fe^{2+} reduction in the presence of the lamellar liquid crystalline system–$C_{12}H_{25}(OCH_2CH_2)_4OH$/water solution under applied magnetic field. Some fraction of spherical particles of 10–15-nm diameter was also present in the product [190].

It was demonstrated that the external magnetic field have an effect on the shape of amorphous iron oxide nanoparticles synthesized via the sonication of iron pentacarbonyl in decalin under ambient atmosphere [191]. Amorphous acicular iron oxide nanoparticles (of 50 nm in length and 5 nm in width) in amount of 20–30% were observed in the product prepared under applied 7 kG magnetic field in contrast to the quasispherical nanoparticles (about 25 nm in diameter) generated via the same procedure in the absence of an applied field. The difference in the shape of nanoparticles was reflected in their magnetic properties – the magnetic moment versus temperature measurements – and Mössbauer spectroscopy revealed a large shift of the blocking temperature of about 70 K toward higher temperatures for the sample obtained under applied magnetic field possibly due to the significant enhancement of the nanoparticle shape magnetic anisotropy [191].

Figure 5.12 demonstrates the effects of applied external magnetic field on the shape of iron oxide nanoparticles photochemically generated at the air–water interface [287].

Nanoparticles were synthesized by UV decomposition of iron pentacarbonyl in stearic acid Langmuir monolayer. The shape of substantial part of nanoparticles was changed from circular plate-like and ring-like

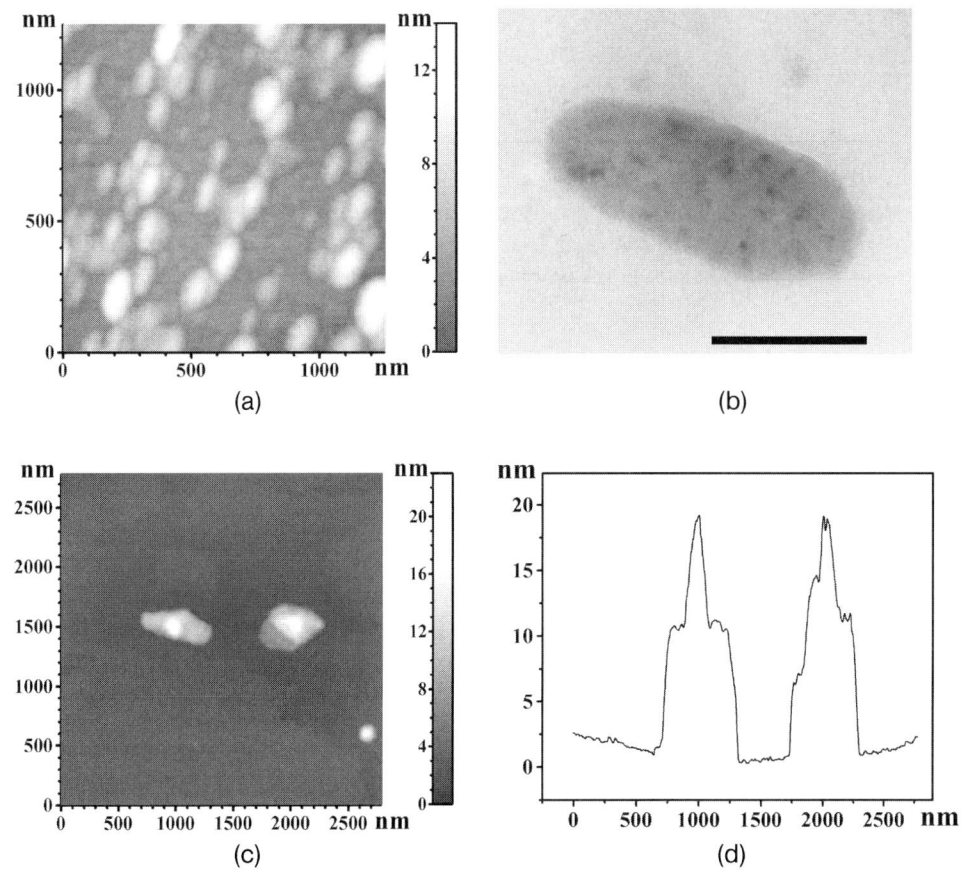

Figure 5.12 Images (a) and (b): AMF and transmission electron micrographs showing typical iron-containing nanoparticle synthesized in Langmuir monolayer under applied external magnetic field ($H = 2 \times 10^3$ Oe) parallel to the plane of monolayer (initial Fe(CO)$_5$/SA ratio was 10:1, UV exposure time 4 min). Image (c): AFM tapping mode topographic images of iron-containing magnetic nanoparticles synthesized in Langmuir monolayer under applied external magnetic field ($H = 2$ kOe) perpendicular to the plane of the monolayer and deposited under conditions as on Figure 5.3 (initial Fe(CO)$_5$/SA ratio was 10:1, UV exposure time 4 min, $T = 294$ K, aqueous subphase pH $= 5.6$, monolayer surface pressure during the synthesis of nanoparticles was $\pi \cong 0$). Image (d): cross-section profile of the image 5.12(c) parallel to the x-axis [75].

(Figures 5.2(a), (b) and (d)) to the field-aligned ellipsoidal when external magnetic field parallel to the plane of precursor monolayer was applied during the nanoparticle synthesis (Figures 5.12(a) and (b)). The evident correlation between the direction of the applied magnetic field and the long axis of anisotropic nanoparticles grown and deposited under the applied field was observed. Figure 5.12(c) presents AFM topographic images of anisotropic iron oxide nanoparticles synthesized in interfacial monolayer under 2 kOe magnetic field applied perpendicular to the plane of precursor monolayer. Characteristic ellipsoidal nanoparticles generated in monolayer under applied field parallel to the monolayer surface (Figure 5.12(a)) were not observed. Figure 5.12(d) shows typical results of cross-section analysis of the image of nanoparticulate monolayer presented in Figure 5.12(c). Characteristic quasipyramidal shape of nanoparticles with central peak can be observed on Figure 5.12(d). The anisotropic shape of iron oxide nanoparticles synthesized under magnetic field can be a result of the effects of anisotropic diffusion of growing magnetic particles and nucleus due to their anisotropic dipole–dipole interactions [192, 193].

The effect of applied magnetic field on morphology of arrays of self-assembled colloidal magnetic nanoparticles formed by drying a drop of nanoparticle suspension on a substrate surface was studied in [38], and the formation of anisotropic 3D arrays of magnetic nanoparticle columns was observed by applying external magnetic field perpendicular to the substrate surface. These data correlate with results presented in Figures 5.12(c) and (d) and indicate the important leading role of applied magnetic field in anisotropic organization and growth of magnetic nanoparticles.

5.3.2
Ligand Effects and Synthesis in Nanostructured Media

The nature of organic ligands and the composition and structure of reaction media have substantial effects on the shape, properties, and organization of synthesized inorganic nanoparticles and nanostructures [46, 194]. Thus, anisotropic inorganic nanoparticles and nanoparticulate structures can be obtained via appropriate control of composition and ligand nature in reaction system during the synthesis of nanoparticles. A relatively large number of organic ligands, including, for example, various fatty acids (in particular, alkanecarboxilic, sulfonic, and phosphonic acids), alkylamine, trioctylphosphine, and alcohols, in particular, poly(ethylene glycol), have been investigated as stabilizing agents for colloid magnetic nanoparticles dispersed in organic media [195–198]. For stabilization of colloid nanoparticles in polar liquids, such as water or ethylene glycol, the electrostatic interactions of surface charges were exploited [46]. In particular, magnetic iron oxide nanoparticle aggregation was prevented by electrostatic stabilization via surface adsorption of citrate ions [195] or hydroxide [199] ions to produce negatively charged colloid nanoparticles. Bilayers of ionic surfactants have also been reported to provide electrostatic stabilization through the outer layer of charged head groups surrounding the particles [200].

Anisotropic ellipsoidal γ-Fe_2O_3 nanoparticles with narrow distribution of sizes and shapes were synthesized successfully [201]. Anisotropic spindle-shaped colloidal hematite α-Fe_2O_3 particles with narrow size distribution were prepared by hydrolysis of $FeCl_3$ in solution at high temperatures. It was found that small additions of phosphate and hypophosphate ions in reaction system before aging resulted in significant effect on the shape of synthesized iron oxide particles. Surface interactions of phosphate ions with nuclei were proposed to determine the following growth mechanism of anisotropic iron oxide particles [202]. Similar observation was reported earlier in [203]. Surface treatment of magnetic nanoparticles can result in substantial changes in their magnetic properties. Thus, the coercivity of γ-Fe_2O_3 nanoparticles increased substantially after treating with polyphosphate anions [204, 205].

Anisotropic disklike iron oxide nanoparticles were obtained via surfactant-mediated thermal decomposition of iron pentacarbonyl following oxidation by oxygen [206].

It was demonstrated that the shape of magnetic nickel nanoparticles is dependent on the nature of organic ligand present during the nanoparticle synthesis. Ni nanorods were obtained when hexadecylamine was present in the reaction media at high concentrations [207]. Also, anisotropic cobalt nanoparticles were prepared via the use of appropriate stabilizing ligands in processes of their colloidal liquid-phase synthesis [59]. Anisotropic disk-shaped cobalt nanoparticles (about 100 nm in width and 15 nm in thickness) were obtained via sonication of aqueous solution of Co^{2+} ions and hydrazine [208]. Lorentz microscopy indicated that these were single-magnetic domain particles with the axis of magnetization located in the (101) plane different by some appreciable angle from the (001) axis.

Anisotropic Fe-based elongated alloy particles protected by alumina coating have been obtained by dehydration and further thermal reduction of undoped and Co-doped goethite precursors. The precursors were obtained by oxidation–precipitation of a mixture of $FeSO_4$ and $Co(NO_3)_2$ in the presence of NaOH and Na_2CO_3. Prior to the thermal treatment, the particles were coated with a homogeneous thin layer of alumina. Values of the coercivity around 1200 Oe have been found in the prepared FeCo metallic nanoparticles protected by the alumina layer [209]. Quasimonodisperse spherical iron nanoparticles with diameters of 2 nm and monodisperse rod-shaped iron nanoparticles with 2 nm width and 11 nm length were synthesized and characterized in [210]. At first, spherical 2-nm-sized Fe nanoparticles were prepared by the thermal decomposition of organometallic precursor ($Fe(CO)_5$) in the presence of a stabilizing surfactant trioctylphosphine oxide (TOPO). Then, to the spherical nanoparticles solution in TOPO, the solution of 2.6 mM $Fe(CO)_5$ in trioctylphosphine was added at 320 °C. The resulting black solution was aged for 30 min at 320 °C. This step was repeated once more, and the resulting reaction mixture was cooled to room temperature to get a black solid. Butanol was added to solubilize the reaction mixture, and the resulting butanol solution was added into excess acetone, resulting in a black precipitate. The precipitate was collected by centrifugation and washed several times with acetone to remove excess surfactant. The precipitate was dissolved in 19 mL of pyridine containing 0.5 g of didodecyldimethylammonium bromide, and the resulting solution was refluxed for 12 h. The precipitate formed during the reflux was removed by centrifugation, and the supernatant was vacuum-dried to yield a black powder. The dry powder was soluble in pyridine, and the solution was kept stable for a week without precipitation under an ambient condition.

Synthesis of inorganic nanophase materials in anisotropic reaction systems can result in correspondingly anisotropic nanoparticles and nanostructures. Such anisotropy of reaction media can be provided by spatial (structural) anisotropy of the matrix used as a chemical nanoreactor to generate nanoparticles. An example of formation of anisotropic plate-like iron oxide nanoparticles in anisotropic interfacial quasi-2D reaction system was presented in Section 5.2.2.

Similar approach can be used for the preparation of acicular nanoparticles and chain nanostructures. In that case, the reaction system should have quasi-1D structural elements. Nanowires and nanotubes of various nature can be prepared via the synthesis and assembly within the ordered pores of appropriate materials. The advantage of the approach is that due to the cylindrical pores of uniform diameter, the monodisperse nanocylinders of the desired material (polymers, metals, semiconductors, and other) with controlled dimensions can be obtained [211]. For example, metallic nanowires were formed inside carbon nanotubes [212, 213]. Carbon nanotubes can also be loaded with magnetic nanoparticles. Thus, Fe nanoparticles [214], Ni nanoparticles [215, 216], Co nanoparticles [217], Fe–Co alloy nanoparticles [218], nanocrystalline iron oxide nanoparticles [219], CoFeO nanowires [220], and FeCo nanowires [221] were formed inside carbon nanotubes.

Magnetic multiwalled carbon nanotubes were filled with γ-Fe_2O_3 nanoparticles and exploited for building up electrochemical biosensor by combining with horseradish peroxidase and assembling in multilayer films with the aid of applied magnetic field [222].

Various magnetic nanowires were electrodeposited into pores of membranes and layers. Typical nanowires fabricated by this method have a diameter in the range of 30–500 nm for a length of the order of 10 µm, and can be composed of a stack of layers of different metals with thicknesses in the nanometer range (multilayered nanowires) [223].

One of the first and most widely used porous template matrix exploited as an anisotropic reaction system for electrochemical generation of inorganic rodlike nanoparticles, nanofibriles, and nanotubes was porous anodic alumina layer with quasi-1D pores unidirectionally aligned in the layer [211, 224]. Magnetic metallic iron nanowires in anodic aluminium oxide porous layer were prepared by electrodeposition using ac electrolysis in an electrolyte composed by $FeSO_4$, boric acid and ascorbic acid at pH = 3 [225]. The diameters of obtained nanowires were in the region 20–220 nm and aspect ratio was up to 60. The magnetic measurements showed that the coercivity of magnetic nanowires was highly anisotropic and dependent on aspect ratio of acicular Fe nanowire-like particles.

Ferromagnetic nanowires were formed by electrodeposition of ferromagnetic metals (Fe, Co, and Ni) into porous anodic alumina templates, with diameters as small as 5 nm. Coercivity as a function of diameter reveals a change of the magnetization reversal mechanism from localized quasicoherent nucleation for small diameters to a localized curlinglike nucleation as the diameter exceeds a critical value determined by the exchange length [226].

Organized arrays of magnetic Ni nanowires were formed by Ni electrodeposition into porous matrix of anodic aluminum oxide films [227]. The effect of the form factor on the magnetic properties of nanowire arrays was observed in those nanostructures. Nanometer scale ordered arrays of magnetic metallic cylindrical nanoparticles with aspect ratio ~50–70 and ultrahigh uniformity were formed via alumina anodization and electrochemical deposition of metal nanophase. Pulse reverse voltage wave forms were used for better control of the nucleation and growth of the anisotropic magnetic particles. The resulting nanoparticles were polycrystalline, and grains were randomly oriented [228].

Quasi-1D porous silicon channel structures with pores oriented perpendicular to the substrate surface with diameter of about 50 nm and length up to 50 μm were exploited for electrochemical generation of high aspect ratio Ni nanowires. This obtained porous silicon/Ni nanowire nanocomposite system exhibited a twofold switching behavior of the magnetization curve at two different magnetic field ranges [229]. Similar porous silicon matrix templates were used for electrodeposition of high aspect ratio quasi-1D metallic nanostructures (nanowires, nanoparticles) in them [230]. The advantage of this porous template matrix is the probability for direct integration of metallic nanostructures with other silicon components in electronic devices because of better physical and electrical interconnection between the nanostructure and the silicon substrate [230]. In particular, single-segment Ni–Fe and Au and two-segment Ni–Fe/Au nanowires of diameter ~275 nm and length up to 100 μm were electrodeposited in the porous silicon templates with ordered and controlled nanometer-size pores, 40 nm and 290 nm in diameter. The formation of both vertical and branched Ni–Fe nanowires along with nanoparticles was observed due to discontinuous growth of nanowires.

Mesoporous silica (100) films with unidirectionally oriented pores were used to prepare magnetic iron nanowires [231]. The films were deposited on SiC/Si(100) substrate by spin-coating technique. All pores were parallel to the substrate surface and aligned radially by centrifugal force with very high level of local alignment (90% on 2 cm^2) of the pores. Magnetic iron nanowires were synthesized in the mesoporous silica channels by intercalation of a hydrophobic iron compound into a hydrophobic part of the silica-surfactant composite followed by decomposition of the complex and additional crystallization of iron in hydrogen flow. It was found that an ordered structure preserved after the chemical modification and synthesized iron nanowires uniformly filled the porous film structure [232]. Mesoporous silica matrix was also used for the formation of cobalt nanowire arrays by incorporation of precursor hydrophobic metal compound $Co_2(CO)_8$ into the hydrophobic part of the silica-surfactant composite. Thermal modification was performed in order to provide a crystallization process of the cobalt nanowires, which were uniform in size and well ordered in the silica matrix. For Co-containing sample annealed at 300 °C, the coercive force at room temperature was found to be 42.2 kA/m at saturation magnetization of 0.5 A m^2/kg [233]. Also chemical modification of anion-substituted layered double hydroxides was used for the preparation of anisotropic magnetic nanocomposites with different morphology and sizes of magnetic metal (Fe and Ni) nanoparticles.

Porous and nanostructured organic matrixes were also used for the synthesis of anisotropic magnetic 1D structures. Magnetite nanostructures grown by electrodeposition in the form of nanocrystalline film (200 nm thick) and nanocylinders (50 nm diameter, 200 nm length) were prepared inside the pores of polycarbonate membranes [234]. The effects of shape magnetic anisotropy, antiphase boundary interactions, and the Verwey transition were studied in the obtained anisotropic nanostructures. Nanoporous polycarbonate and alumina membranes containing uniformly distributed cylindrical pores with monodispersed diameters (varying between 20 and 200 nm) and thicknesses of 6 and 60 µm, respectively, were used for the fabrication of iron oxide particle nanowires via hydrolysis and polymerization of iron salts [235]. Alternatively, the sol–gel technique using metalloorganic compounds as precursors, fibrils, and tubes of different iron oxide phases were obtained inside the membrane nanopores in the same porous matrix. It was found that below the blocking temperature, the magnetic behavior of the nanowires was governed by the dipolar interaction between nearest-neighbor nanoparticles inside the pore, whereas the energy barrier, and therefore the blocking temperature value, was mainly governed by the dipolar interaction between magnetic moments over larger (interpore) distances. Crystalline iron oxide nanotubes exhibited magnetic perpendicular anisotropy due to their magnetocrystalline and shape anisotropy [235].

Polycarbonate membranes were exploited for electrodeposition of Ni/Cu multilayers in the form of nanowires (60–80 nm in diameter and ∼5 µm in length) [236]. Biwires with diameters 200 nm, 400 nm, 1 µm, and 2 µm were prepared by electrodeposition into pores produced by nuclear particle track etching in polycarbonate membranes [237].

Sulfonated mesoporous styrene-divinylbenzene copolymer template was used to prepare size-controllable magnetite nanoparticles by alkaline oxidation of adsorbed ferrous ions [238]. Prolate α-Fe nanoparticles 50–150 nm long and 5 nm in diameter were synthesized via sodium borohydride reduction of ferrous ions in ferrous chloride solution in the presence of the lamellar liquid crystalline structure $C_{12}EO_4$/water [190]. Magnetic nanowires (CoPt and FePt) were synthesized using a virus-based scaffolds of M13 bacteriophage [239]. The incorporation of specific, nucleating peptides into the generic scaffold of the M13 virus coat structure provided a viable template for the directed synthesis of semiconducting and magnetic materials. Removal of the viral template by means of annealing promoted oriented, aggregation-based crystal growth, forming individual crystalline nanowires.

Thus, the shape and physical–chemical properties of colloid magnetic nanoparticles can be dependent on the stabilizing ligand nature thus giving possibility for control of their morphology, colloidal stability, and assembling. Anisotropic acicular magnetic nanoparticles and their ordered ensembles can be formed via synthesis of magnetic nanophase in appropriate nanostructured matrix.

5.3.3
Quasilinear Nanoparticulate Structures

Under applied magnetic field, the formation of chains and 1D structures of colloid magnetic nanoparticles is promoted and the chains are predominantly aligned in the external field direction [25, 46, 240, 241]. The effect of self-organization of colloid magnetic nanoparticles in suspension confined in a thin gap perpendicular to the applied magnetic field with formation of fixed quasiregular arrays of nanoparticulate columns was used for the separation of a large duplex DNA in a microchannel device prepared by soft lithography [242]. The column spacing can be tuned from submicrometer to about 100 mm, by varying cell size and magnetic particle concentration. In a monodisperse ferrofluid emulsion, an applied magnetic field induces a magnetic dipole moment in each droplet. When the dipole–dipole interaction energy exceeds the thermal energy, a phase transition occurs and the fluid of randomly dispersed droplets changes to a solid of nearly equally sized and spaced columns [243].

Interesting patterns formed by Fe_3O_4 nanoparticles in films of kerosene-based ferrofluids under applied parallel and perpendicular magnetic fields were observed [244]. In parallel external magnetic field applied to a ferrofluid film, the ordered quasiperiodic chains of nanoparticles were obtained. When magnetic field was applied perpendicular to the film surface, the column structures were formed and the distances between these columns decreased as the strength of the applied magnetic field increased [244]. Formation of tubes of ferrite colloid nanoparticles stabilized by citrate ions was observed during the evaporation of a drop of dispersion under applied external field [58].

It is known that due to the anisotropic dipole–dipole interactions, colloidal magnetic nanoparticles depending on the particle size and magnetic properties can self-assemble into quasilinear chain structures in zero external magnetic filed [46, 245–247]. The transition from separate particles to randomly oriented linear aggregates and branched chains and networks was observed in zero field in dispersion of magnetic colloid iron nanoparticles with increasing particle size. When dispersion was vitrified in a permanent magnetic field, these chains aligned in the field direction and formed thick elongated structures, indicating lateral attraction between parallel dipole chains [248].

Monodisperse colloid 20 nm Co nanoparticles prepared via thermal decomposition of precursor compound – dicobalt octacarbonyl – in hydrocarbon solvent in the presence of polymeric compound methylmethacrylate-ethylacrylate-vinyl pyrrololidone formed long-chain nanostructures [249]. In colloidal dispersion of 16 nm iron nanoparticles prepared via thermal decomposition of iron pentacarbonyl in the presence of poly(4-vinylpyridine-styrene)/dichlorobenzene, the formation of chain structures of magnetic nanoparticles was also observed [250]. Formation of self-assembled chain-like complex aggregates of anisotropic ferromagnetic Co disk-shaped nanoparticles without applied filed was reported [251, 252]. It was found that superparamagnetic maghemite nanoparticles (~9.2 nm diameter) formed needle-like assemblies of nanoparticle arrays [253].

The effect of magnetic-field-induced formation of linear chain aggregates of magnetic nanoparticles was used for the formation of quasi-1D nanoparticulated template. Colloidally synthesized cobalt nanoparticles in aqueous solution were exploited as sacrificial electroless deposition templates for the synthesis of hollow gold nanospheres [254] by oxidation of Co(0) with Au(3+) salt ($HAuCl_4$). Under applied magnetic field, linearly aggregated cobalt nanoparticles were found to produce long, well-organized gold nanotubes with controllable shape, size, and length, tens of nm in diameter, a few nm in wall thickness, and up to 5 μm in length [255].

In some cases, surfactants can assist the unidirectional aggregation of magnetic nanoparticles. Thus, amorphous spherical Co nanoparticles of about 5–10 nm size were prepared by ultrasonic decomposition of the volatile precursor $Co(NO)(CO)_3$ in decane solution with oleic acid as a stabilizing surfactant [256]. It was found that with time (about a month), Co nanoparticles self-organized into acicular ellipsoidal nanoparticle aggregates of about 1 μm size with aspect ratio of 5:1. The magnetic dipole–dipole interactions were proposed to play a directing role in the formation of elongated aggregates. Also, the role of surfactant was to stabilize the unidirectional aggregation by inhibiting the side interconnection of the magnetic nanoparticles [256].

Self-organization of magnetic Ni nanoparticles (size ~2.5 nm) on very smooth (gold on HOPG) surfaces with the formation of chains was described in [257]. The effect of formation of quasi-1D chains of iron oxide nanoparticles synthesized in Langmuir monolayer via decomposition of iron pentacarbonyl under short UV illumination conditions are illustrated in Figure 5.13.

The external magnetic field can have even more pronounced effect on the magnetic nanoparticulate nanostructure morphology. Figure 5.14 presents TEM images of maghemite nanocrystals deposited on the cooper grid with and without magnetic field [258]. The nanocrystals were synthesized by soft chemistry using functionalized surfactant forming micelles in water. Without applied magnetic field, spherical aggregates differing by various sizes appeared (Figure 5.14(a)). In 0.59 T magnetic field, nanocrystals formed thick striped structures in the direction of the applied field (Figures 5.14(b) and (c)). This

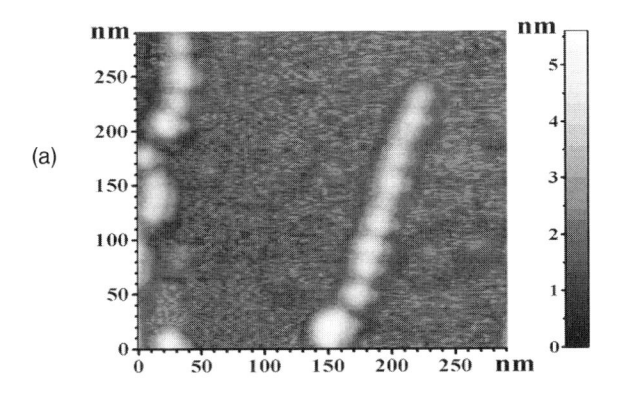

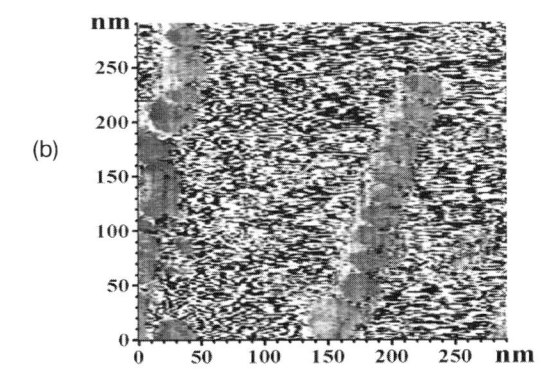

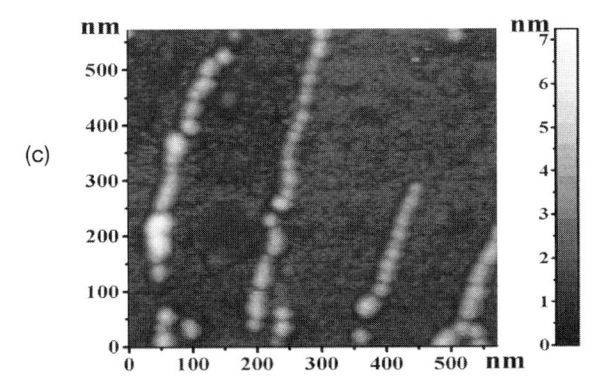

Figure 5.13 AFM tapping mode topographic images of iron-containing nanoparticles synthesized in uncompressed Langmuir monolayer ($\pi \cong 0$) at $T = 21\,°C$ and deposited onto the mica substrate at $\pi = 25$ mN/m using vertical substrate-lifting method. Image (a): chain nanostructures formed via dark incubation of the nanoparticulate monolayer for 4 min after the UV illumination for 6 sec. Image (b): Phase-contrast image corresponding to the image 5.13(a). Image (c): large-scale image of chain-like nanostructures presented in the sample corresponding to the image 5.13(a).

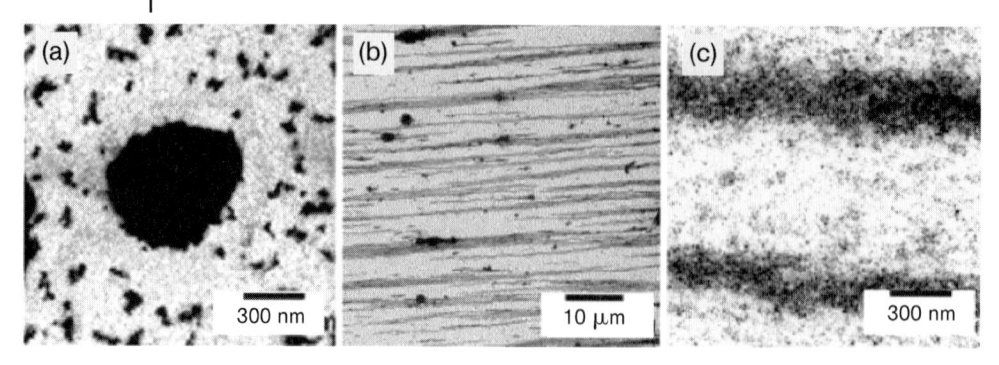

Figure 5.14 Transmission electron microscopy images of maghemite nanocrystals deposited on the copper grid with and without magnetic field. (a) Nanocrystals coated with octanoic acid and deposited without magnetic field. (b) and (c) Coated nanocrystals deposited with magnetic field. The nanocrystals were synthesized by soft chemistry [258].

effect strongly depends on the thickness of surfactant layer on the surface of nanoparticles [258].

Silica-coated cobalt nanoparticles with an average diameter of 32 nm and a silica-shell thickness of about 4 nm were found to organize into chains when driven by a weak external magnetic field [259]. The formation of chains of cobalt nanoparticles is attributed to the magnetic dipole–dipole interaction between neighboring particles. Due to the anisotropic nature of the dipolar interaction, the formation of chains is favored, since the north and south poles of the dipolar nanomagnets attract each other, while particles coming close to each other side by side with the parallel magnetization direction will repel each other, thus favoring the formation of "pearl necklaces."

Polystyrene-coated cobalt nanoparticles were synthesized by the thermolysis of dicobaltoctacarbonyl (Co_2CO_8) in the presence of end-functional polymeric surfactants in refluxing 1,2-dichlorobenzene [260]. Aligned chains of nanoparticles were clearly evident in topographical AFM and MFM images. These hybrid materials were weakly ferromagnetic at room temperature and strongly ferromagnetic at 40 K. Significant enhancement of the magnetic coercivity was observed by aligning nanoparticle chains under a weak magnetic field at 300 K due to the coupling of magnetic dipole moments along the 1D assembly.

5.3.4
Templated Structures

Anisotropic nanoparticulate structures can be formed via binding of magnetic nanoparticles with corresponding anisotropic templating substrates. For that purpose, the synthetic and assembling methods based on adsorption processes can be exploited. Thus, for example, surface decorations of carbon nanotubes with various metal and metal oxide nanoparticles have attracted great interest as an approach to the formation of anisotropic organized nanostructures with advanced electronic, magnetic, adsorption, optical, and mechanical properties (see, for example, [261–267]). Single-walled carbon nanotubes were used as templates for deposition of small magnetic Co clusters [268]. Attachment of magnetic nanoparticles on carbon nanotubes and their soluble derivatives was reported in [269]. Various other magnetic nanoparticles and nanocomposites were attached on carbon nanotubes, including Fe clusters [270], Co and Ni particles [262], Fe–Ni alloy nanoparticles [271], Fe–Co alloy nanoparticles [272], $Fe_xCo_{81-x}Ni_{19}$/CNTs nanocomposites [273], magnetite nanoparticles [274], and Co–B amorphous alloy [275].

Magnetic multiwalled carbon nanotube composites were obtained by homogeneous decoration of iron oxide nanoparticles on or in carbon nanotubes. The method involved the dispersion of the carbon nanotubes in iron pentacarbonyl $Fe(CO)_5$ followed by vacuum thermolysis and subsequent oxidation [274, 276–279]. It was demonstrated that γ-Fe_2O_3 nanoparticles, with controllable sizes varying from 6 to 15 nm and with selectively ferromagnetic or superparamagnetic properties, could be uniformly nucleated on both the inner and outer surfaces of multiwall carbon nanotubes by the vacuum thermolysis of $Fe(CO)_5$ confined in/around nanotubes and subsequent controlled oxidation [280].

Catalytic metallization and formation of Ni nanoshells on carbon nanotubes was reported in [281]. Uniform layers of magnetite/maghemite nanoparticles (6–10 nm in diameter) were deposited onto multiwalled carbon nanotubes by combining the polymer wrapping and LbL assembly techniques [282]. The particle-coated carbon nanotubes were superparamagnetic and aligned at room temperature on a substrate surface by deposition from an aqueous solution in an external magnetic field $B = 0.2$ T.

Magnetic CoPt and FePt nanowires were synthesized by using highly ordered filamentous capsid of the M13 bacteriophage as a template [239]. Linear polymeric molecules, in particular, DNA, can also be used as substrates for the preparation of anisotropic quasilinear arrays of magnetic nanoparticles.

DNA molecules are attractive candidates for building up precisely controlled and reproducible nanostructures due to the unique DNA recognition capabilities and highly selective binding, predictability of interactions, physicochemical and mechanical stability, and synthetic availability of practically any desired nucleotide sequences and lengths. Iron oxide nanoparticle arrays and nanostructures were formed via DNA templating and scaffolding [283, 284].

DNA complexes immobilized on the solid substrate surfaces were used as templates and iron-containing precursor material in the process of formation of organized iron oxide nanoparticulate structures [285]. First, interfacial planar DNA complexes with Langmuir monolayer of amphiphilic polycation were formed via binding of bulk aqueous-phase DNA molecules with floating monolayer and then deposited onto the solid substrates (mica,

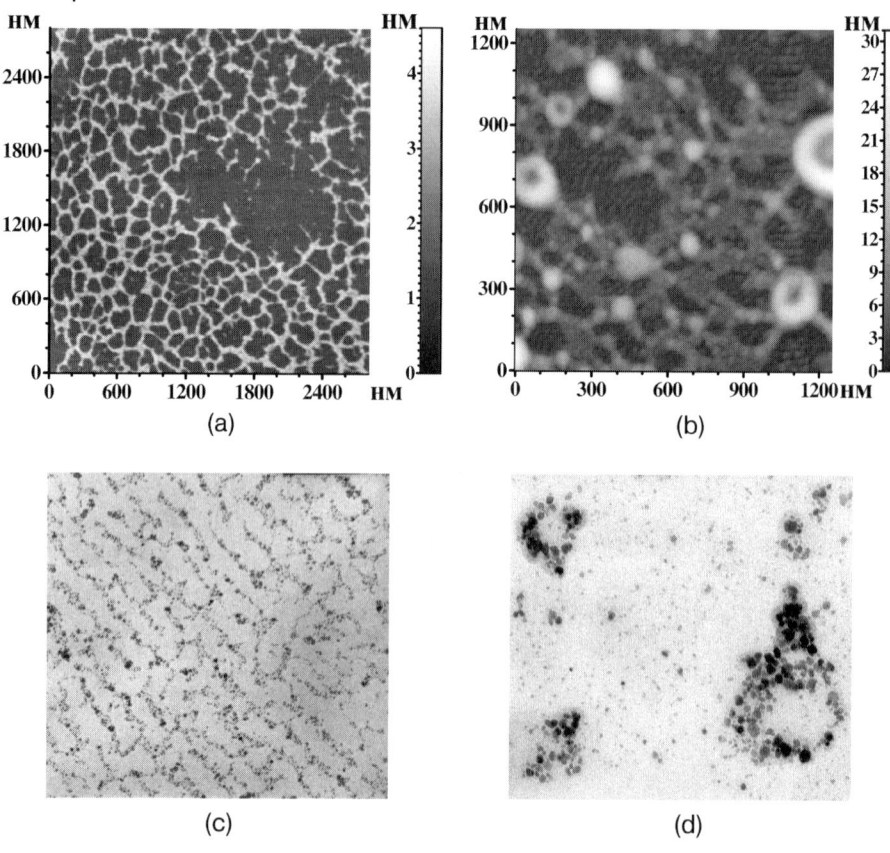

Figure 5.15 Images (a) and (b) AFM topographic images of DNA/PVP-16 complex bilayer LB film formed on mica substrate. Image (a)): Complex was formed under monolayer surface pressure value $\pi \sim 0$. Image (b) surface pressure $\pi = 20$ mN/m. Images (c) and (d): Transmission electron micrographs of magnetite nanoparticulate structures formed via incubation of DNA/PVP-20 complex LB film in suspension of prersinthesized colloidal cationic magnetite nanoparticles. Image (c) corresponds to the film structure shown in image 5.9(a). Image (d) corresponds to the film structure shown in image 5.9(b)). Image size: (c) 1.5 × 1.5 μm; (d) 1 × 1 μm [285].

silicon wafers with natural oxide layer, TEM grids with carbon layer). The ultrastructure of deposited LB films of amphiphilic polycations PVP-16 and PVP-20 was dependent on the corresponding Langmuir monolayer surface pressure and incubation time on the aqueous-phase surface [286, 287]. The effective surface concentration of amphiphilic polycation molecules in Langmuir monolayer interacting with bulk-phase DNA molecules can be easily and controllably changed and adjusted by variation of the monolayer area. Figure 5.15 demonstrates characteristic morphologies of interfacially formed DNA–amphiphilic polycation complexes deposited onto the mica substrate surface. Planar netlike or lattice structures were obtained when DNA molecules interacted with uncompressed PVP-16 Langmuir monolayer (Figure 5.15(a)). Figure 5.15(b) shows the structure of DNA–PVP-16 complexes formed when compressed PVP-16 Langmuir monolayer with quasihomogeneous, positively charged surface interacted with DNA molecules from the aqueous phase. Quasicircular toroidal structures bound by fibers are clearly seen in Figure 5.15(b). Such toroidal morphology is known for native compacted DNA structures and is most common morphology in DNA condensates from aqueous solution [288, 289]. These data demonstrate the probability of controlling the structure of interfacially formed amphiphilic polyelectrolyte complexes with bulk-phase polymeric ligands via controlling the conditions of the interfacial complex formation.

Figures 5.15(c) and (d) show the structure of assemblies of colloidal cationic Fe_3O_4 nanoparticles bound with immobilized DNA/amphiphilic polycation complexes. One can see that the structure of magnetite nanoparticulate assemblies is dependent on the morphology of the DNA/PVP-20 complex used as a binding template. Planar extended netlike nanostructures (Figure 5.15(c)) and massive circular aggregates (Figure 5.15(d)) of magnetic Fe_3O_4 nanoparticles were obtained via interaction of colloidal magnetite nanoparticles with preformed netlike (Figure 5.15(a)) and toroidal (Figure 5.15(b)) immobilized DNA/amphiphilic polycation complexes. The size of circular nanoparticulate aggregates in Figure 5.15(b) is of the order of hundred nanometers and corresponds to the order of magnitude of the mean diameter of toroidal structures shown in Figure 5.15(b).

DNA complexes immobilized on the solid substrate surfaces were used as templates and iron-containing precursor material in the processes of synthesis of iron oxide nanoparticulate structures. Figure 5.16 illustrates the results of the formation of DNA-templated magnetic iron oxide nanoparticulate structures. The use of DNA/Fe^{3+}/PVP-16 complex LB films as precursors resulted in the generation of quasilinear aggregates (Figures 5.16(a) and 5(b)), chain-like arrays of small iron oxide nanoparticles (Figure 5.16(c)), and their linear aggregates (Figure 5.16(d)). Figure 5.16(e) shows typical linear necklace-like aggregates of positively charged presynthesized magnetite nanoparticles (5–7 nm in diameter) in a bulk aqueous phase in the presence of DNA molecules at pH = 3. The morphology of these aggregates is very similar to that presented in Figure 5.16 (b). At neutral pH values, formation of DNA-templated magnetite nanoparticle aggregates was suppressed presumably due to the augmented OH^- binding to iron oxide nanoparticles at those pH values. For comparison, typical disordered aggregates of magnetite nanoparticles formed by drying of a droplet of magnetite colloid suspension on the TEM substrate are shown in Figure 5.16(f). The presented data illustrate the important role of DNA molecules in spatial organization of inorganic nanoparticles generated in DNA complexes.

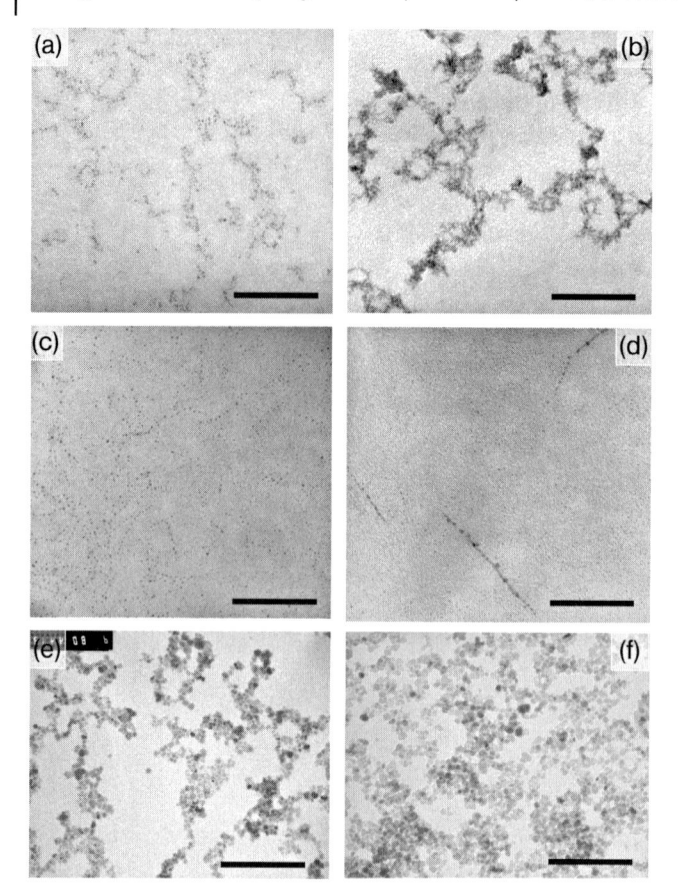

Figure 5.16 Transmission electron micrographs showing arrays of iron oxide nanoparticles synthesized in planar DNA/Fe^{3+}/PVP-16 complex LB films. Images (a), (b), and (d): $NaBH_4$ (5×10^{-4} M) was used as a reductant. Image (c): ascorbic acid (10^{-3} M) was used as a reductant. Image (e): Linear aggregates of magnetite nanoparticles formed in the presence of DNA (10^{-5} M per monomer) in colloid magnetite nanoparticles suspension. Image (f): typical aggregates of magnetite nanoparticles formed by drying a droplet of magnetite colloid suspension on the TEM substrate. Bar size: (a) 150 nm; (b) 170 nm; (c) 130 nm; (d) 100 nm: (e) 170 nm; and (f) 170 nm.

5.3.5
Nanotubes

Magnetic nanotubes may be of interest for magnetic-field-assisted separation, interaction, catalysis and targeting, and drug delivery biomedical applications. Nanotubes of various nature can be prepared via the templated synthesis and assembly within the cylindrical pores of corresponding thin-film materials similar to the preparation of nanowires in anisotropic reaction systems

(Section 5.3.2). The advantage of the approach is that due to the cylindrical pores of uniform diameter, the monodisperse nanocylinders of the desired material (polymers, composites, metals, semiconductors, and other) with controlled dimensions can be obtained. A further advantage of this method is that the nanotube material synthesized within the pores can then be freed from the membrane and used for further investigations or building up of superstructures.

The other way to form nanotubes is by using nanowire template for the formation of nanoshell, following decomposition of the core template similar to the procedure of formation of hollow colloid capsules (Section 5.2.3.2).

The three-step procedure for the preparation of single-crystalline magnetic Fe_3O_4 nanotubes with controllable length, diameter, and wall thickness is described in [290]. Single-crystalline MgO nanowires were first grown on Si/SiO_2 substrates. A conformal layer of Fe_3O_4 was then deposited onto the MgO nanowires using the pulsed laser deposition technique to obtain MgO/Fe_3O_4 core–shell nanowires. Finally, the MgO inner cores of the MgO/Fe_3O_4 core–shell nanowires were selectively etched in $(NH_4)_2SO_4$ solution (10 wt%, pH $\sim$ 6.0) at 80 °C. An etching time of $\sim$1.5 h was typically used to obtain micrometer-long Fe_3O_4 nanotubes with completely etched inner cores.

The LbL assembly technique (see Section 5.2.3) can be used for the preparation of nanotubes in porous matrixes and on nanowire-like templates. The wall thickness and inner diameter of as-prepared nanotubes can be tuned by controlling the composition and number of assembled layers. The length and outer diameter of nanotubes are dependent on the morphology of used templates. From practical point of view, the experimental LbL procedure to form such nanotubes in nanoporous template substrates is substantially simpler than the process of formation of colloid composite capsules via similar LbL technique. Due to that, the composite magnetic nanotubes can compete effectively with conventional colloidal magnetic containers in drugs transport and delivery applications [291].

Multilayer superparamagnetic nanotubes were assembled via the LbL deposition of polyelectrolytes (poly(allylamine hydrochloride) and poly(styrene sulfonate)) and magnetite nanoparticles in the pores of track-etched polycarbonate membranes [292]. Magnetic polypeptide nanotubes were fabricated through the LbL assembly of poly-L-lysine hydrochloride, poly-L-glutamic acid, and magnetic nanoparticles within the inner pores of polycarbonate membrane templates with subsequent removal of the templates. It was demonstrated that such magnetic polypeptide nanotubes can be used as a DNA carrier and manipulated under applied magnetic field. [293].

Ferromagnetic cobalt/polymer composite nanotubes with diameters of 160–450 nm and wall thicknesses of a few nanometers were prepared in self-ordered porous alumina membranes by deposition of a polystyrene or poly-L-lactide layer containing a metalloorganic precursor [294, 295].

Ferromagnetic manganite nanotubes with composition $La_{0.67}Sr_{0.33}MnO_3$ and $La_{0.67}Ca_{0.33}MnO_3$ and external diameters from 100 to 800 nm with walls of around 50 nm thickness were formed and studied in [296]. The nanotube walls were constituted by an assembly of nanoparticles with a non-Gaussian size distribution and a maximum at about 25 nm. It was determined that the crystallites were single magnetic domains with a magnetic dead layer on the surface, which avoided exchange interactions among nanoparticles [296].

Magnetic $NiFe_2O_4$ polycrystalline cubic spinel nanotubes with tailored structural characteristics were fabricated inside the nanochannels of porous anodic aluminum oxide templates by wetting chemical deposition and thermal decomposition using a mixture of Fe nitrate and Ni nitrate as precursors [297].

As was mentioned above in Section 5.3.2, iron oxide nanotubes with magnetic perpendicular anisotropy were prepared in nanoporous polycarbonate and alumina membranes by the sol–gel technique with the use of metalloorganic compounds as precursors [235].

Ordered arrays of iron oxide nanotube of controlled geometry and tunable magnetic properties were formed by atomic layer deposition technique [298]. To form Fe_2O_3 nanolayers in porous anodic alumina membrane matrix, water and the homoleptic dinuclear iron(III) *tert*-butoxide complex $Fe_2(OtBu)_6$ were used. The internal walls of the pores were covered conformally with a smooth layer of Fe_2O_3-yielding arrays of tubes with aspect ratios up to 100 with the growth rate being 0.26 Å/cycle. Reduction of the Fe_2O_3 nanotubes in the Al_2O_3 matrix by 5% H_2/95% Ar at 400 °C converted Fe_2O_3 to Fe_3O_4 [298].

The high aspect ratio of Ni and Co nanotubes were produced in porous membranes with pore diameters of 180, 220, and 260 nm. Ni nanotubes were deposited into the pores of alumina membrane between two layers of TiO_2. The synthetic procedure included the formation of NiO and TiO_2 layers by atomic layer deposition technique, which consists of the sequential deposition of thin layers from two different vapor-phase reactants. Nickeltocene or cobaltocene and H_2O or O_3 were used as precursor reactants for the deposition of a thin meatl oxide layer, which was then reduced into a magnetic Ni layer by annealing at 400 °C under Ar + 5% H_2 atmosphere. The TiO_2 layers were also obtained by atomic layer deposition technique and used for adding a higher stability against oxidation to the Ni and Co nanotubes and for preventing their damage in the etching process [299, 300].

Ferromagnetic carbon-nanotube-supported Ni/NiO nanotubes were formed via initial assembly of catalytic Pt nanoparticles on carbon nanotubes, followed by the deposition of Ni layer [301].

Very long lipid tubules with uniform widths, spirals, and spirals nested in tubular superstructures were formed from a lipid analogue made by acylating 1-(N, N-1-dimethylamino)ethyl-2,3-dihydroxybutyramide with two molecules of pentacosa-10,12-diynoic acid. The high propensity of this molecule for tubule formation is attributed to the large carbonyl dipole in the headgroup. The tubules formed from the pure lipid are hollow and can be gold-plated by vapor deposition without collapsing. The superstructures can also be metallized by adsorption of nickel(II) or copper(II) ions, followed by reduction with borohydride. The nickel-metallized tubules can be oriented in very modest external magnetic fields. The unusual length, stability, and ease of preparation of these tubules may be of use for the construction of functional microstructures [302].

Tubular magnetic nanostructures are highly attractive due to their structural attributes, such as the distinctive inner and outer surfaces, over conventional spherical nanoparticles. Inner voids can be used for capturing, concentrating, and releasing species ranging in size from large proteins to small molecules. Distinctive outer surfaces can be differentially functionalized with environment-friendly and/or probe molecules to a specific target. Due to the combining attractive tubular structure with magnetic properties, the magnetic nanotubes can be a promising candidate for the multifunctional nanomaterials toward biomedical applications, such as targeting drug delivery with MRI capability. In particular, magnetic silica–iron-oxide composite nanotubes were synthesized and demonstrated to be useful in magnetic-field-assisted chemical and biochemical separations, immunobinding, and drug delivery applications [303].

5.4
Patterned, Self-Organized, Composite, and Other Complex Magnetic Nanoparticulate Nanostructures

The fabrication of ordered patterned ensembles of magnetic nanoparticles utilizing a variety of preparation techniques such as different lithography and nanoimprinting techniques, as well as self-assembled and nanotemplate-assisted growth of nanostructures, is an important growing field of modern research. Advanced sample growth and patterning techniques allow one to control and modify shape and size of magnetic particles and the geometry of their ordered arrays. The behavior of ordered arrays of nanoscale magnetic particles is interesting both from a fundamental point of view, and also for applications in magnetic recording media, various magnetoelectronic and optoelectronic devices. Patterning of magnetic nanoparticles and thin films is of particular interest because of the potential applications in device fabrication, such as ultrahigh density data-storage media. The most common feature of these components consists of dots and other ordered arrays, where feature size and homogeneity are important. Although photolithographic methods have been the standard for microelectronic industry, they typically involve complex lithography procedures. Unconventional approaches have been developed in order to overcome various restrictions of the photolithographic methods that include high cost and limitation of accessibility. Photolithographic techniques are typically not suitable for direct patterning of nanoparticles either. New techniques useful for the formation of a number of ordered complex nanoparticulate structures on surfaces have appeared recently including nanopatterning via molding and embossing, using STM and AFM, and by printing, templating, and lithography [304–314, 315]. These techniques were also exploited for fabrication of patterned magnetic nanostructures, which are believed to be useful for creation of new-generation ultrahigh density magnetic recording media with densities of up to 150 Gbit $\times$ cm^{-2} [316]. Formation of complex patterned nanostructures on surfaces is often a multistep process, which includes a combination of physical methods and manipulations with assembling and synthetic chemical procedures.

The approach to form patterned nanostructures on surfaces can be illustrated by multilayer transfer printing approach. In that approach, a polyelectrolyte or composite multilayer film fabricated by the LbL alternate adsorption deposition technique directly onto the surface of a poly(dimethylsiloxane) stamp is transferred to the final substrate surface. It was demonstrated that such approach allows preparation of multicomponent multilayer patterns with differing functionalities onto a surface if combined with a positioning system [317].

Patterned magnetic nanoparticulate structures can also be based on patterned nanoparticulate LB films. Soft lithography method [318] was used for the preparation of patterned LB film structures. Patterned arrays of close-packed monodisperse (11 and 13 nm diameter) γ-Fe$_2$O$_3$ nanoparticles were formed onto solid substrates (silicon wafer) by deposition of LB films of oleic acid-stabilized iron oxide nanoparticles on microdot PDMS stamp and transfer of nanoparticulate layer to the solid substrate via soft lithography [100]. For that, double layers of γ-Fe$_2$O$_3$ nanoparticles were deposited onto the patterned PDMS stamp at a surface pressure of 40 mN/m. This patterned LB film was then transferred onto a silicon wafer, freshly cleaned with acetone and methanol in a sonication bath and dried with a stream of N$_2$, using microcontact printing. It was found that the quality of the original photo master, the replicated PDMS stamps, and the applied pressure were some of the key factors that can affect the shapes of the printed patterns of magnetic nanoparticles. Also, magnetic FePt-patterned microstructures were prepared from LB films of platinum–iron oxide core–shell nanoparticles [102] using a soft lithographic technique.

The combination of physical patterning tools with self-assembly has led to patterned self-assembly of nanoparticles. For example, a silicon wafer with a photoresist film with patterned holes made by UV lithography can be used as a template to direct the self-assembly of magnetic nanoparticles. The nanoparticle dispersion is dropped onto the patterned holes. After removal of the photoresist, a patterned disk of the self-assembled magnetic nanoparticles can be obtained. Alternatively, a focused laser beam has been applied for direct thermal patterning of a self-assembled nanoparticle array. The beam can locally heat a small area of the self-assembled array on an insulating substrate. The heating carbonizes the surfactant around the particles and strengthens the linkage between them. The unexposed particles can be washed away with fresh solvent, leaving patterned islands of exposed nanoparticles [319].

5.4 Complex Magnetic Nanoparticulate Nanostructures

The patterned disks, with an average diameter of 2.0 µm and a height of 250 nm, composed of self-assembled presynthesized FePt nanoparticles (diameter 3 nm) were prepared on silicon substrate surface via deposition of the FePt nanoparticles dispersion on the patterned holes of the photoresist film made by conventional UV-lithography technique. After being heat-treated at 100 °C for 30 min under vacuum condition, the photoresist was stripped out by dipping the sample in acetone [320].

Block copolymer lithography uses self-assembling properties of block copolymers to pattern nanoscale features over large areas. Although the resulting patterns have good short-range order, the lack of long-range order limits their utility in some applications. Lithographically assisted self-assembly method allows ordered arrays of nanostructures to be formed by spin casting a block copolymer over surfaces patterned with shallow grooves. The ordered block copolymer domain patterns are then transferred into an underlying silica film using a single etching step to create a well-ordered hierarchical structure consisting of arrays of silica pillars with 20 nm feature sizes and aspect ratios greater than 3 [321].

Nanosphere lithography technique is widely used for the preparation of organized macroscopic arrays of magnetic nanostructures on surfaces, which is interesting for applications related to ultrahigh density data-storage media. In that method, the ordered compact arrays of micro- or nanospheres are used as templates for deposition of nanophase magnetic material in the holes between neighboring spheres. Self-assembly of nanospheres over macroscopic areas on a water/air interface is facilitated by a combination of electrostatic and capillary forces and allows obtaining nanosphere templates with long-range order stretching over areas greater than 1 cm^2 [322].

Ordered DNA/magnetite nanoparticle patterns on silica wafer surface were prepared using a combination of templating Fe_3O_4 nanoparticles with DNA molecules, LbL deposition technique, and dip-pen nanolithography using AFM [284].

Effect of self-organization of spherical micelles formed by diblock copolymers was used to produce large-scale arrays of size-selected fct FePt nanoparticles. The synthesis of those FePt nanoparticles included their formation in cluster beams and from wet-chemical procedures at elevated temperatures [323]. Local magnetic field gradients (10^6–10^7 T/m) formed in the magnetized composite nanorod structure comprising ferromagnetic (Co–Ni) and diamagnetic (Au) sections were used for sub-100 nm trapping and spatial localization of superparamagnetic 8 nm γ-Fe_2O_3 nanoparticles in the high magnetic-field-gradient regions at the boundaries between the nanorod sections [324]. Interesting ring-like aggregates of magnetic nanoparticles formed via self-assembly were reported in a number of works. Thus, sonochemically synthesized colloidal barium hexaferrite $BaFe_{12}O_{19}$ nanoparticles formed nanorings and intersecting "Olympic ring" structures under drying of nanoparticle suspension [325]. It was observed that ferromagnetic cobalt nanoparticles can assemble spontaneously into nanosized "bracelet-like" nanorings when dispersed in organic solvents containing resorcinarenes as surfactants. Nanoring self-assembly of Co nanoparticles occurred in solution and was directed by magnetic dipolar interactions [326]. Using electron holography technique, the magnetic flux in nanorings formed by self-assembled Co nanoparticles was observed. The rings contained closed magnetic field flux, which can flow in one of two directions, clockwise or counterclockwise in the ring [327].

Self-organization of Mn_{12} cluster molecule magnets into ring structures was observed under cooling evaporation of a CH_2Cl_2 solution of Mn_{12} clusters on highly oriented pyrolitic graphite [328].

Self-assembled rings about 1 µm diameter of $CoPt_3$ nanoparticles (6 nm diameter) were formed in ultrathin polymer films due to phase separation of a binary polymer solution on a water surface with subsequent dewetting of the top layer, and its decomposition into droplets on the surface of the bottom layer. The polymer system was a blend of 50 wt% of a 1% nitrocellulose solution in amyl acetate and 50% of a $CoPt_3$ nanoparticle solution in hexane with hexadecylamine as stabilizing agent [329].

It was demonstrated that magnetite nanocube particles could be assembled into flux-closure rings, which consisted of several to dozens of magnetic nanocubes [330]. The observed nanorings had one-particle annular thickness, and individual magnetic nanocubes were spaced fully together. It was found that only nanocubes, the average size of which was close to 50 nm, could be assembled into rings, while slightly smaller magnetite nanoparticles were aligned in dipolar chains, suggesting that ring self-assembly could be produced by the degradation of dipolar chains that were a metastable structure with respect to rings, and only larger nanocubes with strong magnetic dipoles could overcome the potential barrier for the transformation from chains to rings [330].

Three-dimensional organized nanostructures – crystalline superlattices – formed by iron nanoparticles of cubic shape with edges of 7 nm were observed in [331]. These superlattices were formed in solution, precipitated in high yield, and were redissolved and redeposited as 2D arrays. Highly ordered fcc supracrystals of cobalt nanocrystals were obtained by slow evaporation of a solution of colloid cobalt nanoparticles in hexane [332]. The nanoporous and monodispersed spherical magnetite aggregates with mean size about 100 nm and nanopores less than 3 nm were formed by the assembling of Fe_3O_4 primary nanoparticles (~5 nm diameter) [333].

Complex magnetic nanostructured composite materials are a subject of current and future applied and basic researches because of possibilities to obtain materials with novel properties and multifunctional magnetic materials for novel applications. Nanocomposites of organic materials and inorganic nanoparticles are of great promise as composites for utilization in high-speed and high-capacity optical and magnetic information storage media [334]. Nanoparticle-based composite materials and systems can be classified as composite nanostructured nanoparticles and as matrixes (of organic or

5.4 Complex Magnetic Nanoparticulate Nanostructures

inorganic nature) incorporating inorganic nanoparticles (and, possibly other nanosized components).

Magnetic luminescent nanocomposite particles were prepared composed of Fe_3O_4 magnetic nanoparticle core (~8.5 nm diameter) covered by a number of CdTe quantum dots/polyelectrolyte multilayers. Layer-by-layer assembly technique was used for the deposition of polyelectrolyte and CdTe nanocrystal multilayers onto the surface of magnetite nanoparticles. It was shown that formed nanocomposite particles could be easily separated and collected in an external magnetic field, which makes them of some interest for applications in separation or site-specific transport [335].

Magnetic luminescent nanocomposite colloid multilayer structures were prepared using magnetite nanoparticles of 8.5 nm diameter as a template for the deposition of CdTe quantum dots/polyelectrolyte multilayers. The nanocomposite colloids were able to be easily separated and collected in an external magnetic field [335]. Magnetic colloid particles with polystyrene core and magnetic nanocomposite shell structure were prepared via sequential adsorption of magnetite nanoparticles and polyelectrolyte layers on polystyrene latex colloidal template [336]. Biocompatible magnetically and optically active dual-functional Au–Fe_3O_4 nanoparticles were prepared and demonstrated to be useful for simultaneous magnetic and optical detection as a new type of multifunctional probe for diagnostic and therapeutic applications [337].

Fluorescein isothiocyanate-incorporated silica-coated core–shell superparamagnetic iron oxide colloid nanoparticles with diameters of 50 nm were prepared and used as a bifunctionally magnetic vector that can efficiently label human mesenchymal stem cells via clathrin- and actin-dependent endocytosis with subsequent intracellular localization in late endosomes/lysosomes [338].

Complex core–shell multifunctional nanoparticles possessing magnetic, up-conversion fluorescence and bioaffinity properties were prepared [339]. Magnetic iron oxide core (5–15 nm diameter) was covered with ytterbium and erbium co-doped sodium yttrium fluoride via co-precipitation of the rare-earth metal salts with fluoride in the presence of fluoride, EDTA, and iron oxide nanoparticles. Then magnetic/fluorescent nanoparticles were coated with SiO_2 and functionalized with glutaraldehyde and streptavidin. Specific binding of streptavidin-coated magnetic/fluorescent particles with biotinylated IgG was demonstrated [339].

Composite iron boride-silica core–shell nanoparticles (84 ± 20 nm diameter) with soft ferromagnetic properties were prepared via one-pot aqueous chemical synthesis. Silica-passivated ferromagnetic nanoparticles were mainly composed of polycrystalline iron boride with 4.6% of alpha iron phase [340].

Core–shell colloid magnetic nanoparticles with stabilizing silica coating on magnetic FePt [341] and magnetite template core particles were synthesized [342]. It was found that the superparamagnetic nature of the magnetic nanoparticles remained unchanged with self-assembled bilayer molecular coatings and gave the way for their use in colloid suspension for device applications [343].

Oil in water emulsions with homogeneous fluorescence and high magnetic nanoparticle concentrations in oil droplets were prepared via encapsulation of both magnetic oxide and fluorescent semiconductor nanoparticles in emulsion oil droplets. For a given size and a given quantum dot (QD) concentration, the droplet fluorescence intensity droped sharply as a function of the magnetic nanoparticle concentration [344].

Transition metal–copolymer composite nanospheres with magnetic Ni nanoparticles ($\sim$10 nm in size) immobilized on anionic polymer nanospheres (about 300-nm diameter) were formed using UV irradiation at room temperature [345].

Ferromagnetic nanoparticles (Co) coated with carbon using carbon-arc method exploited for the preparation of fullerenes were reported in [346].

Various magnetic composite colloid particles representing the organic matrix particle (in particular, polymeric latex particle, dendrimers, etc.) decorated or impregnated with magnetic nanoparticles were reported in the literature (see, for example, [347–359]).

Multilayer films on 420-nm colloidal latex particles formed via LbL assembly and containing 12-nm magnetite nanoparticles, polyelectrolytes, glucose oxidase, and silica nanoparticles were prepared and possessed biocatalitic properties [360]. The presence of magnetic nanoparticles in that structure allowed self-stirring of the formed nanobioreactors with a rotating magnetic field and enhanced its productivity.

Novel magnetorheological fluids were developed based on anisotropic colloids – carbon nanotube–iron oxide nanoparticle nanocomposites [361, 362].

Magnetic nanocomposites can be interesting objects for investigating basic magnetic properties of spatially fixed interacting magnetic nanoparticles [363]. Various inorganic and organic matrixes were used for dispersing and incorporating magnetic nanoparticles. Some examples were described in Section 5.3.2, where synthesis of nanoparticles in porous nanostructured media was discussed. Zeolites and amorphous glasslike carbons were impregnated with magnetic iron oxide and Fe_3C nanoparticles using organometallic compounds as ferrocene and iron pentacarbonyl [364].

Anisotropically one-direction conductive composite material can be prepared via magnetic alignment of conductive magnetic particles in nonconductive matrix material such as elastomeric or adhesive polymers. Such composites can contain many field-aligned and laterally isolated chains of magnetic conducting particles [365, 366]. Polymeric nanocomposite material transparent in the visible region at room temperature was obtained via chemical synthesis of γ-Fe_2O_3 nanoparticles in an ion-exchange resin at 60 °C. It was found that ultrasmall γ-Fe_2O_3 nanoparticles were considerably more transparent to visible light than the corresponding bulk iron oxide. The magnetization of the composite was greater by more than an order of magnitude than that of well-known strong room-temperature transparent magnets FeF_3 and $FeBO_3$ [367].

Composite copolymer-magnetic nanoparticle (symmetric polystyrene–polybutylmethacrylate diblock copolymer with incorporated nanoparticles γ-Fe_2O_3 4 nm diameter) thin films were fabricated by spin coating and annealing at 150 °C [368].

Polymeric magnetic composite films were prepared via synthesis of magnetic metallic nanoparticles into porous polymer matrix by reduction of metal ions [369–372].

Nanomaterial with high conductivity and magnetizability was developed based on maghemite/polyaniline nanocomposite [373]. To prepare that material, maghemite nanoparticles (~10 nm in diameter) were coated with 4-dodecylbenzenesulfonic acid, and polyaniline chains were doped with 10-camphorsulfonic acid. The coated nanoparticles and doped polymers were soluble in common solvents, and casting the solutions readily gave freestanding nanocomposite films with nanocluster contents as high as ~50 wt%. The prepared nanocomposites showed high conductivity (82–237 S/cm) and magnetizability (up to ~35 emu/g).

Recent theoretical arguments [374–376] have suggested that synergistic interactions between self-organizing particles and a self-assembling matrix material can lead to hierarchically ordered structures. Mixtures of diblock copolymers and either cadmium selenide- or ferritin-based nanoparticles exhibit cooperative, coupled self-assembly on the nanoscale. In thin films, the copolymers assemble into cylindrical domains, which dictate the spatial distribution of the nanoparticles; segregation of the particles to the interfaces mediates interfacial interactions and orients the copolymer domains normal to the surface, even when one of the blocks is strongly attracted to the substrate. Organization of both the polymeric and particulate entities is thus achieved without the use of external fields [377] opening a simple and general route for fabrication of nanostructured materials with hierarchical order.

Biodegradable composite polymer-based magnetic gels have been synthesized using hydroxypropyl cellulose and maghemite [378]. These magnetic gels have a network of nanoparticles of hydroxypropyl cellulose (30–100 nm) and a homogeneous distribution of nanosized maghemite (7 nm). This has been observed in a STEM micrograph. The surface structure of the gels has been observed by atomic force microscopy, while transmission electron microscopy has shown the distribution of iron oxide in gel nanoparticles. These gels have marked magneto-elastic properties.

Anisotropic nanocomposite magnetic fiber-like structures and materials can be prepared from polymer solutions containing magnetic nanoparticles using electrospinning technique. More than 100 different polymers of natural, synthetic origin, their blends, and composites have been electrospun into different 3D macrostructures. Electrospinning provides opportunities to manipulate and control surface area, fiber diameter, porosity, and pore size of nanofiber matrices. For example, self-assembled magnetic FePt nanoparticles (4 nm diameter) were encapsulated in poly(ε-caprolactone) nanofibers by coaxial electrospinning [379]. The encapsulated array of the discrete FePt nanoparticles can reach as long as 3000 nm along the fiber axis. Assisted by electrostatic interactions, the FePt/PCL composite nanofibers were readily patterned forming a uniaxial alignment over a large area.

Superparamagnetic nanocomposite polymeric nanofibers were produced using electrospinning technique from colloidally stable suspensions of magnetite nanoparticles in polyethylene oxide and polyvinyl alcohol solutions. The magnetite nanoparticles were aligned in columns parallel to the fiber axis direction within the fiber. The polymer/magnetite nanofibers exhibited superparamagnetic behavior at room temperature, and deflected in the presence of an applied magnetic field. The mechanical properties of the nanofibers were maintained or improved after incorporating the magnetite nanoparticles [380].

Polycrystalline nanofibers of $NiFe_2O_4$ with an average diameter of 46 nm were prepared by electrospinning a solution that contained polyvinyl pyrrolidone and alkoxide precursors to nickel and iron oxides, followed by hydrolysis and calcination at 550 °C in air. Significant differences in magnetic properties were observed between these nanofibers and powder nickel ferrite prepared using the conventional sol–gel process, resulted, possibly, from size and morphological differences between the nanofibers and the powders. [381].

Magnetic composites based on silk fiber were prepared via colloid magnetite nanoparticles (10–20 nm diameter) adsorption onto fibrous templates incubated in water or water–methanol solution of magnetic nanoparticles [382].

Figure 5.17 illustrates the possibilities of applying originally prepared novel magnetic nanofilm material (see Section 5.2.4) in the magnetic separation technologies and in modification of fibers. Composite nanofilm material based on magnetite nanoparticles and spermin with bound colloid latex particles is shown in Figure 5.17(a). Composite structure – cotton fiber – with deposited nanofilm material is shown in Figure 5.17(b). Highly dense surface coverage of the fiber with magnetic nanoparticles is clearly seen from the figure.

5.5
Bioinorganic Magnetic Nanostructures

Biogenic magnetic iron-containing nanoparticles were found in many organisms including magnetotactic bacteria, algae, honeybee, migratory ant, termites, salmon, trout, some amphibians, bobolink, pigeon, dolphin, and human and represented iron-containing compounds, mainly oxides [383–385].

Formation of iron oxide nanophases accompanies many neurodegenerative diseases and disruption of iron homeostasis in the brain [384]. Self-organized 1D nanostructures (chains) of iron oxide (magnetite, Fe_3O_4, with admixture of γ-Fe_2O_3) and greigite (Fe_3S_4) magnetic nanoparticles are present in magnetotactic bacteria and play an important physiological role in their space orientation via interaction with geomagnetic field [383].

5.5 Bioinorganic Magnetic Nanostructures

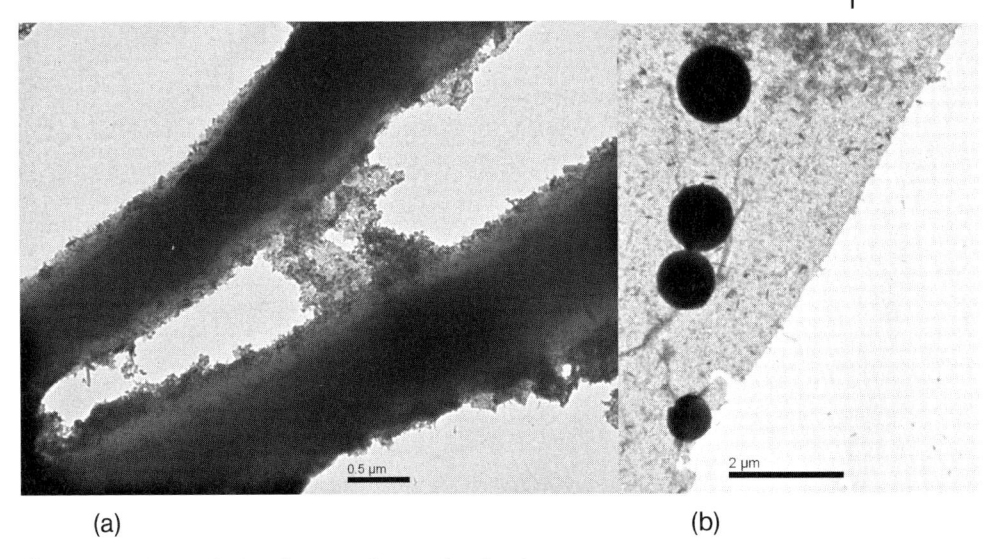

(a) (b)

Figure 5.17 Transmission electron micrographs showing magnetic composite nanofilm material based on spermine/magnetite nanoparticle complex with bound colloid latex particles (b) and deposited on natural cotton fiber (a) [46].

Biogenic and synthetic magnetic nanoparticles (usually superparamagnetic due to their small size) with appropriate surface modification and functionalization have been used in a number of biomedical applications [386–388]. Due to the biocompatibility and good magnetic properties, iron oxide nanoparticles currently dominate in biomedical applications. Among their current applications are magnetic resonance imaging contrast enhancement [389–391], magnetic detection of biomolecular interactions and bioassays [392], preparation of magnetic gels [393], biocompatible films [394], anticancer agents and targeting hyperthermia [395, 396], targetable drug delivery [397–400] and gene delivery (magnetic transfection) [401–403] with carrier localization in a specific area, concentration of trace amounts of specific targets and magnetic separation [404–406], analyses of DNA [407], cell labeling [408] concentration of trace amounts of specific targets, such as bacteria or leukocytes. Multifunctional magnetic Au/Ni nanowires were fabricated by electrochemical deposition in nanoporous templates and were exploited as a type of magnetic carrier that offered significant potential advantages over commercially available magnetic particles [409].

It was shown that magnetite nanoparticles loaded into sperm cells did not affect their motility and functionality (the ability to fertilize the egg) [410]. The redox processes in biological systems with participation of iron oxide nanoparticles are still not well understood, though antioxidant properties of magnetite nanoparticles were reported [411]. Most of these promising applications require iron oxide nanoparticles to possess good chemical stability, narrow particle-size distribution, and uniform morphology.

Magnetic iron oxide nanoparticles for biomedical applications can be prepared and functionalized by a number of chemical techniques [9, 11–15]. Fabrication and study of organized nanostructures with integrated biological and synthetic components can be important for nanotechnological, medical, biotechnological, and many other applications. It can result in the development of new advanced materials, films and biocompatible coatings with high reproducibility, functional specificity, and effectiveness characteristic for biological systems along with functionality, multifunctionality, enhanced processability, and applicability due to the synthetic polymeric and/or inorganic components. Unfortunately, many biomolecules and biomolecular structures are fairly nonrobust, and the engineered integrated nanosystems with immobilized and stabilized bionanocomponents can allow them to be used in real practical applications.

Organized composite magnetic bionanoreactor with LbL-assembled multilayer shells of glucose oxidase, oppositely charged polyelectrolytes, and inorganic nanoparticles (9 nm, 20 nm, and 45 nm diameter silica or 12 nm magnetite) on 420-nm latex particles was described in [360]. The role of magnetic nanoparticles in that construction was to allow self-stirring of the nanoreactors with a rotating applied magnetic field resulting in enhancing its productivity.

Figure 5.18 illustrates the iron oxide nanoparticles generated in biomineralization processes connected with the formation of magnetic iron oxide nanoparticles in DNA complexes in the presence of only reagents of biological nature – ferritin as an iron source, ascorbic acid as a reductant, at physiological pH value (7.5) and ambient conditions [285]. The native double-stranded DNA molecules were immobilized via binding with water-insoluble amphiphilic intercalating dye N, N'-Dioctadecyloxacarbocyanine-4-toluenesulfonate Langmuir monolayer followed by monolayer complex deposition onto the substrate. Also, immobilized DNA/spermine complexes were formed by incubation of the substrate with immobilized DNA/amphiphilic dye layer in the aqueous spermine solution. The obtained immobilized DNA complex samples were first incubated for various time intervals in the aqueous ferritin suspension, and after washing in pure water were placed in the reductant ascorbic acid solution at ambient conditions (with access of air oxygen). TEM investigation of the samples before the placement in ascorbic acid solution did not reveal any considerable quantities of bound unwashed ferritin molecules or any inorganic nanoparticles.

It follows from Figures 5.18(a) and (b) that the described synthetic procedures resulted in generation of iron oxide inorganic nanoparticles characterized by wide size dispersion (numerical analyses give dimensions from 1 to 15 nm with mean diameter about 3 nm) and apparently developed crystalline structure. The typical corresponding electron diffraction picture obtained with those samples is shown as an insert on Figure 5.18. The character of

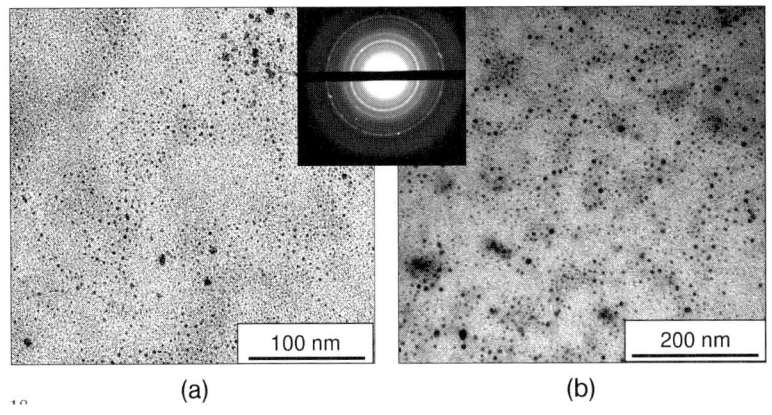

Figure 5.18 Transmission electron micrographs of iron oxide nanoparticles synthesized in immobilized DNA complexes. DNA was immobilized via binding with Langmuir monolayer of amphiphilic intercalating dye N, N′-Dioctadecyloxacarbocyanine-4-toluenesulfonate followed by monolayer deposition onto TEM substrate. Image (a): Iron oxide nanoparticles synthesized via incubation of deposited amphiphilic dye monolayer with bound DNA molecules in ferritin suspension for 1 h followed by incubation in ascorbic acid solution (1 h). Image (b): Iron oxide nanoparticles synthesized via incubation of deposited amphiphilic dye monolayer with bound DNA molecules in spermine solution (1 h) followed by incubation in ferritin suspension for 1 h and then in ascorbic acid solution (1 h), pH = 7.5. The insert in images (a) and (b) shows typical selected area electron diffraction picture obtained in those samples corresponding to Fe_3O_4 nanophase [285].

diffraction is typical for polycrystalline systems of iron oxide (particularly, magnetite) nanoparticles and points to the pronounced crystalline structure of the generated nanoparticles in contrast to the ferritin. In Figure 5.18(a), one can see organized quasilinear chain arrays of small (∼2 nm) nanoparticles which can be a result of attachment of nanoparticles to the linear DNA molecules. Nanoparticles in DNA/spermine complex sample (Figure 5.18(b)) form rather dense aggregates that can be a result of more compact structure of condensed DNA/spermine complexes. Substantial difference in size and structure between ferritin and generated nanoparticles along with their linear arrangement in the DNA complex sample (Figure 5.18(a)) can be a result of formation of observed iron oxide nanoparticles from iron ions transferred from ferritin to linear DNA molecules during incubation of immobilized DNA complexes in ferritin solution. The surface density of synthesized nanoparticles in DNA/spermine complex (Figure 5.18(b)) is substantially higher than that in only DNA sample (Figure 5.18(a)). That result points to the higher Fe cations binding capacity of DNA/spermine complexes in comparison with DNA molecules alone due to the presence of polyamine in the sample. The data obtained give evidence for the possibility to form magnetic iron oxide nanoparticles in DNA and DNA/polyamine complexes in biological systems under appropriate conditions providing source of Fe ions and redox processes. That finding can be related to possible natural biogenic process involving iron metabolism in living organisms.

5.6
Magnetic Properties of Organized Ensembles of Magnetic Nanoparticles

Assembling nanoparticles generates new organized nanostructures with unforeseen collective, intrinsic physical properties. These properties can be exploited for multipurpose applications in information storage and processing, nanoelectronics, spintronics, sensors, and various functional hybrid materials. Magnetic nanostructures, including organized ensembles of magnetic nanoparticles, are particularly interesting class of materials for scientific and technological explorations. Various exciting phenomena such as giant, colossal and tunneling magnetoresistance, interlayer coupling and exchange bias, spin injection, magnetic vortices, etc., have eventually led to the possibility of utilizing electron spin for information processing or spintronics [412, 413]. As a rule, thin film and multilayer systems are considered as the most promising materials for spintronics and, therefore, investigated for prototypical device implementation. However, nanoscale particulate composites are also of great interest; they can cover a broad diversity of materials from biomaterials [414] to superspin glasses [415]. Systems of strong interacting nanoparticles [416] and nanogranular films [417] serve as model systems for the study of ageing, rejuvenation, and memory phenomena.

Highly ordered arrangements of magnetic nanoparticles and organized nanoparticulate structures are interesting systems for study and understanding fascinating effects generated by the coexistence of various driven forces: geometric confinement, structural order, particle–particle and matrix–particle interactions, etc., [2, 418].

In studying ordered arrays of magnetic nanoparticles, it has been found that the most important factor in their magnetism is interparticle long-range dipolar interaction, which decays as $1/r^3$ (neglecting possible dielectric screening), where r is average distance between neighboring particles, that is more slowly in comparison with electrostatic forces. Many new phenomena have been discovered that are essentially related to dipolar interaction effects in ordered nanoparticle assembles, such as sufficient changing of the blocking temperature, remanent magnetization, magnetic coercivity, magnetic anisotropy, etc., [419, 420]. Moreover, the dipolar interaction has been identified as the force of creating and controlling self-organized magnetic nanoparticle systems [25].

Dipolar interactions promote a formation of flux-closed (ring-like) structures in which the magnetic field created by each magnetic particle biases its neighbors. Magnetic ring-like structures are very interesting for magnetic recording purposes since such configuration forces magnetization to be

5.6 Magnetic Properties of Organized Ensembles of Magnetic Nanoparticles

circular, enabling a stable flux closure mode [421]. The ring-like nanostructures can be successfully fabricated by numerous lithography methods [184, 422] and self-assembled processes [423]. In ferrofluids, magnetic particles form various structures, such as separate and branched chains, circles and other ring-like structures, clusters and fractal-like complexes [424–426]. It was shown [427] that in magnetically dilute ferrofluids dipole–dipole interactions between magnetic particles promote a formation of flexible ring-like nanostructures. In weak magnetic fields, these ring-like structures are stable and orient perpendicular to the field. In strong magnetic fields, the rings are destroyed and the particles form linear chains [427].

The use of biological molecules (like DNA) and their molecular-recognition properties to guide the assembly of nanoparticle structures on a solid surface looks attractive [285, 428, 429]. In this case, particles should form essentially rigid structure so that the distances between them are invariant. As it is shown in [430], magnetic properties of such rigid ring-like structures depend strongly on the dipole–dipole and Zeeman magnetostatic interactions, and can be sufficiently modified by the single-particle magnetic anisotropy.

The magnetic properties of ring-like planar nanoparticles were studied theoretically and experimentally in a number of works [422, 423, 430–438]. In nanosized magnetic elements (continuous circular or elliptical metallic dots), magnetic vortex structures appear from a competition between exchange and magnetostatic energies, while anisotropy energy is usually considered to be negligible. In the present work, different ring-like nanoparticle structures are studied. Dipole–dipole (magnetostatic), Zeeman, and anisotropy interactions are the most important in these structures, whereas the exchange energy does not play any significant role. Similar to ring-like continuous nanostructures, nanoparticle structures, studied in our work, show various spin configurations – flux-closed vortexes, "onion"-like, and more complex ones.

In the work [430], numerical calculations of equilibrium state energies and local magnetic fields in planar ring-like nanoparticle structures were performed. The dipole–dipole, Zeeman, and magnetic anisotropy interactions were included into the model. The result of their competition depends on the value of the external magnetic field, magnetic parameters of an individual nanoparticle, size, and shape of the structures. Flux-closed vortexes, single domain, two-domain "onion"-like, "hedgehog"-like, and more complex spin structures can be realized. The critical field, providing a sharp transition from the flux-closed vortex to the "onion"-like state, can be regulated by a variation of the particle magnetization and anisotropy constant, their easy directions, and particle space arrangement.

Let us consider a system of N spherical equivalent magnetic particles (with magnetic moments $\mathbf{m}_i$ and a diameter d) located in vertexes of a regular polygon with a distance R between the polygon center and each particle (Figure 5.19). The total energy W of classical interactions is the sum of Zeeman W_Z,

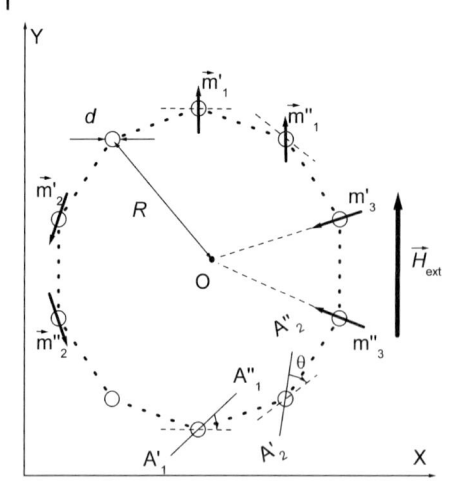

Figure 5.19 Schematic illustration of arrangement of magnetic nanoparticles in vertexes of a regular planar polygon, where R is the distance between the polygon center (O) and its vertex, d is the particle diameter. Easy directions (solid line segments A'_1-A''_1, A'_2-A''_2 as an example) are determined by the angle θ, dotted lines denote the tangents to the circumcircle in the vertexes. Thick vertical arrow shows the direction of the external magnetic field H_{ext}. Pairs of arrows (m'_1; m''_1), (m'_2; m''_2), (m'_3; m''_3) show directions of neighboring magnetic moments in two initial configurations, which were used in the energy minimization procedures [430].

dipole–dipole W_{DD}, and single-particle magnetic anisotropy W_{anis} terms:

$$W_Z = -\sum_{i=1}^{N} \vec{m}_i \vec{H}; \tag{5.7}$$

$$W_{DD} = \sum_{i=1}^{N}\sum_{j=i+1}^{N} \frac{\vec{m}_i \vec{m}_j r_{ij}^2 - 3(\vec{m}_i \vec{r}_{ij})(\vec{m}_j \vec{r}_{ij})}{r_{ij}^5}; \tag{5.8}$$

$$W_{anis} = -K_{anis}\sum_{i=1}^{N} \cos^2(\varphi_i), \tag{5.9}$$

where $\vec{m}_i$ is the in-plane magnetic moment of the ith particle, $\vec{H}$ is the in-plane external magnetic field, $\vec{r}_{ij}$ are differences between radius-vectors of ith and jth particles, K_{anis} is the parameter of the single-particle axial magnetic anisotropy, and $\cos(\varphi_i) = \dfrac{(\vec{m}_i \vec{l}_i)}{|\vec{m}_i| \cdot |\vec{l}_i|}$ is the cosine of the angle between the magnetic moment of ith particle and its easy-axis direction $\vec{l}_i$. In Figure 6.1, the examples of easy directions are shown by segments $B'_1 B''_1$ and $B'_2 B'_2$, which inclined the angle θ with respect to tangents to the incircle. Assuming the values of $|\vec{H}|$, $|K_{anis}|$, $|\vec{r}_{ij}|$, and θ are fixed, the total energy W can be regarded as a function of N parameters, for example, angles α_i between the vectors $\vec{m}_i$ and the X-axis. Using a so-called dynamical version [439] of the conjugate graduate method [440], the minimization procedures were performed for this function with different N.

In order to escape the local minima at the start of calculations, the following various initial spin configurations were used, including the following symmetrical spin positions: (1) parallel spins (see the orientation of m'_1 and m''_1 vectors in Figure 5.19); (2) tangent spins (see the orientation of m'_2 and m''_2 vectors in Figure 5.19), and (3) radial spins (see the orientation of m'_3 and m''_3 vectors in Figure 5.19). Random initial spin positions were also used.

Besides the ideal circle structures, oblate and oblong ellipses as well as circle-like structures without some particles (defects structures) were investigated [430]. In that way, a semicircle and arcs of different length were considered with the same R and d values.

Using the classical formula

$$\vec{H}_{i,\text{DD}} = -\sum_{j=i+1}^{N} \frac{\vec{m}_j r_{ij}^2 - 3\vec{r}_{ij}(\vec{m}_j \vec{r}_{ij})}{r_{ij}^5}, \tag{5.10}$$

dipole–dipole local magnetic fields acting on the nanoparticles and the local field in each point of the system can be easily calculated.

In the calculations, magnetic parameters for magnetite (Fe_3O_4) were used: saturation magnetization $M_s = 480$ emu/cm^3, and the volume anisotropy constant $|K_{\text{anis}}| \approx 10^5$ erg/cm^3 [441]. Magnetite is often used in biomedical nanoparticle applications due to its robustness, low toxicity, and relatively high magnetization. Other parameters for the model were chosen as follows: $d = 2$ nm, $R = 10$ nm. In this case, the maximum number of particles should be equal to $\sim 2\pi R/d \approx 30$.

Magnetic exchange interactions between neighboring particles were not included in the total system assuming that nanoparticles are separated. For real nanoparticles, which are in direct contact, the possible interparticle short-range forces are very complicated and cannot be reduced to a simple isotropic two-spin exchange term.

Spin configurations in circular nanoparticle structures can be rather different. Some of the configurations are shown in Figure 5.20.

Let us consider the case $K_{\text{anis}} = 0$. If $H_{\text{ext}} = 0$, the flux-closed vortex structure has a minimal energy (Figure 5.20(a)). In this case, we get $W_Z \approx 0$ and $W \approx W_{\text{DD}}$. With H_{ext} increased, the spins smoothly turn to the external field direction, but while $H_{\text{ext}} < H_{\text{cr}}$ they remain approximately directed along tangents to the circumcircle and the flux-closed structure is not disturbed significantly. At $H_{\text{ext}} = H_{\text{cr}}$, the vortex structure drastically transforms to "onion"-like structure, which is characterized by the existence of two "domains" separated by a "domain" wall (Figure 5.20(b)). The type of a preferable spin structure is determined by the energy gain, which the system obtains after a transition from one spin configuration to another. If H_{ext}

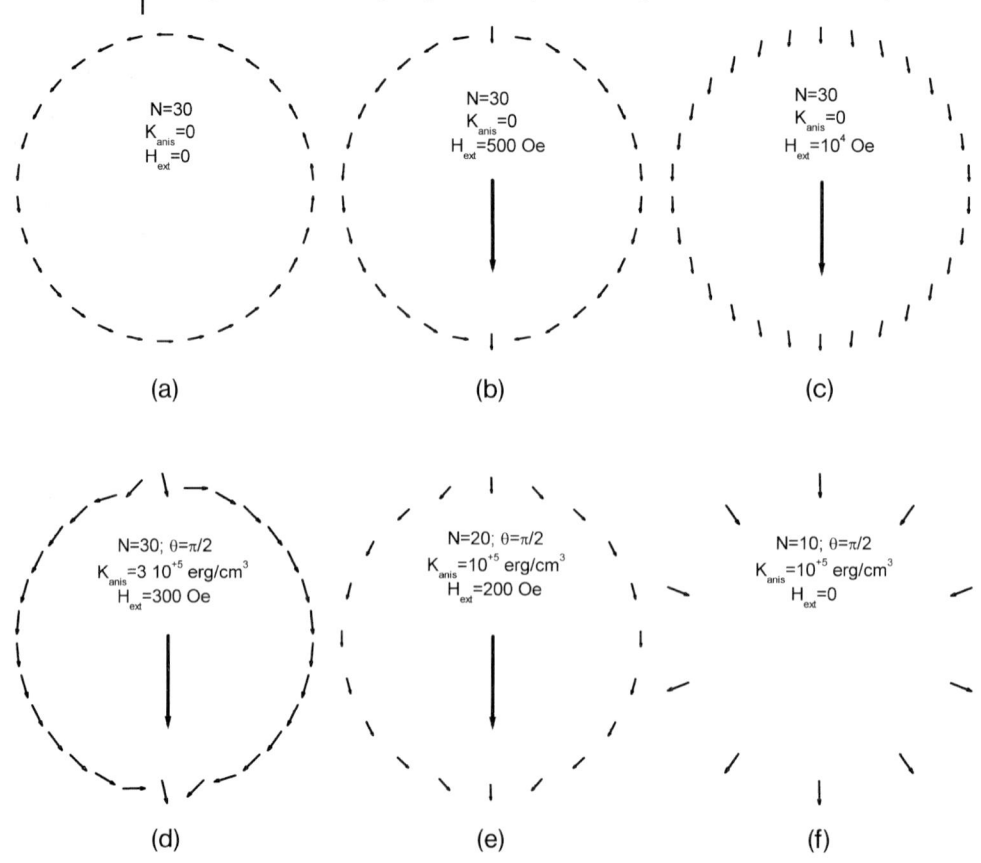

Figure 5.20 Examples of different spin configurations in the circle nanoparticle structure: (a) flux-closed vortex state; (b) two-domain "onion"-like state; (c) intermediate state between homogeneous "single-domain" state and "onion"-like state; (d) and (e) "onion"-like states with extensive "domain boundary"; (f) "hedgehog"-like state. Model parameters N, H_{ext}, K_{anis}, and θ are shown in the figure [430].

increases more, the "onion"-like structure gradually transforms into a single-domain structure, in which all spins are parallel to the external field. An intermediate situation between the uniform single-domain state and the two-domain "onion"-like state is presented in Figure 5.20(c).

More complex spin behavior takes place in case of nonzero magnetic anisotropy. If the value of the angle θ (Figure 5.19) is close to zero, the anisotropy energy stabilizes the flux-closed vortex state (Figure 5.20(a)). A qualitative transformation of the spin structure can be observed for $\theta \approx \pi/2$. Firstly, the "wall" between "onion" domains can become more extended (Figures 5.20(d) and (e)). Secondly, if N is relatively small and the maximum value of $W_{anis,max}$ (which is realized for $\theta \approx \pi/2$) is comparable with the maximum value of $W_{DD,max}$ (which is realized for tangent to the circumcircle spins), the ground-state structure is not flux-closed vortex, but "hedgehog"-like (Figure 5.20(f)) even in case of zero external magnetic field. With increasing $W_{DD,max}/W_{anis,max}$ ratio, the zero-field ground-state structure will become a flux-closed vortex, irrespective of θ value. This happens, e.g., for $N = 20$ and 30, $H_{ext} = 0$, and $K_{anis} \leq 10^5$ erg/cm^3.

The transition from the flux-closed vortex to the "onion"-like state is accompanied by drastic changes in various parameters (Figure 5.21). The Zeeman and dipole–dipole energies undergo sharp transformations at H_{cr}, while the total energy has a kink at this point. Average magnetic field, acting

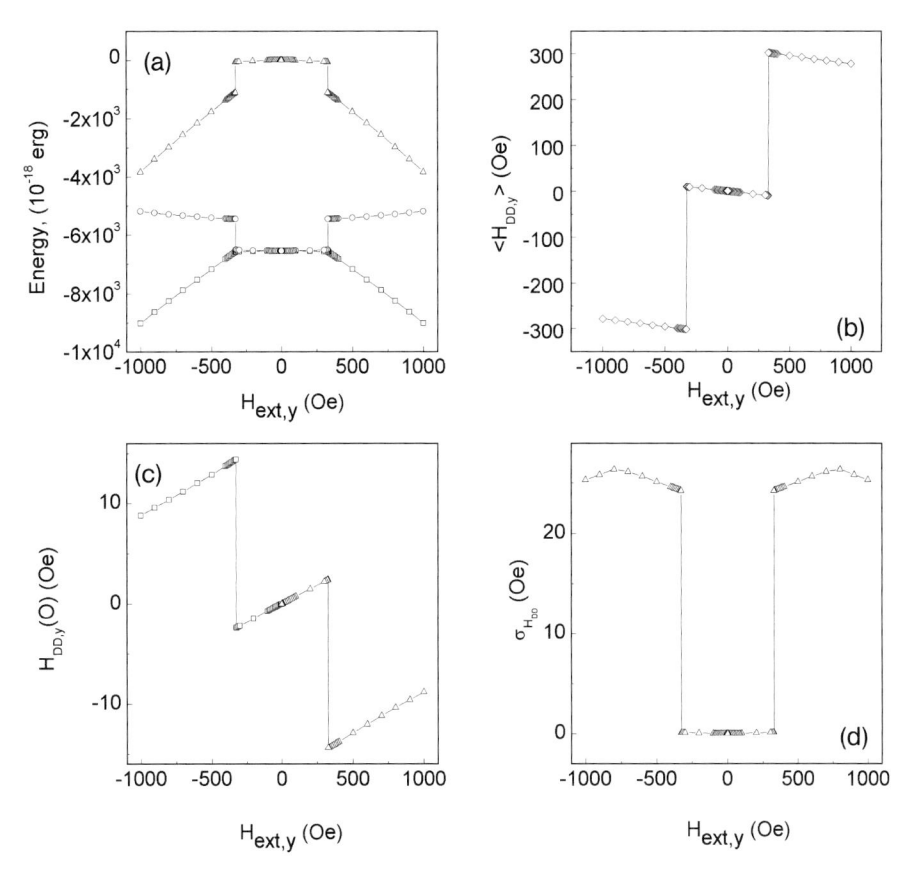

Figure 5.21 The energy (per particle) and local magnetic field as a function of the external magnetic field for the circle nanoparticle structure with $N = 30$, $K_{anis} = 0$ (see Figures 5.20(a)– 5.(c)). (a) squares – W, circles – W_{DD}, triangles – W_Z; (b) average dipole–dipole magnetic field (Y-projection) acting on a particle; (c) dipole–dipole magnetic field (Y-projection) in the center of the polygon; (d) dispersion of the absolute value of the dipole–dipole magnetic field acting on a particle.

on a nanoparticle due to magnetic dipole moments of other particles, and its dispersion are changed abruptly at H_{cr}. The same behavior is typical for the local dipole–dipole magnetic field in the circle center. It is important that the local field dispersion is very small in the flux-closed vortex state, that is, all particles are in nearly the same conditions. This is the feature of the flux-closed state.

The critical field H_{cr} depends on the averaged interparticle distance. In the case of ideal circle structures, the value of H_{cr} increases linearly with averaged $1/r_{ij}^3$ (Figure 5.22(a)). However, for more complex structures, e.g., the circles with a "gate" (Figure 5.22(b)), the critical field changes nonmonotonously with $1/r_{ij}^3$. The "gap" between spins breaks the dipole–dipole interaction chain, typical of the flux-closed vortex state, and makes this state less stable in the external magnetic field (left part of the curve in Figure 5.22(b)). But, H_{cr} grows with increasing $\langle 1/r_{ij}^3 \rangle$ again when the "gate" in the circle becomes sufficiently large (Figure 5.22(b)). It should be noted that the spin configurations of the "defective" circles (Figure 5.22(b)) can be in some way different from those presented in Figure 5.22. However, the critical field can be easily determined by observing the sharp changes in local magnetic fields.

The transformation of the ideal circle into ellipses results in a significant reduction of the critical field (Figure 5.22(c)). For the ellipse stretched along the external magnetic field axes, this reduction is clearly less pronounced than for the ellipse elongated perpendicular to H_{ext}. The reduction of H_{cr} can be explained by a significant increase in the average interparticle distance (compare X-axis scales in Figures 5.22(a) and (c)). Furthermore, the flux-closed vortex configuration should be more stable for the oblong ellipse than for the oblate one, since in the former case the particles with parallel magnetic moments are situated closer and, therefore, effectively magnetize each other.

It is interesting to compare the dipole–dipole energy with a thermal energy $W_T \cong \kappa_B T \approx 5 \times 10^{-14}$ erg ($T = 300$ K). In case of $d = 2$ nm and $R = 10$ nm, the magnetic moment m of the magnetite nanoparticle is equal to ~ 200 μ_B and a maximal value of W_{DD} is estimated as $\sim N \cdot m^2/d^3 \approx 5 \times 10^{-15}$ erg ($N = 30$). Since $m \sim d^6$, then we get $W_{DD} \sim d^3$, assuming N to be a constant. In the case of nanoparticles with $d = 4$ nm, the dipole–dipole and thermal energies will be nearly equal. Therefore, magnetic vortex-like structures, considered in this work, will be stable at room temperature in spite of superparamagnetic fluctuations if $d > 5$–6 nm.

Results of the work [430] showed that without an external magnetic field, when only the dipole–dipole and magnetic anisotropy energies compete, the flux-closed vortex and "hedgehog"-like configurations appear. The former is stable if the anisotropy energy is small or the particles magnetization easy directions are tangent to the circumcircle. The "hedgehog"-like spin configuration is stable if the magnetic anisotropy, is sufficiently large and the particles magnetization easy directions are perpendicular to the circumcircle tangents.

5.6 Magnetic Properties of Organized Ensembles of Magnetic Nanoparticles | 181

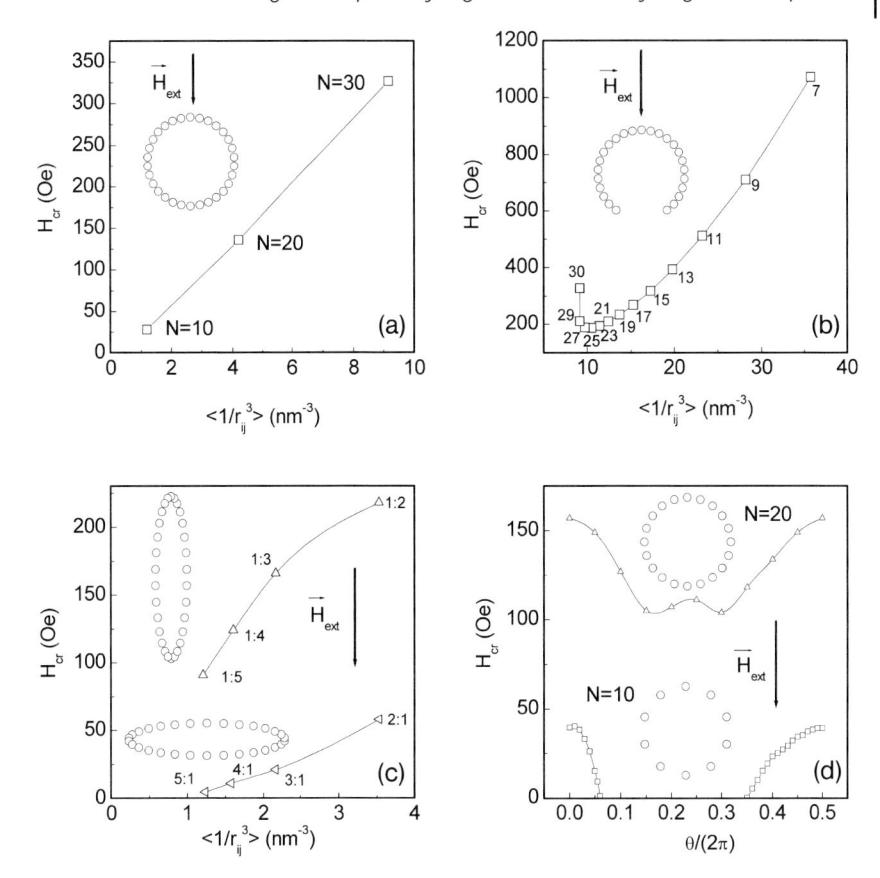

Figure 5.22 Variation of the critical magnetic field H_{cr} depending on the interparticle distances (a)–(c), and the easy direction (d). In cases (a)–(c) H_{cr} is plotted as a function of the average value of $1/r_{ij}^3$, where r_{ij} is the interpaticle distance. The value of r_{ij} was changed (a) by the variation of the particle number ($N = 10; 20; 30$) in the ideal circle structure; (b) by making a symmetrical "gap" in the circle structure, removing some particles, but keeping the distance between neighbor particles constant; the figures in (b) show the number of residuary particles; (c) by distorting of the structure shape from circular to elliptical one, the corresponding ratios of horizontal and vertical semi-axes lengths are indicated in (c), the minimal axis length is equal to R. In the case (d) ideal circle structures with $N = 10, 20$, and 30 are considered, $K_{anis} = 10^5$ erg/cm^3, the meaning of θ is clear from Figure 5.19.

Without single-particle anisotropy, we found three basic spin configurations. These depend on external magnetic field strength and intensity of dipole–dipole interactions. If the external field is high enough, a "single-domain" uniform structure is realized. In this case, magnetic moments are parallel to external magnetic field. With external magnetic field decrease, a two-domain "onion"-like structure gradually appears without sharp changes of magnetic properties. There is a critical value for the magnetic field

(H_{cr}) at which the sharp transition from "onion"-like to vortex configuration occurs.

These studies demonstrated that ring-like nanoparticle structures have rather intricate magnetic properties. Even more interesting can be spherical nanoparticle structures [442].

5.7
Conclusions and Perspectives

In recent years, the developments in the magnetic nanoparticle synthesis methods have shown wide possibilities for investigation and exploiting their unique properties. The assembly of nanoparticles and the formation of organized nanoparticulate nanostructures have shown the potential of numerous applications. Magnetic nanoparticles can be assembled in different dimensions and arrays using a variety of techniques as described in this chapter. Those techniques are usually nonspecific for magnetic nanoparticles (excepting specific magnetic interactions) and can be used for building-up of nanoparticulate structures of various nature. Future efforts will be directed to the building-up of highly ordered defect-free and reproducible nanoparticle assemblies and nanostructures, which is necessary for the development of reliable devices. Different methods of nanoparticle assembly offer different advantages and disadvantages. Relatively simple methods based on evaporation-induced assembly can produce large structures but there is little scope for controlling the interparticle distances or structure of assemblies. Assembly methods based on chemical bonding and conjugations allow tailoring surface properties of the nanoparticles; however, this generally involves careful and complicated tailoring of ligand composition and surface functionalities on nanoparticles, substrates, or template surfaces making these methods more complicated and chemically sophisticated.

Future perspectives in developments of nanoparticulate structures are also connected with creating complex, multicomponent, hybrid, integrated, and, as a result multifunctional nanomaterials and nanosystems. The resulting properties of such nanosystems are substantially dependent on the nanoscale organization of structural and functional nanocomponents and their collective behavior. It shows possibilities for fine tuning and controlling the properties of integrated nanosystems, which is important for applications. To realize these possibilities, the insights into basic mechanisms that govern the nanoscale processes of inorganic phase growth and morphological evolution, interparticle interactions, and structural organization of nanoparticulate systems and nanostructures are necessary. The investigation of these mechanisms and development of novel organized functional and polyfunctional nanomaterials will be the subject of future research.

The described synthetic strategies and methods can be useful for investigation of fundamental mechanisms of nanoscale structure formation, organization, and transformations in complex nanosystems. The methods are relatively simple, rapid, inexpensive, and allow large-scale preparation of organized nanostructures at ambient and ecologically friendly conditions. It makes them promising practical instruments for molecular *nanotechnology* with potential for nanobiotechnological and biomedical applications.

Acknowledgments

This work was supported by Russian Foundation for Basic Research (Grant 08-03-01081).

References

1. *Nanocrystals Forming Mesoscopic Structures*, M.P. Pileni Ed.; Wiley-VCH, Weinheim, **2005**.
2. F. Dumestre, S. Martinez, D. Zitoun, M.-C. Fromen, M.-J. Casanove, P. Lecante, M. Respaud, A. Serres, R.E. Benfield, C. Amiens, B. Chaudret, *Faraday Discuss.*, **2004**, *125*, 265.
3. Y.-W. Jun, J.-S Choi, J. Cheon, *Chem. Commun.*, **2007**, 1203.
4. S. Kinge, M. Crego-Calama, D.N. Reinhoudt, *Chem. Phys. Chem.*, **2008**, *9*, 20.
5. D.L. Leslie-Pelecky, R.D. Rieke, *Chem. Mater.*, **1996**, *8*, 1770.
6. X. Batlle, A. Labarta, *J. Phys. D*, **2002**, *35*, R15.
7. J.L. Dormann, D. Fiorani (Eds.), *Magnetic Properties of FineParticles*, Elsevier, Amsterdam, **1992**.
8. R.H. Kodama, *J. Magn. Magn. Mater.*, **1999**, *200*, 359.
9. S.P. Gubin, Yu.A. Koksharov, G.B. Khomutov, G.Yu. Yurkov, *Russ. Chem. Rev.*, **2005**, *74*, 539.
10. M. Giersig, M. Hilgendorff, *Eur. J. Inorg. Chem.* **2005**, 3571.
11. R.M. Cornell, U. Schwertmann, *Iron Oxides: Structure, Properties, Reactions, Occurrences and Uses*, Willey-VCH, Weinheim, Germany, **2003**.
12. A.K. Gupta, M. Gupta, *Biomaterials*, **2005**, *26*, 3995.
13. D.K. Kim, Y. Zhang, W. Voit, K.V. Rao, J. Kehr, B. Bjelke, M. Muhammed, *Scripta Materialia*, **2001**, *44*, 1713.
14. Y. Zhang, N. Kohler, M. Zhang, *Biomaterials*, **2002**, *23*, 1553.
15. C.C. Berry, A.S.G. Curtis, *J. Phys. D*, **2003**, *36*, R198.
16. M.A. Willard, L.K. Kurihara, E.E. Carpenter, S. Calvin, V.G. Harris, *Int. Mater. Rev.*, **2004**, *49*, 125.
17. U. Jeong, X. Teng, Y. Wang, H. Yang, Y. Xia, *Adv. Mater.*, **2007**, *19*, 33.
18. T. Hyeon, *Chem. Commun.*, **2003**, 927.
19. S.-J. Park, S. Kim, S. Lee, Z.G. Khim, K. Char, T. Hyeon, *J. Am. Chem. Soc.*, **2000**, *122*, 8581.
20. M. Vazquez, C. Luna, M.P. Morales, R. Sanz, C.J. Serna, C. Mijangos, *Phys. B: Conden. Matter*, **2004**, *354*, 71.
21. K. Raj, R. Moskowitz, *J. Magn. Magn. Mater.*, **1990**, *85*, 233.
22. B.M. Berkovsky, V.F. Medvedev, M.S. Krokov, *Magnetic Fluids: Engineering Applications*, Oxford University Press, Oxford, **1993**.
23. G. Bossis, S. Lacis, A. Meunier, O. Volkova, *JMMM*, **2002**, *252*, 224.
24. C. Kormann, H.M. Laun, H.J. Richter, *Int. J. Mod. Phys. B*, **1996**, *10*, 3167.

25. V. Cabuil, *Curr. Opin. Colloid Interface Sci.*, **2000**, *5*, 44.
26. V. Salgueiriño-Maceira, L.M. Liz-Marza'n, M. Farle, *Langmuir*, **2004**, *20*, 6946.
27. J. Shen, J. Kirschner, *Surf. Sci.*, **2002**, *500*, 300.
28. D.K. Lee, Y.H. Kim, C.W. Kim, H.G. Cha, Y.S. Kang, *J. Phys. Chem. B*, **2007**, *111*, 9288.
29. J.F. Smyth, S. Schultz, D.R. Fredkin, D.P. Kern, S.A. Rishton, H. Schmid, M. Cali, T.R. Koehler, *J. Appl. Phys.*, **1991**, *69*, 5262.
30. C. Shearwood, S.J. Blundell, M.J. Baird, J.A.C. Bland, M. Gester, H. Ahmed, H.P. Hughes, *J. Appl. Phys.*, **1994**, *75*, 5249.
31. B.D. Terris, D. Weller, L. Folks, J.E.E. Baglin, A.J. Kellock, H. Rothuizen, P. Vettiger, *J. Appl. Phys.*, **2000**, *87*, 7004.
32. Y.-Z. Huang, D.J.H. Cockayne, J. Ana-Vanessa, R.P. Cowburn, S.-G. Wang, R.C.C. Ward, *Nanotechnology*, **2008**, *19*, 015303.
33. L.J. Heyderman, H.H. Solak, C. David, D. Atkinson, R.P. Cowburn, F. Nolting, *Appl. Phys. Lett.*, **2004**, *85*, 4989.
34. (a) N. Singh, S. Goolaup, A.O. Adeyeye, *Nanotechnology*, **2004**, *15*, 1539;(b) Y. Luo, V. Misra, *Nanotechnology*, **2006**, *17*, 4909.
35. J.I. Martín, J. Nogués, K. Liu, J.L. Vicent, I.K. Schuller, *J. Magn. Magn. Mater.*, **2003**, *256*, 449.
36. T. Fried, G. Shemer, G. Markovich, *Adv. Mater.*, **2001**, *13*, 1158.
37. H. Kodama, S. Momose, N. Ihara, T. Uzumaki, A. Tanaka, *Appl. Phys. Lett.*, **2003**, *83*, 5253.
38. L. Chitu, Y. Chushkin, S. Luby, E. Majkova, A. Satka, J. Ivan, L. Smrcok, A. Buchal, M. Giersig, M. Hilgendorff, *Mater. Sci. Eng. C*, **2007**, *27*, 23.
39. X. Zhang, Y. Cao, S. Kan, Y. Chen, J. Tang, H. Jin, Y. Bai, L. Xiao, T. Li, B. Li, *Thin Solid Films*, **1998**, *327–329*, 568.
40. K. Baba, F. Takase, M. Miyagi, *Opt. Comm.*, **1997**, *139*, 35.
41. R. Shenhar, T.B. Norsten, V.M. Rotello, *Adv. Mater.*, **2005**, *17*, 657.
42. S. Sun, *Adv. Mater.*, **2006**, *18*, 393.
43. M. Nagtegaal, P. Stroeve, W. Tremel, *Thin Solid Films*, **1998**, *327–329*, 571.
44. X.K. Zhao, P.J. Herve, J.H. Fendler, *J. Phys. Chem.*, **1989**, *93*, 908.
45. H. Zeng, R. Skomski, L. Menon, Y. Liu, S. Bandyopadhyay, D.J. Sellmyer, *Phys. Rev. B*, **2002**, *65*, 134426.
46. G.B. Khomutov, Yu.A. Koksharov, *Adv. Colloid Interface Sci.*, **2006**, *122*, 119.
47. M. Shimomura, T. Sawadaishi, *Curr. Opin. Colloid Interface Sci.*, **2001**, *6*, 11.
48. V. Palermo, P. Samori, *Angew. Chem. Int. Ed.*, **2007**, *46*, 4428.
49. Z. Lin, S.J. Granick, *J. Am. Chem. Soc.*, **2005**, *127*, 2816.
50. T.P. Bigioni, X.-M. Lin, T.T. Nguyen, E.I. Corwin, T.A. Witten, H.M. Jaeger, *Nature Mater.*, **2006**, *5*, 265.
51. S. Sun, C.B. Murray, *J. Appl. Phys.*, **1999**, *85*, 4325.
52. M.P. Pileni, *J. Phys. Chem. B*, **2001**, *105*, 3358.
53. S. Sun, C.B. Murray, D. Weller, L. Folks, A. Moser, *Science*, **2000**, *287*, 1989.
54. C. Petit, A. Taleb, M.-P. Pileni, *Adv. Mater.*, **1998**, *10*, 259.
55. J. Legrand, C. Petit, D. Bazin, M.P. Pileni, *Appl. Surf. Sci.*, **2000**, *164*, 186.
56. J. Legrand, A.-T. Ngo, C. Petit, M.-P. Pileni, *Adv. Mater.*, **2001**, *13*, 58.
57. V. Russier, C. Petit, J. Legrand, M.P. Pileni, *Phys. Rev. B*, **2000**, *62*, 3910.
58. M.P. Pileni, Y. Lalatonne, D. Ingert, I. Lisiecki, A. Courty, *Faraday Discussions*, **2004**, *125*, 251.
59. V.F. Puntes, K.M. Krishnan, P. Alivisatos, *Appl. Phys. Lett.*, **2001**, *78*, 2187.
60. V.F. Puntes, P. Gorostiza, D.M. Aruguete, N.G. Bastus, A.P. Alivisatos, *Nature Mater.*, **2004**, *3*, 263.
61. X.A. Teng, H. Yang, *J. Mater. Chem.*, **2004**, *14*, 774.
62. M. Chen, D.E. Nikles, *Nano Lett.*, **2002**, *2*, 211.

63. D.M. Lawson, P.J. Artymiuk, S.J. Yewdall, J.M.A. Smith, J.C. Livingstone, A. Treffry, A. Luzzago, S. Levi, P. Arosio, G. Cesareni, C.D. Thomas, W.V. Shaw, P.M. Harrison, *Nature*, **1991**, *349*, 541.
64. N.D. Chasteen, *J. Struct. Biol.*, **1999**, *126*, 182.
65. E.C. Theil, M. Matzapetakis, X. Liu, *J. Biol. Inorg. Chem.*, **2006**, *11*, 803.
66. I. Langmuir, *J. Am. Chem. Soc.*, **1917**, *39*, 1848.
67. K.B. Blodgett, I. Langmuir, *Phys. Rev.*, **1937**, *57*, 964.
68. G.L. Gaines, *Insoluble Monolayers at Liquid–Gas Interfaces*, Interscience Publishers, New York, **1966**.
69. H. Kuhn, D. Möbius, H. Bucher, Spectroscopy of monolayer assemblies, in A. Weissberger and B.W. Rossiter (Eds.), *Techniques of Chemistry*, Wiley, New York, **1972**.
70. R.H. Tredgold, *Rep. Progr. Phys.*, **1987**, *50*, 1609.
71. G.G. Roberts, *Langmuir–Blodgett Films*, Plenum, NY, **1990**.
72. H.M. McConnell, *Annu. Rev. Phys. Chem.*, **1991**, *42*, 171.
73. H. Möhwald, *Rep. Prog. Phys.*, **1993**, *56*, 653.
74. B.P. Binks, *Adv. Colloid Interface Sci.*, **1991**, *34*, 343.
75. G.B. Khomutov, *Adv. Colloid Interface Sci.*, **2004**, *111*, 79.
76. D.R. Talham, *Chem. Rev.*, **2004**, *104*, 5479.
77. M. Pomerantz, *Surf. Sci.*, **1984**, *142*, 556.
78. G.B. Khomutov, *Macromolecular Symposia*, **1998**, *136*, 33.
79. A.M. Tishin, Yu.A. Koksharov, J. Bohr, G.B. Khomutov, *Phys. Rev. B*, **1997**, *55*, 11064.
80. T.V. Murzina, A.A. Fedyanin, T.V. Misuryaev, G.B. Khomutov, O.A. Aktsipetrov, *Appl. Phys. B*, **1999**, *68*, 537.
81. Yu.A. Koksharov, I.V. Bykov, A.P. Malakho, S.N. Polyakov, G.B. Khomutov, J. Bohr, *Mater. Sci. Eng. C*, **2002**, *22*, 201.
82. G.B. Khomutov, Yu.A. Koksharov, I.L. Radchenko, E.S. Soldatov, A.S. Trifonov, A.M. Tishin, J. Bohr, *Mater. Sci. Eng. C*, **1999**, *8–9*, 299.
83. A. Brugger, Ch. Schoppmann, M. Schurr, M. Seidl, G. Sipos, C.Y. Hahn, J. Hassmann, O. Waldmann, H. Voit, *Thin Solid Films*, **1999**, *338*, 231.
84. N. Bagkar, R. Ganguly, S. Choudhury, P.A. Hassan, S. Sawant, J.V. Yakhmi, *J. Mater. Chem.*, **2004**, *14*, 1430.
85. M. Clemente-Leon, C. Mingotaud, C.J. Gomez-Garcia, E. Coronado, P. Delhaes, *Thin Solid Films*, **1998**, *327–329*, 439.
86. S. Lefebure, C. Ménager, V. Cabuil, M. Assenheimer, F. Gallet, C. Flament, *J. Phys. Chem. B*, **1998**, *102*, 2733.
87. T. Nakaya, Y.-J. Li, K. Shibata, *J. Mater. Chem.*, **1996**, *6*, 691.
88. Y.S. Kang, D.K. Lee, P. Stroeve, *Thin Solid Films*, **1998**, *327–329*, 541.
89. X.K. Zhao, S. Xu, J.H. Fendler, *J. Phys. Chem.*, **1990**, *94*, 2573.
90. F.C. Meldrum, N.A. Kotov, J.H. Fendler, *J. Phys. Chem.*, **1994**, *98*, 4506.
91. J. Yang, X.G. Peng, T.J. Li, S.F. Pan, *Thin Solid Films*, **1994**, *243*, 643.
92. T. Meron, Y. Rosenberg, Y. Lereah, G. Markovich, *J. Magn. Magn. Mater.*, **2005**, *292*, 11.
93. Y.S. Kang, D.K. Lee, C.S. Lee, P. Stroeve, *J. Phys. Chem. B*, **2002**, *106*, 9341.
94. D.K. Lee, Y.S. Kang, C.S. Lee, P. Stroeve, *J. Phys. Chem. B*, **2002**, *106*, 7267.
95. D.K. Lee, Y.H. Kim, Y.S. Kang, P. Stroeve, *J. Phys. Chem. B*, **2005**, *109*, 14939.
96. X.-G Peng, Y. Zhang, J. Yang, B. Zou, L. Xiao, T. Li, *J. Phys. Chem.*, **1992**, *96*, 3412.
97. S.A. Iakovenko, A.S. Trifonov, M. Gicrsig. A. Mamedov, D.K. Nagesh, V.V Hanin, E.S. Soldatov, N.A. Kotov, *Adv. Mater.*, **1999**, *11*, 388.
98. J. Yang, X.-G. Peng, Y. Zhang, H. Wang, T.-J. Li, *J. Phys. Chem.*, **1993**, *97*, 4484.

99. D.C. Lee, D.K. Smith, A.T. Heitsch, B.A. Korgel, *Annu. Rep. Prog. Chem., Sect. C*, **2007**, *103*, 351.
100. Q. Guo, X. Teng, S. Rahman, H. Yang, *J. Am. Chem. Soc.*, **2003**, *125*, 630.
101. K. Kuroishi, M.-P. Chen, Y. Kitamoto, T. Seki, *Electrochimica Acta*, **2005**, *51*, 867.
102. Q. Guo, X. Teng, H. Yang, *Adv. Mater.*, **2004**, *16*(3) 1337.
103. C.H. Yu, N. Caiulo, C.C.H. Lo, K. Tam, S.C. Tsang, *Adv. Mater.*, **2006**, *17*, 2312.
104. M. Wen, K.E, H. Qi, L. Li, J. Chen, Y. Chen, Q. Wu, T. Zhang, *J. Nanoparticle Res.*, **2007**, *9*, 909.
105. Y. Wang, B. Ding, H. Li, X. Zhang, B. Cai, Y. Zhang, *JMMM*, **2007**, *308*, 108.
106. M. Mitsuishi, J. Matsui, T. Miyashita, *Polym. J.*, **2006**, *38*, 877.
107. T. Yamaki, T. Yamada, K. Asai, K. Ishigure, *Thin Solid Films*, **1998**, *327–329*, 586.
108. G.B. Khomutov, A.Yu. Obydenov, S.A. Yakovenko, E.S. Soldatov, A.S. Trifonov, V.V. Khanin, S.P. Gubin, *Mater. Sci. Eng.: C*, **1999**, *8–9*, 309.
109. G.B. Khomutov, *Colloids Surf. A*, **2002**, *202*, 243.
110. G.B. Khomutov, R.V. Gaynutdinov, S.P. Gubin, A.Yu. Obydenov, E.S. Soldatov, A.L. Tolstikhina, A.S. Trifonov, *Mater. Res. Soc. Symp. Proc.*, **2001**, *635*, C4201.
111. P. Jund, S.G. Kim, D. Tomanek, J. Hetherington, *Phys. Rev. Lett.*, **1995**, *74*, 3049.
112. R.K. Iler, *J. Colloid Interface Sci.*, **1966**, *21*, 569.
113. G.L. Gaines Jr., *Thin Solid Films*, **1983**, *99*, 243.
114. M.D. Musick, C.D. Keating, L.A. Lyon, S.L. Botsko, D.J. Pena, W.D. Holliway, T.M. McEvoy, J.N. Richardson, M.J. Natan, *Chem. Mater*, **2000**, *12*, 2869.
115. I. Ichinose, H. Tagawa, S. Mizuki, Yu. Lvov, T. Kunitake, *Langmuir*, **1998**, *14*, 187.
116. N.A. Kotov, I. Dekany, J.H. Fendler, *J. Phys. Chem.*, **1995**, *99*, 13065.
117. E.R. Kleinfeld, G.S. Ferguson, *Science*, **1994**, *265*, 370.
118. D.G. Peiffer, L.E. Nielsen, *J. Appl. Polym. Sci.*, **1979**, *23*, 2253.
119. D.G. Peiffer, *J. Appl. Polym. Sci.*, **1979**, *24*, 1451.
120. G. Decher, J.D. Hong, *Macromol. Chem., Macromol. Symp.*, **1991**, *46*, 321.
121. L. Krasemann, B. Tieke, *Mater. Sci. Eng. C*, **1999**, *8–9*, 513.
122. G. Decher, *Science*, **1997**, *277*, 1232.
123. W. Knoll, *Curr. Opin. Coll. Interface Sci.*, **1996**, *1*, 137.
124. G.M. Halpern, *J. Vac. Sci. Technol.*, **1980**, *17*, 1184.
125. D.G. Peiffer, T.J. Corley, G.M. Halpern, B.A. Brinker, *Polymer*, **1981**, *22*, 450.
126. R. Pommersheim, J. Schrezenmeir and W. Vogt, *Macromol. Chem. Phys.*, **1994**, *195*, 1557.
127. F. Caruso, *Adv. Mater.*, **2001**, *13*, 11.
128. G.B. Sukhorukov, E. Donath, H. Lichtenfeld, E. Knippel, M. Knippel, A. Budde, H. Mohwald, *Colloids Surf. A*, **1998**, *137*, 253.
129. K. Ariga, J.P. Hill, Q. Ji, *Phys. Chem. Chem. Phys.*, **2007**, *9*, 2319.
130. M. Sano, Y. Lvov, T. Kunitake, *Annu. Rev. Mater. Sci.*, **1996**, *26*, 153.
131. S. Tripathy, J. Kumar, H.S. Nalwa (Eds.), *Handbook of Polyelectrolytes and Their Applications*, American Scientific Publishers, Stevenson Ranch, CA, **2002**.
132. P.T. Hammond, *Curr. Opin. Colloid Interface Sci.*, **2000**, *4*, 430.
133. C. Lesser, M. Gao, S. Kirstein, *Mater. Sci. Eng. C*, **1999**, *8–9*, 159.
134. M. Eckle, G. Decher, *Nano Lett.*, **2001**, *1*, 45.
135. S. Dante, Z. Hou, S. Risbud, P. Stroeve, *Langmuir*, **1999**, *15*, 2176.
136. Z. Dai; H. Mohwald, B. Tiersch, L. Dahne, *Langmuir*, **2002**, *18*, 9533.
137. F. Hua, T. Cui, Yu. Lvov, *Langmuir*, **2002**, *18*, 6712.
138. F. Hua, J. Shi, Y. Lvov, T. Cui, *Nano Lett.*, **2002**, *2*, 1219.
139. C.J. Slevin, A. Malkia, P. Liljeroth, M. Toiminen, K. Kontturi, *Langmuir*, **2003**, *19*, 1287.

140. Y. Wang, Z. Tang, M.A. Correa-Duarte, L.M. Liz-Marzan, N.A. Kotov, *J. Am. Chem. Soc.*, **2003**, *125*, 2830.
141. C.A. Constantine, K.M. Gattas-Asfura, S.V. Mello, G. Crespo, V. Rastogi, T.C. Cheng, J.J. DeFrank, R.M. Leblanc, *Langmuir*, **2003**, *19*, 9863.
142. G.M. Lowman, S.L. Nelson, S.M. Graves, G.F. Strouse, S.K. Buratto, *Langmuir*, **2004**, *20*, 2057.
143. J.A. He, R. Mosurkal, L.A. Samuelson, L. Li, J. Kumar, *Langmuir*, **2003**, *19*, 2169.
144. J.F. Hicks, Y. Seok-Shon, R.W. Murray, *Langmuir*, **2002**, *18*, 2288.
145. C.A. Constantine, S.V. Mello, A. Dupont, X. Cao, D. Santos Jr., O.N. Oliveira Jr., F.T. Strixino, E.C. Pereira, T.C. Cheng, J.J. Defrank, R.M. Leblanc, *J. Am. Chem. Soc.*, **2003**, *125*, 1805.
146. E.J. Calvo, R. Etchenique, L. Pietrasanta, A. Wolosiuk, C. Danilowicz, *Anal. Chem.*, **2001**, *73*, 1161.
147. F. Caruso, D. Trau, H. Mohwald, R. Renneberg, *Langmuir*, **2000**, *16*, 1485.
148. G.B. Sukhorukov, M.M. Montrel, A.I. Petrov, L.I. Shabarchina, B.I. Sukhorukov, *Biosens. Bioelectron.*, **1996**, *11*, 913.
149. M.M. Montrel, G.B. Sukhorukov, A.I. Petrov, L.I. Shabarchina, B.I. Sukhorukov, *Sens. Actuators B*, **1997**, *42*, 225.
150. S.H. Sun, S. Anders, H.F. Hamann, J.U. Thiele, J.E.E. Baglin, T. Thomson, E.E. Fullerton, C.B. Murray, B.D. Terris, *J. Am. Chem. Soc.*, **2002**, *124*, 2884.
151. S. Sun, S. Anders, T. Thomson, J.E.E. Baglin, M.F. Toney, H.F. Hamann, C.B. Murray, B.D. Terris, *J. Phys. Chem. B*, **2003**, *107*, 5419.
152. G.A. Held, Hao Zeng, Shouheng Sun, *J. Appl. Phys.*, **2004**, *95*, 1481.
153. T. Thomson, M.F. Toney, S. Raoux, S.L. Lee, S. Sun, C.B. Murray, B.D. Terris, *J. Appl. Phys.*, **2004**, *96*, 1197.
154. S. Anders, M.F. Toney, T. Thomson, J.-U. Thiele, and B.D. Terris, *J. Appl. Phys.*, **2003**, *93*, 7343.
155. T.V. Murzina, A.A. Nikulin, O.A. Aktsipetrov, J.W. Ostrander, A.A. Mamedov, N.A. Kotov, M.A.C. Devillers, J. Roark, *J. Appl. Phys. Lett.*, **2001**, *79*, 1309.
156. O.A. Aktsipetrov, *Colloids Surf. A: Physicochem. Eng. Aspects*, **2002**, *202*, 165.
157. D.A. Gorin, D.O. Grigorev, A.M. Yashchenok, Yu.A. Koksharov, A.A. Neveshkin, A.V. Pavlov, G.B. Khomutov, H. Möhwald, G.B. Sukhorukov, *Proc. SPIE – The Int. Soc. Opt. Eng.*, **2007**, *6536*, 653607.
158. D. Grigoriev, D. Gorin, G.B. Sukhorukov, A. Yashchenok, E. Maltseva, H. Mohwald, *Langmuir*, **2007**, *23*, 12388.
159. A.A. Mamedov, N.A. Kotov, *Langmuir*, **2000**, *16*, 5530.
160. M.A. Correa-Duarte, M. Giersig, N.A. Kotov, L.M. Liz-Marzan, *Langmuir*, **1998**, *14*, 6430.
161. Y. Liu, A. Wang, R.O. Claus., *Appl. Phys. Lett.*, **1997**, *71*, 2265.
162. H.S. Kim, B.H. Sohn, W. Lee, J.-K. Lee, S.J. Choi, S.J. Kwon, *Thin Solid Films*, **2002**, *419*, 173.
163. S. Masayuki. M. Yasuo, H. Yuki, S. Osamu, S. seimei, E. Yasuaki, *Nippon Kagakkai Koen Yokoshu*, **2005**, *85*, 486.
164. M.A. Correa-Duarte, M. Grzelczak, V. Salgueiriño-Maceira, M. Giersig, L. Liz-Marzan, M. Farle, K. Sierazdki, R. Diaz, *J. Phys. Chem. B*, **2005**, *109*, 19060.
165. J.E. Wong, A.K. Gaharwar, D. Muller-Schulte, D. Bahadur, W. Richtering, *J. Magn. Magn. Mater.*, **2007**, *311*, 219.
166. E. Donath, G.B. Sukhorukov, F. Caruso, S.A. Davis, H. Mohwald, *Angew. Chem. Int. Ed.*, **1998**, *37*, 2201.
167. D.V. Volodkin, A.I. Petrov, M. Prevot, G.B. Sukhorukov, *Langmuir*, **2004**, *20*, 3398.
168. D.G. Shchukin, T. Shutava, E. Shchukina, G.B. Sukhorukov, Y.M. Lvov, *Chem. Mater.*, **2004**, *16*, 3446.

169. F. Caruso, M. Spasova, A. Susha, M. Giersig, R.A. Caruso, *Chem. Mater.*, **2001**, *13*, 109.
170. A. Voigt, N. Buske, G.B. Sukhorukov, A.A. Antipov, S. Leporatti, H. Lichtenfeld, H. Baumler, E. Donath, H. Mohwald, *J. Magn. Magn. Mater.*, **2001**, *225*, 59.
171. Z. Lu, M.D. Prouty, Z. Guo, V.O. Golub, C.S.S.R. Kumar, Y.M. Lvov, *Langmuir* **2005**, *21*, 2042.
172. N. Gaponik, I.L. Radtchenko, G.B. Sukhorukov, A.L. Rogach, *Langmuir*, **2004**, *20*, 1449.
173. B. Zebli, A.S. Susha, G.B. Sukhorukov, A.L. Rogach, W.J. Parak, *Langmuir*, **2005**, *21*, 4262.
174. D.A. Gorin, D.G. Shchukin, Yu.A. Koksharov, S.A. Portnov, K. Köhler, I.V. Taranov, V.V. Kislov, G.B. Khomutov, H. Möhwald, G.B. Sukhorukov, *Proc. SPIE*, **2007**, *6536*, 653604.
175. A.A. Mamedov, J. Ostrander, F. Aliev, N.A. Kotov, *Langmuir*, **2000**, *16*, 3941.
176. R. Gunawidjaja, C. Jiang, S. Peleshanko, M. Ornatska, S. Singamaneni, V.V. Tsukruk, *Adv. Funct. Mater.*, **2006**, *16*, 2024.
177. C. Jiang, S. Markutsya, V.V. Tsukruk, *Adv. Mater.*, **2004**, *16*, 157.
178. C. Jiang, S. Markutsya, Y. Pikus, V.V. Tsukruk, *Nature Mater.*, **2004**, *3*, 721.
179. C. Jiang, S. Markutsya, H. Shulha, V.V. Tsukruk, *Adv. Mater.*, **2005**, *17*, 1669.
180. S. Markutsya, C. Jiang, Y. Pikus, V.V. Tsukruk, *Adv. Funct. Mater.*, **2005**, *15*, 771.
181. G.B. Khomutov, Yu.A. Koksharov, *Nanofilm Materials and the Method for Production of Nanofilm Materials*, Patent application RU2006147123, **2006**.
182. J. Hu, T.W. Odom, C.M. Lieber, *Acc. Chem. Res.*, **1999**, *32*, 435.
183. Y. Xia, P. Yang, Y. Sun, Y. Wu, B. Mayers, B. Gates, Y. Yin, F. Kim, H. Yan, *Adv. Mater.*, **2003**, *15*, 353.
184. R.P. Cowburn, M.E. Welland, *Science*, **2000**, *287*, 1466.
185. A. Imre, G. Csaba, L. Ji, A. Orlov, G.H. Bernstein, W. Porod, *Science*, **2006**, *311*, 205.
186. A.K. Bentley, A.B. Ellis, G.C. Lisensky, W.C. Crone, *Nanotechnology*, **2005**, *16*, 2193.
187. I.S. Jacobs, C.P. Bean, *Phys. Rev.*, **1955**, *100*, 1060.
188. M. Ozaki, E. Matijevic, *J. Colloid Interface Sci.*, **1985**, *107*, 199.
189. J.L. Cain, D.E. Nikles, *IEEE Trans. Magn.*, **1996**, *32*, 4490.
190. J. Chen, D.E. Nikles, *IEEE Trans. Magn.*, **1996**, *32*, 4478.
191. T. Prozorov, R. Prozorov, Yu. Koltypin, I. Felner, A. Gedanken, *J. Phys. Chem. B*, **1998**, *102*, 10165.
192. G.B. Khomutov, S.P. Gubin, V.V. Khanin, Yu.A. Koksharov, A.Yu. Obydenov, V.V. Shorokhov, E.S. Soldatov, A.S. Trifonov, *Colloids Surf. A*, **2002**, *198–200*, 593.
193. G.B. Khomutov, S.P. Gubin, Yu.A. Koksharov, V.V. Khanin, A.Yu. Obidenov, E.S. Soldatov, A.S. Trifonov, *Mat. Res. Soc. Symp. Proc.*, **1999**, *577*, 427.
194. C. Burda, X. Chen, R. Narayanan, M.A. El-Sayed, *Chem. Rev.*; **2005**; *105*, 1025.
195. E. Dubois, V. Cabuil, F. Boue, R. Perzynski, *J. Chem. Phys.*, **1999**, *111*, 7147.
196. C. Yee, G. Kataby, A. Ulman, T. Prozorov, H. White, A. King, M. Rafailovich, J. Sokolov, A. Gedanken, *Langmuir*, **1999**, *15*, 7111.
197. A.K. Boal, K. Das, M. Gray, V.M. Rotello, *Chem. Mater.*, **2002**, *14*, 2628.
198. H. Khalil, D. Mahajan, M. Rafailovich, M. Gelfer, K. Pandya, *Langmuir*, **2004**, *20*, 6896.
199. F.A. Tourinho, R. Franck, R.J. Massart, *Mater. Sci.*, **1990**, *25*, 3249.
200. D. Maity, D.C.J. Agrawal, *J. Magn. Magn. Mater.*, **2007**, *308*, 46.
201. C. Salling, S. Schultz, I. McFadyen, M. Ozaki, *IEEE Trans. Magn.*, **1991**, *27*, 5184.
202. M. Osaki, S. Kratohvil, E. Matijevic, *J. Colloid Interface Sci.*, **1984**, *102*, 146.
203. M. Ocana, M. Morales, C.J. Serna, *J. Colloid Interface Sci.*, **1995**, *171*, 85.
204. D.E. Nikles, M.R. Parker, E.M. Crook, T.M. Self, *J. Appl. Phys.*, **1994**, *75*, 5565.

205. F.E. Spada, F.T. Parker, C.Y. Nakamura, A.E. Berkowitz, *J. Magn. Magn. Mater.*, **1993**, *120*, 129.
206. M.F. Casula, Y.-W. Jun, D.J. Zaziski, E.M. Chan, A. Corrias, A.P. Alivisatos, *J. Am. Chem. Soc.*, **2006**, *128*, 1675.
207. N. Cordente, M. Respaud, F. Senocq, M.-J. Casanove, C. Amiens, B. Chaudret, *Nano Lett.*, **2001**, *1*, 565.
208. C.P. Gibson, K.J. Putzer, *Science*, **1995**, *267*, 1338.
209. N.O. Nunez, P. Tartaj, M.P. Morales, R. Pozas, M. Ocana, C.J. Serna, *Chem. Mater.*, **2003**, *15*, 3558.
210. S.-J. Park, S. Kim, S. Lee, Z.G. Khim, K. Char, T. Hyeon, *J. Am. Chem. Soc.*, **2000**, *122*, 8581.
211. C.R. Martin, L.S. Van Dyke, Z. Cai, W. Liang, *J. Am. Chem. Soc.*, **1990**, *112*, 8976.
212. X.P. Gao, Y. Zhang, X. Chen, G.L. Pan, J. Yan, F. Wu, H.T. Yuan, D.Y. Song, *Carbon*, **2004**, *42*, 47.
213. J. Bao, C. Tie, Z. Xu, Z. Suo, Q. Zhou, J. Hong, *Adv. Mater.*, **2002**, *14*, 1483.
214. D. Seifu, Y. Hijji, G. Hirsch, S.P. Karna, *J. Magn. Magn. Mater.*, **2008**, *320*, 312.
215. J. Bao, Q. Zhou, J. Hong, Z. Xu, *Appl. Phys. Lett.*, **2002**, *81*, 4592.
216. D. Gozzi, A. Latini, G. Capannelli, F. Canepa, M. Napoletano, M.R. Cimberle, M. Tropeano, *J. Alloys Compd.*, **2006**, *419*, 32.
217. R. Kozhuharova, M. Ritschel, D. Elefant, A. Graff, A. Leonhardt, I. Mönch, T. Mühl, S. Groudeva-Zotova, C.M. Schneider, *Appl. Surf. Sci.*, **2004**, *238*, 355.
218. R. Kozhuharova, M. Ritschel, D. Elefant, A. Graff, I. Mönch, T. Mühl, C.M. Schneider, A. Leonhardt, *J. Magn. Magn. Mater.*, **2005**, *290–291*, 250.
219. B.K. Pradhan, T. Toba, T. Kyotani, A. Tomita, *Chem. Mater.*, **1998**, *10*, 2510.
220. C. Pham-Huu, N. Keller, C. Estournes, G. Ehret, M.J. Ledoux, *Chem. Commun.*, **2002**, 1882.
221. A.L. Elias, J.A. Rodriguez-Manzo, M.R. McCartney, D. Golberg, A. Zamudio, S.E. Baltazar, F. Lopez-urias, E. Munoz-Sandoval, L. Gu, C.C. Tang, D.J. Smith, Y. Bando, H. Terrones, M. Terrones, *Nano Lett.*, **2005**, *5*, 467.
222. S. Qu, F. Huang, G. Chen, S. Yu, J. Kong, *Electrochem. Comm.*, **2007**, *9*, 2812.
223. A. Ferta, L. Piraux, *J. Magn. Magn. Mater.*, **1999**, *200*, 338.
224. J. Yu, J.Y. Kim, S. Lee, J.K.N. Mbindyo, B.R. Martin, T.E. Mallouk, *Chem. Commun.*, **2000**, 2445.
225. D. AlMawlawi, N. Coombs, M. Moscovits, *J. Appl. Phys.*, **1991**, *70*, 4421.
226. H. Zeng, J. Li, J.P Liu, Z.L. Wang, S. Sun, *Nature*, **2002**, *420*, 395.
227. K.S. Napolskii, A.A. Eliseev, N.V. Yesin, A.V. Lukashin, Yu.D. Tretyakov, N.A. Grigorieva, S.V. Grigoriev, H. Eckerlebe, *Physica E*, **2007**, *37*, 178.
228. M. Sun, G. Zangari, M. Shamsuzzoha, R.M. Metzger, *Appl. Phys. Lett.*, **2001**, *78*, 2964.
229. P. Granitzer, K. Rumpf, H. Krenn, *J. Nanomater.*, **2006**, *2006*, Article ID 18125.
230. S. Aravamudhan, K. Luongo, P. Poddar, H. Srikanth, S. Bhansali, *Appl. Phys. A: Mater. Sci. Process.*, **2007**, *87*, 773.
231. A.A. Eliseev, I.V. Kolesnik, A.V. Lukashin, Y.D. Tretyakov, *Adv. Eng. Mater.*, **2005**, *7*, 213.
232. M.V. Chernysheva, A.A. Eliseev, K.S. Napolskii, A.V. Lukashin, Y.D. Tretyakov, N.A. Grigoryeva, S.V. Grigoryev, M. Wolff, *Thin Solid Films*, **2006**, *495*, 73.
233. M.V. Chernysheva, N.A. Sapoletova, A.A. Eliseev, A.V. Lukashin, Y.D. Tretyakov, P. Goernert, *Pure Appl. Chem.*, **2006**, *78*, 1749.
234. D. Carlier, C. Terrier, C. Arm, J.-Ph. Ansermet, *Electrochem. Solid State Lett.*, **2005**, *8*, 43.
235. L. Suber, P. Imperatori, G. Ausanio, F. Fabbri, H. Hofmeister, *J. Phys. Chem. B*, **2005**, *109*, 7103.
236. L. Wang, K. Yu-Zhang, A. Metrot, P. Bonhomme, M. Troyon, *Thin Solid Films*, **1996**, *288*, 86.

237. K. Hong, F.Y. Yang, K. Liu, D.H. Reich, P.C. Searson, C.L. Chien, F.F. Balakirev, G.S. Boebinger, *J. Appl. Phys.*, **1999**, *85*, 6184.
238. D. Rabelo, E.C.D. Lima, and A.C. Reis, *Nano Lett.*, **2001**, *1*, 105.
239. C. Mao, D.J. Solis, B.D. Reiss, S.T. Kottmann, R.Y. Sweeney, A. Hayhurst, G. Georgiou, B. Iverson, A.M. Belcher, *Science*, **2004**, *303*, 213.
240. D. Wirtz, M. Fermigier, *Langmuir*, **1995**, *11*, 398.
241. A.T. Skjeltorp, *J. Appl. Phys.*, **1985**, *57*, 3285.
242. P.S. Doyle, J. Bibette, A. Bancaud, J.-L. Viovy, *Science*, **2002**, *295*, 2237.
243. J. Liu, E.M. Lawrence, A. Wu, M.L. Ivey, G.A. Flores, K. Javier, J. Bibette, J. Richard, *Phys. Rev. Lett.*, **1995**, *74*, 2828.
244. C.-Y. Hong, I.J. Jang, H.E. Homg, C.J. Hsu, Y.D. Yao, H.C. Yang, *J. Appl. Phys.*, **1997**, *81*, 4275.
245. P.G. de Gennes, P.A. Pincus, *Phys. Condens. Mater.*, **1970**, *11*, 189.
246. J.J. Weis, *Mol. Phys.*, **1998**, *93*, 361.
247. R.W. Chantrell, A. Bradbury, J. Popplewell, S.W. Charles, *J. Appl. Phys.*, **1982**, *53*, 2742.
248. K. Butter, P.H.H. Bomans, P.M. Frederik, G.J. Vroege, A.P. Philipse, *Nature Mater.*, **2003**, *2*, 88.
249. J.R. Thomas, *J. Appl. Phys.*, **1966**, *37*, 2914.
250. C.H. Griffiths, M.P. O'Horo, T.W. Smith, *J. Appl. Phys.*, **1979**, *50*, 7108.
251. V.F. Puntes, K.M. Krishnan, A.P. Alivisatos, *Science*, **2001**, *291*, 2115.
252. V.F. Puntes, D. Zanchet, C.K. Erdonmez, A.P. Alivisatos, *J. Am. Chem. Soc.*, **2002**, *124*, 12874.
253. G.B. Biddlecombe, Y.K. Gun'ko, J.M. Kelly, S.C. Pillai, J.M.D. Coey, M. Venkatesan, A.P. Douvalis, *J. Mater. Chem.*, **2001**, *11*, 2937.
254. A.M. Schwartzberg, T.Y. Olson, C.E. Talley, J.Z. Zhang, *J. Phys. Chem. B*, **2006**, *110*, 19935.
255. A.M. Schwartzberg, Tammy Y. Olson, Chad E. Talley, J.Z. Zhang, *J. Phys. Chem. C*, **2007**, *111*, 16080.
256. V.P. Kurikka, M. Shafi, A. Gedanken, R. Prozorov, *Adv. Mater.*, **1998**, *10*, 590.
257. V. Kislov, B. Medvedev, Yu. Gulyaev, I. Taranov, V. Kashin, G.B. Khomutov, M. Artemiev, S. Gurevich, *Int. J. Nanosci.*, **2007**, *6*, 373.
258. J. Richardi, L. Motte, M.P. Pileni, *Curr. Opin. Colloid Interface Sci.*, **2004**, *9*, 185.
259. V. Salgueiriño-Maceira, M.A. Correa-Duarte, A. Hucht, M. Farle, *J. Magn. Magn. Mater.*, **2006**, *303*, 163.
260. B.D. Korth, P. Keng, I. Shim, S.E. Bowles, C. Tang, T. Kowalewski, K.W. Nebesny, J. Pyun, *J. Am. Chem. Soc.*, **2006**, *128*, 6562.
261. B.C. Satishkumar, E.M. Vogl, A. Govindaraj, C.N.R. Rao, *J. Phys. D: Appl. Phys.*, **1996**, *29*, 3173.
262. L.M. Ang, T.S.A. Hor, G.Q. Xu, C.H. Tung, S.P. Zhao, J.L.S. Wang, *Carbon*, **2000**, *38*, 363.
263. B. Rajesh, T.K. Ravindranathan, J.-M. Bonard, B. Viswanathan, *J. Mater. Chem.*, **2000**, *10*, 1757.
264. B. Xue, P. Chen, Q. Hong, J. Lin, L.T. Kuang, *J. Mater. Chem.*, **2001**, *11*, 2378.
265. M.A. Correa-Duarte, L.M. Liz-Marzan, *J. Mater. Chem.*, **2006**, *16*, 22.
266. J. Wei, J. Ding, X. Zhang, D. Wu, Z. Wang, J. Luo, K. Wang, *Mater. Lett.*, **2005**, *59*, 322.
267. J. Sun, L. Gao, *J. Electroceram.*, **2006**, *17*, 91.
268. T.W. Odom, J.-L. Huang, C.L. Cheung, C.M. Lieber, *Science*, **2000**, *290*, 1549.
269. V. Georgakilas, V. Tzitzios, D. Gournis, D. Petridis, *Chem. Mater.*, **2005**, *17*, 1613.
270. D.L. Peng, X. Zhao, S. Inoue, Y. Ando, K. Sumiyama, *J. Magn. Magn. Mater.*, **2005**, *292*, 143.
271. H.-Q. Wu, Y.-J. Cao, P.-S. Yuan, H.-Y. Xu, X.-W. Wei, *Chem. Phys. Lett.*, **2005**, *406*, 148.
272. H.-Q. Wu, P.-S. Yuan, H.-Y. Xu, D.-M. Xu, B.-Y. Geng, X.-W. Wei, *J. Mater. Sci.*, **2006**, *41*, 6889.

273. D.-M. Xu, H.-Q. Wu, Q.-Y. Wang, Q. Wang, B. Niu, Z.-M. Hu, *J. Funct. Mater.*, **2007**, *38*, 1777.
274. F. Stoffelbach, A. Aqil, C. Jerome, R. Jerome, C. Detrembleur, *Chem. Commun.*, **2005**, *36*, 4532.
275. Z.-J. Liu, Z. Xu, Z.-Y. Yuan, W. Chen, W. Zhou, L.-M. Peng, *Mater. Lett.*, **2003**, *57*, 1339.
276. F. Tan, X. Fan, G. Zhang, F. Zhang, *Mater. Lett.*, **2007**, *61*, 1805.
277. C. Huiqun, Z. Meifang, L. Yaogang, *J. Solid State Chemistry*, **2006**, *179*, 1208.
278. C. Gao, W. Li, H. Morimoto, Y. Nagaoka, T. Maekawa, *J. Phys. Chem. B*, **2006**, *110*, 7213.
279. L. Jiang, L. Gao, *Chem. Mater.*, **2003**, *15*, 2848.
280. X. Fan, F. Tan, G. Zhang, F. Zhang, *Mater. Sci. Eng. A*, **2007**, *454–455*, 37.
281. M. Grzelczak, M.A. Correa-Duarte, V. Salgueiriño-Maceira, B. Rodríguez-González, J. Rivas, L.M. Liz-Marzán, *Angew Chem Int Ed Engl.*, **2007**, 7026–7030.
282. M. Correa-Duarte, M. Grzelczak, V. Salgueiriño-Maceira, M. Giersig, L. Liz-Marzan, M. Farle, K. Sierazdki, R. Diaz, *Phys. Chem. B*, **2005**, *109*, 19060.
283. G.B. Khomutov, M.N. Antipina, A.N. Sergeev-Cherenkov, A.A. Rakhnyanskaya, M. Artemyev, D. Kisiel, R.V. Gainutdinov, A.L. Tolstikhina, and V.V. Kislov, *Int. J. Nanosci.*, **2004**, *3*, 65.
284. D. Nyamjav, A. Ivanisevic, *Biomaterials*, **2005**, *26*, 2749.
285. G.B. Khomutov, in: *Nanomaterials for Application in Medicine and Biology*, M. Giersig, G.B. Khomutov, (Eds.), Springer, Dordrecht, The Netherlands, **2008**, 39.
286. M.N. Antipina, R.V. Gainutdinov, A.A. Rachnyanskaya, A.L. Tolstikhina, T.V. Yurova, and G.B. Khomutov, *Surf. Sci.*, **2003**, *532–535*, 1025.
287. G.B. Khomutov, V.V. Kislov, R.V. Gainutdinov, S.P. Gubin, A.Yu. Obydenov, S.A. Pavlov, A.N. Sergeev-Cherenkov, E.S. Soldatov, A.L. Tolstikhina, A.S. Trifonov, *Surf. Sci.*, **2003**, *532–535*, 287.
288. L.C. Gosule, J.A. Schellmann, *Nature*, **1976**, *259*, 333.
289. T.H. Eickbush, E.N. Moudrianakis, *Cell*, **1978**, *13*, 295.
290. Z. Liu, D. Zhang, S. Han, C. Li, B. Lei, W. Lu, J. Fang, C. Zhou, *J. Am. Chem. Soc.*, **2005**, *127*, 6.
291. Z. Liang, A.S. Susha, A. Yu, F. Caruso, *Adv. Mater.*, **2003**, *15*, 1849.
292. D. Lee, R.E. Cohen, M.F. Rubner, *Langmuir*, **2007**, *23*, 123.
293. Q. He, Y. Tian, Y. Cui, H. Mohwald, J. Li, *J. Mater. Chem.*, **2008**, *18*, 748.
294. K. Nielsch, F.J. Castaño, C.A. Ross, R. Krishnan, *J. Appl. Phys.*, **2005**, *98*, 034318.
295. K. Nielsch, F.J. Castano, S. Matthias, W. Lee, C.A. Ross, *Adv. Eng. Mater.*, **2005**, *7*, 217.
296. J. Curiale, R.D. Sanchez, H.E. Troiani, A.G. Leyva, P. Levy, *Appl. Surf. Sci.*, **2007**, *254*, 368.
297. F. Li, L. Song, D. Zhou, T. Wang, Y. Wang, H. Wang, *J. Mater. Sci.*, **2007**, *42*, 7214.
298. J. Bachmann, J. Jing, M. Knez, S. Barth, H. Shen, S. Mathur, U. Gosele, K. Nielsch, *J. Am. Chem. Soc.*, **2007**, *129*, 9554.
299. M. Daub, M. Knez, U. Goesele, K. Nielsch, *J. Appl., Phys.*, **2007**, *101*, 09J111.
300. M. Knez, A. Kadri, C. Wege, U. Goesele H. Jeske, K. Nielsch, *Nano Lett.*, **2006**, *6*, 1172.
301. M. Grzelczak, M.A. Correa-Duarte, V. Salgueirino-Maceira, B. Rodriguez-Gonzalez, J. Rivas, L.M. Liz-Marzan, *Angew. Chem. – Int. Ed.*, **2007**, *46*, 7026.
302. G. Wang, R.I. Hollingsworth, *Langmuir*, **1999**, *15*, 6135.
303. S.J. Son, J. Reichel, B. He, M. Schuchman, S.B. Lee, *J. Am. Chem. Soc.*, **2005**, *127*, 7316.
304. B.D. Gates, Q. Xu, M. Stewart, D. Ryan, C.G. Willson, G.M. Whitesides, *Chem. Rev.*, **2005**, *105*, 1171.
305. M. Wirtz, C.R. Martin, *Adv. Mater.*, **2003**, *15*, 455.

306. S. Liu, R. Maoz, G. Schmid, J. Sagiv, *Nano Lett.*, **2002**, *2*, 1055.
307. M. Ben Ali, T. Ondarcuhu, M. Brust, C. Joachim, *Langmuir*, **2002**, *18*, 872.
308. H. Zhang, K.-B. Lee, Z. Li, C.A. Mirkin, *Nanotechnology*, **2003**, *14*, 1113.
309. H.S. Shin, H.J. Yang, Y.M. Jung, S.B. Kim, *Vib. Spectrosc.*, **2002**, *29*, 79.
310. Q. Guo, X. Teng, H. Yang, *Adv. Mater.*, **2004**, *16*, 1337.
311. Z.M. Fresco. J.M.J. Frechet, *J. Am. Chem. Soc.*, **2005**, *127*, 8302.
312. B.D. Gates, Q. Xu, M. Stewart, D. Ryan, C.G. Willson, Whitesides G.M., *Chem. Rev.*, **2005**, *105*, 1171.
313. J.E. Barton, C.L. Stender, T.W. Odom, *Acc. Chem. Res.*, **2006**; *39*, 249.
314. D.D. Awschalom, D.P. DiVincenzo, *Phys. Today*, **1995**, 43.
315. A.D. Kent, S. von Molnar, S. Gider, D.D. Awschalom, *J. Appl. Phys.*, **1994**, *76*, 6656.
316. C.A. Ross, *Ann. Rev. Mater. Res.*, **2001**, *31*, 203.
317. J. Park, P.T. Hammond, *Adv. Mater.*, **2004**, *16*, 520.
318. Y. Xia, G.M. Whitesides, *Angew. Chem. Int. Ed.*, **1998**, *37*, 550.
319. Z.R. Dai, S. Sun, Z.L. Wang, *Nano Lett.* **2001**, *1*, 443.
320. M. Chen, D.E. Nikles, H. Yin, S. Wang, J.W. Harrell, S.A. Majetich, *J. Magn. Magn. Mater.*, **2003**, *266*, 8.
321. J. Cheng, W. Jung, C.A. Ross, *Phys. Rev. B*, **2004**, *70*, 064417.
322. S.M. Weekes, F.Y. Ogrin, W.A. Murray, P.S. Keatley, *Langmuir* **2007**, *23*, 1057.
323. A. Ethirajan, U. Wiedwald, H.-G. Boyen, B. Kern, L. Han, A. Klimmer, F. Weigl, G. Kastle, P. Ziemann, K. Fauth, R.J. Cai, R.J. Behm, A. Romanyuk, P. Oelhafen, P. Walther, J. Biskupek, U. Kaiser, *Adv. Mater.*, **2007**, *19*, 406.
324. A.R. Urbach, J.C. Love, M.G. Prentiss, G.M. Witesides, *J. Am. Chem. Soc.*, **2003**, *125*, 12704.
325. V.P. Kurikka, M. Shafi, I. Felner, Y. Mastai, A Gedanken, *J. Phys. Chem. B*, **1999**, *103*, 3358.
326. S.L. Tripp, Stephen V. Pusztay, Alexander E. Ribbe, Alexander Wei, *J. Am. Chem. Soc.*, **2002**, *124*, 7914.
327. S.L. Tripp, R. Dunin-Borkowski, A. Wei, *Angew. Chem. Int. Ed.*, **2003**, *42*, 5591.
328. J. Gómez-Segura, O. Kazakova, J. Davies, P. Josephs-Franks, J. Veciana, D. Ruiz-Molina, *Chem. Commun.*, **2005**, 5615.
329. L.V. Govor, J. Parisi, G.H. Bauer, *Z. Naturforsch*, **2003**, *58a*, 392.
330. Y. Xiong, J. Ye, X. Gu, Q.-W. Chen, *J. Phys. Chem. C*, **2007**, *111*, 6998.
331. F. Dumestre, B. Chaudret, C. Amiens, P. Renaud, P. Fejes, *Science*, **2004**, *303*, 821.
332. I. Lisiecki, P.-A. Albouy, M.-P. Pileni, *Adv. Mater.*, **2003**, *15*, 712.
333. Y. Zhu, W. Zhao, H. Chen, J. Shi, *J. Phys. Chem. C*, **2007**, *111*, 5281.
334. B.J. Lemaire, P. Davidson, J. Ferré, J.P. Jamet, P. Panine, I. Dozov, J.P. Jolivet, *Phys. Rev. Lett.*, **2002**, *88*, 1255071.
335. X. Hong, J. Li, M. Wang, J. Xu, W. Guo, J. Li, Y. Bai, T. Li, *Chem. Mater.*, **2004**, *16*, 4022.
336. F. Caruso, A.S. Susha, M. Giersig, H. Mohwald, *Adv. Mater.*, **1999**, *11*, 950.
337. C. Xu, J. Xie, D. Ho, C. Wang, N. Kohler, E.G. Walsh, J.R. Morgan, Y.E. Chin, S. Sun, *Ang. Chem. – Int. Ed.*, **2008**, *47*, 173.
338. C.-W. Lu, Y. Hung, J.-K. Hsiao, M. Yao, T.-H. Chung, Y.-S. Lin, S.-H. Wu, S.-C. Hsu, H.-M. Liu, C.-Y. Mou, C.-S. Yang, D.-M. Huang, Y.-C. Chen, *Nano Lett.*, **2007**, *7*, 149.
339. H. Lu, G. Yi, S. Zhao, D. Chen, L.-H. Guo, J. Cheng, *J. Mater. Chem.*, **2004**, *14*, 1336.
340. C. Saiyasombat, N. Petchsang, I.M. Tang, J.H. Hodak, *Nanotechnology*, **2008**, *19*, 85705.
341. V. Salgueiriño-Maceira, M.A. Correa-Duarte, M. Farle, *Small*, **2005**, *1*, 1073.
342. F.G. Aliev, M.A. Correa-Duarte, A. Mamedov, J.W. Ostrander, M. Giersig, L.M. Liz-Marzan, N.A. Kotov, *Adv. Mater.*, **1999**, *11*, 1006.

343. L. Fu, V.P. Dravid, D.L. Johnson, *Appl. Surf. Sci.*, **2001**, *181*, 173.
344. S.K. Mandal, N. Lequeux, B. Rotenberg, M. Tramier, J. Fattaccioli, J. Bibette, B. Dubertret, *Langmuir*, **2005**, *21*, 4175.
345. B. Cheng, Y.R. Zhu, W.Q. Jiang, C.Y. Wang, Z.Y. Chen, *J. Chem. Res. (S)*, **1999**, 506.
346. E.M. Brunsman, R. Sutton, E. Bortz, S. Kirkpatrick, K. Midelfort, J. Williams, P. Smith, M.E. McHenry, S.A. Majetich, J.O. Artman, M. De Graef, S.W. Staley, *J. Appl. Phys.*, **1994**, *75*, 5882.
347. S.P. Gubin, G.Yu. Yurkov, N.A. Kataeva, *Inor. Mater.*, **2005**, *41*, 1017.
348. H. Xu, L. Cui, N. Tong, H. Gu, *J. Am. Chem. Soc.*, **2006**, *128*, 15582.
349. X.G. Li, S. Takahashi, K. Watanabe, Y. Kikuchi, M. Koishi, *Nano Lett.*, **2001**, *1*, 475.
350. K. Landfester, L.P. Ramírez, *J. Phys.: Condens. Matter*, **2003**, *15*, S1345.
351. V. Holzapfel, M. Lorenz, C.K. Weiss, H. Schrezenmeier, K. Landfester, V. Mailander, *J. Phys.: Condens. Matter*, **2006**, *18*, S2581.
352. Y. Deng, L. Wang, W. Yang, S. Fu, A. Elaïssari, *J. Magn. Magn. Mater.*, **2003**, *257*, 69.
353. K. Wormuth, *J. Colloid Interface Sci.*, **2001**, *241*, 366.
354. J.W.M. Bulte, T. Douglas, B. Witwer, S.-C. Zhang, E. Strable, B.K. Lewis, H. Zywicke, B. Miller, P. van Gelderen, B.M. Moskowitz, I.D. Duncan, J.A. Frank., *Nature Biotechnology*, **2001**, *19*, 1141.
355. N.S. Kommareddi, M. Tata, V.T. John, G.L. McPherson, M.F. Herman, *Chem. Mater.*, **1996**, *8*, 801.
356. H. Shiho, Y. Manabe, N. Kawahashi, *J. Mater. Chem.*, **2000**, *10*, 333.
357. A.Yu. Men'shikova, B.M. Shabsel's, Yu.O. Skurkis, K.S. Inkin, N.A. Chekina, S.S. Ivanchev, *Russ. J. Gen. Chem.*, **2007**, *77*, 354.
358. M. Chu, X. Song, D. Cheng, S. Liu, J. Zhu, *Nanotechnology*, **2006**, *17*, 3268.
359. D. Horák, E. Petrovský, A. Kapička, T. Frederichs, *J. Magn. Magn. Mater.*, **2007**, *311*, 500.
360. M. Fang, P.S. Grant, M.J. McShane, G.B. Sukhorukov, V.O. Golub, Y.M. Lvov, *Langmuir*, **2002**, *18*, 6338.
361. H. Pu, F. Jiang, *Nanotechnology*, **2005**, *16*, 1486.
362. S. Samouhos, G. McKinley, *J. Fluids Eng., Trans. ASME*, **2007**, *129*, 429.
363. S. Mørup, *Europhys. Lett.*, **1994**, *28*, 671.
364. L.N. Mulay, D.W. Collins, A.W. Thomson, P.L. Walker, *J. Organomet. Chem.*, **1979**, *178*, 217.
365. S. Jin, R.C. Sherwood, J.J. Mottine, T.H. Tiefel, R.L. Opila, *J. Appl. Phys.*, **1988**, *64*, 6008.
366. S. Jin, T.H. Tiefel, R. Wolfe, R.C. Sherwood, J.J. Mottine, *Science*, **1999**, *255*, 446.
367. R.F. Ziolo, E.P. Giannelis, B.A. Weinstein, M.P. O'Horo, B.N. Ganguly, V. Mehrota, M.W. Russell, D.R. Huffman, *Science*, **1992**, *257*, 219.
368. V. Lauter-Pasyuk, H.J. Lauter, D. Ausserre, Y. Gallot, V. Cabuil, B. Hamdoun, E.I. Kornilov, *Physica B*, **1998**, *248*, 243.
369. S.V. Stakhanova, E.S. Trofimchuk, N.I. Nikonorova, A.V. Rebrov, A.N. Ozerin, A.L. Volynskii, N.F. Bakeev, *Polym. Sci., Ser. A*, **1997**, *39*, 229.
370. N.I. Nikonorova, E.V. Semenova, V.D. Zanegin, G.M. Lukovkin, A.L. Volynskii, N.F. Bakeev, *Polym. Sci.*, **1992**, *34*, 711.
371. T. Kimura, H. Ago, M. Tobita, S. Ohshima, M. Kyotani, M. Yumura, *Adv. Mater.*, **2002**, *14*, 1380.
372. T. Hayashi, S. Hirono, M. Tomita, S. Umemura, *Nature*, **1996**, *381*, 772.
373. B.Z. Tang, Y. Geng, Q. Sun, X.X. Zhang, X. Jing, *Pure Appl. Chem.*, **2000**, *72*, 157.
374. A.C. Balazs, *Curr.Opin. Colloid Interface Sci.*, **2000**, *4*, 443.
375. J.Y. Lee, Z. Shou, A.C. Balazs, *Phys. Rev. Lett.*, **2003**, *91*, 136103.
376. J.Y. Lee, Z. Shou, A.C. Balazs, *Macromolecules*, **2003**, *36*, 7730.
377. P. Mansky, Y. Liu, E. Huang, T.P. Russell, C. Hawker, *Science*, **1997**, *275*, 1458.
378. J. Chatterjee, Y. Haik, C.J. Chen, *Colloid Polym. Sci.*, **2003**, *281*, 892.

379. T. Song, Y. Zhang, T. Zhou, C.T. Lim, S. Ramakrishna, B. Liua, *Chem. Phys. Lett.*, **2005**, *415*, 317.
380. M. Wang, H. Singh, T.A. Hatton and G.C. Rutledge, *Polymer*, **2004**, *45*, 5505.
381. D. Li, T. Herricks, Y. Xia, *Appl. Phys. Lett.*, **2003**, *83*, 4586.
382. E.L. Mayes, F. Vollrath, S. Mann, *Adv. Mater.*, **1998**, *10*, 801.
383. *Iron Biominerals*, R.B. Frankel, R.P. Blakemore (Eds.), Plenums, New York, **1991**.
384. I. Safarik, M. Safarikova, *Chemical Monthly*, **2002**, *133*, 737.
385. B. Gilbert, J.F. Banfield, *Rev. Mineral. Geochem.*, **2005**, *59*, 109.
386. V. Salgueiriño-Maceira, M.A. Correa-Duarte, *Adv. Mater*, **2007**, *19*, 4131.
387. Q.A. Pankhurst, J. Connolly, S.K. Jones, J. Dobson, *J. Phys. D*, **2003**, *36*, R167.
388. D. Schuler, R.B. Frankel, *Appl. Microbiol. Biotechnol.*, **1999**, *52*, 464.
389. Y.-X.J. Wang, S.M. Hussain, G.P. Krestin, *Eur. Radiol.*, **2001**, *11*, 2319.
390. D. Portet, B. Denizot, E. Rump, J.-J. Lejeune, P. Jallet, *J. Colloid Interface Sci.* **2001**, *238*, 37.
391. E.X. Wu, H. Tang, *NMR Biomed.*, **2004**, *17*, 478.
392. T. Osaka, T. Matsunaga, T. Nakanishi, A. Arakaki, D. Niwa, H. Iida, *Anal. Bioanal. Chem.*, **2006**, *384*, 593.
393. M. Zrínyi, L. Barsi, A. Büki, *Polym. Gels Networks*, **1997**, *5*, 415.
394. C. Albornoz, S.E. Jacobo, *J. Magn. Magn. Mater.*, **2006**, *305*, 12.
395. D.C.F. Chan, D.B. Kirpotin, P.A. Bunn, Jr., *J. Magn. Magn. Mater.*, **1993**, *122*, 374.
396. A. Jordan, R. Scholz, P. Wust, H. Schirra, S. Thomas, H. Schmidt, R. Felix, *J. Magn. Magn. Mater.*, **1999**, *194*, 185.
397. A. Senyei, K. Widder, G. Czerlinski, *J. Appl. Phys.*, **1978**, *49*, 3578.
398. T. Kubo, T. Sugita, S. Shimose, Y. Nitta, Y. Ikuta, T. Murakami, *Int. J. Oncol.*, **2001**, *18*, 121.
399. Ch. Alexiou, A. Schmidt, R. Klein, P. Hulin, Ch. Bergemann, W. Arnold, *J. Magn. Magn. Mater.*, **2001**, *252*, 363.
400. A.S. Lubbe, C. Bergemann, J. Brock, D.G. McClure, *J. Magn. Magn. Mater.* **1999**, *194*, 149.
401. F. Scherer, M. Anton, U. Schillinger, J. Henke, C. Bergemann, A. Kruger, B. Gansbacher, C. Plank, *Gene Therapy*, **2002**, *9*, 102.
402. C. Plank, U. Schillinger, F. Scherer, C. Bergemann, J.-S. Rémy, F. Krotz, M. Anton, J. Lausier, J. Rosenecker, *Biol. Chem.*, **2003**, *384*, 737.
403. Z.P. Xu, Z.Q. Hua, G.Q. Lu, A.B. Yu, *Chem. Eng. Sci.*, **2006**, *61*, 1027.
404. O. Olsvik, T. Popovic, E. Skjerve, K.S. Cudjoe, E. Hornes, J. Ugelstad, M. Uhlen, *Clinical Microbiol. Rev.*, **1994**, *7*, 43.
405. W. Kemmner, G. Moldenhauer, P. Schlag, R. Brossmer, *J. Immunol. Methods*, **1992**, *147*, 197.
406. S. Bucak, D.A. Jones, P.E. Laibinis, T.A. Hatton, *Biotechnol. Prog.*, **2003**, *19*, 477.
407. L. Josephson, J.M. Perez, R. Weissleder, *Angew. Chem., Int. Ed.*, **2001**, *40*, 3204.
408. A.K. Gupta, A.S.G. Curtis, *Biomaterials*, **2004**, *25*, 3029.
409. D.H. Reich, M. Tanase, A. Hultgren, L.A. Bauer, C.S. Chen, G.J. Meyer, *J. Appl. Phys.*, **2003**, *93*, 7275.
410. S.B.-D. Makhluf, R. Qasem, S. Rubinstein, A. Gedanken, H. Breitbart, *Langmuir*, **2006**, *22*, 9480.
411. D.G. Shchukin, A.A. Patel, G.B. Sukhorukov, Y.M. Lvov, *J. Am. Chem. Soc.*, **2004**, *126*, 3374.
412. I. Žutić, J. Fabian, S.D. Sarma, *Rev. Mod. Phys.*, **2004**, *76*, 323.
413. S.D. Bader, *Rev. Mod. Phys*, **2006**, *78*, 1.
414. *Biomedical Nanostructures*, K.E. Gonsalves, C.R. Halberstadt, C.T. Laurencin, L.S. Nair (Editors), John Wiley & Sons, Hoboken, New Jersey, **2008**.
415. D. Parker, V. Dupuis, F. Ladieu, J.-P. Bouchaud, E. Dubois, R. Perzynski, E. Vincent, *Phys. Rev. B*, **2008**, *77*, 104428.

416. M. Sasaki, P.E. Jönsson, H. Takayama, H. Mamiya, *Phys. Rev. B*, **2005**, *71*, 104405.
417. W. Kleemann, O. Petracic, Ch. Binek, G.N. Kakazei, Yu.G. Pogorelov, J.B. Sousa, S. Cardoso, P.P. Freitas, *Phys. Rev. B.*, **2001**, *63*, 134423.
418. R. Birringer, H. Wolf, C. Lang, A. Tschöpe, A. Michels, *Z. Phys. Chem.*, **2008**, *222*, 229.
419. P. Poddar, T. Telem-Shafir, T. Fried, G. Markovich, *Phys. Rev. B.*, **2002**, *66*, 060403.
420. M.S. Seehra, H. Shim, P. Dutta, A. Manivannan, J. Bonevich, *J. Appl. Phys.*, **2005**, *97*, 10J509.
421. J.-G. Zhu, Y. Zheng, G.A. Prinz, *J. Appl. Phys.*, **2000**, *87*, 6668.
422. J. Aizpurua, P. Hanarp, D.S. Sutherland, M. Käll, G.W. Bryant, F.J. García de Abajo, *Phys. Rev. Lett.*, **2003**, *90*, 057401.
423. S.L. Tripp, S.V. Pusztay, A.E. Ribble, A. Wei, *J. Am. Chem. Soc.*, **2002**, *124*, 7914.
424. P. Jund, S.G. Kim, D. Tománek, J. Hetherington, *Phys. Rev. Lett.*, **1995**, *74*, 3049.
425. W. Wen, F. Kun, K.F. Pál, D.W. Zheng, K.N. Tu, *Phys. Rev. E*, **1999**, *59*, R4758.
426. A. Ghazali, J.-C. Lévy, *Phys. Rev. B*, **2003**, *67*, 064409.
427. J.Y. Cheng, W. Jung, C.A. Ross, *Phys. Rev. B*, **2004**, *70*, 064417.
428. L.M. Demers, S.J. Park, T.A. Taton, Z. Li, C.A. Mirkin. *Angew. Chem. Int. Ed.*, **2001**, *40*, 3071.
429. D. Gerion, W.J. Parak, S.C. Williams, D. Zanchet, C.M. Micheel, A.P. Alivisatos, *J. Am. Chem. Soc.*, **2002**, *124*, 7070.
430. Yu.A. Koksharov, G.B. Khomutov, E.S. Soldatov, D. Suyatin, I. Maximov, L. Montelius, P. Carlberg, *Thin Solid Films*, **2006**, *515*, 731.
431. R. Scomski, *J. Phys.: Condens. Matter*, **2003**, *15*, R841.
432. S.A. Wolf, D.D. Awschalom, R.A. Buhrman, J.M. Daughton, S. von Molnár, M.L. Roukes, A.Y. Chtchelkanova, D.M. Treger, *Science*, **2001**, *294*, 1488.
433. R.P. Cowburn, *J. Phys. D: Appl. Phys.*, **2000**, *33*, R1.
434. J.G. Zhu, Y.F. Zheng, G.A. Prinz, *J. Appl. Phys.* **2000**, *87*, 6668.
435. W. Jung, F.J. Castaño, C.A. Ross, R. Menon, A. Patel, E.E. Moon, H.I. Smith, *J. Vac. Sci. Technol. B*, **2004**, *22*, 3335.
436. R.P. Cowburn, D.K. Koltsov, A.O. Adeyeye, M.E. Welland, D.M. Tricker, *Phys. Rev. Lett.*, **1999**, *83*, 1042.
437. R.P. Boardman, H. Fangohr, S.J. Cox, A.V. Goncharov, A.A. Zhukov, P.A.J. de Groot, *J. Appl. Phys.*, **2004**, *95*, 7037.
438. T. Okuno, K. Mibu, T. Shinjo, *J. Appl. Phys.*, **2004**, *95*, 3612.
439. *Numerical Calculations for Theorist Physicists*, V.A. Il'ina, P.K. Silaev (Editors), Moscow-Izhevsk, Institute of Computer Research (in Russian), **2003**.
440. *Numerical Recipes*, W.H. Press, S.A. Teukolsky, W.T. Vetterling, B.P. Flannery (Editors), Cambridge University Press, Cambridge, **1995**.
441. C.M. Sorensen, in: *Nanoscale Materials in Chemistry*, K.J. Klabunde (Editor), John Wiley & Sons, New York, **2003**, 169.
442. P.V. Melenev, V.V. Rusakov, Yu.L. Raikher, *Pisma J. Thech. Phys. (Russ.)*, **2008**, *34*, 50.

6
Magnetism of Nanoparticles: Effects of Size, Shape, and Interactions
Yury A. Koksharov

6.1
Introduction

Magnetism of nanoparticles is an area of intense development that touches many fields including material science, condensed matter physics, biology, medicine, planetary science, and so on [1–5]. Nanoscale magnetic materials are of interest for applications in ferrofluids, high-density magnetic storage, high-frequency electronics, high-performance permanent magnets, magnetic refrigerants, etc. Small magnetic particles exhibit many unique phenomena such as superparamagnetism [6], quantum tunneling of magnetization [7], enhanced magnetic coercivity [8]. Due to their advanced magnetic properties, certain magnetic nanoparticles (e.g., CoPt, FePt) are of high interest for future high-density recording media [9–12]. Magnetic nanoparticles are also used in medicine and biotechnology [13–17]. For example, iron oxide colloids have a low toxicity and show good biocompatibility which makes them suitable in various areas of medicine, like drug delivery systems and hyperthermia treatment of cancer.

Natural magnetic nanoparticles are everywhere [18]: in human brain [19], in bacteria, algae, birds, ants, bees, etc. [20–22], in soils and lacustrine sediments [23], in meteorites [24, 25], and in interstellar space [26, 27]. There are magnetic materials that exist, probably, only in nanoparticle form. The best knowing example is ferrihydrite – widespread iron oxyhydroxide [28, 29]. So, magnetic nanoparticles are not the Homo sapience invention. However, human beings can visualize and manipulate nanoparticles, as well as synthesize those with required properties.

In case of bulk defect-free materials, their intrinsic magnetic properties (e.g., saturation magnetization M_S, coercive force H_C, and Curie temperature T_C) depend only on chemical and crystallographic structure. The size and shape of studied bulk samples are not crucially important; for example, M_S, H_C, and T_C values of small and big cobalt samples are all equal. Magnetic nanoparticles show a wide variety of unusual magnetic properties

Magnetic Nanoparticles. Sergey P. Gubin
© 2009 WILEY-VCH Verlag GmbH & Co. KGaA, Weinheim
ISBN: 978-3-527-40790-3

as compared to the respective bulk materials. Magnetic characteristics of nanoparticles are strongly influenced by so-called finite-size and surface effects. Their relevance increases with decreasing particle size. Finite-size effects result, in the strict sense of the word, from quantum confinement of the electrons. Surface effects can be related, in simplest case, to the symmetry breaking of the crystal structure at the boundary of each particle, but can be also due to different chemical and magnetic structures of internal ("core") and surface ("shell") parts of a nanoparticle. Below we will see how magnetic properties of nanoparticles depend on their size, shape, and environment, including interparticle and particle–matrix interactions.

6.2
Magnetism of Nanoparticles in the View of Atomic and Solid State Physics

The magnetic characteristics of atoms and relatively small molecules, containing up to several tens of atoms, can be *a priori* calculated by quantum chemistry methods. In macroscopic objects, a number of atoms are very large ($>10^{23}$ particles), and therefore one should use specific methods of solid state physics, based on symmetry analysis, statistical and thermodynamic approaches, etc. Nanoparticles contain from several hundreds up to $\approx 10^5$ atoms [2]. So they take a place where quantum chemistry and solid state physics meet together. It is not surprising that when researchers attempt to explain the properties of nanoparticles, they use tools and conceptions from very different fields of chemistry and physics. In some cases, nanoparticles are considered as large molecules (or clusters) with discrete energy or space structure. We found the examples of the quantized energy states of electrons and holes in models of quantum dots [30] and small metallic particles [31]. The discrete structure approach relates often to magnetism. The examples are various atomic-scale models in which magnetic nanoparticles are realized as assembles of strongly interacting, though separate, magnetic moments [32–36].

In other situations, nanoparticles are treated as continuum medium of very small size and deviations from "bulk" behavior are explained taking into account mainly finite-size effects. The representation of a magnetic nanoparticle as single-domain hard ferromagnetic body is the example of simple continuous (macroscopic) approximation in which explicit atomic structure of nanoparticles is inessential [37–39]. Such approach is related to micromagnetic theory [40], which merges classical electrodynamics of continuous media, some branches of condensed matter physics, and various phenomenological concepts. In the micromagnetic theory, a magnetic sample is described by a set of macroscopic variables, e.g., the directional cosines of the magnetization vector within the sample [41].

6.3
Magnetic Finite-Size Effects and Characteristic Magnetic Lengths. Single-Domain Particles

Before answering the question: "How particle size affects magnetic phenomena?" let us briefly recall basics of magnetism phenomena in macroscopic specimens. Magnetic materials are all around us [42]. Understanding their properties and development, their applications underlie much of today's scientific and engineering work. Despite its centuries-old history, magnetism is still a science field with great deal of puzzles. Most of pure stable chemical elements (79 of the 103) carry an atomic moment in the atomic ground state. However, among pure elements in polyatomic state, only O, Cr, Fe, Mn, Co, Ni, and some rare earth elements show magnetic ordering. In the case of substances consisting of more than one type of atom, no simple approach allows one to predict whether a given substance will be magnetic. For example, YFe_2Si_2 is nonmagnetic, although most iron-based compounds are ferro- or ferrimagnetic. On the other hand, Cu_2MnAl and $MnBi$ are ferromagnetic, although all constituent elements in metal state are nonferromagnetic: copper and bismuth are diamagnetic, aluminum is paramagnetic, and manganese is antiferromagnet below 100 K [42].

The fundamental source of material magnetism is magnetic moments of electrons which constitute electron shells of atoms and form electron structure of molecules and crystals [43]. As a rule, if the electron shell of an isolated atom is not closed, the atom has nonzero magnetic moment. In molecules and especially in condensed matter, interatomic interactions have significant action upon electron and magnetic properties. If magnetism of individual atoms is preserved, as it takes place, for example, in iron oxides, interactions between localized magnetic moments can result in magnetic ordering. In metals, like Fe, Co, Ni, and $ZrZn_2$, the magnetic moment of delocalized electrons can also be a reason of magnetic ordering. It is a standard practice to analyze magnetic properties of solids in terms of these two approaches – models of "localized moments" and "itinerant electrons" [44]. In real materials, both mechanisms can be important to a greater or lesser extent. There is also an approach (the so-called spin fluctuations theory) that tries to create unified picture of magnetism [45].

A classification of substances by their magnetic properties includes "weak"-magnetic (diamagnetic and paramagnetic) and "strong"-magnetic (ferromagnetic, ferrimagnetic, antiferromagnetic, etc.) materials. The carriers of magnetic moment are usually sketched out as magnetic dipoles (see, e.g., arrows in Figure 6.1). Dipole magnetic moment per unit volume is called the magnetization, M. The diamagnetic and paramagnetic materials have zero magnetization at any temperature in the absence of the external magnetic field.

Zero-field (spontaneous) magnetization takes place only in "strong"-magnetic materials below the Curie (or Néel) temperature due to long-range

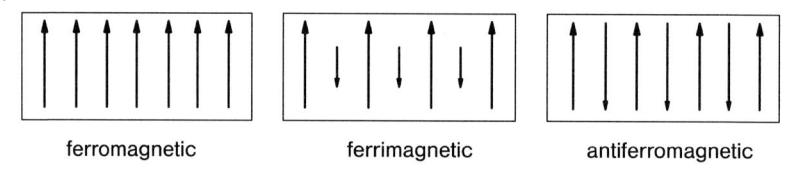

Figure 6.1 Schematic illustrating the arrangements of magnetic dipoles for the most common types of "strong"-magnetic materials. Note, that such arrangements take place below the ordering temperature regardless of the absence or presence of an external magnetic field (H).

ordering of magnetic dipoles. The origin of this ordering lies in the quantum-mechanical exchange forces. While ferromagnetic has only one magnetic lattice composed of parallel dipoles, antiferromagnetics and ferrimagnetics can be considered as superposition of two oppositely directed magnetic sublattices (Figure 6.1). Since the sublattices in antiferromagnetic crystal are identical, then their total magnetic moment vanishes. As it seems from Figure 6.1 ferro- and ferrimagnetic specimens should have very large total magnetic moment even in the absence of an external magnetic field. Actually, this is not the case. The reason is the so-called domain structure of ferromagnets.

The typical experimental magnetization curve of ferromagnets (Figure 6.2) shows paradoxical situation: (1) in some cases, the maximal (saturation) magnetization are attained by the application of a very weak magnetic field (0.01 Oe in Figure 6.2) and (2) it is possible for the magnetization of the same specimen to be zero in very small (nearly zero) applied field. The first fact is most amazing since, for example, in the paramagnetic salt, effectively only one magnetic moment in 10^9 is "oriented" by a field of 0.01 Oe [46], so that the distribution of magnetic moment directions remains essentially random and total specimen magnetization is very far from saturation. This apparent paradox can be resolved by Weiss [47] who explained the principal aspects of ferromagnetism by means of two assumptions: the existence of the strong ($\sim 10^7$ Oe) internal molecular field, which aligns magnetic moments, and the existence of domain structure, which reduces the total magnetic moment of a ferromagnetic specimen. The explanation of the molecular field in terms of exchange forces was contributed by Heisenberg [48], and the explanation of the origin of domains in terms of magnetic field energy was given by Landau and Lifshitz [49].

Figure 6.3 illustrates bulk ferromagnetic specimens consisting of a number of small regions, each spontaneously magnetized to saturation. These regions are called "domains." The boundaries between domains are called domain walls. The domain walls must not be regarded as infinitely thin surfaces but rather as zones of transition of finite thickness in which the magnetization gradually changes from the direction on one side to that on the other [50].

The domain structure can be favorable energetically if the decrease of the magnetostatic energy, which is due to magnetic field around a specimen,

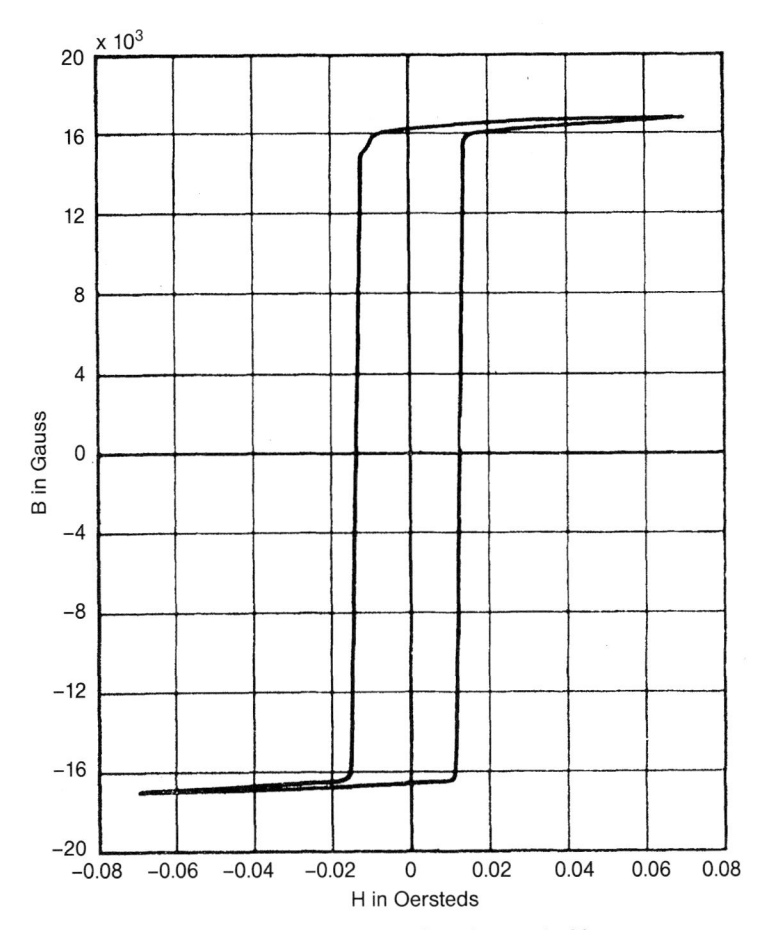

Figure 6.2 Experimental magnetization curve of single crystal of ferromagnetic iron [46]. Note that $B = \mu_0(H + M) \approx \mu_0 M$.

dominates the increase in the exchange energy relating to domain walls. In the absence of an applied magnetic field, the demagnetized state is the stable state in large ferromagnetic crystals. In a demagnetized specimen, the directions of magnetization of the individual domains are distributed at random among various possible directions, so that the magnetic flux circuit lies almost entirely within the specimen (Figure 6.3(a)). The magnetization directions of domains are determined mainly by the so-called crystalline anisotropy. The energy of atom in crystals depends on orientation of the magnetic moment with respect to crystallographic axis. The directions for which the energy has minimum (maximum) value are called directions (axes) of easy (hard) magnetization. As a result, the saturation magnetization in bulk ferromagnets like Fe, Co, Ni is not a simple scalar magnitude but depends on the orientation of their crystallographic axes in the external magnetic field.

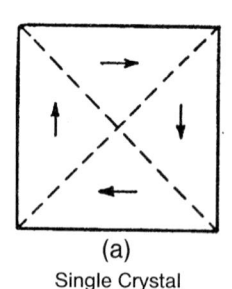

Figure 6.3 Schematic domain arrangements for zero resultant magnetic moment in a single crystal (a) and in a polycrystalline specimen (b). The domain structure of the polycrystalline specimen has been drawn for simplicity as if each crystallite contained only a single domain; this is not usually the case [46].

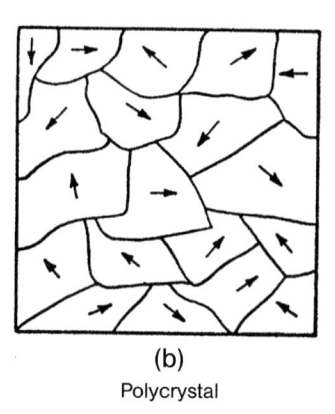

(a) Single Crystal

(b) Polycrystal

As the specimen is subjected to larger and larger fields, at first the magnetization remains along directions of easy magnetization, but domains magnetized close to the field direction grow at the expense of those magnetized away from the field direction (domain walls displacement). Eventually, only the directions of easy magnetization near the field direction are occupied. At still higher fields, the magnetization rotates toward the field direction and the magnetic saturation can be reached after all.

When at extremely high applied fields, the magnetization approaches the saturation magnetization value M_S and the field is decreased, the magnetization does not follow, in general, the initial magnetization curve obtained during the increase. When the field is decreased to zero, in general, a nonzero (remanent) magnetization M_r survives. If the field is applied in the reverse direction, the magnetization is equal to zero at a field $(-H_C)$, where H_C is called the coercive force. When the negative field increases still further, the magnetization reaches the saturation value again. This irreversible behavior of magnetization versus external magnetic field is called hysteresis [51]. An example of a hysteresis curve is given in Figure 6.4. In general, experimentally observed hysteresis is a nonequilibrium (in thermodynamical sense) phenomenon complicated by the influence of the ferromagnet's real structure (e.g., defects in single crystals, morphology in polycrystals and composites, etc). The realization of a specific domain pattern

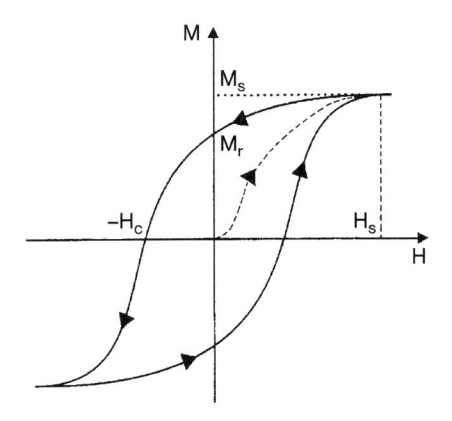

Figure 6.4 Typical ferromagnetic hysteresis curve. M_S is the saturation magnetization, H_S is the saturation field, M_r is the remanent magnetization, and H_C is the coercive force.

depends essentially on the magnetic history. Due to "impurities" (or defects) acting as pinning centers, at temperatures much below the Curie temperature, domain walls propagate very slowly and the equilibrium distribution of domains can only be reached in very long times. Naturally, domain wall dynamics becomes faster at temperatures near the Curie temperature.

The coercive force is the most sensitive property of ferromagnetic materials which is subject to control, and is one of the most important criteria in the selection of ferromagnetic materials for practical applications. The essential difference between material for permanent magnets and material for transformer cores lies in the coercive force, which may range from the value of 10^4 Oe in NdFeB and SmCo to the value of 0.01 Oe in NiFe [52]. Hence, the coercive force in bulk ferromagnetics may be varied over a range of 10^6. The coercive force of small ferromagnetic particles is observed to increase, as the particle size decreases, at least until extremely small sizes are reached (see Section 6.9).

The increasing of the coercive force with reducing of the particle size has been considered as substantial evidence of real existence of single-domain particles [46]. Indeed, if no domain boundaries are formed, then only magnetization changes in a specimen occur through spin rotation. Spin rotation is opposed by the anisotropy forces, which are usually much greater than the local forces opposing movement of a domain boundary. With decreasing particle size, it may therefore be expected that the coercive force will increase. However, in particles with size below some characteristic value D_{SD} (see below), one could observe the decrease in the coercive force with decreasing particle size if the temperature is higher than the so-called blocking temperature T_B (see Section 6.5).

The multidomain state is energetically favorable if the energy consumption for the formation of the domain walls is lower than the difference between the magnetostatic energies of the single-domain and multidomain states. As the dimensions of the specimen are diminished, the relative contributions of the various energy terms to the total energy of ferromagnetic specimen

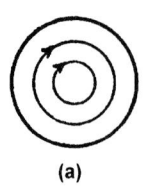

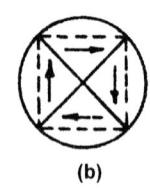

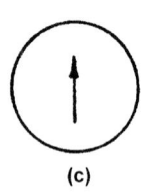

Figure 6.5 Types of magnetization in small sphere. (a) Low anisotropy; (b) high anisotropy in a cubic crystal; and (c) single-domain particle of uniaxial crystal [54].

are changed, and the surface energy of domain walls becomes more important than the magnetostatic volume energy. There is a specimen critical size D_{SD}, at which it is favorable energetically to do away with the domain boundaries [53]. A particle of ferromagnetic material, below a critical particle size, would consist of a single magnetic domain (so-called single-domain particle). The configuration of the magnetization inside a single-domain particle depends strongly on the magnetic anisotropy and particle's shape. If the ferromagnetic ellipsoidal (or spherical) particle has a negligibly small crystalline anisotropy, the atomic magnetic moment may be expected to point along closed rings (Figure 6.5(a)) [54], so that total magnetic moment of the particle is equal to zero. If the crystalline anisotropy is relatively large most of the atomic magnetic moments may be expected to lie along easy directions (Figure 6.5(b)). For example, in case of the strong uniaxial anisotropy, the single-domain ellipsoidal particle has uniform magnetization (along unique easy axis) and can act as a permanent magnet (Figure 6.5(c)).

Charles Kittel [46] presented an order-of-magnitude estimate by comparing the energy necessary to create a domain wall with the reduction of the magnetostatic energy during the creation of a domain structure. Kittel's critical radius for uniform magnetization state is valid for cubic crystals with saturation magnetization M_S given by

$$R = 9\gamma_w/(\pi M_S^2),$$

where $\gamma_w = 2(AK)^{1/2}$ is the surface energy of a Bloch wall in an infinite material with low anisotropy, A is the exchange stiffness constant, and K is the anisotropy constant. Note that the values A, K, and M_S^2 represent the exchange, anisotropy, and dipolar volume energies, respectively [55].

The combination of the various energy parameters introduces characteristic length scales, for example, the exchange length $l_{ex} = (A/M_S)^{1/2}$, the Bloch wall thickness $l_w = (A/|K_1|)^{1/2}$. The relative ordering of the quantities l_{ex}, l_w, and the particle size l is of great importance for the critical properties of the nanoparticle. Roughly speaking, for $l \ll l_w < l_{ex}$, the particle is dominated by the exchange interaction and the magnetic behavior will resemble that described by Stoner and Wohlfarth [37]. For elongated particles, l_{ex} approximately indicates the particle radius, above which nonuniform magnetization processes become important during magnetization reversal [41].

The size of a ferromagnetic particle, below which domain wall formation is unstable, can be obtained from the comparison of the energy of the particle without domain wall and with domain wall. Estimates of these energies are $NVM_S^2/2$ (N is the smallest demagnetizing field factor and V the volume of the particle) and $\gamma_w V^{2/3}$, respectively.

This gives the characteristic length, $D_{SD} = \gamma_w/M_S^2$, where the coefficient $2/N$, related to the particle shape, is often taken equal to unity. Particles of size $l < D_{SD}$ are usually named single domain [55]. Note that this is a steady description to be applied at equilibrium state and zero Kelvin only (no temperature excitations).

In the view of the condensed matter physics, finite-size effects [4] are originated due to cutting off of characteristic length (exchange length, domain size, etc, [56, 57]), resulting from the geometric limitation of particle volume. For example, macroscopic ferromagnetic single-crystal materials have well-defined T_C values depending exclusively on their composition [58]. As a specimen characteristic dimension d approach nanosize values and since the magnetic correlation length diverges at T_C, the correlated fluctuating magnetic moments in a volume are influenced by the finite size of the specimen. The T_C is then reduced as

$$[T_C(d) - T_C(\infty)]/T_C(\infty) = \pm(d/d_0)^{-\lambda},$$

where $T_C(\infty)$ is the bulk Cure temperature, λ is related to a correlation length exponent, and d_0 is an order of the characteristic microscopic dimension [59]. Hence, if the particle size decreases, we can anticipate that the Curie temperature should also decrease. However, changing of crystallographic parameters or composition, both in the particle core or its surface layer, can mask and even inverse this effect [60–62].

Surface effects are due to the lack of translational symmetry at outer boundaries of a particle, reduced coordination number, and broken magnetic exchange bonds of surface atoms [63]. Decreasing the particle size gradually increases the ratio of surface spins to the total number of spins. For instance, in a maghemite (γ-Fe_2O_3) particle of radius about 4 nm, 50% of atoms lie on the surface [64]. In a certain sense, surface effects can be considered as a sort of finite-size effects since the surface influence is most significant in smallest nanoparticles and should vanish for very large particles. Physical properties of the surface and the core of a nanoparticle can differ very much. Hence, the competition (or cooperation) between surface and core magnetic subsystems determines magnetic parameters of nanoparticle as a whole. So then, magnetic nanoparticles can have a complex (not uniform) structure. Often the presentation of a magnetic nanoparticle as the single super-large magnetic moment (super-moment) is the oversimplification.

Finite-size and surface effects can drastically change magnetic properties of nanoparticle in comparison with corresponding "bulk" counterpart [65, 66]. Let us imagine starting with a bulk single crystal spherical sample of ferromagnetic material at a temperature much lower than the magnetic ordering temperature (T_C or T_N) of the bulk material. We also assume that we are in thermodynamic equilibrium at each step in the size reduction process. In that case, the specimen has an ordered magnetic structure with equilibrium domain structure, which is consistent with applied field and the sample's shape and orientation in the applied field. As the specimen size decreases and becomes comparable to equilibrium domain sizes of the bulk material (10^6 to 1 nm, depending on the intrinsic material properties), the domain configuration or domain microstructure must significantly modify and adjust itself to the new equilibrium configuration of each new size. At this point, particle shape and surface start to play a marked role.

Below some critical size D_{SD}, the nanoparticle will become single domain, rather than multidomain. This is an important boundary that dramatically affects the particle's magnetic properties. The particle no longer sustains a domain wall and now consists of an essentially uniformly magnetized core carrying a net magnetic super-moment frozen along one of its magnetocrystalline easy directions. At $D > D_{SD}$, one could magnetize or demagnetize the particle under changing applied field simply by moving the domain walls, a relatively low barrier energy process. At $D < D_{SD}$, the only way to change the magnetization of a sample containing spatially fixed single-domain particles is for the applied field to overcome the magnetocrystalline barrier by causing a uniform rotation of the strongly exchange coupled moments, that is, of the particle super-moment. This is of course a simplified picture (the magnetization and the rotation are never perfectly uniform) but it is often close enough to reality to be very useful [37]. Although at $D \sim D_{SD}$, the particle surface acts as a significant perturbation, the surface region of a particle still plays a supporting role in formation of particle's magnetic characteristics, while the inner part (core) of the particle plays dominant role.

It should be noted that many estimations of the critical size of the crossover from the single to the multidomain state are often debatable [67]. Especially those which are based on comparing the energy of the single-domain particle with roughly calculated energy of an arbitrarily chosen magnetization structure for the multidomain state. Brown [68] emphasized that all such approaches cannot even establish the existence of a single-domain particle, let alone evaluate its size. When a particular magnetization structure is used for comparing the energy, it is impossible to know whether another configuration, not considered in that calculation, may not have a still lower energy than that of all the configurations which are considered. Even a calculation which is done rigorously and without approximation can yield only an upper bound to the size below which the particle must be uniformly magnetized. It proves that it is energetically favorable for the particle not to be uniformly magnetized above a certain size, but then it is impossible to be sure that a more sophisticated configuration will not have a lower energy even for a lower size. It thus takes a rigorous lower bound to prove that the process of finding more and more sophisticated configurations will end at a finite particle size, namely that there exists a size below which the particle indeed becomes a single domain.

Using analytical micromagnetics, Brown [68] developed a rigorous calculation of the critical size for the spherical particle. He established the upper and lower limits of the relevant energies at which the energies of the uniform state and the lowest energy of the nonuniform state become equal. Brown provided such a lower bound for a spherical particle with uniaxial anisotropy by proving all the necessary mathematical rigor that the energy of a uniformly magnetized ferromagnetic sphere is smaller than that of all possible magnetization configurations in that sphere, provided its radius R is smaller than R_0.

$$R_0 = 3.6055 \{2A/(4\pi M_S^2)\}^{1/2}.$$

If $R < R_0$, the lowest energy state is the single-domain configuration with uniform magnetization. Brown [68] also calculated two upper limits. The lowest energy state of a ferromagnetic sphere is definitely not that of a uniform magnetization, whenever its radius R is larger than the smaller of these two entities:

$$R_{C1} = 4.5292 \cdot (2A)^{1/2}/(4\pi M_S^2 - 5.6150 K_1)^{1/2}$$

$$R_{C2} = (9/8) \cdot \{2A(8\pi\sigma M_S^2 + K_1)\}^{1/2} / \{(3\sigma - 2)M_S^2\},$$

where σ is a numerical factor equal to 0.785398, K_1 is the absolute value of the first-order magnetocrystalline (uniaxial) anisotropy constant. It should be noted that formula for R_{C1} is valid only as long as the expression under the square root in that equation is positive. Hence, it gives the upper bound for low-anisotropy materials, where the energy of the curling or vortex state is compared with the energy of the uniform magnetization state. In high-anisotropy materials, the expression for R_{C1} can become meaningless and only R_{C2} should be used. Besides, in case of large K_1, the two-domain structure instead of the curling competes with the uniform magnetization state. Let us designate $K_1/(4\pi M_S^2)$ as ξ, R_{C1}/R_0 as ξ_1, R_{C2}/R_0 as ξ_2.

Of the two upper bounds for R_C, the lower is R_{C1} when $\xi < 0.1768$ and R_{C2} when $\xi > 0.1768$. When $\xi = 0.1768$, $\xi_1 = \xi_2 = 14.7$, and when $\xi \gg 1$, $\xi_2 = 11.0 \xi$.

For low anisotropy, the critical radius can be calculated as $0.5 \cdot (R_0 + R_{C1}) = 4.07(2A/4\pi)^{1/2}/M_S$ with accuracy about 11% [69]. It must be emphasized, though, that these results apply strictly to the case of a spherical shape only. Note also that the Brown's method estimated the size, but did not give the type of configuration which would have the lowest energy just above that critical size.

The values of critical radii R, R_0, R_{C1}, R_{C2}, l_{ex}, l_w for some typical magnetic materials are listed in the Table 6.1. It should be noted that since the exchange constant A is not known with the sufficient accuracy, these values should be considered as evaluative.

At present, the problem of critical radii of a spherical soft-magnetic nanoparticle seems to be solved. Three-dimensional computations [70, 71] showed clearly that the lowest energy remanent state of a ferromagnetic sphere is the collinear (saturated) state below a certain size and curling (vortex) configuration above that size. The question that remains unanswered is if any realistic particles can be approximated by the shape of an ideally smooth sphere. Even in the finite-difference computations, it was quite difficult to incorporate enough structural elements to approximate a sufficiently smooth sphere. Whether this shape can be made of atoms remains an open question. It should also be borne in mind that there can exist a lower bound to the validity of micromagnetic theory, which is not exactly known but which may well be within the realms of achievable particle size. We cannot exclude that the whole micromagnetic approach must be modified in case of very small particles. For example, the Heisenberg approximation, on which micromagnetics is based and which assumes strictly nonitinerant electrons, may break down in the limit of extremely small particles [72].

Many others questions exist, even if the micromagnetic theory gives correct picture for the size-depending magnetic behavior of nanoparticles. For example, what is the magnetic structure for particles of intermediate size between the upper and lower limits of the critical size (R_0 and $R_{C1,2}$ in case of spherical particles)? Kákay and Varga [71] studied remanent states of sphere-like particles in order to establish the critical radius for the transition from the single-domain to vortex configuration. A hard-axis-oriented vortex state has been found as a transition state between the monodomain and easy-axis-oriented vortex magnetization states.

It should bear in mind that since the magnetic parameters (the saturation magnetization and the crystalline anisotropy) are sensitive to temperature and external magnetic field, then critical radii can also be temperature and field dependent [73].

6.4
Shape Effects

In general, the lowest energy state of a magnetic particle depends on its size, shape, and the strength and character of its anisotropy. The shape of nanoparticles can influence its magnetic properties in different ways. For example, classical electrodynamics teaches us that the homogenous magnetization is achievable only for ellipsoidal bodies. Hence, the ideal single-domain particle has to be ellipsoidal. Distortions of particle shape can induce additional anisotropy, stabilizing (or destabilizing) the single-domain state. Small deviations from uniformity in the magnetization field within the nanoparticles can play an essential role in determining its magnetic properties (susceptibility, anisotropy, hysteresis features, etc.) [74]. The surface effects can be also shape-dependent since the relative number of surface atoms depends on the particle shape [75].

6.4 Shape Effects

The rigorous upper and lower bounds for the critical nanoparticle radii, which Brown obtained for a sphere, have been extended, to some extent, to the case of a prolate spheroid [76, 77]. Again, it was rigorously proved that the saturated (single-domain) state has the lowest energy of all possible configurations, below a certain size. Numerical computations of the lowest-energy state just above that size showed a cylindrical wall structure. The latter is quite similar to the case of the curling in a sphere. Therefore, the conclusion for prolate spheroids is similar to that for the sphere. The theoretical results are unique, clear-cut, and rigorous, but it is not clear at all if they apply to any realistic particle which does not have the shape of an ideally smooth ellipsoid. Various imperfections can strongly affect nanoparticle magnetic properties. In that case, studies of nanoparticles of ideal shapes may become purely academic [78].

Experiments using transmission electron microscope show that the shape of real magnetic nanoparticles can markedly deviate from the ellipsoidal one [79]; they can be, for example, cubic [80] or triangle [81]. Ferromagnetic nanoparticles of ideal geometrical shape (parallelepiped, cylinder, triangular prism, etc) can be now obtained by means of electron beam or imprint lithography [82] or by colloidal chemical synthetic procedures [79, 83]. To what extent does the properties of a small ferromagnetic particle of nonellipsoidal external shape differ from those of an ellipsoidal one? The results of numerical simulations [41, 84–89] show that the quasiuniform magnetization is the lowest energy state for particles of various external shapes at small enough sizes. In case of cubic nanoparticle, this quasiuniform magnetization is called a flower state [84] because the magnetization in the cube corners spreads outwards like the petal.

Independently of the particle shape, if its size is small compared to the lengths l_w and l_{ex}, the exchange coupling dominates over all other interaction mechanisms. Each volume element of the particle is tightly coupled to every other part of the sample. Large deviations of the magnetization from the uniform state are energetically unfavorable since the associated expense of short-range interaction energy cannot be balanced by the long-range magnetostatic interaction. Therefore, a first approximation to the magnetization configuration of all particles, which are small compared to the exchange length, is the uniform state. The degree by which the equilibrium state differs from the uniform magnetization state is a function of the size of the particle. The numerical calculations show that the uniform state may be used as a good approximation for particle of arbitrary shape which size is less than 20% of the least of l_{ex} and l_w. For bigger particles, the full structure of the nonuniform magnetization field must be taken into account and the particle shape becomes important. For example, a perfectly uniformly magnetized cube (if it existed) would possess a nonuniform demagnetization field. Hence, the uniform state of magnetization is never an exact equilibrium state for cubic particles. For cubic particles with the uniaxial anisotropy, various nonuniform magnetic states were found: symmetrical and twisted vortex (if the anisotropy is small) or two- and multidomain structures (in case of large anisotropy) [87].

To consider the relative number of surface atoms depending on the particle shape, the shape factor α can be introduced as the ratio of the surface area of a nonspherical nanoparticle S' to that of a spherical nanoparticle S, where both of the nanoparticle have identical volume, that is, $\alpha = S'/S$ [75].

On the basis of this definition, the geometry of nonspherical nanoparticle can be described by the particle size and the shape factor. The particle size is defined as the diameter of the corresponding spherical nanoparticle, and the difference between spherical nanoparticles and nonspherical nanoparticles is well described by the shape factor. The shape factor of spherical nanoparticles equals 1, which means that the spherical shape is included in the definition of shape factor. The shape factor of nonspherical nanoparticles is always larger than 1 and, therefore, shape factor can approximately describe the shape difference between spherical and nonspherical nanoparticles. Unfortunately, in that way introduced, the shape factor allows, in principle, that particles with different shapes may have the identical shape factor.

One more example of the shape effects on nanoparticle magnetic properties is the sensitivity of dipole–dipole (magnetostatic) interactions between particles to their shapes. The long-range magnetostatic interaction is often treated simplistically as if each nanoparticle were a pure dipole. The explicit consideration of the shape-conditional demagnetization factor of nanoparticles is not an easy task. Recently, a new effective method within the framework of a Fourier space approach was developed [90] for the dipolar interaction energy between nanoparticles of arbitrary shape and magnetization state. This approach takes into account the particle shape anisotropy without resorting to any approximation and obtained explicit results for interacting cylinders of variable aspect ratios (disks and rods) and special magnetization states (in-plane and axial). The main advantages of this approach, compared to the standard methods, are an easy and compact mathematical treatment of shape anisotropy, the accuracy of the results, and the availability of advanced fast Fourier transform algorithms, which can efficiently evaluate the various real space quantities.

6.5
Superparamagnetism

Although real nanoparticles can have a complex (not uniform) magnetic structure, as a rule an assembly of noninteracting single-domain isotropic particles behaves like classical paramagnetic matter with very high ($\sim 10^3 - 10^4\ \mu_B$) effective magnetic moment μ per "atom" particle. If we have this assembly at a temperature T in an applied field H, and assume that it has achieved thermodynamic equilibrium, there will be a Boltzmann distribution of the orientations of μ with respect to H. The average assembly moment in the direction of the field is equal to

$$\langle\mu\rangle = \mu \cdot \Lambda(\mu H/kT) = \mu \cdot (\coth(\mu H/kT) - kT/\mu H), \qquad (6.1)$$

where Λ is the Langevin function and k_B is Boltzmann's constant.

Since Eq. (6.1) is typical of classical paramagnetic and μ is much larger than any atomic magnetic moment (e.g., Gd^{3+} has greatest effective magnetic moment $\approx 8\ \mu_B$), the behavior of magnetization in that way has been called "superparamagnetism" [39]. Equation (6.1) can be used for experimental determination of magnetic moment and average magnetization $M_S = \mu/V$ of superparamagnetic nanoparticles (V is the particle volume).

Equation (6.1) is valid if the thermal energy at the temperature of the experiment is sufficient to equilibrate the magnetization of an assembly in a time short compared with that of the experiment [6]. It is possible in case of negligible anisotropy energy. However, real single-domain particles are usually not isotropic in their properties, but have anisotropic contributions to their energy arising from the shape of the particle, imposed external stresses due to both environment and lattice deformation in particle surface layer, and the crystalline structure itself (magnetocrystalline anisotropy).

The anisotropy energy of a single-domain particle is proportional, in a first approximation, to its volume V [91]. For most simple case of the uniaxial anisotropy, the associated energy barrier E_{an}, separating easy magnetization directions, is proportional to KV, where K is the anisotropy constant. With decreasing particle size, the anisotropy energy decreases, and for a particle size lower than a characteristic value, it may become so low as to be comparable to or lower than the thermal energy $k_B T$. This implies that the energy barrier for magnetization reversal may be overcome, and then the total magnetic moment of the particle can thermally fluctuate, like a single atomic magnetic moment in a paramagnetic material. The total magnetic moment of the particle may be freely rotated, whereas the moments within the particle remain magnetically coupled (ferromagnetically or antiferromagnetically). In that case, an assembly of nanoparticles can quickly approach to thermal equilibrium if external magnetic field or temperature changes.

The actual magnetic behavior of nanoparticle assemble depends on the value of the measuring time (τ_m) of the specific experimental technique with respect to the relaxation time (τ) associated with the overcoming of the energy barriers. In case of noninteracting single-domain nanoparticles with the uniaxial magnetic symmetry, the relaxation time is given by the Arrhenius law:

$$\tau = \tau_0 \exp(E_a/k_B T),$$

where E_a is the activation energy. Depending on the context, it is also known as the Néel or Néel–Brown relaxation law (see below).

If we consider single-domain particles with uniaxial anisotropy and uniform magnetization, and suppose that the particle magnetic moment reverses in external magnetic field by the coherent rotation, the energy barrier to thermal activation is given by

$$E_a = KV(1 - H/H_{sw})^m,$$

where H_{sw} is the switch field, m is phenomenological constant that depends on the angle ψ between the magnetic field H and anisotropy axis. If the field is parallel to the axis (aligned particles) then $m = 2$ and $H_{sw} = H_K$, where $H_K = 2K/M_S$ is the anisotropy field [4]. For randomly oriented particles, $m \approx 3/2$ [92].

Experiments on isolated, magnetic particles at low temperatures have confirmed the Néel–Brown theoretical approach to thermal activation [93]. However, at high temperatures, e.g., in a temperature range between room temperature and T_C, relaxation times may deviate by many orders of magnitude. These deviations are mainly due to the fact that the magnetic nanoparticle itself is a thermodynamic system, with a temperature-dependent magnetization and anisotropy energy. Consequently, the relevant energy barrier E_a is temperature dependent [35].

The value of τ_0 is typically in the range 10^{-13}–10^{-9} s [91]. If $\tau_m \gg \tau$, the relaxation appears to be so fast that a time average of the magnetization orientation is observed in the experimental time window, and the assembly of particles behaves like a paramagnetic system (superparamagnetic state). Certainly, there is not hysteresis in the superparamagnetic state. On the contrary, if $\tau_m \ll \tau$, the relaxation appears so slow that thermodynamical nonequilibrium properties are observed (blocked state). The blocking temperature T_B, separating the two states, is defined as the temperature at which $\tau_m = \tau$. Near T_B, changes of magnetization reorientation occur with relaxation times comparable with the time of a measurement; the result is an observable lag of magnetization changes behind field changes. This phenomenon is called magnetic after-effect or magnetic viscosity. Well below T_B the thermal agitation can be neglected and static magnetization is calculated with help of minimizing total system free energy. This is the well known Stoner–Wohlfarth calculations [37], which showed that the blocking state is closely related to hysteresis, because in certain magnetic field ranges there are two or more minima, and thermally activated transitions between them are neglected.

The exact value of T_B is in some ways blurry due to particle inequality and rather arbitrary choice of τ_m. Also the blocking temperature T_B is not uniquely defined since the values of τ_m depend on the experimental technique. For example, in Mössbauer spectroscopy, the timescale τ_m is of the order of the nuclear Larmor precession time, i.e., typically a few nanoseconds, while in static magnetization measurements, the timescale is typically of the order of 1 s. Therefore, the same nanoparticles can be in the superparamagnetic state in static magnetization experiments and in the blocked state in Mössbauer spectroscopy experiments [91]. In any case, the blocking temperature T_B increases with increasing of a nanoparticle size and for a given size increases with decreasing measuring time. The highest possible value of T_B is represented by the Curie (or Néel) temperature, at which the magnetic moments within each particle decouple.

The interesting question is under what conditions an assembly of single-domain particles can achieve thermal equilibrium in a time short compared with that of an experiment. The simplest means of approaching equilibrium is by rotation of the individual particle body as whole, as would occur if the magnetic nanoparticles were suspended in a liquid medium [94]. Such mechanisms are most important in ferrofluids [95]. In that case, the main factors determining the rate of approach to equilibrium are the viscosity of the suspension medium and the effective particle volume [96]. Evidently, this relaxation mechanism will not be available in solids (except for, probably, some specific polymers [97]). In 1949, Néel [38] pointed out that if a single-domain particle were small enough, thermal fluctuations could cause its direction of magnetization to undergo a sort of Brownian rotation – thermally activated reversal of magnetization (above-mentioned coherent reversal). He derived the conditions under which an assembly of such particles could come to thermal equilibrium in a given time. This relaxation calculation has been generalized by Brown [98] with results that are in essential agreement with Néel's theory. The detailed theoretical analysis of the Brown's approach to the superparamagnetic relaxation was performed in [99].

The experiments on water-based magnetic fluids containing magnetic iron oxide nanoparticles [100] indicate that the time dependences of Brownian and Brown–Néel relaxation are clearly different. This is due to the relatively narrow distribution of Brownian relaxation times compared to the extremely wide distribution of Brown–Néel relaxation times. The presence of a Brown–Néel relaxation signal in ferrofluids gives evidence that at least for a fraction of particles the Brownian mechanism is inhibited. This is the case, e.g., for magnetic nanoparticles fixed to a solid phase or unstable aggregated magnetic fluids containing sediments. Since the relaxation signal yields information about the underlying relaxation process, it can be consequently utilized to distinguish between particles bound to a solid phase and free particles. This is an important feature for the analysis of biological binding processes.

There is one more mechanism of magnetic relaxation in nanoparticles and molecular clusters – the quantum tunneling. This mechanism was suggested by Bean and Livingston for spontaneous magnetization reversal of small particles [39], and developed theoretically by a number of authors [101–103]. The quantum tunneling is relevant at rather low temperatures, where $\exp(-E_a/k_B T)$ is negligible and thermal fluctuations become ineffective, while the quantum tunneling barrier is low or narrow enough to allow macroscopic quantum tunneling of the super-moment between its easy axis orientations. Evidence of such tunneling has been observed by the electron paramagnetic resonance technique in molecular clusters [104] and iron-oxide nanoparticles [105].

6.6
Surface Effects

Let us continue decreasing the particle size below the critical single-domain size D_{SD}. In a particle of radius 4 nm, about 50% of atoms lie on the surface. It is reasonable to suppose that at some characteristic size D_{SR}, the roles of particle's core and surface region become comparable. Now the surface region essentially modifies magnetization configuration of the particle and has effect on its magnetic characteristics. When $D < D_{SR}$, the concept of a well-defined super-moment breaks down and we need to use rather discrete models with individual atomic (ionic) magnetic moments. In real particles, the surface region thickness (and, correspondingly, D_{SR}) is very sensitive to particle shape distortion, surface roughness, surface impurities, defects (like vacancies), compositional inhomogeneities, surface chemical bonds with environment, etc. Hence, it is not possible create a column for even approximated values of D_{SR} in Table 6.1. However, to our best knowledge, no experiment shows that $D_{SR} > D_{SD}$. Reviews of surface effects and their importance in dealing with real magnetic nanoparticles have been given by Kodama [33], Battle and Labarta [4], and in the recent book [106].

If we continue to decrease the particle size below D_{SR}, we eventually reach the objects, where it may be more relevant to speak of polynuclear molecules than of nanoparticles. These are, for example, molecular clusters of only several tens paramagnetic cations [107, 108]. Crystals of molecular clusters can be elegant model systems of identical magnetic nanoparticles (e.g., [109]). When one reaches these sizes, any change in cluster size (i.e., by as little as one paramagnetic cation or one coordinating anion) can cause dramatic changes in the magnetic properties. In case of antiferromagnetic coupling, clusters with odd numbers of paramagnetic cations have large residual super-moments whereas clusters with even numbers of paramagnetic cations do not. Similarly, substitution of a single coordinating (i.e.,

Table 6.1 Critical radii and characteristic length scales for typical ferro- and ferrimagnets.

	M_S (emu/cm^3)	K_1 (10^4 erg/cm^3)	A (10^{-6} erg/cm)	R (nm)	R_0 (nm)	R_{C1} (nm)	R_{C2} (nm)	l_{ex} (nm)	l_w (nm)
Iron	1710[a]	45[a]	1.0[a]	13.1	8.4	11.0	116.5	5.9	14.9
Iron	1714[b]	47[b]	2.1[b]	19.4	12.1	15.9	168.5	8.5	21.1
Cobalt	1430	430[a]	1.0[a]	58.1	15.8	51.4	146.0	7.0	4.8
Nickel	483	4.5[a]	1.0[a]	52.1	29.8	39.1	412.9	20.7	47.1
γ-Fe$_2$O$_3$	350[c]	−4.6[c]	0.1[c]	31.7	13.0	17.9	181.0	9.0	14.7
Fe$_3$O$_4$	480[c]	−11.0[c]	0.1[c]	26.1	9.5	13.4	132.3	6.6	9.5

[a] [67].
[b] [71].
[c] [41].

superexchange) anion can change a key magnetic exchange bond from being antiferromagnetic to being ferromagnetic or vice versa. Also, with such small clusters, the magnetic exchange bond topology becomes critical and, as with low dimensional materials, the concept of a bulk magnetic ordering temperature becomes irrelevant. This is because single-ion magnetic exitations become the energetically preferred, even at the lowest temperatures, relative to cooperative magnetic excitations such as super-moment and superparamagnetic fluctuations [110]. In this limit, therefore, depending on temperature, magnetocrystalline anisotropy, exchange strength, and exchange bond topology, the concept of a super-moment looses its usefulness because fluctuations in the super-moment magnitude become comparable to the fluctuations in super-moment direction. The magnetism of molecular clusters is better described in terms of exchange-modified paramagnetism than in term of superparamagnetism [111].

Thus, with the decreasing size of a magnetic particle, which remains sufficiently large to not be turned into the molecular cluster, the surface effects are believed to become more and more pronounced. A simple argument based on the estimation of the fraction of surface atoms shows that for a particle of spherical shape and diameter D (in units of the lattice spacing), this fraction is an appreciable number of order $6/D$. Regarding the fundamental property of magnetic particles, the magnetic anisotropy, the role of surface atoms is augmented by the fact that these atoms in many cases experience surface anisotropy that by far exceeds the bulk anisotropy.

The magnetocrystalline anisotropy reflects the symmetry of the neighbors of each atom. The large perturbations to the crystal symmetry at surfaces should lead to magnetocrystalline anisotropy of different magnitude and symmetry for surface sites. In fact, surface anisotropy has a crystal-field nature and it comes from the symmetry breaking at the boundaries of the particle. As was suggested by Néel [112] and microscopically shown in [113], the leading contribution to the anisotropy is due to pairs of atoms and can be written as

$$H_A = (1/2) \Sigma L_{ij}(\mathbf{m}_i \mathbf{e}_{ij}),$$

where $\mathbf{m}_i$ is the reduced magnetization of the ith atom, $\mathbf{e}_{ij}$ are unit vectors directed from the ith atom to its neighbors, and L_{ij} is the pair anisotropy coupling that depends on the distance between atoms. This equation describes in a unique form both the bulk anisotropy including the effect of elastic strains and the effect of missing neighbors at the surface that leads to the surface anisotropy. Surface atoms experience large anisotropy of order L due to the broken symmetry of their crystal environment – the so-called Néel surface anisotropy.

A model for describing the combined effect of reduced coordination and surface anisotropy in ionic materials was developed by Kodama et al. [32]. They assumed that the pairwise exchange interactions have the same magnitude for bulk and surface atoms, but that the total exchange interaction is less for surface atoms because of their lower coordination. They also postulated the existence of a fraction of extra broken exchange bonds due to oxygen vacancies or bonding with ligands other than oxygen at the surface. They treated the surface anisotropy as uniaxial, with the axis defined by the dipole moment of the neighboring ions. Since at the surface, some of the neighbors are missing, the total dipole electric field acting on the surface atom is nonzero and directed approximately normal to the surface. Hence, in this model, the easy axis is approximately radial.

Surface atoms can make a contribution to the effective volume anisotropy decreasing as $1/D$ with the particle's linear size: $K_{V,\text{eff}} = K_V + K_S/D$, as was observed in a number of experiments [114, 115]. Many other experimental results for metallic and oxide particles indicate that the anisotropy of fine particles increases as the volume is reduced because of the contribution of what is known as surface anisotropy. For example, the anisotropy per unit volume increases by more than one order of magnitude for 1.8 nm fcc Co particles [116] being 3×10^7 erg/cm^3 compared with the bulk value of 2.7×10^6 erg/cm^3. An anisotropy value of one order of magnitude larger than the preceding case has been reported for Co particles embedded in a Cu matrix [117] as well as in polymeric matrix [118].

Illustrative example of surface effects in magnetic nanoparticle FePt is presented by Labaye et al. [119]. Authors considered the effect of the surface anisotropy on an isolated single-domain spherical nanoparticle using atomic Monte Carlo simulation of the low-temperature spin ordering. The particles studied composed of magnetic atoms forming a simple cubic array with parameter a with six nearest neighbors, except at the surface. One sphere has radius $R = 6a$ and contains 925 atoms, with a surface-to-volume ratio of 0.38; the other has radius $R = 15a$ and contains 14,328 atoms, with a surface-to-volume ratio of 0.16. A classical magnetic moment is associated with each atom and the total energy is minimized for a given configuration. The nanoparticle core was defined as the region where every atom has six nearest neighbors and the nanoparticle surface as the outer shell where every atom has at least one dangling bond.

The magnetic system was described by a Heisenberg-type Hamiltonian which includes terms representing the nearest-neighbor exchange interaction, the bulk anisotropy, and the surface anisotropy:

$$H_i = -\Sigma J_{ij}\mathbf{S}_i\mathbf{S}_j - K_V(\mathbf{S}_i \cdot \mathbf{z})^2 - K_S(\mathbf{S}_i \cdot \mathbf{n})^2, \qquad (6.2)$$

where J_{ij} are the nearest-neighbor exchange coupling constants with different values depending whether the sites belong to the bulk (J_b) or to the surface (J_s), $\mathbf{S}_i$ and $\mathbf{S}_j$ are the spins on the i and j sites, K_V is the bulk anisotropy for all the sites belonging to the volume, and K_S is the surface anisotropy for all the sites belonging to the surface. It was supposed that the bulk anisotropy is uniaxial along the Oz axis (corresponding vector $\mathbf{z}$ in Eq. (6.2)) and the surface anisotropy is normal (corresponding vector $\mathbf{n}$ in Eq. (6.2)) to the surface at every site, in a direction calculated from the surface neighbor positions. Because of the lower coordination of the sites in the surface, $J_b \neq J_s$ in general, but in simulation, authors assumed $J_b = J_s = 10$ (ferromagnetic coupling), $K_V = 20$, and K_S ranging between 20 and 2000. If the unit of J is equivalent to 0.5 K, $J_b = 10$ corresponds to a Curie temperature $T_C = 725$ K, and $K_V = 20$ corresponds to a bulk anisotropy of about 9 MJ/m^3. These values are close to those of FePt nanoparticles [120].

For $K_S = K_V = 20$, the calculated spin structure is essentially collinear along Oz, both in the core and at the surface of the nanosphere (Figure 6.6(a)). Slight fluctuations in spin direction can be due to thermal effects. A different spin configuration is apparent for $K_S = 200$ as it is shown in Figure 6.6(b), the increased surface anisotropy now tries to impose a radial orientation for the surface spins, a tendency which propagates into the core via the exchange coupling and competes with the bulk magnetocrystalline anisotropy which favors the spins to be aligned along Oz. This competition yields a "throttled" structure that has the surface spins oriented inward for the upper hemisphere; they reverse progressively at the equator and become oriented outward for the lower hemisphere. The surface spin reversal from inward to outward creates the throttling of the core spin configuration. Nevertheless, the mean orientation of the core spins remains the Oz axis as imposed by the bulk

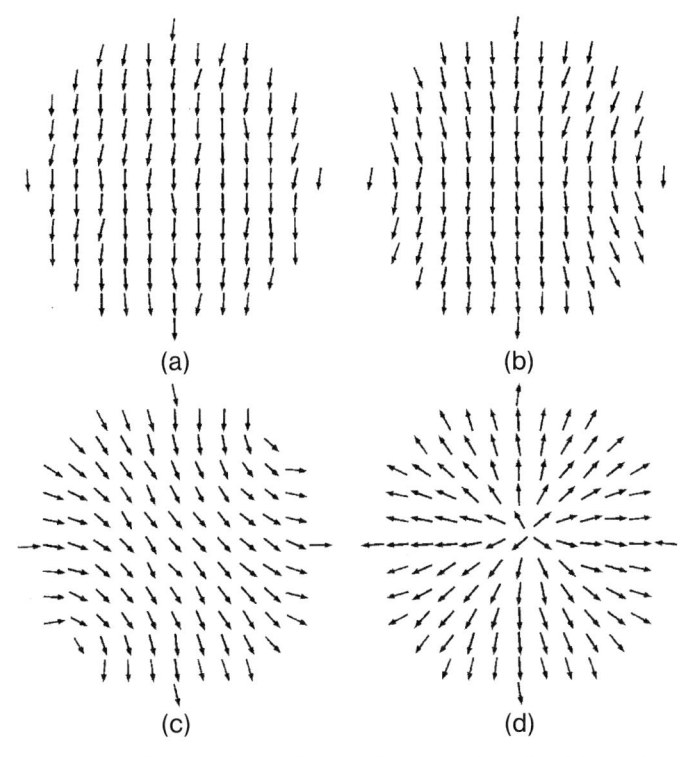

Figure 6.6 The central plane for a smaller nanosphere for $K_S/K_V = 1$ (a), 10 (b), 40 (c), and 60 (d). The Oz axis is vertical [119].

anisotropy. For $K_S = 400-1000$, the spin structure remains throttled but the reversal of the surface spins takes place only over few atomic spacings, creating vortex-type reversal centers. The spins of the nanosphere core become canted away from the Oz axis and tend to lie perpendicular to the line joining the two vortex centers (Figure 6.6(c)). For $K_S > 1200$, all the spins are radially oriented, either outward or inward the center of the nanoparticle, giving rise to a "hedgehog"-type spin structure (Figure 6.6(d)).

Figure 6.7 shows the spin configurations obtained for the larger particle with the same values of K_S. There is the same tendency to form a throttled state, with vortex centers at the surface. The crossover to the hedgehog state is not reached within the range of parameters we have explored. Authors supposed that this may be because the size of the particle exceeds the domain wall width $(2J/K_V)a = 10a$.

Analogous behavior was found in the work [36] where the effect of surface anisotropy upon the magnetic structure of ferrimagnetic maghemite nanoparticles was studied with the help of three-dimensional classical Heisenberg–Hamiltonian and a Monte Carlo–Metropolis approach. Results reveal throttle structure as surface anisotropy increases, as well as a marked

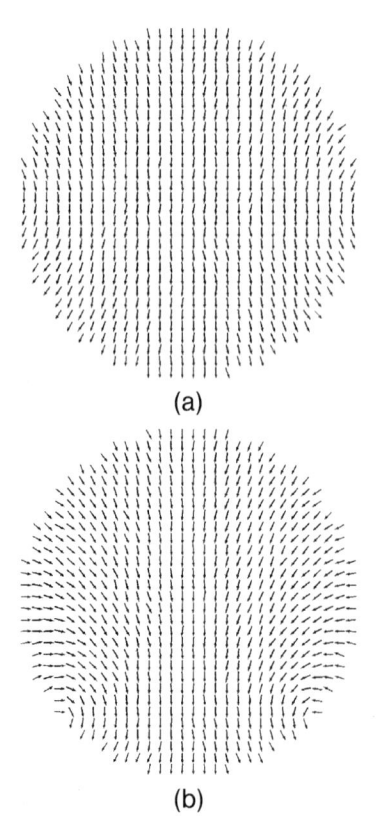

Figure 6.7 The central plane for a larger nanosphere for $K_S/K_V = 10$ (a) and 60 (b). The bulk anisotropy axis is vertical [119].

6.6 Surface Effects

decrease of the Curie temperature of the considered nanoparticle with that obtained of a bulk maghemite.

Some *ab initio* band calculations predict enhancement of magnetic moment per atom in thin films compared to the bulk value [121], and it seems that we could anticipate the same in the surface layers of nanoparticles. However, the reduction of the saturation magnetization, M_S, is a common experimental observation in many fine-particle systems [91]. In early models, this fact was interpreted by postulating the existence of a dead magnetic layer originated by the demagnetization of the surface spins, which caused a reduction of M_S because of its paramagnetic behavior [122]. A random canting of the surface spins caused by competing antiferromagnetic interactions between sublattices was proposed by Coey [123] to account for the reduction of M_S in maghemite ferrimagnetic particles. He found that even a magnetic field of 5 T was not enough to align all spins in the field direction for particles of 6 nm in size. The existence of canted spins was verified in different nanoparticle assemblies of ferrimagnetic oxides like maghemite, $NiFe_2O_4$, $CoFe_2O_4$, $CuFe_2O_4$ by Mössbauer spectroscopy, polarized and inelastic neutron scattering, and ferromagnetic resonance technique [4]. However, the origin of the lack of full alignment of spins in fine particles of ferrimagnetic oxides is an object of continuing discussion, but up to the moment, no clear conclusions have been established. So, the original suggestion of Coey [123] that spin canting occurs preferentially at the particle surface has been supported by means of Mössbauer spectroscopy in particles of maghemite [124]. At the same time, from other studies also based on Mössbauer spectroscopy, it has been suggested that spin canting is not a surface effect, but rather a finite-size effect which is uniform throughout the whole volume of the particle [125]. Spin canting has not been observed yet in metallic ferromagnetic nanoparticles, a fact which supports the hypothesis that the competition of frustrated antiferromagnetic interactions is in the origin of the spin misalignment [4].

Monte Carlo computer calculations show [126] that the effects of surface anisotropy on the spin configuration of magnetic nanoparticles should be most evident in those of ferromagnetic actinide compounds such as US and rare-earth metals and alloys with Curie points below room temperature and less significant for 3d ferromagnets and their alloys, with the possible exception of FePt [119].

In consideration of surface effects on nanoparticle magnetization, it should take into account that for ferromagnetic metal nanoparticles, pure finite-size effects are expected to enhance the M_S value with respect to the bulk, in contrast to ferrimagnetic oxides. Thus, metal atoms at the surface present a higher magnetic moment due to the band narrowing caused by the lack of orbital overlap [4, 33]. Indeed, it was reported about enhanced (25–30% as compared with a bulk value) atomic magnetization of cobalt nanoparticles [65, 114, 118]. However, for most metallic nanoparticles, the negative contribution of the surface even at low temperature often leads to a noticeable reduction of the magnetization for sufficient small particles which should be ascribed to the relative increase of the contribution from impurities and oxides at the surface layer [127].

On the other hand, the surface anisotropy makes the surface layer magnetically harder than the core of the particle. At low temperature, this can result in the marked magnetization enhancement, especially in amorphous nanoparticles [128]. Another interesting surface-related magnetic phenomenon is the so-called exchange bias (see Section 6.7), which appears in nanoparticles with ferromagnetic core and antiferromagnetic surface layer [129] or in nanoparticles embedded in either a paramagnetic or an antiferromagnetic matrix [130].

6.7
Matrix Effects

Naturally, matrix effects can be considered as a specific sort of surface effects. The magnetic exchange coupling induced at the interface between ferromagnetic and antiferromagnetic systems can provide an extra source of anisotropy, leading to magnetization stability. This principle was demonstrated [130] for ferromagnetic cobalt nanoparticles of about 4 nm in diameter that were embedded in either a paramagnetic or an antiferromagnetic matrix. Whereas the cobalt cores lost their magnetic moment at 10 K in the first system, they remained ferromagnetic up to about 290 K in the second. This behavior is ascribed to the specific way ferromagnetic nanoparticles couple to an antiferromagnetic matrix.

The system studied by Skumryev et al. [130] consisted of Co nanoparticles embedded either in a paramagnetic matrix (C or Al_2O_3) or in an antiferromagnetic matrix (CoO). The samples were grown by sequentially depositing a layer of matrix material of thickness 15–20 nm followed by a layer of nanoparticles. The Co–CoO core–shell nanoparticles were produced by gas condensation of sputtered atoms. As revealed by transmission electron microscopy, the Co particles had a roughly spherical core of $d_{core} < 3$–4 nm in diameter and a face-centered cubic (fcc) structure. The shell was fcc CoO with a thickness of about $t_{shell} < 1$ nm. The Co–CoO core–shell nanoparticles exhibit a log-normal distribution with a mean $d_{core} + t_{shell} < 4.7$ nm and a standard deviation of 1.1 nm. The average in-plane and out-of plane interparticle distances were around 12 nm, significantly larger than the diameter of the nanoparticles. As the maximum dipolar field between two nanoparticles in contact amounts to 0.07 T, interparticle dipolar interactions cannot be the source of the effects.

The temperature dependence of the magnetic moment under an applied magnetic field (0.01 T) was measured after field cooling (FC) and after zero-field cooling (ZFC) (Figure 6.8). As stated in Section 6.3, the blocking temperature (T_B) is the temperature at which superparamagnetism sets in. Below T_B, the FC and ZFC magnetization curves are split, whereas above

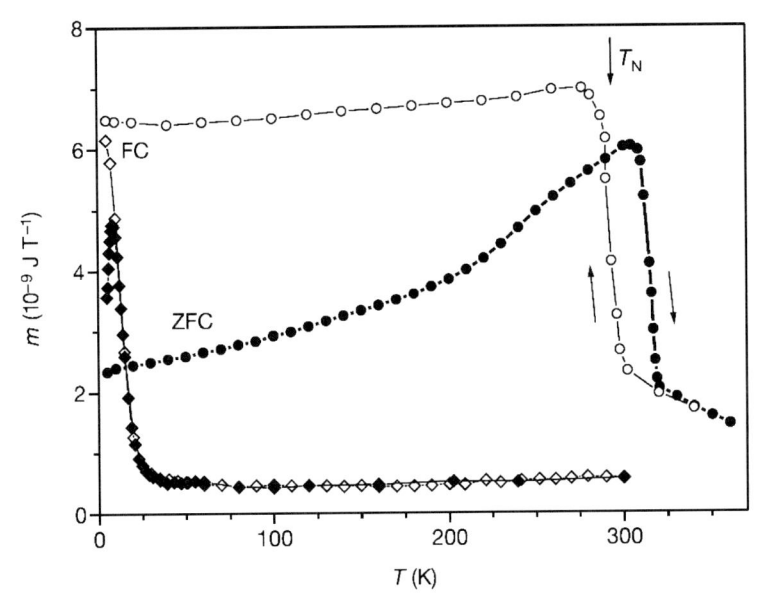

Figure 6.8 Magnetic moments of 4-nm Co–CoO particles versus temperature. The zero-field-cooled (filled symbols) and field-cooled (open symbols) magnetic moment. Particles were embedded in a paramagnetic (Al$_2$O$_3$) matrix (diamonds) or in an antiferromagnetic (CoO) matrix (circles). The Néel temperature of CoO is indicated by an arrow [130].

T_B, they coincide as the remanence and coercivity have vanished. For 4-nm particles embedded in an Al$_2$O$_3$ (paramagnetic) matrix, $T_B < 10$ K. Similar results were obtained for a C matrix. Increasing the particle size to about 7 nm resulted in an enhancement of T_B up to 2 K, that is qualitatively in accord with the well-known relation $T_B = KV/25k_B$, where V is the particle volume [91].

The hysteresis loops of both systems (with Al$_2$O$_3$ and C matrices) measured below T_B are characterized by a small value of coercivity (0.02 T) and no loop shift on the magnetic field axis for samples field cooled in fields up to 5 T. It is known [131] that exchange bias at ferromagnetic–antiferromagnetic interfaces is characterized by coercivity enhancement, revealing induced uniaxial or multiaxial anisotropy, and hysteresis loop shift along the field axis after FC, revealing unidirectional anisotropy (see Figure 6.9(b) and (c)).

The lack of exchange bias in the nonmagnetic matrix systems can tentatively be attributed to the small thickness of the antiferromagnetic CoO shell. When such Co–CoO core–shell particles were deposited without dilution in a paramagnetic matrix and hence in close contact with each other, they were found to exhibit all the features of an exchange-biased system – exchange bias field of about 0.92 T and enhanced coercivity (0.39 T in the ZFC case and 0.59 T in the FC one) at 4.2 K. The bias field decreased as temperature increased,

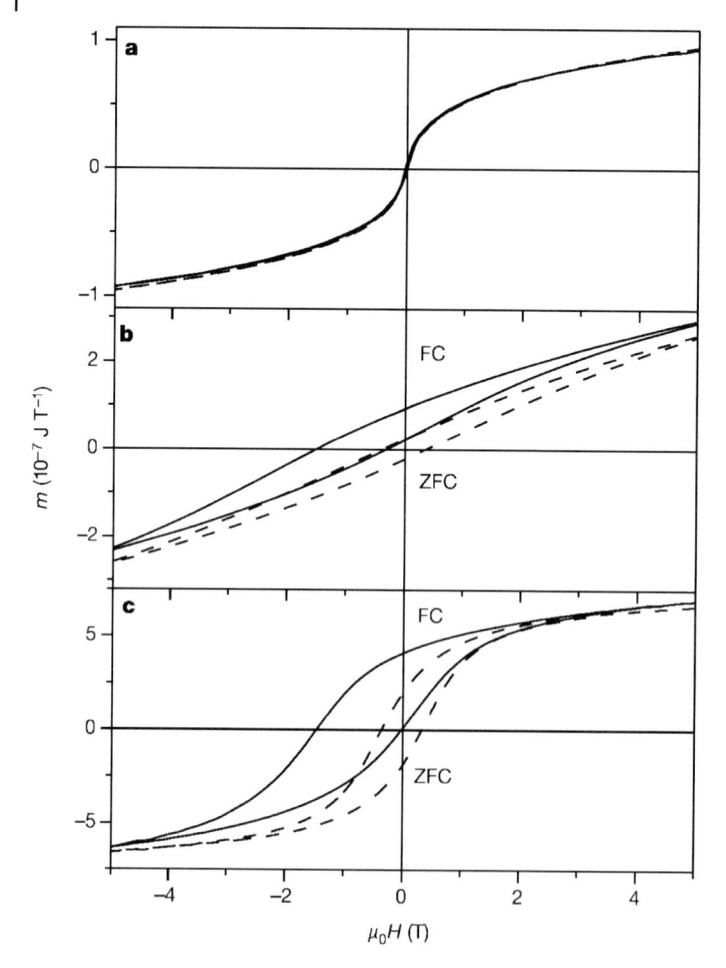

Figure 6.9 Hysteresis loops at 4.2 K of 4-nm Co–CoO particles embedded in different matrices. Data are shown after ZFC (dashed lines) and FC (5 T; solid lines) procedures. (a) Embedded in a paramagnetic (Al_2O_3) matrix, (b) a compacted matrix, and (c) an AFM (CoO) matrix [130].

and vanished at about 180 K. This difference in the behavior of isolated and compacted Co–CoO core–shell particles emphasizes the important role of interparticle coupling in stabilizing not only the ferromagnetism of the particle core but also the antiferromagnetism of the particle shell. It should be noted that the stabilization of ferromagnetic nanoparticle magnetization through interparticle magnetic interactions is not of interest for magnetic recording [132].

The magnetic behavior of the isolated Co–CoO particles changes markedly when, instead of being embedded in a paramagnetic matrix, they are embedded in an antiferromagnetic CoO matrix of similar thickness. As can be seen from the ZFC–FC curves, the Co nanocores remain ferromagnetic up to the Néel temperature, T_N, of CoO ($T_N < 290$ K). This indicates that an extra anisotropy is induced such that $KV \gg k_B T$. In this case, the nanocore moments are prevented from flipping over the energy barrier for all temperatures below T_N of CoO, and thus the nanoparticles remain magnetically stable below this temperature. A hysteresis loop typical of Co–CoO nanoparticles embedded in a CoO matrix is shown in Figure 6.9(c).

Below T_N, it is a characteristic of an exchange-biased system, exhibiting an exchange bias field of 0.74 T and enhanced coercivity of 0.76 T at 4.2 K. Both the coercive force H_C and the remanent magnetization M_R remain much larger than zero for $T < T_N$. In contrast, the Co nanoparticles embedded in a paramagnetic matrix have zero M_R and H_C for $T > T_B = 10$ K.

A behavior similar to the one described above for Co_{core}/CoO_{shell} particles was also observed for pure Co nanoparticles (i.e., without CoO shell) embedded in a CoO matrix. However, because of the poorer quality of the interface between the antiferromagnetic matrix and the Co ferromagnetic nanoparticles, all the effects associated with exchange anisotropy were less pronounced.

Results of this remarkable work [130] clearly demonstrated how the coupling of ferromagnetic particles with an antiferromagnetic matrix can be a source of a large effective anisotropy, leading to a considerable increase of the nanoparticle blocking temperature and therefore to higher thermal stability which is desirable for the magnetic recording applications.

Another important matrix effect relates to magnetoresitance phenomena. Berkowitz et al. [133] have observed giant magnetoresitance effect in heterogeneous thin film Cu–Co alloys, consisting of ultrafine Co-rich particles in a Cu-rich matrix. It was found that the magnetoresistance (MR) scales inversely with the average particle diameter. This behavior was successfully modeled by including spin-dependent scattering at the interfaces between the particles and the matrix, as well as the spin-dependent scattering in the Co-rich particles. The giant magnetoresistance (GMR) effects were also found in many other binary metallic systems (Co–Ag, Fe–Ag, Fe–Cu, Fe–Au) [134].

6.8
Interparticle Interaction Effects

At temperatures much greater than the blocking temperature when magnetic relaxation is fast on the timescale of the experiment, magnetic nanoparticles exhibit superparamagnetic behavior which weakly depends on interparticle interactions. However, at lower temperatures, interactions between the nanoparticles become important and can have a significant influence on their dynamics.

The dipole–dipole interaction is the most important type of magnetic interactions in nanoparticle systems due to two main features: long-range character and rather large value of the typical magnetic moment of an individual nanoparticle (10^3–10^4 Bohr's magnetons). The dipole–dipole interaction is present in all magnetic spin systems, but in concentrated magnetic materials, other interaction mechanisms, like the exchange interaction, dominate. For example, the dipolar coupling between magnetic ions in paramagnetic salts corresponds to very low characteristic temperatures in the range of 0.01–0.1 K [135].

If nanoparticles are in close contact, the exchange interactions are pronounced also, especially in case of antiferromagnetic particles for which total magnetic particle moment is reduced in comparison with ferro- and ferrimagnetic ones [136]. If nanoparticles are embedded in electrically conducting matrix, Ruderman–Kittel–Kasuya–Iosida (RKKI) mechanism is also possible [4]. For superparamagnetic nanoparticles in ferrofluids or solid polymer matrix, where there is not direct contact between the particles, exchange and RKKI interaction mechanisms can be discarded.

Dipolar interactions are markedly anisotropic and can favor ferromagnetic or antiferromagnetic alignments of the magnetic moments, depending on geometry. Therefore, the dipolar interaction tends to frustrate [137]. Because of random distribution of particle easy axes and positions in matrix, nanoparticle systems with dipolar interactions are anticipated to demonstrate spin-glass properties [138, 139]. The complex interplay between various sources of magnetic disorder determines the state of the nanoparticle assembly and its dynamical properties. Magnetic interactions modify the energy barrier coming from the anisotropy contributions of each particle and, in the limit of strong interactions, their effects become dominant and individual energy barriers can no longer be considered, only the total energy of the assembly being a relevant magnitude. In this limit, relaxation is governed by the evolution of the system through an energy landscape with a complex hierarchy of local minima similar to that of spin glasses. It is worth noticing that in contrast with the static energy barrier distribution arising only from the anisotropy contribution, the reversal of one-particle moment may change the energy barriers of the assembly, even in the weak interaction limit. Therefore, the energy barrier distribution may evolve as the magnetization relaxes.

The first attempt to introduce interactions in the Néel–Brown model was made by Shtrikmann and Wohlfarth [140]. They predicted a Vogel–Fulcher law for the relaxation time in the weak interaction limit ($T_B \gg T_0$), of the form

$$\tau = \tau_0 \exp(KV/k_B[T - T_0]),$$

where T_B is the blocking temperature and T_0 is an effective temperature which accounts for the interaction effects. Using a mean field approximation, the energy barrier E_a in the Néel–Brown formula for τ can be replaced by $V(K + H_{int}M)$. Here K corresponds to the magnetocrystalline anisotropy constant, H_{int} is the mean interaction field, and $VH_{int}M$ is the contribution of the interaction energy to the barrier. In the first approximation, it is possible to replace the interaction field H_{int} by a statistical mean value equal to $(H^2_{int}MV)/k_BT$. If the relation $KV \gg VH_{int}M$ is valid then

$$KV(1 + VH_{int}M/KV) \approx KV/(1 - VH_{int}M/KV)$$

and

$$E_a/k_BT \approx KV/[k_BT - (H_{int}MV)^2/KV] = KV/k_B[T - T_0].$$

Therefore, the temperature T_0 increases with the interaction strength:

$$T_0 = (1/k_B)(H_{int}MV)^2/KV.$$

The more general approach was developed by Dormann et al. [141]. This model reproduces the variation of the blocking temperature T_B as a function of the observation time window of the experiment in a range of time covering eight decades [142]. The increase of T_B with the strength of the dipolar interactions (e.g., increasing particle concentration or decreasing particle distances) has been predicted by the Shtrikmann–Wohlfarth and Dormann–Bessais–Fiorani models and confirmed experimentally [143] as well as in Monte Carlo numerical calculations [144].

In the work [145], direct control of the magnetic interaction between iron oxide nanoparticles was performed through dendrimer-mediated self-assembly. Positively charged superparamagnetic iron oxide nanoparticles were assembled using a series of anionic polyamidoamine dendrimers. The resulting assemblies featured systematically increasing average interparticle spacing over a 2.4 nm range with increasing dendrimer generation. This increase in spacing modulated the collective magnetic behavior by effective lowering of the dipolar coupling between particles. The dependence of blocking temperature on interparticle spacing was found to deviate from a inverse cubic dependence $T_B \sim 1/d^3$ [144] and point toward a much more dramatic interdependence ($T_B \sim 1/d^6$) at close interparticle spacing and a weaker ($T_B \sim 1/d^{0.6}$) correlation at larger spacings. On the other hand, the inverse cubic size dependence for T_B was found to be valid for monodisperse magnetite particles (5 nm of diameter) in p-xylene matrix [146].

It should have in view that the analysis of interparticle interactions in an actual particle assembly is an extremely complex task since these systems are usually characterized by several degrees of disorder, e.g., topological disorder, volume distribution, random distribution of easy axes [147]. Probably, no model can cover the extreme complexity of real nanoparticle assemblies. Therefore, any model includes more or less drastic simplifications which can always be discussed *a priori*, and some of them contested. For example, Mørup and Hansen [148] called in question of the validity of the Dormann–Bessais–Fiorani model, insisting on decreasing of the activation energy due to dipole–dipole interactions. Experimental confirmations of such point of view are listed in Refs. [148, 149]. Mørup and Hansen suggested that increase of T_B, observed experimentally, can be explained be the transition of superparamagnetic particles to collective ordered state, like spin-glass state.

Iglesias and Labarta [150] studied a model systems of single-domain nanoparticles with dipolar interactions using so-called $T\ln(\tau/\tau_0)$ method (Monte Carlo simulation). Authors believe that their results can reconcile the contradictory explanations of Dormann–Bessais–Fiorani and Mørup–Hansen models concerning the variation of the blocking temperature T_B with particle concentration in terms of energy barrier models. According to [150], for weak interactions (diluted systems), the effective energy barrier distributions shift toward lower E_a values with respect to the noninteracting case and become wider as the strength of the dipolar interaction increases. Therefore, for weakly interacting systems, the energy barriers relevant to the observation time window decrease with increasing interaction and consequently the same behavior is expected for T_B. This corresponds to the observations by Mørup et al. in Mössbauer experiments on maghemite nanoparticles [148, 149]. In dense systems (strong interactions), the energy barrier distributions become decreasing functions of energy with increasing contribution of quasizero barriers as the strength of the dipolar interaction increases. When interparticle interactions are strong enough to dominate over the disorder induced by the distribution of anisotropy axes, the dynamic effects are ruled out by an effective energy distribution that broadens toward higher energies as the strength of the dipolar interaction increases. Consequently, an increase in the blocking temperature is expected as observed in Refs. [151, 152].

When interparticle interactions are sufficiently strong, the blocking processes (related to individual particles) are no longer independent. Dipole–dipole interactions can induce spin-glass-like freezing of particle moments. Spin-glass-like behavior of nanoparticle assemblies has been observed experimentally [153], including frequency-dependent temperature of AC susceptibility peak, various slow dynamics phenomena, and memory effects (the relaxation of the remanent magnetization, aging waiting time dependence of observables, etc). Many researchers have observed a slowing down of the response to the static magnetic field with time spent at a constant temperature [142, 154] and with temperature changes [155]. However, due to the polydispersity of nanoparticles and the superparamagnetic blocking at low temperatures, it is very difficult to disentangle single particle effects from collective effects. Interesting, that numerous spin-glass-like features such as the AC susceptibility peak at a frequency-dependent temperature and the relaxation of the remanent magnetization are also encountered in noninteracting nanoparticle systems and some bulk disordered magnetic materials.

The spin-glass state is characterized by a true thermodynamic phase transition, occurring at a well-defined temperature and accompanied by a critical slowing down of the relaxation time according to a power law and a critical divergence of the nonlinear susceptibility. The full understanding of spin-glass phase nature is still open to question. The droplet and hierarchical models have dominated the discussion. In the droplet model, the system has two pure equilibrium spin configurations which are related by global spin reversal. In the configuration, the relative orientations of spins at distances longer than a characteristic length (overlap length, L) are quite sensitive to small temperature changes. In the hierarchical model, a multivalley structure is hierarchically organized on the free energy surface, and the valleys merge with increasing temperature. For this reason, the hierarchical model predicts that the aging is fully initialized by raising the temperature, while the results of the aging are held during temporary cooling. Some researchers have striven to construct a new model based on the idea that the two models are mutually supplemented [156] because both the models described above cannot explain some features observed by the recent experiments [156, 157].

In general, the low-temperature phase in spin-glass states is nonergodic, with a hierarchical structure of the energy valleys in the phase space, as revealed by the ageing time dependence of the magnetic relaxation. Aging effects on the magnetic relaxation yield clear evidences of the existence of many minima in the phase space. However, this may also occur in disordered systems which do not show a thermodynamic phase transition.

On the contrary, the critical slowing down of the relaxation time and the divergence of the nonlinear susceptibility are characteristic of a phase transition. The occurrence of these latter properties allows one to distinguish spin-glass features that result from homogeneous freezing leading to a phase transition and from spin-glass-like features corresponding, in general, to inhomogeneous freezing without a phase transition. Some results indicating the existence of a phase transition have been reported by using frozen magnetic fluids containing uniform particles [158, 159]. These papers have shown a critical slowing down of the relaxation and a divergence of the nonlinear susceptibility at finite temperature T_g.

However, as far as the behavior of interacting nanoparticles is concerned, it is not yet completely clear at present whether it is similar to a homogeneous freezing or to an inhomogeneous blocking, the occurrence of the two processes possibly depending on the strength of the interactions.

6.9
Nanoparticles of Typical Magnetic Materials: Illustrative Examples

Now we are going to discuss illustrative experimental data related mainly to magnetic nanoparticles of common ferromagnetic materials (iron, cobalt, and their compounds). The work [160] explores in details the magnetic and structural properties of iron nanoparticles with median diameter in the range $d_0 = 5$–20 nm. The particle size distribution was a log-normal distribution with the geometric standard deviation σ lies in the range $1.1 < \sigma < 1.3$.

$$f(d) = 1/(2\pi d^2 \sigma^2)^{1/2} \exp(-(\ln(d) - \ln(d_0))/2\sigma^2).$$

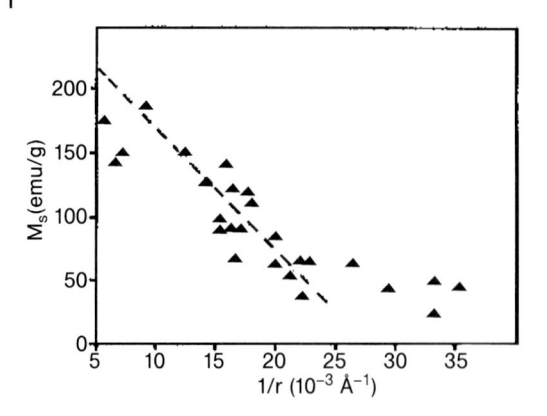

Figure 6.10 Saturation magnetization M_S as a function of the inverse of the total radius. The dashed line is the regression fit to the data, the slope is related to the shell thickness [160].

The particles were not pure metallic but rather α-Fe/Fe$_2$O$_3$(Fe$_3$O$_4$) mixture resulted due to the passivation process.

The saturation magnetization value M_S for all particles was found markedly smaller than that for the bulk α-Fe (220 emu/g) (Figure 6.10). The value of magnetization below 80 emu/g (the bulk saturation magnetization for magnetite) observed experimentally in smaller particles may be due to the domination of nonmagnetic surface layers (dead layers) around the particles or due to spin canting at the outer particle layers [161]. A similar behavior for the magnetization was observed in ferrite nanoparticles [162, 163]. Other possible reasons of reducing the saturation magnetization is bad nanoparticle crystallinity, especially in particle surface layer [164]. The decrease in magnetization with decreasing particle size is related to the higher surface to volume ratio in the smaller particles resulting in a much higher contribution from the surface oxide layer. Authors [160] estimated the atom fractions of metallic and oxidized Fe in nanoparticles and concluded that for smallest particles only approximately 10% of iron atoms are in metallic state. The internal particle core is metallic and, therefore, has larger magnetization than oxidized layer. The oxide surface layer has a thickness of 1–2 nm. Probably, these nanoparticles should be called rather iron-based (containing) than iron.

The temperature dependence of particle magnetization fits satisfactorily the Bloch's law

$$M = M_S(1 - BT^b),$$

where B and b are the Bloch's constant and exponent, respectively. This is explained by the spin-wave theory [165] which holds quite well at temperature $T \ll T_C$ (780 °C for iron) [58]. The value of Bloch's constant for bulk Fe is 3.3×10^{-6} K$^{-3/2}$. For iron-based nanoparticles, B was found to be larger by an order of magnitude [165, 166] (Figure 6.11). A similar increase of B as compared with the bulk value was observed in granular Fe:SiO$_2$ solids [167].

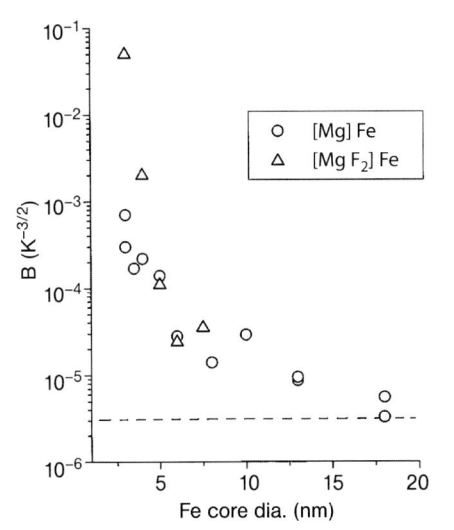

Figure 6.11 Bloch exponent as a function of iron crystallite size in two different matrix. Dashed line is the bulk value [166].

The Bloch exponent determined experimentally can both decrease [166] or increase [168] from the bulk value 3/2 with decreasing particle size, or be approximately equal to it [167]. The reason of this discrepancy is not clear now, but matrix effects are, probably, very important [166].

Increasing of B in the Bloch's law implies a stronger temperature dependence of magnetization in nanoparticles. Theories [169] have shown that the fluctuations are larger for surface magnetic moments than for interior those. As a result, the value of B of the surface atoms is about 2–3.5 times larger than that of the "core" atoms. This naturally explains the larger values of B in smaller particles because the fraction of atoms present on the surface is much higher in the case of smaller particles, compared to the bigger ones.

It seems true that the reduced coordination at the surface will cause the spins at the surface to be more susceptible to thermal excitation, which leads to larger magnetization temperature dependences. However, the problem of the temperature dependence of the nanoparticle magnetization is not so easy because of the possible lattice softening of the small particles [170] (and therefore softening of the spin waves) and the geometrical size effect limiting the value of the spin wavelength [171].

The coercive force of the nanoparticles was found to depend strongly on particle size. Kneller and Luborsky [172] showed the decrease in the coercive force with decreasing particle size due to thermal effects in superparamagnetic particles. If $R_{sp} > R$, then the size dependence of the coercive force follows the law:

$$H_C = (2K/M_S)(1 - (R_{sp}/R)^{3/2}), \quad (6.3)$$

where R_{sp} is the radius of particle which becomes superparamagnetic (in static magnetization experiments) above the room temperature. This radius can be calculated by the relation $KV_{sp} = 25\,kT$, where $T = 300$ K and $V_{sp} = (4/3)\pi R_{sp}^3$. For iron, this critical radius is equal approximately to 12 nm, for hcp cobalt ≈ 4 nm [39].

The size dependence (6.3) was confirmed, for example, for FePt nanoparticles embedded in carbon matrix (Figure 6.12) [173]. The nanoparticles were formed by annealing sputtered FePt/C multilayer precursors.

At low temperatures, where thermal effects are negligible, a difference from (6.3) size dependence can be observed as shown in Figure 6.13 for uniform spherical nanoparticles $Co_{80}Ni_{20}$ with diameters between 18 and 540 nm [127]. At 10 K, the coercivity decreases monotonically with increasing size over the whole range of sizes. Such behavior of H_C might be related to the larger surface to volume ratio in the smaller particles. This could result in a higher effective magnetic anisotropy with large contributions, from the surface and, probably, metallic–oxide interface anisotropy.

At room temperature, a maximum coercivity of $Co_{80}Ni_{20}$ nanoparticles is observed at a critical diameter, d_C, so that two size ranges can be distinguished. Above a critical value, d_C, of around 40 nm, the coercivity decreases as particle size increases, whereas for the finest particles with sizes below d_C, the coercive field decreases as the particle diameter decreases. This variation of the H_C

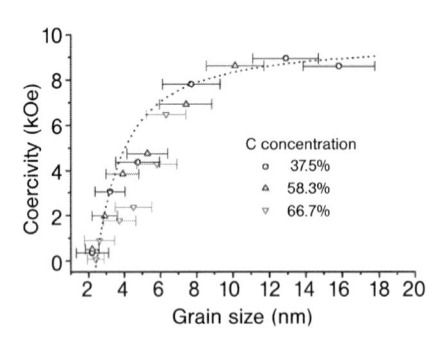

Figure 6.12 Room temperature coercivity of FePt nanoparticles versus grain size at $T = 300$ K. Symbols correspond to different carbon concentration. The dashed line is plotted using Eq. (6.3) [173].

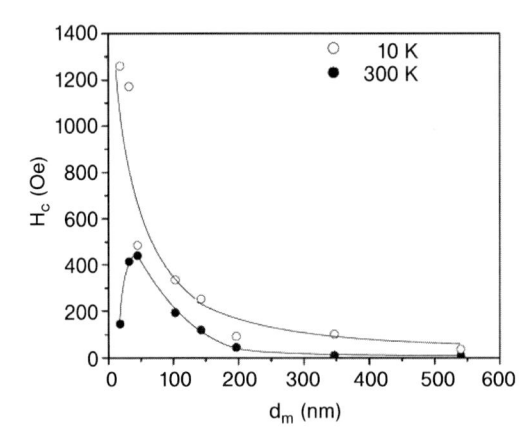

Figure 6.13 Dependence of coercive force on the median diameter at two different temperatures (solid curves are just a guide to the eye) [127].

values can indicate different mechanisms of magnetization reversal as a function of the particle size. For example, the largest particles can behave as magnetic multidomain particles and the magnetization reversal occurs by wall motion, whereas the smallest particles can behave as single-domain particles where the reversal of the magnetization occurs by coherent spin rotation. The reduction of coercivity below d_C is ascribed to thermal effects; however, even for $d = 18$ nm, these thermal effects are not strong enough to reach the superparamagnetic limit where coercivity and remanence vanish.

In should be noted that the coercive force is closely related with the crystalline magnetic anisotropy (see Eq. (6.3)). Many attempts were made to control large crystalline magnetic anisotropy by varying preparation methods. For example, it was reported [174] that coexistence of hematite and maghemite brings about substantial increase in the coercivity. This kind of materials was prepared, for example, by careful annealing of conventionally precipitated hydrous oxide.

Very high coercive force (about 1000 Oe at room temperature), observed in Co nanoparticles on the surface of polytetrafluoroethylene microgranules [175], can result from the "core–shell" structure that could significantly influence on the effective magnetic anisotropy [176].

It is important that the coercivity of the Fe-based nanoparticles could not be explained by assuming the average values of magnetization and magnetocrystalline anisotropy for α-Fe and γ-Fe$_2$O$_3$/Fe$_3$O$_4$ and using the Stoner and Wohlfarth model for noninteracting single-domain particles with shape or magnetocrystalline anisotropy [160]. These particles behave as if they are markedly more anisotropic.

Gangopadhyay et al. [160] made attempts to calculate the value of the effective anisotropy constant K from the magnetization data, using the law of approach to saturation [54, 177]:

$$M = M_S(1 - b/H^2) + \chi \cdot H,$$

where b is a function of M_S and K, and χ is the high-field susceptibility.

From Eq. (6.3), the temperature dependence of the coercive force can be obtained:

$$H_C = H_{C0}(1 - (T/T_B)^{1/2}) = 2K/M_S - (2/M_S)(25 k_B K/V)^{1/2} T^{1/2}, \tag{6.4}$$

where V is the median particle volume. The formula (6.4) also can be used to find K.

Both techniques gave values of K of the order of 10^6 erg/cm^3. Thus, the experimental value of K is larger by an order of magnitude than the bulk Fe and Fe-oxide values. Enhanced values of K have also been reported for various Fe-based nanoparticles [167, 178]. The origin of such large effective anisotropy could be partly due to the "core–shell" particle morphology where the oxide coating is believed to interact strongly with the Fe core ("interface effect") and partly due to the marked contribution of the surface anisotropy ("surface effect") which are expected in nanoparticles.

6.10
Antiferromagnetic Nanoparticles

In nanoparticles with an antiferromagnetically ordered core, the surface spins are expected to dominate the measured magnetization because of their lower coordination and uncompensated exchange couplings. This in turn leads to a large magnetic moment per particle and modified superparamagnetism. Considerable variations of the magnetic properties with change in particle sizes are expected because of the associated changes in the relative number of surface spins. Antiferromagnetic nanoparticle systems where detailed magnetic studies have been reported include NiO [179], ferritin [180], hematite [181], ferrihydrites [182, 183], transition-metal monoxides MnO [184], CoO [185], NiO [186, 187], CuO [188, 189], and $DyPO_4$ [190].

The small magnetic moments of antiferromagnetic nanoparticles make the analysis of magnetization data much less straightforward than the analysis of data for ferro- and ferrimagnetic nanoparticles. This is because the isothermal magnetization curves of particles with very small moments are often far from saturation in the superparamagnetic regime. Furthermore, the anisotropy may be large compared to the Zeeman energy, such that the Langevin function is not a good approximation to the magnetization curves. In most antiferromagnetic nanoparticles, pronounced exchange bias phenomena were found [129].

The size dependence of the magnetic properties of antiferromagnetic nanoparticles differs in several ways from that of ferromagnetic and ferrimagnetic nanoparticles [136]. The magnetic moment of a nanoparticle of a ferromagnetic or ferrimagnetic material is basically determined by the particle volume and the magnetization, which may be similar to the bulk value, although surface effects and defects often result in a (slightly) smaller magnetization. In contrast, the magnetic moment of an antiferromagnetic nanoparticle is mainly a result of imperfections or finite-size effects, e.g., different numbers of spins in the sublattices, which lead to an uncompensated moment and a related increase of the saturation magnetization with decreasing particle size. Numerous magnetization studies of antiferromagnetic nanoparticles have shown that both the initial susceptibility and the magnetization in large applied fields are considerably larger than in the corresponding bulk materials.

In experimental studies of the magnetization of antiferromagnetic nanoparticles, the presence of impurities can be crucial [191]. Even tiny amounts of strongly magnetic phases, which may not be visible in x-ray diffraction measurements, may dominate the magnetization of the samples. During the preparation of many antiferromagnetic nanoparticles, impurity phases can be difficult to avoid. For example, when preparing CoO nanoparticles, the samples may be contaminated with ferromagnetic metallic Co or antiferromagnetic Co_3O_4 with a lower Néel temperature. In samples of hematite nanoparticles, a few per cent of ferrihydrite, which also is antiferromagnetic [182], can give a large contribution to the magnetization [192].

6.11
Semiconductor Magnetic Nanoparticles

6.11.1
Magnetism of Intrinsic and Diluted Magnetic Semiconductors

Ferromagnetism in semiconductors and insulators are rare; the most known ferromagnetic semiconductors are chalcogenides, EuX (X = O, S, and Se) ($T_C < 70$ K), and CdCr$_2$X$_4$ (X = S and Se) ($T_C < 142$ K) [193, 194]. Diluted magnetic semiconductors (DMSs) – also referred to as semimagnetic semiconductors – are semiconducting materials in which a fraction of the host cations (usually nonmagnetic) can be substitutionally replaced by magnetic 3d (transition metal) or 4f (rare earths) ions. Transition metals that have partially filled d states (Sc, Ti, V, Cr, Mn, Fe, Co, Ni, and Cu) and rare earth elements that have partially filled f states (e.g., Eu, Gd, Er) have been used as magnetic atoms in DMS. It is supposed that the partially filled d states or f states contain unpaired electrons which are responsible for them to exhibit magnetic behavior.

There are many mechanisms that could be responsible for magnetic ordering. In DMS materials, the delocalized conduction band electrons and valence band holes interact with the localized magnetic moments associated with the magnetic atoms. Generally, when cations of the host are substituted by 3d transition metal ions, the resultant electronic structure is influenced by strong hybridization of the 3d orbitals of the magnetic ion and mainly the p orbitals of the neighboring host anions. This hybridization gives rise to the strong magnetic interaction between the localized 3d spins and the carriers in the host valence band.

The most extensively studied materials in the early period of this field of this type are II–VI compounds (such as CdTe, ZnSe, CdSe, CdS) doped with transition metal ions T (e.g., T = Mn), substituting their original cations. These alloys are designated as $A^{II}_{1-x}T_xA^{VI}$ or (II,T)VI. The extensive body of research on (II,Mn)VI alloys has been reviewed by Furduna and Kossut [195, 196]. The low critical temperatures and to some extent the difficulty in doping these II–VI-based DMSs made these materials less attractive for applications.

The conventional III–V semiconductors, on the other hand, have been widely used for high-speed electronic devices and optoelectronic devices. The discovery of hole-mediated ferromagnetism in (Ga,Mn)As [197] and heterostructures based on it paved the way for a wide range of possibilities for integrating magnetic and spin-based phenomena with the mainstream microelectronics and optoelectronics as well as taking advantage of the already established fabrication processes. The highest Curie temperature T_C reported in (Ga,Mn)As grown by molecular beam epitaxy, however, is 170 K, which sets T_C higher than room temperature as the major challenge for GaAs-based DMS. The basic idea behind creating these novel ferromagnetic materials is simple [198]: based on the lower valence of Mn, substituting Mn in a

(III, V) semiconductor for the cations (the III elements) can dope the system with holes; beyond a concentration of 1%, there are enough induced holes to mediate a ferromagnetic coupling between the local $S = 5/2$ magnetic moments provided by the strongly localized $3d^5$ electrons of Mn and a ferromagnetic ordered state can ensue. The simplicity of the model hides within it many physical effects present in these materials such as metal–insulator transitions, carrier-mediated ferromagnetism, disorder physics, and many others. Conceptual difficulties of charge transfer insulators and strongly correlated disordered metals are combined in these materials with intricate aspects of heavily doped semiconductors and semiconductor alloys, such as Anderson–Mott localization, defect formation by self-compensation mechanisms, spinodal decomposition, and the breakdown of the virtual crystal approximation [199].

Several extended papers cover the experimental properties of (III,Mn)V DMSs, particularly (Ga,Mn)As and (In,Mn)As, interpreted within the carrier-mediated ferromagnetism model [200–202]. Theoretical predictions based on this model for a number of properties of bulk DMSs and heterostructures have been reviewed by Dietl [203, 204]. A detailed description of wide band-gap oxide DMSs can be found also in [205].

GaN and ZnO have attracted intense attention in searching for high T_C ferromagnetic DMS materials since Dietl et al. [206] predicted that GaN- and ZnO-based DMSs could exhibit ferromagnetism above room temperature upon doping with transition elements such as Mn (on the order of 5% or more) in p-type (on the order of 10^{20} cm^{-3}) materials. This in simple terms is in part due to the strong p–d hybridization, which involves the valence band in the host, owing to small nearest neighbor distance and small spin dephasing spin–orbit interaction [207]. Das et al. [208] showed by first-principles calculations that Cr-doped GaN can be ferromagnetic regardless whether the host GaN is of the form of bulk crystal or clusters. These types of predictions set off a flurry of intensive experimental activity for transition metal doped GaN and ZnO as potential DMS materials with applications in spintronics.

Considerable interest has focused also on achieving room-temperature ferromagnetism in many other transition metal-doped oxide semiconductors, e.g., TiO_2 [209], SnO_2 [210], hematite [211], and CuO [212].

The fast and promising development of high-temperature ferromagnetic semiconductors is challenged now on both experimental and theoretical fronts. Unfortunately, many experimental results are controversial. For example, researchers have identified various properties of Mn-doped ZnO. Fukumura et al. [213] found a spin-glass behavior; Tiwari et al. [214] observed paramagnetism; Jung et al. [215] observed ferromagnetism with a T_C of 45 K; Sharma et al. found ferromagnetic ordering at 425 K [216], while Lue et al. [217] found antiferromagnetism. In Co-doped TiO_2 films grown by pulsed laser deposition, both ferro- and antiferromagnetic behavior were found [218].

These experimental results suggest that the various magnetic properties have different origins, both intrinsic and extrinsic. Different composition and microstructures are obtained, depending on the synthesis methods and postannealing processes used. Therefore, a systematic investigation on the synthesis, microstructure, and magnetic properties is necessary for a better understanding of the magnetic phenomena in DMS oxides. Probably, the experimental depiction of these systems can be adequate only with using of element-specific characterization tools with nanoscale spatial resolution [199]. Moreover, some magnetologists think that in case of DMS, we can face with some new aspects of magnetism physics, the so-called d^0 ferromagnetism [219].

6.11.2
Unusual Magnetism of Magnetic Semiconductors and Role of Nanosize Effects

It is worth to distinguish concentrated magnetic semiconductors, which are intrinsically semiconducting ferromagnetic or ferrimagnetic compounds, such as europium monochalcogenides, chromium chalcospinels, or garnets, from the DMS where an established nonmagnetic semiconductor such as GaAs or InSb is made ferromagnetic by doping with 3d atoms.

The intrinsically ferromagnetic semiconductors were intensively investigated in the 1960 s and 1970 s; however, the transition temperatures in these compounds are rather much below the room temperature. Like room-temperature superconductivity, the dream is to create a useful room-temperature ferromagnetic semiconductor. However, the problem with superconductivity is to bring the critical temperature up above the room temperature, whereas dilute magnetic oxides already appear to be high-temperature ferromagnets, but it is not so clear that they are ferromagnetic semiconductors [219].

What is surprising in the fact that nonmagnetic semiconductors and insulators such as ZnO or TiO_2 become ferromagnetic well above room temperature when they are doped with a few percent of a transition-metal cation such as V, Cr, Mn, Fe, Co, or Ni? The strangest fact is that the ferromagnetism is present at concentrations that lie far below the percolation threshold associated with the nearest-neighbor cation coupling.

The value of the percolation threshold is approximately $2/Z$, where Z is the cation coordination number. Typical values of Z in oxides are 6 or 8, so the percolation threshold is normally greater than 10%. Reports of ferromagnetism in thin films include TiO_2 with V, Cr, Fe, Co, or Ni; SnO_2 with V, Cr, Mn, Fe, or Co; and ZnO with Ti, V, Cr, Mn, Fe, Co, Ni, or Cu. The 3d dopant concentrations are generally below 10%. Nanoparticles of some of these systems are also reported to be ferromagnetic, but well-crystallized, bulk material does not usually order magnetically [220]. Moreover, the average moment per transition-metal cation approaches (or even exceeds) the spin-only moment at low concentrations of magnetic cations. And finely, it falls progressively as x increases toward the percolation threshold [220].

There are different conceivable ways of making a magnetic semiconductor [221]. Magnetic ions can interact with conduction electrons, to produce an n-type semiconductor with a spin-split conduction band. This is the situation in the europium chalcogenides. Alternatively, there may be a spin-split valence band, which is the case, for example, in p-type III–V materials such as $(Ga_{1-x}Mn_x)As$. The Mn doping creates a hole in the valence band, so carrier concentrations are enormous by normal semiconductor standards. A third possibility, in heavily defective material is a spin-split impurity band. This is a model of some relevance for dilute magnetic oxides [222]. Another approach is to take an established ferromagnetic or ferrimagnetic compound like magnetite and render it semiconducting by appropriate doping [223].

The highest Curie temperature achieved for a well-established dilute ferromagnetic semiconductor is 173 K for p-type $(Ga_{0.92}Mn_{0.08})As$ [224]. But 173 K is far from room temperature, and further still from 500 K, the Curie temperature required for ferromagnetic materials to be incorporated into useful devices, with a normal range of operating temperature (-50 to $+120\,°C$).

However, many experiments show that high-temperatures ferromagnetism (HTF) in DMS can really exist. Where are magnetic moments and how they coupled together? A key question is whether or not the ferromagnetism exists in DMS due to the interaction of magnetic dopant atoms with carriers in a spin-split band. Coey supposed that conventional ideas of magnetism in oxides including carrier-induced ferromagnetism are unable to account for the reported examples of HTF in thin films and nanoparticles of DMS. The shortcoming of current theories is principally quantitative. The influential hole-mediated p–d exchange model of Dietl *et al.* is able to envisage Curie temperatures just above room temperature for p-type ZnO doped with 5% Mn [206]. The impurity band exchange model, where the "impurity" band is due to defect states in the crystal, and F-center exchange model, where the dopant ions are coupled via an intervening charged defect, both suffers from insufficient interaction strength and are unable to deliver a Curie temperature close to room temperature.

Coey supposed that HTF in DMS is not due to the spins of the 3d dopant cations but rather due to some defects. It is quite possible that the relevant defects are not point defects, but extended planar defects associated with film surfaces and interfaces, or nanocrystalline surfaces.

There is also a hypothesis that ferromagnetism is as a universal feature of nanoparticles of the otherwise nonmagnetic oxides [225]. Room-temperature ferromagnetism has been observed in the nanoparticles (7–30 nm) of nonmagnetic oxides such as CeO_2, Al_2O_3, ZnO, In_2O_3, and SnO_2. Interesting, that the saturated magnetic moments in undoped CeO_2 and Al_2O_3 nanoparticles are comparable to those observed in corresponding transition metal-doped wide-band semiconducting oxides. The other oxide nanoparticles show lower values of magnetization but with a clear hysteretic behavior.

To check the role of size effects, authors [225] studied the bulk samples obtained by sintering the nanoparticles at high temperatures in air or oxygen and found them diamagnetic. The origin of ferromagnetism in nanoparticles of nonmagnetic oxides may be due to the exchange interactions between localized electron spin moments resulting from oxygen vacancies at the surfaces of nanoparticles. If this hypothesis is correct then all metal oxides in nanosize form would exhibit room-temperature ferromagnetism. The ferromagnetism associated with oxygen vacancies can give also a possible clue to understand the HTF in the thin films of dilute magnetic semiconducting oxides [226].

Thomas Dietl [227] has suggested interesting model explaining above-mentioned puzzle about the origin of ferromagnetic response in DMS, in which an average concentration of magnetic ions is below the percolation limit for the nearest-neighbor coupling and, at the same time, the free carrier density is too low to mediate an efficient long-range exchange interaction. He argued that coherent nanocrystals with a large concentration of magnetic constituent account for high apparent Curie temperatures detected in a number of DMS. It is well-known that phase diagrams of a number of alloys exhibit a solubility gap in a certain concentration range. This may lead to a spinodal decomposition into regions with a low and a high concentration of particular constituents. If the concentration of one of them is small, it may appear in a form of coherent nanocrystals embedded in the majority component. The strong tendency to form nonrandom alloy in the case of DMS was proved by theoretical calculations [228].

In view of typically limit solubility of magnetic atoms in semiconductors, it may be expected that such a spinodal decomposition is a generic property of a number of DMS. Owing to the high concentration of the magnetic constituent, the nanocrystal forms in this way order magnetically at a relatively high temperature, usually much greater than the room temperature. For example, bulk (Zn,Cr)Se and (Zn,Cr)Te, which show distinct superparamagnetic behavior, are to be viewed rather as an ensemble of ferromagnetic particles than a uniform magnetic alloy [227]. Interestingly, that not only ferromagnetic or ferrimagnetic nanocrystals possess a nonzero magnetic moment but also nanocrystals in which antiferromagnetic interactions dominate can also show a nonzero magnetic moment due to the presence of uncompensated spins at their surface, whose relative magnitude grows with decreasing nanocrystal size [229].

Summarizing, nanoparticles seem able to furnish the clue to the explanation of very unusual properties of magnetic semiconductors and point the way to solving the various challenges, like problems of high-temperature ferromagnetism in semiconductors [227] and metal-less organic materials [230], as well as anomaly magnetism of "nonmagnetic" materials [225, 231]. Surprisingly, it can be partly possible due to their nonideal in comparison with bulk counterparts crystalline structure.

6.12
Some Applications of Magnetic Nanoparticles

6.12.1
High-Density Information Magnetic Storage

The technology of magnetic recording has begun time long ago [232]. The concept of magnetic recording is to use current "write head" that generates a magnetic field changing the magnetization of closely spaced magnetic elements (granules, particles, or their groups) in magnetic medium. The data are recovered by the generation of an output signal in the "read head" by sensing the magnetization in the recording medium, using the Faraday's electromagnetic induction or magnetoresitance phenomena. Modern magnetic recording system stores digital data, in which case, the current supplied to the write head is in the form of pulses encoded to represent the digital data ("1" or "0"). The digital data are recorded in the magnetic thin film media as transitions between the two possible states of the magnetization. The transition region between the oppositely directed directions of the magnetization is similar to that between magnetic domains. In conventional hard disk media, data are stored as magnetization patterns in a film consisting of small, weakly coupled magnetic grains, each of which behaves as a single-domain magnetic particle. To obtain an acceptable signal-to-noise ratio (SNR), each bit is written over an area of tens or hundreds of magnetic grains. To increase the data recording density, the grains need to be made smaller, but this leads to superparamagnetic behavior in which thermal energy ($\sim kT$) can reverse the magnetization direction of a grain. Estimates for acceptable thermal stability in hard disk media are that the grains begin to exhibit thermal instability when the ratio of thermal energy kT (k is Boltzmann's constant and T the temperature) to magnetic energy KV (K is the magnetic anisotropy and V the grain volume) must be in the range 40–80, depending on the grain size distribution, intergranular coupling, saturation magnetization, and other properties of the medium. Grain diameters and film thicknesses are currently in the range of 10–20 nm, which leaves little opportunity for decreasing grain volume while retaining thermal stability. Decreases in grain volume could be compensated by increases in K, but for high values of K, it becomes more difficult for the recording head to produce sufficient field to write the data on the medium. This contradiction is called the "superparamagnetic limit" [132, 233]. As shown above in Section 6.7, the magnetic exchange coupling induced at the interface between ferromagnetic and antiferromagnetic systems can provide an extra source of anisotropy, leading to magnetization stability [130].

Besides the superparamagnetic limit, there is other trouble with increasing of magnetic recording density. In usual high-density media, each bit cell contains of the order of 100 grains. Transition noise, originating from irregularities in the magnetization transitions, and increased by collective reversal of groups of grains, dominates the overall SNR of the system. Both the SNR and the minimum width of the transition depend on the grain size of the medium.

One way to overcome the SNR limitation is recording on well-organized bit-patterned media (BPM). The fundamental idea of BPM is that one grain represents one bit so that the entire volume of the bit resists the effect of thermal agitation and higher recording density can be achieved. Investigations of a BPM recording system have shown that recording densities greater than 1 Tb/in.^{2} should be possible [234, 235]. Note that for the recording density of 1 Tb/in.^{2}, the pattern size will be 12.5 × 12.5 nm with an interval of 12.5 nm, when the square bit is used. Recently, it was shown that atomic beam epitaxy of Co on Au(788) is capable of producing particle density of about 26 Tb/in.^{2} with the required uniformity of magnetic properties and absence of dipolar interactions [236].

The realization of suitable patterned media is a serious problem [237]. Top-bottom methods like nanoimprinting or electron-beam lithography are slow and high-cost fabrication processes. The development of the so-called self-organized magnetic array using solution phase synthesis and self-assembly of FePt nanoparticles have generated tremendous interest in recent years due to their low fabrication cost and the highest achievable density [120]. The high magnetic anisotropy and corrosion resistance make these nanoparticles very attractable for the next generations of magnetic recording media and advanced permanent magnets [238]. Recently, a salt-annealing method has been used to produce nearly monodisperse $L1_0$ FePt nanoparticles with stable room-temperature magnetism down to sizes of 3 nm [239]. However, prevention of particle aggregation during annealing, need in aligning of magnetic easy axis, degradation processes after synthesis remain yet serious hurdles for applications.

An alternative bottom-up approach is to use self-assembled porous templates for the growth of magnetic arrays [240]. For example, magnetic nanowire arrays have been produced by electrochemical deposition in porous anodic alumina (PAA) templates [241, 242]. However, high aspect ratio of the nanowires is not desirable for high-density recording due to structural inhomogeneities, incoherent magnetization reversal, and limited writability.

Recently, $L1_0$-ordered CoPt nanocolumns with 100 nm spacing were obtained by electrochemical deposition in PAA with postdeposition annealing [243]. Magnetic and semiconductor nanodot arrays can also be fabricated by physical vapor deposition using ultrathin PAA templates as masks [244, 245].

Kim et al. [240] reported about the development of $L1_0$-structured FePt nanodot (>18 nm in size) arrays with perpendicular anisotropy, high coercivity, and extremely high density. They were fabricated by physical vapor deposition using ultrathin PAA as masks, followed by rapid thermal annealing. There are several advantages of this systems comparing to early available materials: the periodicity of 25 nm leading to extremely high density 1.2×10^{12} dots/in.^{2}, large coercivity (up to 15 kOe) originating from the high magnetocrystalline anisotropy 10^7 erg/cm^3 of $L1_0$-structured FePt, perpendicular easy axis orientation, and ease of fabrication without epitaxy. Hence, this nanoparticle arrays possess essential features desirable for future high-density recording media.

Recently, there has been a significant new emerging technology for fast memory devices – the magnetic random access memory (MRAM). The MRAM device is a possible replacement for the familiar semiconductor memories used in modern computers – dynamic and static random access memory (DRAM and SRAM). The MRAM technology combines a magnetic storage technology together with semiconductor metal-oxide semiconductor devices to result in fast and high-density data memory devices. The technology on which the magnetic part of MRAM is based is an extension of the technology used in magnetic-recording devices identified as the magnetic tunneling junction [246].

Many prototypes of MRAM devices include the ring-shaped memory element forcing magnetization to be circular, enabling a very stable flux closure mode [247]. Flux closure state of magnetic "bits" can also minimize crosstalk between them. Microsize cobalt magnetic rings were fabricated by lithography [248, 249]. However, these structures are far from the physical size limit of flux closure.

In the work by Tripp *et al.* [250], single-walled Co nanoparticle rings were obtained that can support stable flux-closed states at room temperature. Off-axis electron holography was used to visualize magnetic flux with nanometer spatial resolution. The possible formation mechanism of these bracelet-like rings is the dipole-directed self-assembly. These nanoparticle rings were less than 100 nm in diameter that is well below the limit of conventional lithography. Self-assembled nanorings have intriguing possibilities for integration with nano-patterned surfaces, with application in spintronic devices (see Section 6.12.3).

6.12.2
Traditional and New Applications of Ferrofluids

Ferrofluids (also called magnetic fluids, magnetic nanofluids, superparamagnetic colloids) are colloidal suspensions of surfactant-coated magnetic nanoparticles in a liquid medium [251]. The first reported attempts at producing ferromagnetic liquids by dispersing ferromagnetic particles in a carrier fluid were by Gowan Knight in 1779 [252]. He attempted to produce a ferromagnetic liquid using essentially the same technique as that used at the present time by dispersing ferromagnetic particles in carrier liquids, in this case, iron filings in water. After several hours mixing, the water contained a suspension of small particles but the mixture did not have long-term stability. Bitter [253] produced a colloid for magnetic domain studies that consisted of a colloidal suspension of magnetite in water. The particle size was around 10^3 nm. Suspensions containing smaller particles (20 nm) have been prepared by Elmore [254, 255]. Bitter and Elmor used obtained magnetic media for visualization of scattering fields of permanent magnets and developed a "magnetic pattern" method, which became classics of magnetism [256, 257].

These early investigations had no continuation for a quarter of a century up to the middle of the 1960s, when a sharp interest in magnetic fluids arose again. This interest is produced primarily by a great number of potential technical applications of ferrofluids and is based on modern development of chemistry and chemical technology [258]. Stable ferromagnetic liquids were first prepared by Papell [259] for the National Aeronautics and Space Administration. These liquids consisted of ferrite particles in a nonconducting liquid carrier. Other ferromagnetic liquids containing the more magnetic metal particles of iron, cobalt, and nickel dispersed in similar nonconducting carriers are currently being developed. The preparation of ferromagnetic liquids containing iron, nickel–iron, and gadolinium particles in a metallic carrier liquid have been investigated by Shepherd and Popplewell [260]. Now a wide spectrum of magnetic materials (Fe, Co, Ni, FePt, CoPt, various ferrites) has been synthesized as superparamagnetic nanoparticles with narrow size distribution [261, 262]. The possibility of magnetic control of flow and properties of ferrofluids have led to the development of a wide variety of possible applications of these liquids in various fields from mechanical engineering to biomedical employment [263, 264].

The stability of ferrofluids (especially in field gradients) is the more demanding factor [265]. The stability assumes that the particles remain small, in other words that they do not agglomerate. But these are small dipoles, and the dipolar interactions as well as the attractive van der Waals forces between particles tend to cause them to agglomerate. The thermal energy needed to oppose the agglomeration of dipolar origin has the same order of magnitude as that which opposes sedimentation. However, agglomeration of van der Waals origin is irreversible since the energy required to separate two particles, once agglomerated, is very large. Consequently, it is necessary to find a way of preventing the particles from getting too close to each other. This can be done by coating the particles with a polymer (surfactant) layer to isolate one from the other. These are surfaced ferrofluids. Also the stabilization can be reached by electrically charging the particles, which will then repel because of the Coulomb interaction. These are the ionic ferrofluids.

In magnetic fluids, the features of magnetism and fluid behavior are combined in one medium. They remain liquid when highly magnetized, even in the most intense applied magnetic fields. However, the ferrofluids are distinguished from ordinary fluids by the body and surface forces that arise, yielding new fluid mechanical phenomena. For example, these fluids show a pronounced increase of viscosity in the presence of moderate magnetic fields with strengths of the order of several tens of millitesla. Classically, this effect is explained by the hindrance of the free rotation of magnetic particles – with a magnetic moment spatially fixed in the particle – in a shear flow due to magnetic torques trying to align the particles' magnetic moments with the magnetic field direction [266, 267]. Later in some models, it was assumed that the ferrofluid can be described in a first approximation as a bidisperse system containing a large fraction of small particles not contributing to the magnetoviscous effects and a small fraction of relatively large particles able to form agglomerates. This assumption was clearly validated by the fact that the extraction of large particles from the fluid leads to a significant reduction of the magnetoviscous effect. Taking into account the interaction between the magnetic particles, a quantitative description of the magnetoviscous effects was obtained, showing that a small fraction of large particles in the fluid forms chains dominating the rheological properties of the fluids in the presence of magnetic fields [268].

Ordinarily ferrofluids exhibit the Newtonian rheology, with stress proportional to rate of strain, albeit the coefficient of viscosity increases in applied magnetic field. Surprisingly, recent theoretical analysis predicts and experimental investigation demonstrates that under appropriate conditions with a time-varying field, the viscosity of ferrofluids exhibits a substantial reduction, that is, a negative viscosity component [269, 270].

Ferrofluids have a variety of other unusual properties [271]. For example, being placed in nonhomogeneous magnetic field, ferrofluid changes the shape and density. The idea of controlling the properties of ferrofluids with a magnetic field has led to many innovative applications [272]. For example, the easy localization of ferrofluids by a magnetic field leads to the development of a novel class of sealing materials. The basic seal components consist of ferrofluid, a permanent magnet, two pole pieces, and a magnetically permeable shaft. The magnetic structure, completed by the stationary pole pieces and the rotating shaft, concentrates magnetic flux in the radial gap under each pole. When ferrofluid is applied to the radial gap, it assumes the shape of a "liquid ring" and produces a hermetic seal. Such magnetic sealants have several advantages over conventional mechanical sealants: low viscous drag, 1% torque transmission, high-speed capability, noncontaminating, long and reliable life, very good leak tightness, wide temperature range. Hence, ferrofluids can be effectively used as rotating shaft seals in satellites, and in a variety of machines including 1% leak-free, no-wear vacuum rotary seal for use in the manufacture of semiconductor wafers, vacuum processing applications, centrifuges, computer hard disk drives that prevent contamination, increased memory capacity, and improved processing throughput. Ferrofluid-based hermetic seal has been developed for smooth operation of airborne targeting cameras, which are used in harsh environments of military aircrafts. A leak-free seal (without any dripping of the sealant) for hydrocarbon and gas handling applications has been developed, which is being used in a variety of fans, blowers, and vertical pumps. Certain devices (accelerometers and inclinometers) are based on the principle that a film of ferrofluid can be held fixed by magnets, but remains deformable under the influence of acceleration or gravity. The deformation of the fluid leads to a modification of the electrical permeability of the medium between the measuring coils.

Certain applications exploit the fact that a coating of ferrofluid can be held, in a tube for instance, by permanent magnets. This film of ferrofluid can ensure lubrication or improve the flow of a fluid, and consequently the transport of heat. The latter can also be modified by controlling convection by a magnetic field, since a field acts on the viscosity, and a field gradient and temperature affect the magnetization, hence the forces involved in the process of convection. Ferrofluids have enabled loudspeaker manufacturers to improve speaker performance. In this application, which takes advantage of the ferrofluid's heat transfer property, the fluid is magnetically positioned in the gap of the driver assembly. This protects the speaker driver from thermal failure.

Ferrofluid-based viscous dampers that improve the performance of stepper motors are being used extensively in automation processes such as lens grinding, robotics, machine tools, and disk read/write head actuators. Inertial shock absorbers use the lift or levitation exerted by a ferrofluid on a body, magnetic or nonmagnetic. This effect is controlled by the magnetic field. For loudspeakers of the moving coil type, for example, the ferrofluid stabilizes the movement of the coil thereby reducing sound distortion. Other shock absorbers make use of the fact that the viscosity of the ferrofluid is modified by a magnetic field: their damping ability can be adjusted to the load or the roughness of the terrain by controlling viscosity using the field.

Magnetic fields can artificially impart high specific gravity in ferrofluids. This property is exploited for separating mixtures of industrial scrap metals such a titanium, aluminum, and zinc and for sorting diamonds. Another promising application is to use the magnetic fluid inks for high-speed, inexpensive, and silent printers. Magnetic inks are being used in printing paper currency in United States because the genuineness of the currency can be easily checked with a magnet.

Ferrofluids are inexpensive and relatively easily synthesized magneto-optic materials that offer attractive optical and magneto-optical characteristics. When a field is applied, the ferrofluid is magnetized and becomes optically active. When the polarizer and analyzer are cross positioned, light may pass through them, while in zero field it cannot. This phenomenon can be exploited in magnetic field detectors or in light modulators, and may also be used in a more indirect way to produce a viscometer. The principle is to dilute ferrofluid in a liquid whose viscosity is to be measured, and to determine the law governing the variation of the transmitted light over time after the magnetic field has been abruptly cut off. This law is exponential, and its time constant is directly related to the viscosity of the fluid.

In recent years, the investigators found a variety of ordered structures in the ferrofluid under magnetic fields [273, 274]. For example, the ordered structures like chains (or columns) generate various magneto-optic effects, such as birefringence, field-dependent transmittance. It is worth mentioning that the ordered hexagonal structure of columns in the ferrofluid thin film under perpendicular magnetic fields results in the novel magnetochromatics.

This phenomenon implies that the ferrofluid thin film with the ordered hexagonal structure acts as a two-dimensional tunable grating [275].

Summarizing, ferrofluids exhibit the possibility to widely modify their properties and to control their flow by moderate magnetic fields. This fact creates a research field which is actually strongly developing. Magnetic fluids are also an interesting model object for the investigation of the influence of interparticle interaction on the various physical properties [276] and for finding of a connection between the structural and macroscopic properties of nanosystems.

6.12.3
Magnetic Nanoparticles and Spintronics

In a narrow sense, spintronics refers to spin electronics, the phenomena of spin-polarized transport in metals and semiconductors. Until recently, the spin of the electron was ignored in charge-based electronics [277, 278]. Spintronics (spin transport electronics or spin-based electronics), where it is not only the electron charge but also the electron spin that carries information, offers opportunities for a new generation of devices combining standard microelectronics with spin-dependent effects that arise from the interaction between spin of the carrier and the magnetic properties of the material. The goal of spintronics is to find effective ways of controlling electronic properties, such as the current or accumulated charge, by spin or magnetic field, as well as of controlling spin or magnetic properties by electric currents or gate voltages. The ultimate goal is to make practical device schemes that would enhance functionalities of the current charge-based electronics. An example is a spin field-effect transistor, which would change its logic state from ON to OFF by flipping the orientation of a magnetic field [279].

In more broad sense, spintronics is a study of spin phenomena in solids, in particular metals and semiconductors and semiconductor heterostructures. Such studies characterize electrical, optical, and magnetic properties of solids due to the presence of equilibrium and nonequilibrium spin populations, as well as spin dynamics. These fundamental aspects of spintronics give us important insights about the nature of spin interactions in solids: spin-orbit, hyperfine, or spin exchange couplings. In spintronics, we also learn about the microscopic processes leading to spin relaxation and spin dephasing, microscopic mechanisms of magnetic long-range order in semiconductor systems, topological aspects of mesoscopic spin-polarized current flow in low-dimensional semiconductor systems, or about the important role of the electronic band structure in spin-polarized tunneling.

Traditional approaches using spin are based on the alignment of a spin (either "up" or "down") relative to a reference (an applied magnetic field or magnetization orientation of the ferromagnetic film). Device operations then proceed with some quantity (electrical current) that depends in a predictable way on the degree of alignment. Adding the spin degree of freedom to conventional semiconductor charge-based electronics or using the spin degree of freedom alone will add substantially more capability and performance to electronic products. The advantages of these new devices would be nonvolatility, increased data processing speed, decreased electric power consumption, and increased integration densities compared with conventional semiconductor devices.

Spintronics originates in the discovery of the giant magnetoresistive (GMR) effect – the large decrease of the sample's resistivity under application of an external magnetic field [280–282]. Albert Fert and Peter Grünberg were awarded the 2007 Nobel Prize for the discovery GMR phenomenon.

Initially GMR was observed in artificial thin-film materials composed of alternate ferromagnetic and nonmagnetic layers [280]. The resistance of the material is lowest when the magnetic moments in ferromagnetic layers are aligned and highest when they are antialigned. The current can either be perpendicular to the interfaces or can be parallel to the interfaces. New materials operate at room temperatures and exhibit substantial changes in resistivity when subjected to relatively small magnetic fields (1 to 10 Oe).

GMR spin valve read heads performed a revolution in computer hard drives. Although some alternative configurations have been proposed, nearly all commercial GMR heads use the spin valve format as originally proposed by IBM [283]. The MR of spin valves has increased markedly from about 5% in early heads to 20% today. As hard drive storage densities approach 1 Gbits/in.^{2}, sensor stripe widths are approaching 0.1 mm and current densities are becoming very high.

GMR was also observed in granular magnetic composites, which typically consists of nanometer-sized magnetic particles embedded in a nonmagnetic metallic host [133, 284]. In granular solids, the volume concentration p of magnetic particles affects the MR. At low concentrations, the MR increases with p as a result of the increase in the concentration of magnetic scattering centers. At $p \approx 25\%$, the MR reaches a maximum. As p is further increased, the MR decreases and approaches a very small value. The GMR effect vanishes when all the magnetic particles are connected. GMR effect has a close relationship with the local magnetization of ferromagnetic particles, for example, sensitive to the relative orientation between the electron spin and the magnetic dipole moments in the magnetic particles [284, 285].

In granular films, GMR originates from the spin-dependent scattering of conduction electrons at the interface between the ferromagnetic particles and nonmagnetic matrix as well as within the ferromagnetic particles. The external field rotates the moments of the magnetic entities, the scattering potentials for the electrons are modified, and eventually the resistivity of the sample changes. It is therefore expected that the field dependence of the MR reflects the modifications to the micromagnetic configuration as the field varies in strength. In the early works [284], it was demonstrated that the MR is determined solely by the long-range magnetic order of the granular metal, as the universal parabolic dependence of the MR on the sample magnetization (M) dictates. However, flattening of the MR–M observed for weak fields (1–10 Oe) has been subsequently attributed to the existence of short-range ferromagnetic order [286] and the grain-size distribution [287].

Therefore, the particle size and its distribution in granular films are important to improve GMR. Magnetic particles in granular films can be superparamagnetic, single-domain ferromagnetic, and multidomain ferromagnetic particles, and they can display different magnetic properties under applied fields, giving rise to different effects on the spin-dependent scattering related to GMR. Some experiments and theories [288, 289] indicate that the smaller the particle size, the larger the ratio of surface to volume, resulting in stronger spin-dependent scattering of conduction electrons. It seems that GMR should increase monotonically with decreasing particle size. However, it was found in others works [134, 290–292] that there is no monotonic relationship between particle size and GMR. For granular films with low-volume particle fraction, all the magnetic particles are so small that they almost behave as superparamagnetic, therefore leading to low GMR. With increasing volume fraction, particles gradually become larger, causing the increase of the fraction of single-domain particles and then the improvement of GMR. With further increasing volume fraction, the partial particles will grow or touch each other to become multidomain particles, giving rise to the decrease of GMR due to disadvantageous effect of multidomain particles on GMR [293]. So, GMR can reach a peak value at some average particle size. Very likely, the single-domain ferromagnetic particles play a key role in GMR and the multidomain particles play a secondary role in GMR in case of nanoparticulated films.

Tunneling spectroscopy technique allows use magnetic nanoparticles as model systems for the study of the interaction of transport and magnetism [7, 294]. However, underlying technological challenges make the experimental progress in this field much slower than the theoretical one [295]. Probably, in the near future, the most interesting results in exploring of magnetic nanoparticles for spintronics applications could be related to using them as testing area for searching of magnetic materials with new advanced properties.

6.13
Final Remarks

The history of solid state physics (and magnetism as the part of this branch of physics) has demonstrated us the evolution from ideal ordered systems (perfect crystals) to nonideal disordered systems (amorphous solids, glasses, etc.). Concurrently, the interests of physicists have shifted from three-dimensional to low-dimensional systems: two- (e.g., ultrathin films) and one-dimensional (e.g., various quasilinear magnetics) systems. It is natural that now this evolution leads to zero-dimensional objects (nanoparticles) for which the structural defects are inevitable and intrinsic. It is very likely that the progress in understanding of magnetic nanoparticles and their practical applications will depend on ability of scientists to take proper account for numerous complex conditions that determine, effect, and change fascinating nanoparticle features.

References

1. S.P. Gubin, Yu.A. Koksharov, G.B. Khomutov, G.Yu. Yurkov, *Russ. Chem. Rev.*, **2005**, *74*, 539.
2. *Nanoscale Materials in Chemistry*, K.J. Klabunde, Editor; John Wiley & Sons, New York, **2001**.
3. *Nanoparticles. From Theory to Application*, G. Schmid, Editor; Willey-VCH, Weinheim, **2004**.
4. X. Battle, A. Labarta, *J. Phys D.: Appl. Phys.*, **2002**, *35*, R15.
5. D.G. Rancourt, *Rev. Mineral. Geochem.*, **2001**, *41*, 7.
6. S. Jacobs, C.P. Bean, in *Magnetism*, G.T. Rado, H. Suhl, Editors; Academic, New York, **1963**, 271.
7. S. Guéron, M.M. Deshmukh, E.B. Myers, D.C. Ralph, *Phys. Rev. Lett.*, **1999**, *83*, 4148.
8. S.H. Liou, S. Huang, E. Klimek, R.D. Kirby, Y.D. Yao, *J. Appl. Phys.*, **1999**, *85*, 4334.
9. M. Chen, D.E. Nikles, *J. Appl. Phys.*, **2002**, *91*, 8477.
10. I. Panagiotopoulos, S. Stavroyiannis, D. Niarchos, J.A. Christodoulides, G.C. Hadjipanayis, *J. Appl. Phys.*, **1999**, *87*, 4358.
11. S. Sun, *Adv. Mater.*, **2006**, *18*, 393.
12. T. Thomson, B.D. Terris, M.F. Toney, S. Raoux, J.E.E. Baglin, S.L. Lee, S. Sun, *J. Appl. Phys.*, **2004**, *95*, 6738.
13. C. Xu, S. Sun, *Polym. Int.*, **2007**, *56*, 821.
14. O.V. Salata, *J. Nanobiotechnol.*, **2004**, *2*, 3.
15. S.M. Moghimi, A.C. Hunter, J.C. Murray, *FASEB J.*, **2005**, *19*, 311.
16. P. Tartaj, M.P. Morales, S. Veintemillas-Verdaguer, T. González-Carreño, C.J. Serna, *J. Phys. D: Appl. Phys.*, **2003**, *36*, R182.
17. *Bionanotechnology: Lessons from Nature*, D.S. Goodsell, Wiley-Liss, New Jersey, **2004**.
18. M.F. Hochella Jr., S.K. Lower, P.A. Maurice, R.L. Penn, N. Sahai, D.L. Sparks, B.S. Twining, *Science*, **2008**, *319*, 1631.
19. J.L. Kirschvink, A. Kirschvink-Kobayashi, B.J. Woodford, *Proc. Natl. Acad. Sci.*, **1992**, *89*, 7683.
20. *Magnetic Orientation in Animals*, R. Wiltschko, W. Wiltschko, Springer, Berlin, **1995**.
21. J.L. Kirschvink, *Bioelectromagnetics*, **1989**, *10*, 239.
22. I. Stokroos, L. Litinetsky, J.J.L. van der Want, J.S. Ishay, *Nature*, **2001**, *411*, 654.
23. J.A. Tarduno, *Geophys. Res. Lett.*, **1995**, *22*, 1337.
24. J.P. Bradley, H.Y. Sween Jr., R.P. Harvey, *Meteorite Planet Sci.*, **1998**, *33*, 765.
25. P.R. Buseck, R.E. Dunin-Borkowski, B. Devouard, R.B. Frankel, M.R. McCartney, P.A. Midgley, M. Pósfai, M. Weyland, *Proc. Natl. Acad. Sci.*, **2001**, *98*, 13490.
26. A.A. Goodman, D.C.B. Whittet, *Astrophys. J.*, **1955**, *455*, L181.
27. J.L. Kirschvink, A.T. Maine, H. Vali, *Science*, **1997**, *275*, 1629.
28. J.L. Jambor, J.E. Dutrizac, *Chem. Rev.*, **1998**, *98*, 2549.
29. F.M. Michel, L. Ehm, S.M. Antao, P.L. Lee, P.J. Chupas, G. Liu, D.R. Strongin, M.A.A. Schoonen, B.L. Phillips, J.B. Parise, *Science*, **2007**, *316*, 1726.
30. Y. Alhassid, *Rev. Mod. Phys.*, **2000**, *72*, 895.
31. W.F. Halperin, *Rev. Mod. Phys.*, **1986**, *58*, 533.
32. R.H. Kodama, A.E. Berkowitz, E.J. McNiff, Jr., S. Foner, *Phys. Rev. Lett.*, **1996**, *77*, 394.
33. R.H. Kodama, *J. Magn. Magn. Mater.*, **1999**, *2*, 359.

34. G. Füzi, G. Kádar, *Physica B*, **2004**, *343*, 293.
35. U. Nowak, O.N. Mryasov, R. Wieser, K. Guslienko, R.W. Chantrell, *Phys. Rev. B.*, **2005**, *72*, 172410.
36. J. Restrepo, Y. Labaye, J.M. Greneche, *Physica B*, **2006**, *384*, 221.
37. E.C. Stoner, E.P. Wohlfarth, *Philos. Trans. R. Soc. A*, **1948**, *240*, 599.
38. L. Néel, *Ann. Geophys.*, **1949**, *5*, 99.
39. C.P. Bean, J.D. Livingstone, *J. Appl. Phys.*, **1959**, *30*, 120S.
40. *Micromagnetics*, W.F. Brown, Jr., Editor; Wiley, New York, **1963**.
41. M.E. Shabes, *J. Magn. Magn. Mater.*, **1991**, *95*, 249.
42. *Magnetism*, É. du Trémolet, D. Gignoux, M. Schlenker, Editors; **2005**, Springer, Berlin, vols. 1 and 2.
43. *Simple Models of Magnetism*, R. Skomski, Oxford University Press, Oxford, **2008**.
44. *Fundamentals of Magnetism*, M. Getzlaff, Springer, Berlin, Heidelberg, **2008**.
45. *Spin Fluctuations in Itinerant Electron Magnetism*, T. Moriya, Springer, Heidelberg, **1985**.
46. C. Kittel, *Rev. Mod. Phys.*, **1949**, *21*, 541.
47. P. Weiss, *J. de Phys.*, **1907**, *6*, 661.
48. W. Heisenberg, *Z. für Physik*, **1928**, *49*, 619.
49. L. Landau, E. Lifshitz, *Physik. Zeits. Sovjetunion*, **1935**, *8*, 153.
50. L. Néel, *Adv. Phys.*, **1955**, *4*, 191.
51. *Hysteresis in Magnetism – for Physicists, Material Scientists, and Engineers*, G. Bertotti, Academic Press, New York, **1998**.
52. R. Scomski, *J. Phys.: Condens. Matter*, **2003**, *15*, R841.
53. J. Frenkel, J. Dorfman, *Nature*, **1930**, *126*, 274.
54. *The physical principles of magnetism*, A.H. Morrish, John Willey & Son, New York, **1965**.
55. B. Barbara, *Solid State Sci.*, **2005**, *7*, 668.
56. C.L. Dennis, R.P. Borges, L.D. Buda, U. Ebels, J.F. Gregg, M. Hehn, E. Jouguelet, K. Ounadjela, I. Petej, I.L. Prejbeanu, M.J. Thornton, *J. Phys.: Condens. Matter.*, **2002**, *14*, R1175.
57. R. Skomski, *J. Magn. Magn. Mater.* **2004**, *272–276*, 1476.
58. *Introduction to Solid State Physics*, C. Kittel, Wiley, New York, **1965**.
59. M.N. Barber, in *Phase Transitions and Critical Phenomena*, C. Domb, J.L. Lebowitz, Editors, vol. 8, Academic Press, New York, **1983**.
60. M. Zheng, X.C. Wu, B.S. Zou, Y.J. Wang, *J. Magn. Magn. Mater.*, **1998**, *183*, 152.
61. C.A. Grimes, J.L. Horn, G.G. Bush, J.L. Allen, P.C. Eklund, *IEEE Trans. Magn.*, **1997**, *33*, 3736.
62. T.S. Vedantam, J.P. Liu, H. Zeng, S. Sun, *J. Appl. Phys.*, **2003**, *93*, 7184.
63. E. Tronc, A. Ezzir, R. Cherkaoui, C. Chanéac, M. Noguès, H. Kachkachi, D. Fiorani, A.M. Testa, J.M. Grenèche, J.P. Jolivet, *J. Magn. Magn. Mater.* **2000**, *221*, 63.
64. H. Kachkachi, M. Noguès, E. Tronc, D.A. Garanin, *J. Magn. Magn. Mater.* **2000**, *221*, 158.
65. J.P. Chen, C.M. Sorensen, K.J. Klabunde, G.C. Hadjipanayis, *Phys. Rev. B.*, **1995**, *51*, 11527.
66. D.V. Talapin, E.V. Shevchenko, H. Weller, in *Nanoparticles. From Theory to Application*, G. Shmid, Editor; Willey-VCH, Weinheim, **2004**, 199.
67. A. Aharoni, *J. Appl. Phys.*, **2001**, *90*, 4645.
68. W.F. Brown, Jr., *Ann. N. Y. Acad. Sci.* **1969**, *147*, 463.
69. W.F. Brown, Jr., *J. Appl. Phys.*, **1968**, *39*, 993.
70. D.R. Fredkin, T.R. Koehler, *J. Appl. Phys.*, **1990**, *67*, 5544.
71. A. Kákay, L.K. Varga, *J. Appl. Phys.*, **2005**, *97*, 083901.
72. A. Aharoni, *IEEE Trans. Magn.*, **1993**, *29*, 2596.
73. O. Popov, M. Mikhov, *phys. stat. sol. (a)*, **1990**, *118*, 289.
74. R.P. Cowburn, *J. Phys. D: Appl. Phys.*, **2000**, *33*, R1.
75. W.H. Qi, M.P. Wang, Q.H. Lui, *J. Mater. Sci.*, **2005**, *40*, 2737.
76. A. Aharoni, *J. Appl. Phys.*, **1988**, *63*, 5879.

77. A. Aharoni, *J. Appl. Phys.*, **1988**, *64*, 3330.
78. A. Aharoni, *IEEE Trans. Magn.*, **1986**, *22*, 478.
79. A.-H. Lu, E.L. Salabas, F. Schüth, *Angew. Chem. Int. Ed.*, **2007**, *46*, 1222.
80. C.P. Gräf, R. Birringer, A. Michels, *Phys. Rev. B.*, **2006**, *73*, 212401.
81. Y. Li, M. Afzaal, P. O'Brien, *J. Mater. Chem.*, **2006**, *16*, 2175.
82. S.Y. Chou, P.R. Krauss, W. Zhang, L. Guo, L. Zhuang, *J. Vac. Sci. Technol. B*, **1997**, *15*, 2897.
83. T. Hyeon, *Chem. Commun.*, **2003**, 927.
84. M.E. Schabes, H.N. Bertram, *J. Appl. Phys.*, **1988**, *64*, 1347.
85. N.A. Usov, S.E. Peschany, *J. Magn. Magn. Mater.*, **1994**, *130*, 275.
86. N.A. Usov, S.E. Peschany, *J. Magn. Magn. Mater.*, **1994**, *135*, 111.
87. W. Rave, K. Fabian, A. Hubert, *J. Magn. Magn. Mater.*, **1998**, *190*, 332.
88. A.J. Newell, R.T. Merrill, *J. Appl. Phys.*, **1998**, *84*, 4394.
89. D.K. Koltsov, R.P. Cowburn, M.E. Welland, *J. Appl. Phys.*, **2000**, *88*, 5315.
90. M. Beleggia, S. Tandon, Y. Zhu, M. De Graef, *J. Magn. Magn. Mater*, **2004**, *278*, 270.
91. J.L Dormann, D. Fiorani, E. Tronc, *Advan. Chem. Phys.*, **1997**, *XCVIII*, 283.
92. R.D. Kirby, M. Yu, D.J. Selmeyer, *J. Appl. Phys.*, **2000**, *87*, 5696.
93. W. Wernsdorfer, E.B. Orozco, K. Hasselbach, A. Benoit, B. Barbara, N. Demoncy, A. Loiseau, H. Pascard, D. Mailly, *Phys. Rev. Lett.*, **1997**, *78*, 1791.
94. R.E. Rosensweig, *Ann. Rev. Fluid Mech.*, **1987**, *19*, 437.
95. A.M. Konn, P. Laurent, P. Talbot, M. Le Floc'h, *J. Magn. Magn. Mater.*, **1995**, *140–144*, 367.
96. A.B. Pakhomov, Y. Bao, K.M. Krishnan, *J. Appl. Phys.*, **2005**, *97*, 10Q305.
97. J. Tejada, X. Zhang, E. Kroll, X. Bohigas, R.F. Ziolo, *J. Appl. Phys*, **2000**, *87*, 88.
98. W.F. Brown, Jr., *Phys. Rev.*, **1963**, *130*, 1677.
99. W.T. Coffey, D.S.F. Crothers, J.L. Dormann, L.J. Geoghegan, C. Kennedy, W. Wernsdorfer, *J. Phys.: Condens. Matter.*, **1998**, *10*, 9093.
100. R. Kötitz, W. Weitschies, L. Trahms, W. Semmler, *J. Magn. Magn. Mater.*, **1999**, *201*, 102.
101. E.M. Chudnovsky, L. Gunther, *Phys. Rev. Lett.*, **1988**, *60*, 661.
102. A. Garg, G.-H. Kim, *Phys. Rev. Lett.*, **1989**, *63*, 2512.
103. P.C.E. Stamp, E.M. Chudnovsky, B. Barbara, *Int. J. Mod. Phys. B*, **1992**, *6*, 1355.
104. S. Takahashi, R.S. Edwards, J.M. North, S. Hill, N.S. Dalal, *Phys. Rev. B*, **2004**, *70*, 094429.
105. N. Noginova, T. Weaver, E.P. Giannelis, A.B. Bourlinos, V.A. Atsarkin, V.V. Demidov, *Phys. Rev. B*, **2008**, *77*, 014403.
106. *Surface Effects in Magnetic Nanoparticles*, D. Fiorani, Editor; Springer, Berlin, **2005**.
107. A. Cornia, D. Gatteschi, R. Sessoli; *Coord. Chem. Rev.*, **2001**, *219–221*, 573.
108. J.R. Long, in *Chemistry of Nanostructured Materials*; P. Yang, Editor; World Scientific Publishing, Hong Kong, **2003**.
109. A.-L. Barra, P. Debrunner, D. Gatteschi, Ch.E. Schulz, R. Sessoli, *Europhvs. Lett.*, **1996**, *33*, 133.
110. D.G. Rancourt, *Sol. Stat. Commum.*, **1986**, *58*, 433.
111. R.L. Carlin, *Magnetochemistry*, Springer, Berlin, **1986**.
112. L. Néel, *J. Physique Radium*, **1954**, *15*, 255.
113. R.H. Victora, J.M. MacLaren, *Phys. Rev. B*, **1993**, *47*, 11583.
114. M. Respaud, J.M. Broto, H. Rakoto, A.R. Fert, L. Thomas, B. Barbara, M. Verelst, E. Snoeck, P. Lecante, A. Mosset, J. Osuna, T. Ould Ely, C. Amiens, B. Chaudret, *Phys. Rev. B*, **1998**, *57*, 2925.

115. C. Chen, O. Kitakami, S. Okamoto, Y Shimada, *J. Appl. Phys.*, **1999**, *86*, 2161.
116. J.P. Chen, C.M. Sorensen, K.J. Klabunde, G.C. Hadjipanayis, *J. Appl. Phys.*, **1994**, *76*, 6316.
117. B.J. Hickey, M.A. Howson, D. Greig, N. Wiser, *Phys. Rev. B*, **1996**, *53*, 32.
118. S.P. Gubin, Yu.I. Spichkin, Yu.A. Koksharov, G.Yu. Yurkov, A.V. Kozinkin, T.I. Nedoseikina, M.S. Korobova, A.M. Tishin, *J. Magn. Magn. Mater.*, **2003**, *265*, 234.
119. Y. Labaye, O. Crisan, L. Berger, J.M. Greneche, J.M.D. Coey, *J. Appl. Phys.*, **2002**, *91*, 8715.
120. S. Sun, C.B. Murray, D. Weller, L. Folks, A. Moser, *Science*, **2000**, *287*, 1989.
121. H.C. Siegman, *J. Phys.: Condens. Matter.*, **1992**, *4*, 8395.
122. A.E. Berkowitz, W.J. Shuele, P.J. Flanders, *J. Appl. Phys.*, **1968**, *39*, 1261.
123. J.M.D. Coey, *Phys. Rev. Lett.*, **1971**, *27*, 1140.
124. K. Haneda, A.H. Morrish, *J. Appl. Phys.*, **1988**, *63*, 4258.
125. M.P. Morales, C.J. Serna, F. Bodker, S. Morup, *J. Phys.: Condens. Mater*, **1997**, *9*, 5461.
126. L. Berger, Y. Labaye, M. Tamine, J.M.D. Coey, *Phys. Rev. B*, **2008**, *77*, 104431.
127. C. Luna, M.P. Morales, C.J. Serna, M. Vázquez, *Nanotechnology*, **2003**, *14*, 268.
128. E. De Biasi, R.D. Zysler, C.A. Ramos, H. Romero, *Physica B*, **2002**, *320*, 203.
129. J. Nogués, J. Sort, V. Langlais, V. Skumryev, S. Suricach, J.S. Mucoz, M.D. Bary, *Phys. Rep.*, **2005**, *422*, 65.
130. V. Skumryev, S. Stoyanov, Y. Zhang, G. Hadjipanayis, D. Givord, J. Nogués, *Nature*, **2003**, *423*, 850.
131. J. Nogués, I.K. Schuller, *J. Magn. Magn. Mater.*, **1999**, *192*, 203.
132. C.A. Ross, *Annu. Rev. Mater. Res.*, **2001**, *31*, 203.
133. A.E. Berkowitz, J.R. Mitchell, M.J. Carey, A.P. Young, S. Zhang, F.E. Spada, F.T. Parker, A. Hutten, G. Thomas, *Phys. Rev. Lett.*, **1992**, *68*, 3745.
134. J.Q. Wang, G. Xiao, *Phys. Rev. B*, **1994**, *49*, 3982.
135. *EPR of Transition Metal Ions*, A. Abraham, B. Bleaney, Clarendon, Oxford, **1970**.
136. S. Mørup, D.E. Madsen, C. Frandsen, C.R.H. Bahl, M.F. Hansen, *J. Phys.: Condens. Matter.*, **2007**, *19*, 213202.
137. K. De'Bell, A.B. MacIsaac, J.P. Whitehead, *Rev. Mod. Phys.*, **2000**, *72*, 225.
138. S.P. Gubin, I.D. Kosobudskii, *Russ. Chem. Rev.*, **1983**, *52*, 766.
139. I.D. Kosobudskii, L.V. Kashkina, S.P. Gubin, G.A. Petrakovskii, V.P. Piskorskii, N.M. Svirskaya, *Polym. Sci. USSR*, **1985**, *27*, 768.
140. S. Shtrikmann, E.P. Wohlfarth, *Phys. Lett. A*, **1981**, *85*, 467.
141. J.L Dormann, L. Bessais, D. Fiorani, *J. Phys. C: Solid State Phys.*, **1988**, *21*, 2015.
142. J.L. Dormann, R. Cherkaoui, L. Spinu, M. Nogués, L. Lucari, F. D'Orazio, D. Fiorani, A. García, E. Tronc, J.P. Jolivet, *J. Magn. Magn. Mater.*, **1998**, *187*, L139.
143. V.M. Rotello, *J. Am. Chem. Soc.*, **2005**, *127*, 9731.
144. D. Kechrakos, K.N. Trohidou, *Appl. Phys. Lett.*, **2002**, *81*, 7574.
145. B.L. Frankamp, A.K. Boal, M.T. Tuominen, V.M. Rotello, *J. Am. Chem. Soc.*, **2005**, *127*, 9731.
146. C.J. Bae, S. Angappane, J.-G. Park, Y. Lee, J. Lee, K. An, T. Hyeon, *Appl. Phys. Lett.*, **2007**, *91*, 102502.
147. J.L. Dormann, D. Fiorani, E. Tronc, *J. Magn. Magn. Mater.* **1998**, *183*, L255.
148. S. Mørup, M.F. Hansen, *J. Magn. Magn. Mater.*, **1998**, *184*, 262.
149. S. Mørup, E. Tronc, *Phys. Rev. Lett.*, **1994**, *72*, 3278.
150. O. Iglesias, A. Labarta, *Phys. Rev. B.*, **2004**, *70*, 144401.
151. W. Luo, S.R. Nagel, T.F. Rosenbaum, R.E. Rosensweig, *Phys. Rev. Lett.*, **1991**, *67*, 2721.
152. F. Luis, F. Petroff, J.M. Torres, L.M. García, J. Bartolomé, J. Carrey,

A. Vaurès, *Phys. Rev. Lett.*, **2002**, *88*, 217205.
153. M. Sasaki, P.E. Jönsson, H. Takayama, H. Mamiya, *Phys. Rev. B*, **2005**, *71*, 104405.
154. T. Jonsson, J. Mattsson, C. Djurberg, F.A. Khan, P. Nordblad, P. Svedlindh, *Phys. Rev. Lett.*, **1995**, *75*, 4138.
155. H. Mamiya, I. Nakatani, T. Furubayashi, *Phys. Rev. Lett.*, **1999**, *82*, 4332.
156. K. Jonason, E. Vincent, J. Hammann, J.P. Bouchaud, P. Nordblad, *Phys. Rev. Lett.*, **1998**, *81*, 3243.
157. E. Vincent, J.P. Bouchaud, J. Hammann, F. Lefloch, *Philos. Mag. B*, **1995**, *71*, 489.
158. C. Djurberg, P. Svedlindh, P. Nordblad, M.F. Hansen, F. Bødker, S. Mørup, *Phys. Rev. Lett.*, **1997**, *79*, 5154.
159. T. Jonsson, P. Svedlindh, M.F. Hansen, *Phys. Rev. Lett.*, **1998**, *81*, 3976.
160. S. Gangopadhyay, G.C. Hadjipanayis, B. Dale, C.M. Sorensen, K.J. Klabunde, V. Papaefthymiou, A. Kostikas, *Phys. Rev. B*, **1992**, *45*, 9778.
161. F.E. Luborsky, *J. Appl. Phys.*, **1958**, *29*, 309.
162. H. Yamada, M. Takano, M. Kiyama, J. Takada, T. Shinjo, K. Watanabe, *Adv. Ceram.*, **1985**, *16*, 169.
163. O. Kubo, T. Ido, H. Yokoyama, Y. Koike, *J. Appl. Phys.*, **1985**, *57*, 4280.
164. M.P. Morales, M. Andres-Vergés, S. Veintemillas-Verdaguer, M.I. Montero, C.J. Serna, *J. Magn. Magn. Mater.*, **1999**, *203*, 146.
165. F. Bloch, *Z. Physik*, **1930**, *61*, 206.
166. D. Zhang, K.J. Klabunde, C.M. Sorensen, G.C. Hadjipanayis, *Phys. Rev. B*, **1998**, *58*, 14167.
167. G. Xiao, C.L. Chien, *J. Appl. Phys.*, **1987**, *51*, 1280.
168. S. Linderoth, L. Balcells, A. Laborta, J. Tejada, P.V. Hendriksen, S.A. Sethi, *J. Magn. Magn. Mater.*, **1993**, *124*, 269.
169. D.L. Mills, *Commun. Solid State Phys.*, **1971**, *4*, 28.
170. K. Haneda, A.H. Morrish, *Nature*, **1979**, *282*, 186.
171. O. Shtnjo, O. Shigematsu, N. Hosaito, T. Iwasaki, O. Takai, *J. Appl. Phys.*, **1982**, *21*, L220.
172. E.F. Kneller, F.E. Luborsky, *J. Appl. Phys.*, **1963**, *34*, 656.
173. J.A. Christodoulides, M.J. Bonder, Y. Huang, Y. Zhang, S. Stoyanov, G.C. Hadjipanayis, A. Simopoulos, D. Weller, *Phys. Rev. B*, **2003**, *68*, 054428.
174. Y. Ichiyanagi, T. Uozumi, Y. Kimishima, *Trans. Mater. Res. Soc. Jpn.*, **2001**, *26*, 1097.
175. G.Yu. Yurkov, D.A. Baranov A.V. Kozinkin, Yu.A. Koksharov, T.I. Nedoseikina, O.V. Shvachko, S.A. Moksin, S.P. Gubin, *Inorg. Mater.*, **2006**, *42*, 1012.
176. E. Eftaxias, K.N. Trohidou, *Phys. Rev. B.*, **2005**, *71*, 134406.
177. R. Gans, *Ann. Phys.*, **1932**, *15*, 28.
178. C.-R. Lin, R.-K. Chiang, J.-S. Wang, T.-W. Sung, *J. Appl. Phys.*, **2006**, *99*, 08N710.
179. M.S. Seehra, H. Shim, P. Dutta, A. Manivannan, J. Bonevich, *J. Appl. Phys.*, **2005**, *97*, 10J509.
180. C. Gilles, P. Bonville; K.K.W. Wong, S. Mann, *Eur. Phys. J. B*, **2000**, *17*, 417.
181. L. Suber, D. Fiorani, P. Imperatori, S. Foglia, A. Montone, R. Zysler, *Nanostruct. Mater.*, **1999**, *11*, 797.
182. M.S. Seehra, V.S. Babu, A. Manivannan, J.W. Lynn, *Phys. Rev. B.*, **2000**, *61*, 3513.
183. M.S. Seehra, A. Punnoos, *Phys. Rev. B.*, **2001**, *64*, 132410.
184. S. Sako, Y. Umemura, K. Ohshima, M. Sakai, S. Bandow, *J. Phys. Soc. J. Appl. Phys.*, **1995**, *65*, 280.
185. L. Zhang, D. Xue, C. Gao, *J. Magn. Magn. Mater.*, **2003**, *267*, 111.
186. S.A. Makhlouf, F.T. Parker, F.E. Spada, A.E. Berkowitz, *J. Appl. Phys*, **1997**, *81*, 1561.
187. S.A. Makhlouf,_, H. Al-Attar, R.H. Kodama, *Solid State Commun.*, **2008**, *145*, 1.
188. A. Punnoose, H. Magnone, M.S. Seehra, J. Bonevich, *Phys. Rev. B.*, **2001**, *64*, 174420.

189. X.G. Zheng, C.N. Xu, K. Nishikubo, K. Nishiyama, W. Higemoto, W.J. Moon, E. Tanaka, E.S. Otabe, *Phys. Rev. B.*, **2005**, *72*, 014464.
190. T.G. Sorop, M. Evangelisti, M. Haase, L.J. de Jongh, *J. Magn. Magn. Mater.*, **2003**, *272–276*, 1573.
191. J.T. Richardson, D.I. Yiagas, B. Turk, K. Forster, M.V. Twigg, *J. Appl. Phys.*, **1991**, *70*, 6977.
192. F. Bødker, M.F. Hansen, C.B. Koch, K. Lefmann, S. Mørup, *Phys. Rev. B.*, **2000**, *61*, 6826.
193. B.T. Matthias, R.M. Bozorth, J.H. Val Vleck, *Phys. Rev. Lett.*, **1961**, *7*, 160.
194. P.K. Baltzer, H.W. Lehmann, M. Robbins, *Phys. Rev. Lett.*, **1965**, *15*, 493.
195. J.K. Furduna, *JAP*, **1988**, *64*, R29.
196. *Diluted Magnetic Semiconductors*, J.K. Furdyna, J. Kossut, Academic Press, New York, **1988**.
197. H. Ohno, *J. Magn. Magn. Mater.*, **1999**, *2*, 110.
198. J. Sinova, T. Jungwirth, in *Frontiers in Magnetic Materials*, A.V. Narlikar, Editor; Springer, Berlin, Heidelberg, New York, **2005** 185.
199. T. Dietl, *J. Appl. Phys.*, **2008**, *103*, 07D111.
200. A.H. MacDonald, P. Schiffer, N. Samarth, *Nature Mater.*, **2005**, *4*, 195.
201. F. Matsukura, H. Ohno, T. Dietl, in *Handbook of Magnetic Materials*, K.H.J. Buschow, Editor; Elsevier, Amsterdam, **2002**, 14 1.
202. T. Jungwirth, Jairo Sinova, J. Mašek, J. Kušera, A.H. MacDonald, *Rev. Mod. Phys.*, **2006**, *78*, 809.
203. T. Dietl, *Semicond. Sci. Technol.*, **2002**, *17*, 377.
204. T. Dietl, in *Advances in Solid State Physics*, B. Kramer, Editor; Springer, Berlin, **2003** 413.
205. T. Fukumura, H. Toyosaki, Y. Yamada, *Semicond. Sci. Technol.*, **2005**, *20*, S103.
206. T. Dietl, H. Ohno, F. Matsukura, J. Cibert, D. Ferrand, *Science*, **2000**, *287*, 1019.
207. C. Liu, F. Yun, H. Morkoc, *J. Mater. Sci: Mater. Electron.*, **2005**, *16*, 555.
208. G.P. Das, B.K. Rao, P. Jena, *Phys. Rev. B*, **2004**, *69*, 214422.
209. J.E. Jaffe, T.C. Droubay, S.A. Chambers, *J. Appl. Phys.*, **2005**, *97*, 073908.
210. A. Punnoose, J. Hays, *J. Appl. Phys.*, **2005**, *97*, 10D321.
211. S.A. Chambers, T.C. Droubay, C.M. Wang, K.M. Rosso, S.M. Heald, D.A. Schwartz, K.R. Kittilsvted, D.R. Gamelin, *Materials Today*, **2006**, *9*, 28.
212. G.N. Rao, Y.D. Yao, J.W. Chen, *J. Appl. Phys.*, **2007**, *101*, 09H119.
213. T. Fukumura, Z. Jin, M. Kawasaki, T. Shono, T. Hasegawa, S. Koshihara, H. Koinuma, *Appl. Phys. Lett.*, **2001**, *78*, 958.
214. A. Tiwari, C. Jin, A. Kvit, D. Kumar, J.F. Muth, J. Narayan, *Solid State Commun.*, **2002**, *121*, 371.
215. S.W. Jung, S.-J. An, G.-C. Yi, C.U. Jung, S.I. Lee, S. Cho., *Appl. Phys. Lett.*, **2002**, *80*, 4561.
216. P. Sharma, A. Gupta, K.V. Rao, F.J. Owens, R. Sharma, R. Ahuja, J.M.O. Guillen, B. Johansson, G.A. Gehring, *Nature Mater.*, **2003**, *2*, 673.
217. J. Luo, J.K. Liangb, Q.L. Liu, F.S. Liu, Y. Zhang, B.J. Sun, G.H. Rao, *J. Appl. Phys.*, **2005**, *97*, 086106.
218. A. Sandhu, *The Advanced Semiconductor Magazine*, **2005**, *18*, 32.
219. J.M.D. Coey, *Curr. Opin. Solid State Mater. Sci.*, **2006**, *10*, 83.
220. J.M.D. Coye, *Solid State Sci.*, **2005**, *7*, 660.
221. J.M.D Coey, S. Sanvito, *J. Phys. D.*, **2004**, *37*, 988.
222. J.M.D. Coey, M. Venkatesan, C.B. Fitzgerald, *Nature Mater.*, **2005**, *4*, 173.
223. A. Bandyopadhay, J. Velev, W.H. Butler, S.K. Sarker, O. Begone, *Phys. Rev. B*, **2004**, *69*, 174429.
224. T. Jungwirth, K.Y. Wang, J. Mašek, K.W. Edmonds, J. König, J. Sinova et al., *Phys. Rev. B*, **2005**, *72*, 165204.
225. A. Sundaresan, R. Bhargavi, N. Rangarajan, U. Siddesh, C.N.R. Rao, *Phys. Rev. B*, **2006**, *74*, 161306.
226. J.M.D. Coey, *J. Appl. Phys.*, **2005**, *97*, 10D313.
227. T. Dietl, *Physica E*, **2006**, *35*, 293.

228. M. van Schilfgaarde, O.N. Mryasov, *Phys. Rev. B*, **2001**, *63*, 233205.
229. K.N. Trohidou, X. Zianni, J.A. Blackman, *phys. stat. sol. (a)*, **2002**, *189*, 305.
230. T. Makarova, in *Frontiers in Magnetic Materials*, A.V. Narlikar, Editor; Springer, Berlin, Heidelberg, New York, **2005**.
231. E. Rodiner, *Chem. Soc. Rev.*, **2006**, *35*, 583.
232. R.M. White, *J. Magn. Magn. Mater.*, **2001**, *226–230*, 2042.
233. D.E. Speliotis, *J. Magn. Magn. Mater.*, **1999**, *193*, 29.
234. X. Shen, S. Hernandez, R.H. Victora, *IEEE Trans. Magn.*, **2008**, *44*, 163.
235. H.J. Richter, A.Y. Dobin, K. Gao, O. Heinonen, R.J. Van de Veerdonk, R.T. Lynch, J. Xue, D.K. Weller, P. Asselin, M.F. Erden, R.M. Brockie, *IEEE Trans. Magn.*, **2006**, *42*, 2255.
236. N. Weiss, T. Cren, M. Epple, S. Rusponi, G. Baudot, S. Rohart, A. Tejeda, V. Repain, S. Rousset, P. Ohresser, F. Scheurer, P. Bencok, H. Brune, *Phys. Rev. Lett.*, **2005**, *95*, 157204.
237. A. Kikitsu, Y. Kamata, M. Sakurai, K. Naito, *IEEE Trans. Magn.* **2007**, *43*, 3685.
238. R. Skomski, J.P. Liu, C.B. Rong, D.J. Sellmyer, *J. Appl. Phys.*, **2008**, *103*, 07E139.
239. C.B. Rong, D.R. Li, V. Nandwana, N. Poudyal, Y. Ding, Z.L. Wang, H. Zeng, J.P. Liu, *Adv. Mater.*, **2006**, *18*, 2984.
240. C. Kim, T. Loedding, S. Jang, H. Zeng, Z. Li, Y. Sui, D.J. Sellmyer, *Appl. Phys. Lett.*, **2007**, *91*, 172508.
241. H. Zeng, M. Zheng, R. Skomski, D.J. Sellmyer, Y. Liu, L. Menon, S. Bandyopadhyay, *J. Appl. Phys.*, **2000**, *87*, 4718.
242. H. Zeng, R. Skomski, L. Menon, Y. Liu, D.J. Sellmyer, S. Bandyopadhyay, *Phys. Rev. B*, **2002**, *65*, 134426.
243. N. Yasui, A. Imada, T. Den, *Appl. Phys. Lett.*, **2003**, *83*, 3347.
244. K. Liu, J. Nogués, C. Leighton, H. Masuda, K. Nishio, I.V. Roshchin, I.K. Schuller, *Appl. Phys. Lett.*, **2002**, *81*, 4434.
245. Y. Lei, W.K. Chim, *Chem. Mater.*, **2005**, *17*, 580.
246. R.L. Comstock, *J. Mater. Sci: Mater. Electron.*, **2002**, *13*, 509.
247. J.-G. Zhu, Y. Zheng, G.A. Prinz, *J. Appl. Phys.*, **2000**, *87*, 6668.
248. J. Rothman, M. Kläui, L. Lopez-Diaz, C.A.F. Vaz, A. Bleloch, J.A.C. Bland, Z. Cui, R. Speaks, *Phys. Rev. Lett.*, **2001**, *86*, 1098.
249. S.P. Li, D. Peyrade, M. Natali, A. Lebib, Y. Chen, U. Ebels, L.D. Buda, K. Ounadjela, *Phys. Rev. Lett.*, **2001**, *86*, 1102.
250. S.L. Tripp, R.E. Dunin-Borkowski, A. Wei, *Angew. Chem. Int. Ed.*, **2003**, *42*, 5591.
251. *Ferrohydrodynamics*, R.E. Rosensweig, Cambridge University Press, Cambridge, **1985**.
252. S.W. Charles, J. Popplewell, in *Ferromagnetic materials*, E.P. Wohlfarth, Editor; North-Holland, Amsterdam, **1980**, *2* 509.
253. F. Bitter, *Phys. Rev.*, **1932**, *41*, 507.
254. W.C. Elmor, *Phys. Rev.*, **1938**, *54*, 309.
255. W.C. Elmor, *Phys. Rev.*, **1938**, *54*, 1092.
256. N.H. Yeh, *IEEE Trans. Magn.*, **1980**, *MAG-16*, 979.
257. R.D. Weiss, J. Schifter, L. Borduz, K. Raj, *J. Appl. Phys.*, **1985**, *57*, 4274.
258. M.I. Shliomis, Yu.L. Raikher, *IEEE Trans. Magn.*, **1980**, *MAG-16*, 237.
259. S.S. Papell, US Patent Specification, **1964**, 3215572.
260. P.J. Shepherd, J. Popplewell, *Phil. Mag.*, **1971**, *23*, 239.
261. M.P. Pileni, *Adv. Funct. Mater.*, **2001**, *11*, 323.
262. U. Jeong, X. Teng, Y. Wang, H. Yang, Y. Xia, *Adv. Mater.* **2007**, *19*, 33.
263. *Magnetic Fluids, Engineering Applications*, B.M. Berkovsky, V.F. Medvedev, M.S. Krakov, Editors; Oxford University Press, Oxford, New York, **1993**.
264. *Magnetic Fluids and Applications Handbook*, B.M. Berkovsky, V. Bashtovoy, Editors; Begell House, New York, **1993**.

265. C. Scherer, A.M.F. Neto, *Brazil. J. Phys.*, **2005**, *35*, 718.
266. V.M. Zaitsev, M.I. Shliomis, *J. Appl. Mech. Tech. Phys.*, **1969**, *10*, 24.
267. M.I. Shliomis, *Soviet Phys. Uspekhi (Engl. Transl.)*, **1974**, *17*, 153.
268. *Magnetiviscous Effects in Ferrofluids*, S. Odenbach, Springer, Berlin, Heidelberg, **2002**.
269. R.E. Rosensweig, *Science*, **1996**, *271*, 614.
270. M.I. Shliomis, K.I. Morozov, *Phys. Fluids*, **1994**, *6*, 2855.
271. E. Blums, A. Cebers, M.M. Maiorov, Editor, *Magnetic Fluids*, W. de Gruyter, Berlin, New York, **1997**.
272. K. Raj, R. Moskowitz, *J. Magn. Magn. Mater.*, **1990**, *85*, 233.
273. V. Cabuil, *Curr. Opin. Colloid Interface Sci.*, **2000**, *5*, 44.
274. G.B. Khomutov, Yu.A. Koksharov, *Adv. Colloid Interface Sci.*, **2006**, *122*, 119.
275. H.E. Horng, C.-Y. Hong, S.Y. Yang, H.C. Yang, *J. Phys. Chem. Solids*, **2001**, *62*, 1749.
276. S. Odenbach, *J. Phys.: Condens. Matter*, **2004**, *16*, R1135.
277. S.A. Wolf, D.D. Awschalom, R.A. Buhrman, J.M. Daughton, S. von Molnár, M.L. Roukes, A.Y. Chtchelkanova, D.M. Treger, *Science*, **2001**, *294*, 1488.
278. *Spin Electronics*, M. Ziese, M.J. Thornton, Editors;, Springer, Berlin, Heidelberg, **2001**.
279. J. Fabian, A. Matos-Abiague, C. Ertler, P. Stano, I. Žutić, *Acta Physica Slovaca*, **2007**, *57*, 565.
280. P. Grünberg, R. Schreiber, Y. Pang, M.B. Brodsky, H. Sowers, *Phys. Rev. Lett.* **1986**, *57*, 2442.
281. M.N. Baibich, J.M. Broto, A. Fert, *Phys. Rev. Lett.*, **1988**, *61*, 2472.
282. G. Binasch, P. Grünberg, F. Saurenbach, W. Zinn, *Phys. Rev. B*, **1989**, *39*, 4828.
283. C. Tsang, C. Tsang, R.E. Fontana, T. Lin, D.E. Heim, V.S. Speriosu, B.A. Gurney, M.L. Williams, *IEEE Trans. Magn.*, **1994**, *30*, 3801.
284. J.Q. Xiao, J.S. Jiang, C.L. Chien, *Phys. Rev. Lett.*, **1992**, *68*, 3749.
285. R.L. White, *IEEE Trans. Magn.*, **1992**, *28*, 2482.
286. P. Allia, M. Knobel, P. Tiberto, F. Vinai, *Phys. Rev. B*, **1995**, *52*, 15.
287. E.F. Ferrari, F.C.S. da Silva, M. Knobel, *Phys. Rev. B*, **1999**, *59*, 8412.
288. S. Zhang, *Appl. Phys. Lett.*, **1992**, *61*, 1855.
289. S. Honda, M. Nawate, M. Tanaka, T. Okada, *J. Appl. Phys.*, **1997**, *82*, 764.
290. C. Wang, X. Xiao, H. Hu, Y. Rong, T.Y. Hsu, *Physica B*, **2007**, *392*, 72.
291. C.L. Chien, J.Q. Xian, J. Samuel Jiang, *J. Appl. Phys.* **1993**, *73*, 5309.
292. F. Parent, J. Tuaillon, L.B. Stern, V. Dupuis, B. Prevel, A. Perez, P. Melinon, G. Guiraud, R. Morel, A. Barthélémy, A. Fert, *Phys. Rev. B*, **1997**, *55*, 3683.
293. A. Hütten, J. Bernardi, C. Nelson, et al., *Phys. Sol. (A)*, **1995**, *150*, 171.
294. M.M. Deshmukh, S. Kleff, S. Guéron, E. Bonet, A.N. Pasupathy J. von Delft, D.C. Ralph, *Phys. Rev. Lett.*, **2001**, *87*, 226801.
295. P. Seneor, A. Bernand-Mantel, F. Petroff, *J. Phys.: Condens. Matter.*, **2007**, *19*, 165222.

7
Electron Magnetic Resonance of Nanoparticles: Superparamagnetic Resonance
Janis Kliava

7.1
Introduction

Nanomagnetism is both a fundamental and applied challenge. The magnetic properties observable on a macroscopic scale are due to a very large number of atoms and, therefore, are inexistent or very different on a microscopic scale – for a single atom or molecule or for a cluster of a few atoms. The nanometric scale, intermediate between the macro- and microscopic ones, reveals new "exotic" properties, still poorly understood and hence poorly controlled during the nanoparticle synthesis. Meanwhile, as long as the advent of nanomaterials is revolutionizing the technology, understanding and mastering their properties are of paramount importance. In particular, superparamagnetic systems consisting of magnetically ordered nanoparticles imbedded in a diamagnetic matrix attract great attention, and a correlation between the physical properties of the nanoparticles and the matrix is one of the hottest problems of the physics of nanosystems, e.g., see [1–6].

When fine magnetically ordered nanoparticles are dispersed in a diamagnetic matrix, a specific type of magnetic behavior, called *superparamagnetism*, is observed [7]. Understanding this phenomenon made a very important contribution to the fundamentals of magnetism and laid the foundation for the development of new materials for high-density information storage [8].

The physical properties of magnetic nanoparticles are determined by both magnetic nature and morphology (size and shape). A number of experimental techniques, such as static magnetic, optical, magnetooptical, rheological measurements, Mössbauer spectroscopy, electron microscopy, X-ray diffraction, small-angle neutron scattering, etc., have been applied to determine the magnetic characteristics of nanoparticles in different superparamagnetic systems and to evaluate their size distribution [8–13].

The electron magnetic resonance (EMR) of nanoparticles in ferrofluids, glasses, and other superparamagnetic systems has been extensively studied [14–52]. In order to distinguish this particular type of magnetic resonance,

Magnetic Nanoparticles. Sergey P. Gubin
© 2009 WILEY-VCH Verlag GmbH & Co. KGaA, Weinheim
ISBN: 978-3-527-40790-3

on the one hand, from the electron paramagnetic resonance (EPR) of diluted ions or other "paramagnetic centres" and, on the other hand, from the ferromagnetic or antiferromagnetic resonance (FMR, AFMR) in bulk magnetic materials, we refer to it as the *superparamagnetic resonance* (SPR). The generic term of EMR encompasses all these resonances.

The EMR *per se* and, *a fortiori*, in combination with other experimental techniques, provides a powerful tool of studying the physical properties of magnetic nanoparticles owing to its sensitivity to both the magnetic state and the morphological characteristics.

Meanwhile, the situation with the SPR spectra is somewhat paradoxical: at first sight they look very simple while the underlying theoretical analysis is very complex. Attempts to extract meaningful information from a visual inspection of a SPR spectrum usually fail because such a spectrum is clearly overparametrized. Indeed, most often this spectrum looks simply as a single slightly asymmetric line or, at the best, as a superposition of two lines (the "two-line pattern"). At the same time, the resonance magnetic field of a given nanoparticle includes anisotropic contributions, such as the magnetocrystalline anisotropy field depending on the physical nature of the particle and the demagnetizing field depending on the particle shape. In a disordered superparamagnetic system, nanoparticles are oriented more or less at random, so that the angular dependence of their resonance field results in orientation broadening of the spectra. Moreover, the observed spectral shape, in fact, is the superposition of a great number of contributions from individual nanoparticles, each characterized by its own size-dependent intrinsic lineshape. In this situation, computer simulations of the SPR spectra become unavoidable.

Several theoretical approaches to computer-assisted analysis of the SPR spectra of magnetic nanoparticles have been put forward [17, 20–23, 25, 29, 30, 38, 39, 42, 44, 53]. We have worked out a relatively simple though rather general and physically meaningful approach [22, 25, 29, 34] based on the general expression of the SPR spectrum outlined in Section 7.2. In Sections 7.3 and 7.4, we consider the calculation of the resonance magnetic field and of the lineshape, respectively.

The hallmark of the SPR is the superparamagnetic narrowing overviewed in Section 7.5. At elevated temperatures, thermal fluctuations of nanoparticle magnetic moments severely reduce both the angular anisotropy of resonance magnetic fields and the intrinsic linewidths, and particularly narrow resonance spectra are observed. The narrowing is more pronounced for smaller particle size. At lower temperatures the thermal fluctuations are gradually frozen out (the blocking phenomenon) and the resonance spectra become very broad (with a linewidth comparable to the resonance field). This means that the usual assumption of narrow resonance lines with Lorentzian-type intrinsic lineshapes fails. Only a few broad-lineshape expressions are quoted in the literature; moreover, some of them seem to be erroneous. In Section 7.4, we will remedy the lack of a systematic analysis of broad resonance lineshapes engendered by different phenomenological equations of motion. We believe that such analysis may be of general interest to people concerned with magnetic resonance spectroscopy.

A particularly interesting feature of the SPR, appearing in variable temperature studies, is a correlation between the apparent resonance magnetic field and the spectra width. As the temperature is lowered, the SPR spectra broaden in a very spectacular way and simultaneously shift to lower magnetic fields. Such behavior has been reported for numerous superparamagnetic systems [17, 19, 26, 32, 33, 39, 49, 51, 52, 54–56]. In Section 7.4, we will also show that such behavior can be well described in the framework of the general model taking into account low-temperature freezing of the fluctuations of orientations of the magnetic moments and including the linewidth expression resulting from the Landau–Lifshitz damped precession equation.

In the majority of papers dealing with the analysis of the SPR spectra, a spherical shape of the superparamagnetic particles has been assumed. Meanwhile, the anisotropy of the particle shapes plays an important role, e.g., in determining the magnetic birefringence of ferrofluids [10]. The nonsphericity of the magnetic nanoparticles is still more pronounced in partially devitrified glasses, in which case considerable statistical distributions of the particle shapes occur, as well. In Section 7.6, we will provide a general analysis relating the nanoparticle morphological characteristics to their SPR spectra.

Finally, in Section 7.7 we will illustrate the application of the above-mentioned theoretical developments to a novel class of superparamagnetic materials, exemplified by partially devitrified oxide glasses with paramagnetic dopants [22, 25, 29, 34, 39, 43–46, 50, 52]. Such glasses, on the one hand, are characterized by a nonlinear magnetic field dependence of the magnetization, in some cases with hysteresis and magnetic saturation. On the other hand, for certain compositions, they are transparent in the visible and near-infrared spectral range. Such a combination of properties makes these glasses particularly promising for various technical applications, in particular, for new magneto-optical data storage and spin electronics devices.

7.2
Superparamagnetic Resonance Spectrum in a Disordered System

An assembly of single-crystal ferromagnetic particles with the size of the order of some nanometers (nanoparticles) isolated in a diamagnetic matrix forms a superparamagnetic system. Nanoparticles of relatively greater sizes may incorporate several magnetic domains; however, below ca. 10 nm they are typically single-domain particles [8]. Such nanoparticle (neglecting surface and disorder effects) can be thought of as a single giant magnetic moment formed by an exchange interaction between all individual spins; this magnetic moment, $\mu = MV$, is a product of the magnetization M (not necessarily the same as in the bulk material) and the particle volume V, and we denote its norm by μ.

Consider a statistical assembly of single-domain magnetic nanoparticles whose characteristics vary from one particle to another. In the laboratory frame each nanoparticle is described by a random vector $\boldsymbol{\xi}$ whose components include magnetic parameters (in the simplest case, the norm of the magnetization M and the first-order anisotropy constant K_1), morphological (size and shape) characteristics, and the orientation of $\boldsymbol{\mu}$ defined by a unit vector $\boldsymbol{u}_\mu = \boldsymbol{\mu}/\mu$. We denote by $f_V(\boldsymbol{\xi}, \boldsymbol{u}_\mu)$ the *volume fraction* (see Section 7.6) of such particles. The contribution to the magnetic resonance from each nanoparticle, $F(\boldsymbol{\xi}, \boldsymbol{u}_\mu, B, \boldsymbol{u}_B)$, is proportional to the derivative of the imaginary part of the *dynamic* magnetic susceptibility. It depends on $\boldsymbol{\xi}$, $\boldsymbol{u}_\mu$ as well as on the norm B, and the orientation of the static magnetizing field $\boldsymbol{B}$ of the spectrometer, defined by a unit vector $\boldsymbol{u}_B = \boldsymbol{B}/B$. The magnetic resonance spectrum of this assembly is a weighted sum of contributions from individual particles with given characteristics, calculated as follows:

$$P(B, \boldsymbol{u}_B) = \int_{\boldsymbol{u}_\mu} \int_{\boldsymbol{\xi}} f_V(\boldsymbol{\xi}, \boldsymbol{u}_\mu) F(\boldsymbol{\xi}, \boldsymbol{u}_\mu, B, \boldsymbol{u}_B) d\boldsymbol{\xi} \, d\boldsymbol{u}_\mu. \tag{7.1}$$

In a particular but quite widespread case of macroscopically isotropic superparamagnetic systems (e.g., powders or glasses), the resonance spectrum becomes orientation independent:

$$P(B, \boldsymbol{u}_B) = \int_0^{2\pi} \int_0^{\pi} \int_{\boldsymbol{u}_\mu} \int_{\boldsymbol{\xi}} f_V(\boldsymbol{\xi}, \boldsymbol{u}_\mu) F(\boldsymbol{\xi}, \boldsymbol{u}_\mu, B, \boldsymbol{u}_B) d\boldsymbol{\xi} \, d\boldsymbol{u}_\mu \sin \vartheta \, d\vartheta \, d\varphi, \tag{7.2}$$

where ϑ are φ are, respectively, the polar and the azimuthal angles of $\boldsymbol{B}$ in the laboratory frame.

In theoretical modeling and, in particular in computer fitting magnetic resonance spectra, it is often convenient to separate the notion of resonance line broadening from that of the distribution of resonance magnetic fields. In this instance, $F(\boldsymbol{\xi}, \boldsymbol{u}_\mu, B, \boldsymbol{u}_B)$ is considered as an "intrinsic lineshape" $F(B, B_r, \Delta_B)$ arising from nanoparticles with given parameters and characterized by a resonance field B_r and an "intrinsic linewidth" Δ_B. The SPR spectrum of an assembly of such nanoparticles can be computed as follows [29]:

$$P(B) = \int_0^{2\pi} \int_0^{\pi} \int_{\boldsymbol{\xi}} f_V(\boldsymbol{\xi}) F[B, B_r(\boldsymbol{\xi}, \vartheta, \varphi), \Delta_B(\boldsymbol{\xi}, \vartheta, \varphi)] \sin \vartheta \, d\boldsymbol{\xi} \, d\vartheta \, d\varphi, \tag{7.3}$$

where $f_V(\boldsymbol{\xi})$ is the volume fraction of particles with a given $\boldsymbol{\xi}$.

If we assume in addition that the magnetic parameters of the nanoparticles have fixed values for all particles and throughout the volume of each particle, the components of ξ will include only morphological characteristics of the nanoparticles.

In the literature concerning the SPR linewidth and, more generally, the magnetic resonance in powders, glasses, or fluids, there has been much confusion between the observed, e.g., peak-to-peak width of the spectral features and the true intrinsic linewidth [17, 21, 26]. In the approach based on Eq. (7.3), the spectra "broadening" due to the spread of the resonance magnetic fields $B_r(\xi, \vartheta, \varphi)$ is explicitly taken into account, so that the intrinsic linewidth $\Delta_B(\xi, \vartheta, \varphi)$ accounts only for the contribution to the SPR spectra from particles of a given size, shape, and orientation. The corresponding broadening in most cases is not obvious from a visual inspection of the experimental resonance spectra and can only be deduced from accurate computer simulations.

7.3
Resonance Magnetic Field

For simplicity, we consider a single-domain nanoparticle of ellipsoidal form with the principal axes x, y, and z. The free energy of such particle to first order in the magnetic symmetry can be expressed by the following equation, e.g., see [57]:

$$E = -\boldsymbol{\mu} \cdot \boldsymbol{B} + K_1 V \Phi(\boldsymbol{u}_\mu) + \frac{1}{2}\frac{\mu_0}{V}\boldsymbol{\mu} \cdot \boldsymbol{N} \cdot \boldsymbol{\mu}. \tag{7.4}$$

The first term on the right-hand side of Eq. (7.4) is the Zeeman energy. The second term is the (first-order) magnetocrystalline anisotropy energy depending on the orientation of the unit vector $\boldsymbol{u}_\mu$ (see the previous section) with respect to the magnetic symmetry axes and of the corresponding constant K_1. The third term is the magnetostatic energy, with $\boldsymbol{N}$ the demagnetizing tensor whose eigenvalues, the demagnetizing factors, are related by [58]

$$N_x + N_y + N_z = 1. \tag{7.5}$$

The $\Phi(\boldsymbol{u}_\mu)$ function depends on the magnetic symmetry [57], and it can be expressed in terms of the directional cosines $l_{\mu_x}, l_{\mu_y}, l_{\mu_z}$ or the polar and azimuthal angles α and β of $\boldsymbol{\mu}$. Namely, in the cubic symmetry

$$\Phi(\boldsymbol{u}_\mu) = l_{\mu_x}^2 l_{\mu_y}^2 + l_{\mu_x}^2 l_{\mu_z}^2 + l_{\mu_y}^2 l_{\mu_z}^2 = \frac{1}{4}(\sin^2 2\alpha + \sin^4 \alpha \sin^2 2\beta) \tag{7.6}$$

and in the axial symmetry

$$\Phi(\boldsymbol{u}_\mu) = 1 - l_{\mu_z}^2 = \sin^2 \alpha. \tag{7.7}$$

The equilibrium orientation of $\boldsymbol{\mu}$ minimises the value of E, i.e., at equilibrium $\boldsymbol{\mu}$ is parallel to the *effective* magnetic field defined as

$$\boldsymbol{B}_{\text{eff}} = -\frac{\partial E}{\partial \boldsymbol{\mu}} = -\frac{1}{\mu}\nabla_u E, \tag{7.8}$$

where the gradient ∇_u is taken in the space of the spherical unit vectors $\boldsymbol{u}_\mu$, $\boldsymbol{u}_\alpha$, $\boldsymbol{u}_\beta$. In the absence of external magnetic field, in a spherical particle $\boldsymbol{\mu}$ is directed along an easy magnetization axis, while in the strong external field case the Zeeman coupling dominates over all other interactions and the direction of $\boldsymbol{\mu}$ is close to that of $\boldsymbol{B}_{\text{eff}}$.

Meanwhile, from the quantum mechanical viewpoint all three components of $\boldsymbol{\mu}$ cannot be simultaneously determined. The semiclassical description of this property is given, in the absence of damping, by the equation of free gyromagnetic precession of $\boldsymbol{\mu}$ around the direction of $\boldsymbol{B}_{\text{eff}}$:

$$\dot{\boldsymbol{\mu}} = \gamma \boldsymbol{\mu} \wedge \boldsymbol{B}_{\text{eff}}, \qquad (7.9)$$

where $\gamma < 0$ is the gyromagnetic ratio defined as $\gamma = \frac{1}{2} ge/m$ (e is the electron charge, m is the electron mass, and g is the electronic g-factor). The norm of $\boldsymbol{\mu}$ is assumed to be constant.

According to Eq. (7.9) $\boldsymbol{\mu}$ precesses at the angular frequency $\omega_{\text{precession}} = -\gamma B_{\text{eff}}$. In a magnetic resonance experiment one applies perpendicularly to B an oscillating magnetic field b of an angular frequency ω (the magnetic component of an electromagnetic wave, typically of the microwave range and of amplitude $b \ll B$), and the resonance occurs at the angular frequency $\omega_r = \omega_{\text{precession}}$. In most EMR spectrometers, however, ω is kept constant and the magnetizing field B is linearly swept, so that the resonance is observed at the field value $B_r = B_{\text{eff}}$.

In the case of damped precession, instead of a fixed resonance field, a resonance line of definite shape is recorded. This will be considered in detail in the next section.

Several alternative procedures of calculating the resonance field have been reported [24, 28, 57, 59, 60]. Most often, the well-known approach first suggested by Smit and Beljers [59] is used for this purpose. More recently, an alternative approach has been put forward by Baselgia et al. [60]. Unfortunately, these authors carried out their derivation in the Cartesian coordinate system and only at the last stage transformed their result to spherical coordinates, more suitable for practical applications. Below we outline their approach by consistently using spherical coordinates.

In the spherical basis with unit vectors $\boldsymbol{u}_\mu$, $\boldsymbol{u}_\alpha$, $\boldsymbol{u}_\beta$ Eq. (7.9) becomes

$$\mu(\dot{\alpha}\boldsymbol{u}_\alpha + \sin\alpha\,\dot{\beta}\boldsymbol{u}_\beta) = \gamma\mu(B_\alpha \boldsymbol{u}_\beta - B_\beta \boldsymbol{u}_\alpha), \qquad (7.10)$$

where B_α and B_β are the corresponding components of $\boldsymbol{B}_{\text{eff}}$. Taking into account Eq. (7.8) and the expression of gradient on the unit sphere

$$\nabla = \left(\frac{\partial}{\partial \alpha}, \frac{1}{\sin\alpha}\frac{\partial}{\partial \beta}\right), \qquad (7.11)$$

from Eqs. (7.8) and (7.10) we get

$$\mu\dot{\alpha} = \frac{\gamma}{\sin\alpha}\frac{\partial E}{\partial \beta}; \qquad \mu\dot{\beta}\sin\alpha = -\gamma\frac{\partial E}{\partial \alpha}. \qquad (7.12)$$

7.3 Resonance Magnetic Field

In order to express the partial derivatives in Eqs. (7.12), we consider the variation of the particle free energy in a neighborhood of equilibrium, $\delta E = E(\boldsymbol{\mu}_0 + \delta\boldsymbol{\mu}) - E(\boldsymbol{\mu}_0)$, where $\boldsymbol{\mu}_0$ is the magnetic moment at equilibrium, and expand it in a bivariate Taylor series:

$$\delta E = \nabla_0 E \cdot \delta\boldsymbol{\mu} + \frac{1}{2}\delta\boldsymbol{\mu}^T \cdot \nabla^2{}_0 E \cdot \delta\boldsymbol{\mu}, \tag{7.13}$$

where

$$\delta\boldsymbol{\mu} = \mu \begin{pmatrix} d\alpha \\ \sin\alpha\, d\beta \end{pmatrix}, \tag{7.14}$$

$\delta\boldsymbol{\mu}^T$ is the transpose of $\delta\boldsymbol{\mu}$ and both the gradient ∇ and the Hessian (and *not* the Laplacian) ∇^2 are taken over the surface of the unit sphere. The subscript 0 here and below indicates that the corresponding derivatives and trigonometric functions are calculated at equilibrium. In spherical coordinates, e.g., see [61]:

$$\nabla^2 = \begin{pmatrix} \dfrac{\partial^2}{\partial\alpha^2} & \dfrac{1}{\sin\alpha}\dfrac{\partial^2}{\partial\beta\partial\alpha} - \dfrac{\text{ctg}\,\alpha}{\sin^2\alpha}\dfrac{\partial}{\partial\beta} \\ \dfrac{1}{\sin\alpha}\dfrac{\partial^2}{\partial\alpha\partial\beta} - \dfrac{\text{ctg}\,\alpha}{\sin^2\alpha}\dfrac{\partial}{\partial\beta} & \dfrac{1}{\sin^2\alpha}\dfrac{\partial^2}{\partial\beta^2} + \text{ctg}\,\alpha\dfrac{\partial}{\partial\alpha} \end{pmatrix}. \tag{7.15}$$

Calculating δE from Eq. (7.13) and comparing with the expression

$$\delta E(\boldsymbol{\mu}) = \frac{\partial E}{\partial\alpha}\delta\alpha + \frac{\partial E}{\partial\beta}\delta\beta, \tag{7.16}$$

we get

$$\frac{\partial E}{\partial\alpha} = \left(\frac{\partial^2 E}{\partial\alpha^2}\right)_0 \delta\alpha + \left(\frac{\partial^2 E}{\partial\alpha\partial\beta} - \text{ctg}\,\alpha\frac{\partial E}{\partial\beta}\right)_0 \delta\beta$$

$$\frac{\partial E}{\partial\beta} = \left(\frac{\partial^2 E}{\partial\alpha\partial\beta} - \text{ctg}\,\alpha\frac{\partial E}{\partial\beta}\right)_0 \delta\alpha + \left(\frac{1}{\sin\alpha}\frac{\partial^2 E}{\partial\beta^2} + \cos\alpha\frac{\partial E}{\partial\alpha}\right)_0 \sin\alpha\,\delta\beta. \tag{7.17}$$

To simplify the subsequent formulae we use the following notation (not to be confused with that used by some other authors, e.g., in Refs. [57, 59, 60]):

$$E_{\alpha\alpha} = \left(\frac{\partial^2 E}{\partial\alpha^2}\right)_0$$

$$E_{\beta\beta} = \left(\frac{1}{\sin^2\alpha}\frac{\partial^2 E}{\partial\beta^2} + \text{ctg}\,\alpha\frac{\partial E}{\partial\alpha}\right)_0$$

$$E_{\alpha\beta} = \left(\frac{1}{\sin\alpha}\frac{\partial^2 E}{\partial\beta\partial\alpha} - \frac{\text{ctg}\,\alpha}{\sin\alpha}\frac{\partial E}{\partial\beta}\right)_0. \tag{7.18}$$

Substituting Eqs. (7.17) in Eqs. (7.12) yields a pair of linear equations for two angular deviations from equilibrium, and these equations are homogenized by seeking the solution in the form $\delta\alpha = \delta\alpha_{max} e^{i\omega t}$; $\delta\beta = \delta\beta_{max} e^{i\omega t}$:

$$\left(E_{\alpha\beta} - i\frac{\omega}{\gamma}\mu\right)\delta\alpha_{max} + E_{\beta\beta}\sin\alpha\,\delta\beta_{max} = 0$$

$$\frac{1}{\sin\alpha} E_{\alpha\alpha} \delta\alpha_{max} + \left(E_{\alpha\beta} + i\frac{\omega}{\gamma}\mu\sin\alpha\right)\delta\beta_{max} = 0. \quad (7.19)$$

The condition for existence of nontrivial solutions of the latter system yields, cf. Ref. [60]:

$$\omega^2 = \frac{\gamma^2}{\mu^2}\left(E_{\alpha\alpha}E_{\beta\beta} - E_{\alpha\beta}^2\right). \quad (7.20)$$

From the latter equation the resonance magnetic field B_r is determined by iteration.

In the earlier approach of Smit and Beljers [59, see also 57] the first derivatives in Eq. (7.18) were omitted, resulting in

$$\omega^2 = \frac{\gamma^2}{\mu^2 \sin^2\alpha}\left[\frac{\partial^2 E}{\partial\alpha^2}\frac{\partial^2 E}{\partial\beta^2} - \left(\frac{\partial^2 E}{\partial\beta\partial\alpha}\right)^2\right]_0. \quad (7.21)$$

From the mathematical viewpoint Eq. (7.21) is quite accurate and fully equivalent to Eq. (7.20); indeed, at equilibrium the first derivatives of E with respect to the spherical angles vanish. Nevertheless, the form Eq. (7.20) is more convenient, since it avoids mixing the different free energy terms and, in contrast to Eq. (7.21), it remains valid also in the particular case $\alpha = 0$ [60]. Indeed, we have checked that computer simulations of the magnetic resonance spectra of nanoparticles based on Eq. (7.20) provide better results in comparison with those using Eq. (7.21).

In the important special case of a strong applied magnetic field, when in Eq. (7.1) the Zeeman energy dominates over the magnetocrystalline anisotropy and magnetostatic energies, Eqs. (7.20) and (7.21) become linear with respect to the resonance field B_r and the latter can be expressed as, e.g., see [57]:

$$B_r = B_0 + B_a + B_d. \quad (7.22)$$

where $B_0 = -\omega/\gamma$, and the magnetocrystalline anisotropy field B_a and the demagnetizing field B_d arise, respectively, from the last two terms of Eq. (7.4). These fields bring about an angular dependence of the resonance spectra. To first order in the parameter K_1/M one gets in cubic symmetry

$$B_a^{cub} = 2\frac{K_1^{cub}}{M}\left[5\left(l_{B_x}^2 l_{B_y}^2 + l_{B_x}^2 l_{B_z}^2 + l_{B_y}^2 l_{B_z}^2\right) - 1\right]$$

$$= -\frac{5}{4}\frac{K_1^{cub}}{M}\left(\cos 4\vartheta + \frac{3}{5} - 2\sin^4\vartheta\sin^2 2\varphi\right) \quad (7.23)$$

and in axial symmetry

$$B_a^{ax} = \frac{K_1^{ax}}{M}(1 - 3l_{B_z}^2) = \frac{K_1^{ax}}{M}(1 - 3\cos^2\vartheta), \tag{7.24}$$

where K_1^{cub} and K_1^{ax} are, respectively, the cubic and axial first-order anisotropy constants, $l_{B_x}, l_{B_y}, l_{B_z}$ are the directional cosines and ϑ and φ are the spherical angles of B defined with respect to the magnetic symmetry axes.

For nanoparticles of the form of a general (three-axes) ellipsoid the demagnetizing field is expressed as

$$B_d = B_0 + \frac{3}{2}\mu_0 M \big(N_x \sin^2\tilde{\vartheta}\cos^2\tilde{\varphi} + N_y \sin^2\tilde{\vartheta}\sin^2\tilde{\varphi}$$
$$+ N_z \cos^2\tilde{\vartheta} - \frac{1}{3}\big), \tag{7.25}$$

and for nanoparticles of the form of an ellipsoid of revolution, setting $N_x = N_y = N_\perp$, $N_z = N_\parallel$, Eq. (7.25) simplifies to

$$B_d = B_0 + \frac{1}{2}\mu_0 M (N_\parallel - N_\perp)(3\cos^2\tilde{\vartheta} - 1), \tag{7.26}$$

where $\tilde{\vartheta}$ and $\tilde{\varphi}$ are the polar and azimuthal angles of B with respect to the principal axis of the shape ellipsoid.

7.4 Resonance Lineshapes

7.4.1 Damped Gyromagnetic Precession

7.4.1.1 Definitions
In the previous section the free gyromagnetic precession of the particle magnetic moment μ around the effective field was considered. More generally, the motion of μ can be characterized as damped precession described by a phenomenological equation of the general form:

$$\dot{\mu} = \gamma \mu \wedge B_{eff} + \dot{\mu}_{damping}. \tag{7.27}$$

The first term on the right-hand side of Eq. (7.27) accounts for the free precession of μ around the effective magnetic field B_{eff}, cf. Eq. (7.9). In the magnetic resonance conditions, B_{eff} includes the static field B "seen" by the particle (the applied magnetizing field as well as internal fields, namely the demagnetizing field and the magnetocrystalline anisotropy field, see Section 7.3) and the microwave magnetic field b. We choose a coordinate system where the static field and the corresponding static magnetization M

are along the z-axis, and the microwave magnetic field lies in the perpendicular plane, $b = (b_x \quad b_y)$.

The second term on the right-hand side of Eq. (7.27) describes the damping (relaxation) of the motion as a torque that tends to align the magnetization with its equilibrium orientation and accounts for homogeneous broadening of the magnetic resonance lines. The physical mechanisms of the relaxation are quite complicated [62], therefore some phenomenological forms of this term are used by different authors. Below we consider successively the forms used in Bloch–Bloembergen [63, 64], modified Bloch [65, 66], Gilbert [67], Landau–Lifshitz [68] and Callen [69] equations and carry out a comparative analysis of the resonance lineshapes resulting from these equations (see also [70]). However, we must warn the reader that, although these equations are the most well-known and the most currently used ones, they are just some particular cases from "the myriad conceivable forms" [71, 72].

The phenomenological equations are resolved using the Polder tensor method, e.g., see [62 p. 586]. The magnetization related to b, $m = (m_x \quad m_y)$, is expressed as

$$m = \begin{pmatrix} \chi & -i\kappa \\ i\kappa & \chi \end{pmatrix} \tag{7.28}$$

where χ and κ describe, respectively, the response to the x and y components of b.

In the case of a circularly polarized microwave field $b^\pm = b_0 e^{\pm i\omega t}$ where the $\pm$ signs stand for the two polarization directions, the dynamic susceptibility is given by

$$\chi_\pm = \chi \mp \kappa = \chi'_\pm - i\chi''_\pm. \tag{7.29}$$

The two polarization directions are usually referred to as "resonant," or "Larmor" and "nonresonant", or "anti-Larmor" polarizations. For a linearly polarized microwave field $b = b_0 \cos \omega t$ applied along the x-axis, the complex susceptibility is

$$\chi = \chi' - i\chi'' = \frac{1}{2}(\chi_+ + \chi_-) = \frac{1}{2}\left[\chi'_+ + \chi'_- - i(\chi''_+ + \chi''_-)\right]. \tag{7.30}$$

The magnetic resonance absorption is proportional to the imaginary part of the dynamic susceptibility.

Below, we list analytical forms of the resonance lineshape for the linear polarization and both circular polarization directions of the microwave radiation. One can easily check that in all above-quoted cases $\chi''(B) = \frac{1}{2}[\chi''_+(B) + \chi''_-(B)]$. Whenever possible, $\chi''(B)$ has been normalized to unity: $\int_{-\infty}^{\infty} \chi''(B) dB = 1$.

Two different cases of magnetic behavior of the system have been considered, see Figure 7.1:

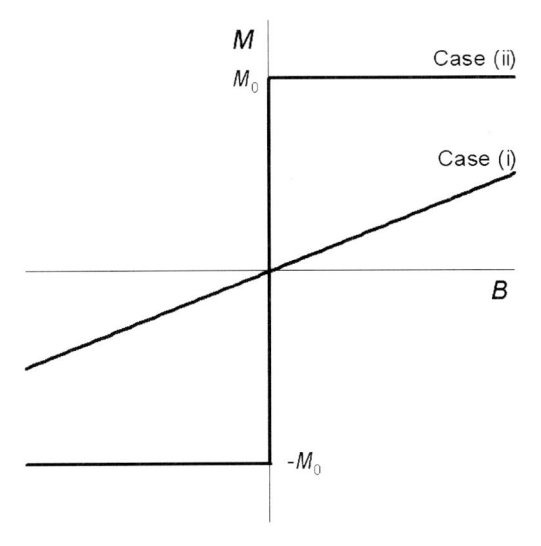

Figure 7.1 Two different cases of magnetic behaviour of the system.

(i) that of a linear paramagnet characterized by static magnetization directly proportional to the static magnetic field, $M_0 = \chi_0 B$, where χ_0 is the static susceptibility;
(ii) that of a perfect soft ferromagnet characterized by a stepwise dependence $M_0(B) = M_0 \operatorname{sgn}(B)$ with $M_0 = \text{const}$.

In the expressions given below the magnetizing field is supposed to be much stronger than the microwave magnetic field, therefore, strictly speaking these expressions fail in the immediate vicinity of $B = 0$; B_r denotes the "true" resonance magnetic field introduced in the Section 7.3 and the linewidth parameter Δ_B is defined as the half-width at half-height in the narrow-line limit $\Delta_B \ll B_r$.

The different absorption lineshapes are displayed in Figure 7.2 for the linear polarization and the "resonant" circular polarization. The corresponding derivative-of-absorption lineshapes in the linear polarization case (the most current experimental situation) are shown in Figure 7.3.

7.4.1.2 The Bloch–Bloembergen Equation

Bloembergen adapted to ferromagnetic resonance the Bloch's nuclear magnetic resonance equation [64]. In a compact form, this equation can be written as follows:

$$\dot{\mu} = \gamma \mu \wedge B_{\text{eff}} - \frac{\mu - \delta_{iz} M_0 V}{T}, \qquad (7.31)$$

where $T = (T_2 \ T_2 \ T_1)$ and $\delta_{iz} = (0 \ 0 \ 1)$; T_1 and T_2 are referred to, respectively, as the spin-lattice and spin-spin relaxation times. We denote the linewidth by $\Delta_B = 1/|\gamma| T_2$.

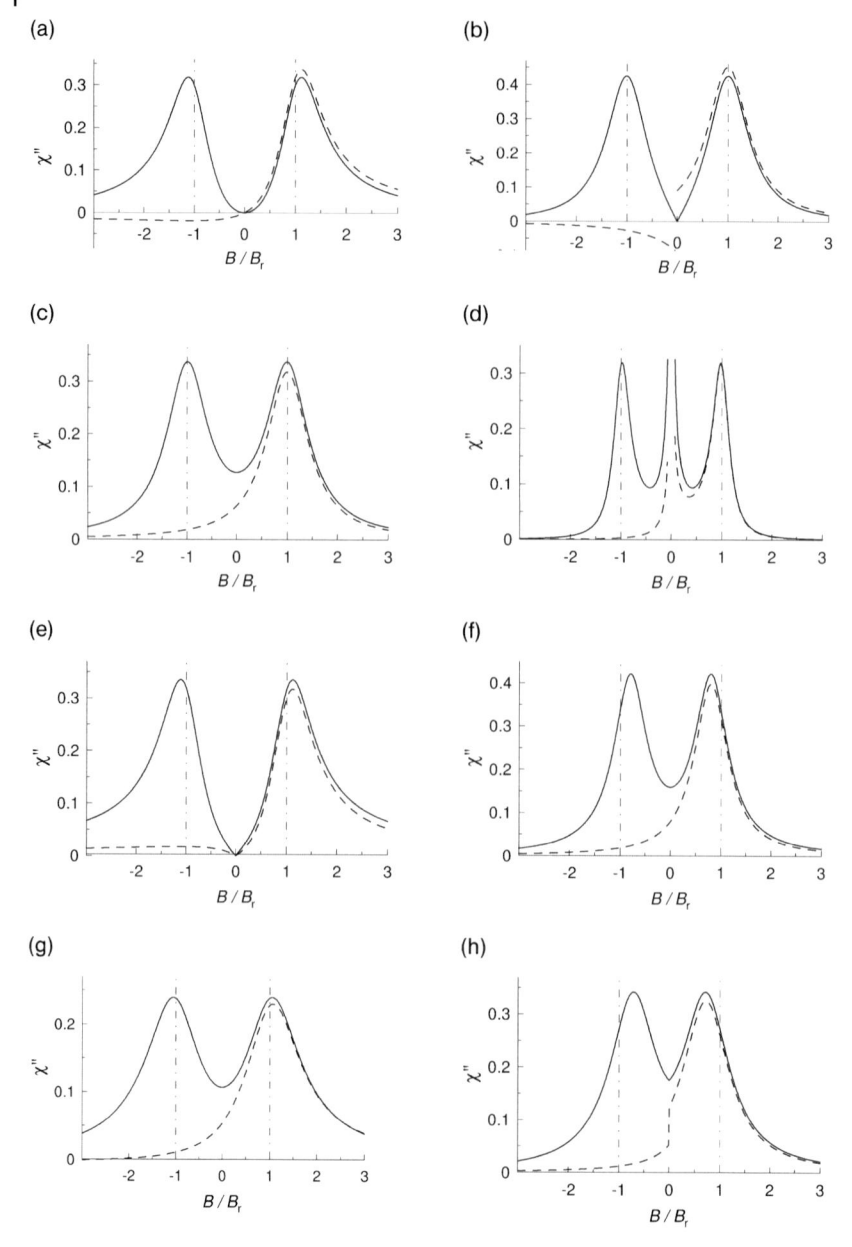

Figure 7.2 Magnetic resonance *absorption* lineshapes obtained with the following equations: (a) Bloch–Bloembergen, case (i), Eq. (7.32); (b) Bloch–Bloembergen, case (ii), Eq. (7.34); (c) modified Bloch, case (i), Gilbert, case (ii), Landau–Lifshitz, case (i), Eq. (7.37); (d) modified Bloch, case (ii), Eq. (7.40); (e) Gilbert, case (i), Eq. (7.43); (f) Landau–Lifshitz, case (ii), Eq. (7.49); (g) Callen, case (i), Eq. (7.52); (h) Callen, case (ii) Eq. (7.54). The linewidth ratio is $\varepsilon = \Delta_B/B_r = 1/2$ in all cases and $\eta = \delta_B/B_r = 1/5$ for the Callen lineshape. Full line: linear polarization; dashed line: right circular polarization [70].

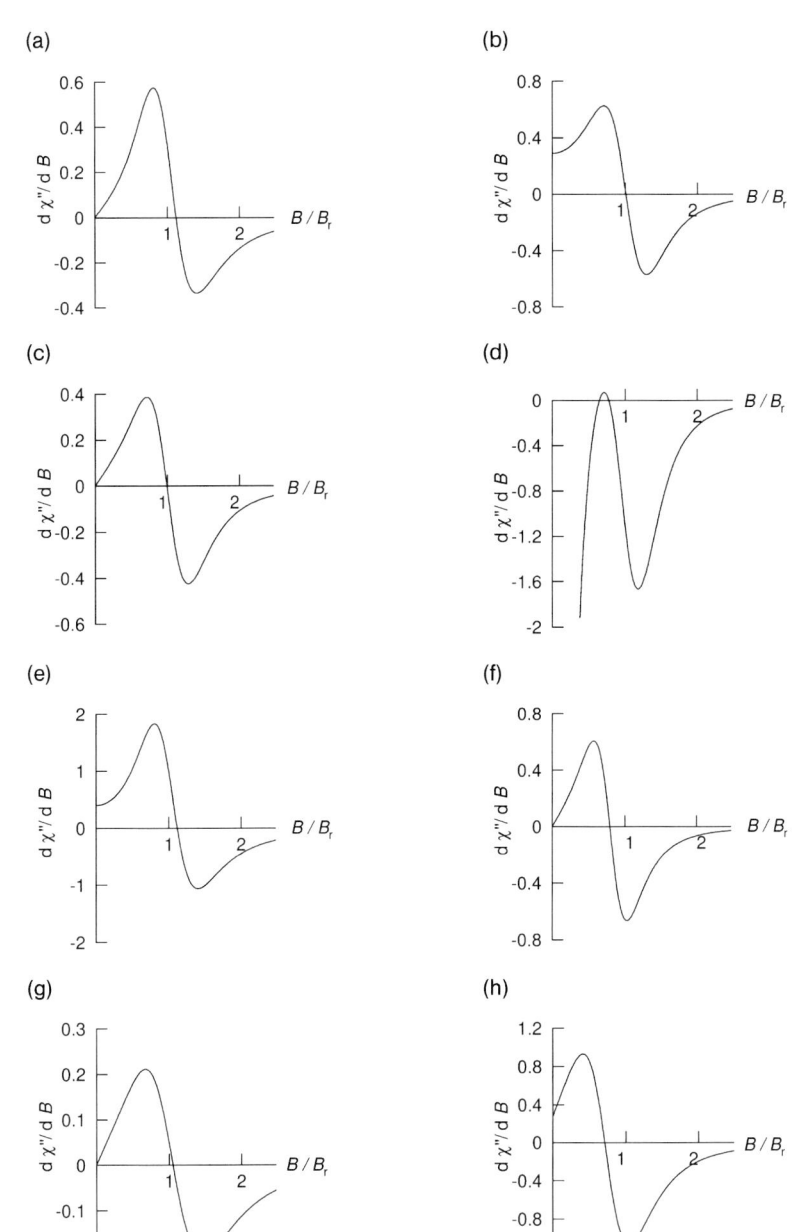

Figure 7.3 Magnetic resonance derivative-of-absorption lineshapes obtained with different equations of motion (linear polarization): (a) Bloch–Bloembergen, case (i); (b) Bloch–Bloembergen case (ii); (c) modified Bloch, case (i), Gilbert, case (ii), Landau–Lifshitz, case (i); (d) modified Bloch, case (ii); (e) Gilbert, case (i); (f) Landau–Lifshitz, case (ii); (g) Callen, case (i); (h) Callen, case (ii). The linewidth ratio is $\varepsilon = \Delta_B/B_r = 1/2$ in all cases and $\eta = \delta_B/B_r = 1/5$ for the Callen lineshape [70].

In case (i) the following lineshapes are obtained, see Figures 7.2 and 7.3(a):

$$\chi''(B) = \frac{2}{\pi} \frac{B^2 \Delta_B}{[(B - B_r)^2 + \Delta_B^2][(B + B_r)^2 + \Delta_B^2]}, \quad (7.32)$$

$$\chi''_\pm(B) \propto \frac{B}{(B \mp B_r)^2 + \Delta_B^2}. \quad (7.33)$$

Note that while the $\chi''(B)$ form of Eq. (7.32) is normalized to unity, those of Eqs. (7.33) are divergent.

In case (ii) the normalized lineshapes are, see Figures 7.2 and 7.3(b):

$$\chi''(B) = \frac{2}{\pi} \frac{B_r |B| \Delta_B}{\arctan\dfrac{B_r}{\Delta_B} [(B - B_r)^2 + \Delta_B^2][(B + B_r)^2 + \Delta_B^2]}, \quad (7.34)$$

$$\chi''_\pm(B) = \pm \frac{1}{2} \frac{\operatorname{sgn} B}{\arctan\dfrac{B_r}{\Delta_B}} \frac{\Delta_B}{(B \mp B_r)^2 + \Delta_B^2}. \quad (7.35)$$

7.4.1.3 The Modified Bloch Equation

The Bloch–Bloembergen equation in the preceding form is unsatisfactory in at least two aspects. First, it predicts that no absorption occurs in the absence of the magnetizing field, while such zero-field absorption can be observed experimentally. Second, it leads to the absurd conclusion that far from resonance, negative absorption of circularly polarized microwaves should be observed, cf. [73 p. 152], as illustrated in Figures 7.2(a) and (b).

In order to avoid these inconsistencies, the Bloch–Bloembergen equation is sometimes modified in such a way that longitudinal relaxation takes place along the direction of the effective magnetic field and lateral relaxation occurs at right angles to it [66]:

$$\dot{\boldsymbol{\mu}} = \gamma \boldsymbol{\mu} \wedge \mathbf{B}_{\text{eff}} - \frac{\boldsymbol{\mu} - M_0 V \mathbf{B}_{\text{total}}/B}{T}. \quad (7.36)$$

In case (i) the following normalized lineshapes are obtained, see Figures 7.2 and 3(c):

$$\chi''(B) = \frac{1}{\pi} \frac{(B^2 + B_r^2 + \Delta_B^2) \Delta_B}{[(B - B_r)^2 + \Delta_B^2][(B + B_r)^2 + \Delta_B^2]}, \quad (7.37)$$

$$\chi''_\pm(B) = \frac{1}{\pi} \frac{\Delta_B}{(B \mp B_r)^2 + \Delta_B^2}. \quad (7.38)$$

Note that for $B = 0$ in the case of linearly polarized radiation one gets

$$\chi''(0) = \frac{1}{\pi} \frac{\Delta_B}{B_r^2 + \Delta_B^2} = \frac{1}{\pi} \frac{|\gamma| T_2}{1 + \omega^2 T_2^2} \quad (7.39)$$

in accordance with the Debye formula for zero-field absorption [66, 74].

In case (ii) the resonance signal shapes are as follows (they are divergent at $B = 0$), see Figures 7.2 and 3(d):

$$\chi''(B) \propto \frac{\left(B^2 + B_r^2 + \Delta_B^2\right) B_r \Delta_B}{|B|\left[(B - B_r)^2 + \Delta_B^2\right]\left[(B + B_r)^2 + \Delta_B^2\right]}, \tag{7.40}$$

$$\chi_\pm''(B) \propto \frac{B_r}{|B|} \frac{\Delta_B}{(B \mp B_r)^2 + \Delta_B^2}. \tag{7.41}$$

7.4.1.4 The Gilbert Equation

Gilbert [67] suggested an equation of motion with a relaxation rate proportional to the total variation rate of the magnetic moment, $\dot{\boldsymbol{\mu}}$:

$$\dot{\boldsymbol{\mu}} = \gamma \boldsymbol{\mu} \wedge \boldsymbol{B}_{\text{eff}} + \frac{\alpha_G}{\mu} \boldsymbol{\mu} \wedge \dot{\boldsymbol{\mu}} \tag{7.42}$$

with $\alpha_G > 0$ a dimensionless damping constant.

Defining the linewidth parameter by $\Delta_B = \alpha_G B_r$, in case (i) one gets, see Figures 7.2, and 7.3(e):

$$\chi''(B) \propto \frac{\left(B^2 + B_r^2 + \Delta_B^2\right)|B|}{\left[(B - B_r)^2 + \Delta_B^2\right]\left[(B + B_r)^2 + \Delta_B^2\right]}, \tag{7.43}$$

$$\chi_\pm''(B) \propto \frac{|B|}{(B \mp B_r)^2 + \Delta_B^2}. \tag{7.44}$$

These lineshapes cannot be normalized since the corresponding integrals are divergent.

In case (ii) the lineshapes are exactly the same as those obtained with the modified Bloch equation in case (i); see Eqs. (7.37) and (7.38) and Figures 7.2 and 3(c).

7.4.1.5 The Landau–Lifshitz Equation

Landau and Lifshitz [68] suggested a damping term with the relaxation rate proportional to the precession component of $\dot{\boldsymbol{\mu}}$. Their equation is currently written in two different notations:

$$\dot{\boldsymbol{\mu}} = \gamma \boldsymbol{\mu} \wedge \boldsymbol{B}_{\text{eff}} - \frac{\lambda}{\mu^2} \nabla \boldsymbol{\mu} \wedge (\boldsymbol{\mu} \wedge \boldsymbol{B}_{\text{eff}})$$

$$= \gamma \boldsymbol{\mu} \wedge \boldsymbol{B}_{\text{eff}} + \frac{\alpha_{\text{LL}} \gamma}{\mu} \boldsymbol{\mu} \wedge (\boldsymbol{\mu} \wedge \boldsymbol{B}_{\text{eff}}), \tag{7.45}$$

where $\lambda > 0$ is a phenomenological damping factor and α_{LL} is a dimensionless constant.

Equations. (7.42) and (7.45) are often considered to be equivalent and sometimes even referred to as the "Landau–Lifshitz–Gilbert equation"

[75 p. 425]. Indeed, it can be shown, e.g., see [73 p.153; [57] p. 57; [75, 76]], that Eqs. (7.42) and (7.45) can be transformed into each other by suitable redefinition of parameters. Below we reproduce this demonstration in order to correct some errors contained in the original papers. Replacing $\dot{\mu}$ on the right-hand side of Eq. (7.42) with the entire right-hand side of the same equation results in

$$\dot{\mu} = \gamma \mu \wedge B_{\text{eff}} + \frac{\alpha_G}{\mu} \mu \wedge \left(\gamma \mu \wedge B_{\text{eff}} + \frac{\alpha_G}{\mu} \mu \wedge \dot{\mu} \right)$$

$$= \gamma \mu \wedge B_{\text{eff}} + \frac{\gamma \alpha_G}{\mu} \mu \wedge (\mu \wedge B_{\text{eff}}) - \alpha_G{}^2 \dot{\mu}, \tag{7.46}$$

where we have taken into account that, as long as the length of the μ vector remains constant, the scalar product $\mu \cdot \dot{\mu}$ vanishes. Equation (7.46) is further rewritten as

$$\dot{\mu} = \frac{\gamma}{1 + \alpha_G{}^2} \mu \wedge B_{\text{eff}} + \frac{\alpha_G \gamma}{(1 + \alpha_G{}^2) \mu} \mu \wedge (\mu \wedge B_{\text{eff}}). \tag{7.47}$$

It is seen that Eq. (7.47) transforms to Eq. (7.45) by the following substitution:

$$\frac{\gamma}{1 + \alpha_G{}^2} \longrightarrow \gamma; \qquad \alpha_G \longrightarrow \alpha_{\text{LL}}. \tag{7.48}$$

Notwithstanding the mathematical equivalence, the underlying physical models of the Gilbert and Landau–Lifshitz equations are quite different, as has been shown, e.g., in [71, 72]. Indeed, it is evident from the above demonstration that contributions of the precession and damping torques are defined in a different way in Eqs. (7.42) and (7.45). In the Landau–Lifshitz equation the damping torque is always perpendicular to the precession torque while in the Gilbert equation it is perpendicular to the time derivative of magnetization. In the limit of very high damping the evolution of magnetization predicted by these equations is quite opposite, namely, $\dot{\mu} \to \infty$ with the Landau–Lifshitz equation and $\dot{\mu} \to 0$ with the Gilbert equation. Obviously, the two equations become physically equivalent in the limit of low damping.

In case (i), with the linewidth parameter denoted by $\Delta_B = \lambda/|\gamma|\chi_0$, we obtain exactly the same normalized lineshape expressions as with the Gilbert equation case (ii) and with the modified Bloch equation case (i); see Eqs. (7.37) and (7.38) and Figures 7.2 and 3(c).

In case (ii), by setting $\Delta_B = \alpha_{\text{LL}} B_r$, the following normalized lineshapes are obtained, see Figures 7.2 and 3(f):

$$\chi''(B) = \frac{1}{\pi} \frac{B_r{}^2 \left[\left(B_r{}^2 + \Delta_B{}^2 \right) B^2 + B_r{}^4 \right] \Delta_B}{\left[B_r{}^2 (B - B_r)^2 + B^2 \Delta_B{}^2 \right] \left[B_r{}^2 (B + B_r)^2 + B^2 \Delta_B{}^2 \right]}, \tag{7.49}$$

$$\chi_\pm''(B) = \frac{1}{\pi} \frac{B_r{}^2 \Delta_B}{B_r{}^2 (B \mp B_r)^2 + B^2 \Delta_B{}^2}. \tag{7.50}$$

Errors in the analogous expression derived in Ref. [77] have been discussed elsewhere [70].

7.4.1.6 The Callen Equation

The Callen [69] dynamical equation with damping has been obtained using a quantum mechanical approach by quantizing the spin waves into magnons. It can be written as follows:

$$\dot{\boldsymbol{\mu}} = \gamma \boldsymbol{\mu} \wedge \mathbf{B}_{\text{eff}} - \frac{\lambda_C}{\mu^2} \nabla \boldsymbol{\mu} \wedge (\boldsymbol{\mu} \wedge \mathbf{B}_{\text{eff}}) - \alpha_{C_2} \boldsymbol{\mu}$$

$$= \gamma \boldsymbol{\mu} \wedge \mathbf{B}_{\text{eff}} + \frac{\alpha_{C_1} \gamma}{\mu} \boldsymbol{\mu} \wedge (\boldsymbol{\mu} \wedge \mathbf{B}_{\text{eff}}) - \alpha_{C_2} \boldsymbol{\mu}, \quad (7.51)$$

where the functions α_{C_1} and α_{C_2} include the magnon statistics. One can see that in this equation the form of the first damping term coincides with that of Landau–Lifshitz, Eq. (7.45) while the second damping term has the same form as the Bloch–Bloembergen one, Eq. (7.31) in the case of lateral relaxation, if one sets $\alpha_{C_2} = T_2^{-1}$.

In case (i), setting $\Delta_B = \lambda_C/|\gamma|\chi_0$ and $\delta_B = \alpha_C/|\gamma|$, we obtain the following normalized lineshapes, see Figures 7.2 and 3(g):

$$\chi''(B) = \frac{1}{\pi} \frac{\left[B_r^2 + (\Delta_B + \delta_B)^2\right]\Delta_B + B^2(\Delta_B + 2\delta_B)}{\left[(B - B_r)^2 + (\Delta_B + \delta_B)^2\right]\left[(B + B_r)^2 + (\Delta_B + \delta_B)^2\right]}, \quad (7.52)$$

$$\chi''_\pm(B) = \frac{1}{\pi B_r} \frac{B_r \Delta_B \pm B \delta_B}{(B \mp B_r)^2 + (\Delta_B + \delta_B)^2}. \quad (7.53)$$

In case (ii), with $\Delta_B = \alpha_{C_1} B_r$ and $\delta_B = \alpha_{C_2}/|\gamma|$, the normalized lineshapes are, see Figures 7.2 and 3(h):

$$\chi''(B) = \frac{1}{N} \frac{B_r^2 \left[(B_r^2 + \Delta_B^2)(B^2 \Delta_B + 2B_r|B|\delta_B) + B_r^2(B_r^2 + \delta_B^2)\Delta_B\right]}{\left[(B - B_r)^2 B_r^2 + (|B|\Delta_B + B_r\delta_B)^2\right]\left[(B + B_r)^2 B_r^2 + (|B|\Delta_B + B_r\delta_B)^2\right]}, \quad (7.54)$$

$$\chi''_\pm(B) = \frac{1}{N} \frac{B_r^2 (\Delta_B \pm \operatorname{sgn} B \, \delta_B)}{B_r^2 (B \mp B_r)^2 + (|B|\Delta_B + B_r\delta_B)^2}, \quad (7.55)$$

where

$$N = \pi - \arctan \frac{B_r^2 + \Delta_B \delta_B}{B_r(\Delta_B - \delta_B)} + \arctan \frac{B_r^2 - \Delta_B \delta_B}{B_r(\Delta_B + \delta_B)}. \quad (7.56)$$

In the EPR studies of diluted ions one usually deals with relatively low damping rates. In this instance all the equations considered above result in a Lorentzian lineshape:

$$\chi''(B) = \frac{1}{\pi} \frac{\Delta_B}{(B - B_r)^2 + \Delta_B^2}, \quad (7.57)$$

with a relatively narrow linewidth $\Delta_B \ll B_r$ (in the case of the Callen equation Δ_B should be replaced by $\Delta_B + \delta_B$). In this approximation the distinction between the responses to the linear and right-polarized microwave radiation disappears. In contrast, in the case of ferromagnetic or low-temperature superparamagnetic resonance of nanoparticles one often deals with broad resonance lines corresponding to high damping. In this case the resonance conditions are considerably altered and the actual lineshape essentially depends on the form of the damping term.

7.4.2
Linewidths and Apparent Shift of the Resonance Field

One can see for some lineshapes in Figures 7.2 and 7.3 that as the resonance linewidth increases, the apparent resonance field (corresponding to the absorption maximum) undergoes a considerable shift. In most cases considered above the analytical expressions of this shift as a function of the linewidth can be obtained. These expressions are summarized in Table 7.1 where B_{max} is the maximum of the resonance absorption, corresponding to zero of the experimentally recorded derivative-of-absorption line and $\varepsilon = \Delta_B/B_r$.

For the modified Bloch lineshape in case (ii) only relatively narrow linewidth can be considered because of the divergence at $B = 0$. Therefore, in the

Table 7.1 Apparent shifts of the resonance magnetic field for different lineshapes.

Equation	B_{max}/B_r	Notes
Bloch–Bloembergen (i) Gilbert (i) Equations (7.32) and (7.43)	$\sqrt{1+\varepsilon^2}$	
Bloch–Bloembergen (ii) Equation (7.34)	$\sqrt{\frac{2}{3}\sqrt{1+\varepsilon^2+\varepsilon^4}+1-\varepsilon^2}$	
Modified Bloch (i) Gilbert (ii) Landau–Lifshitz (i) Equation (7.37)	$\sqrt{2\sqrt{1+\varepsilon^2}-1-\varepsilon^2}$	
Modified Bloch case (ii) Equations (7.40), and (7.41)	$\frac{2}{3}+\frac{1}{3}\sqrt{1-3\varepsilon^2} \approx 1-\frac{1}{2}\varepsilon^2$	$\varepsilon \le \sqrt{3}/3$
Landau–Lifshitz case (ii) Equation (7.49)	$\dfrac{\sqrt{2\sqrt{1+\varepsilon^2}-1-\varepsilon^2}}{1+\varepsilon^2}$	
Callen case (i) Equation (7.52)	$\sqrt{\dfrac{2(1+\eta/\varepsilon)\sqrt{[1+(\varepsilon+\eta)^2](1+\eta^2)}-1-(\varepsilon+\eta)^2}{1+2\eta/\varepsilon}}$	$\eta = \delta_B/B_r$

resonance range the responses to the linear and right-polarized microwave radiation are very close in shape.

Figure 7.4 shows graphs of B_{max}/B_r as a function of ε for some lineshapes. One can see in the left-hand part of this figure that as ε increases, the apparent resonance positions for the Bloch–Bloembergen lineshape cases (i), (ii) and Gilbert lineshape case (i) shift toward high fields. In contrast, those of the modified Bloch case (i), Gilbert case (ii), and Landau–Lifshitz lineshape cases (i), (ii) shift downward, this tendency being particularly pronounced for the latter case. (For the Callen lineshape case (ii), the shift of the resonance is still more striking, as one can see from Figures 7.2 and 3) The latter tendency is most interesting since it corresponds to that observed with the experimental SPR spectra. In the right-hand part of Figure 7.4, we have shown the apparent shift of the resonance field for the Callen lineshape case (i) for different values of η/ε (the case with $\eta/\varepsilon = 0$ coincides with curve (c) in the left-hand part of this figure).

To illustrate the application of the above approach, in Figure 7.5 we show computer-generated SPR spectra together with the corresponding experimental spectra of nanoparticles in a sol–gel silica glass for two measurement temperatures. (The whole temperature dependence of the SPR in this glass will be considered in Section 7.6.) As temperature decreases, one

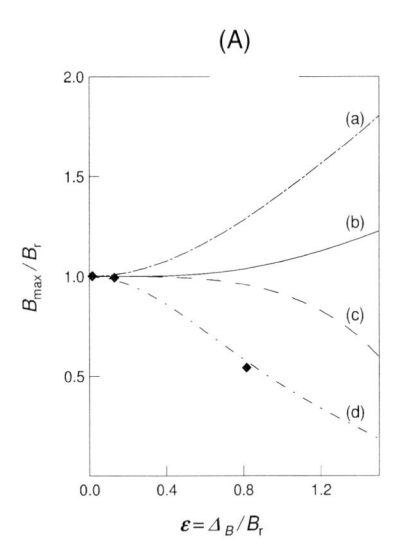

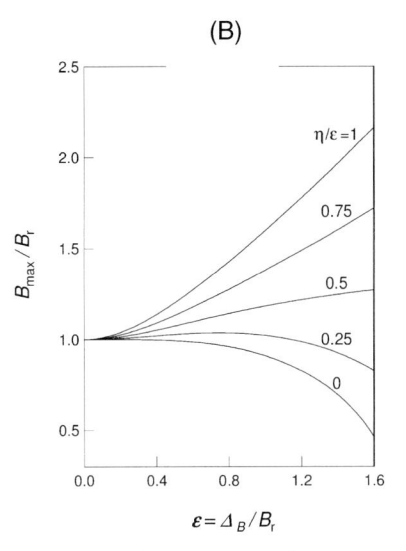

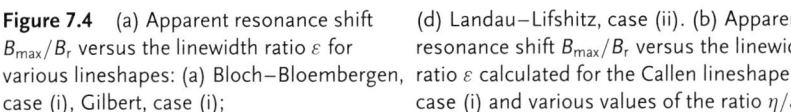

Figure 7.4 (a) Apparent resonance shift B_{max}/B_r versus the linewidth ratio ε for various lineshapes: (a) Bloch–Bloembergen, case (i), Gilbert, case (i); (b) Bloch–Bloembergen, case (ii); (c) modified Bloch, case (i), Gilbert, case (ii), Landau–Lifshitz, case (i); (d) Landau–Lifshitz, case (ii). (b) Apparent resonance shift B_{max}/B_r versus the linewidth ratio ε calculated for the Callen lineshape case (i) and various values of the ratio η/ε. The results of computer simulations of the experimental spectra (see the text) are indicated by the symbols & [70].

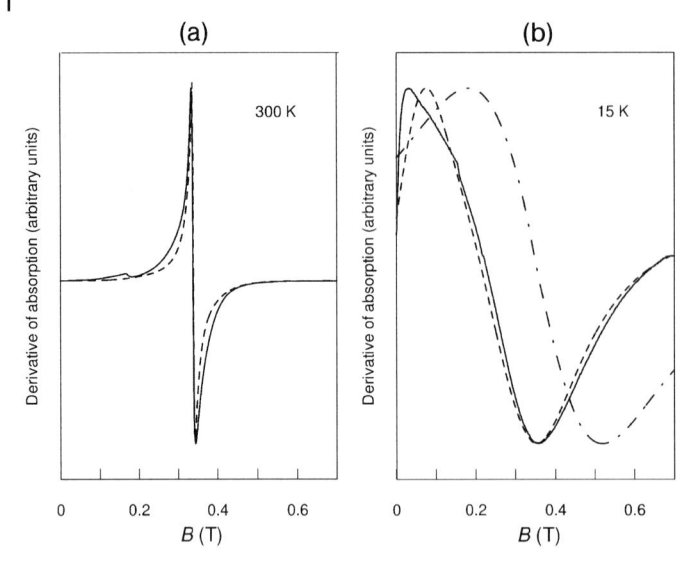

Figure 7.5 Simulations of the experimental SPR spectra (full lines) at 300 K (a) and 15 K (b). The dashed and dash-dotted lines are the best-fit computer-generated SPR spectra for the Landau–Lifshitz case (ii) and Lorentzian lineshapes, respectively. (At 300 K the two simulated spectra practically coincide.) [70].

observes a decrease of the resonance-field value and a concomitant increase of the linewidth. The spectra were calculated for two different lineshapes, the Lorentzian, see Eq. (7.57), and the Landau–Lifshitz case (ii), see Eq. (7.49), At 300 K the two computer-generated spectra practically coincide while at 15 K only the spectrum calculated with the Landau–Lifshitz lineshape, in contrast to that calculated with the Lorentzian lineshape, is in good accordance with the experimental spectrum. In Figure 7.4(a), we have plotted the resonance field versus the intrinsic linewidth values extracted from the computer simulations. One can see that the experimental points closely fit the Landau–Lifshitz curve.

7.4.3
Angular Dependence of the Linewidth

In an anisotropic system not only the resonance field B_r but also the intrinsic linewidth Δ_B can depend on the orientation of the magnetized field **B**. The linewidth anisotropy can readily be obtained for a given equation of damped precession of μ. As an example, below we consider the case of the Landau–Lifshitz equation (7.45). With the latter, we have, cf. Eq. (7.10):

$$\mu(\dot{\alpha} u_\alpha + \sin\alpha \dot{\beta} u_\beta) = \gamma\mu(B_\alpha u_\beta - B_\beta u_\alpha)$$
$$- \frac{\alpha_{LL}\gamma}{\mu}\mu^2(B_\alpha u_\alpha + B_\beta u_\beta), \qquad (7.58)$$

and following the same procedure as in Section 7.3, instead of Eqs. (7.19) we get a new pair of equations with the parameters defined in Eqs. (7.18):

$$\left(E_{\alpha\beta} + \alpha_{LL} E_{\alpha\alpha} - i\frac{\omega}{\gamma}\mu\right)\delta\alpha_{max} + (E_{\beta\beta} + \alpha_{LL} E_{\alpha\beta})\sin\alpha\delta\beta_{max} = 0$$

$$(E_{\alpha\alpha} - \alpha_{LL} E_{\alpha\beta})\delta\alpha_{max} + \left(E_{\alpha\beta} - \alpha_{LL} E_{\beta\beta} + i\frac{\omega}{\gamma}\mu\right)\sin\alpha\delta\beta_{max} = 0. \quad (7.59)$$

The determinant of the system (7.59) can be zero only if ω is complex, in which case it satisfies the following equation, cf. [57 p. 67]:

$$\omega^2 - 2i\omega\Delta_\omega - \omega_r^2 = 0, \qquad (7.60)$$

where ω_r is the angular frequency of resonance and Δ_ω is the linewidth parameter. By identification we get

$$\Delta_\omega = -\frac{1}{2}\frac{\alpha_{LL}\gamma}{\mu}(E_{\alpha\alpha} + E_{\beta\beta}) \qquad (7.61)$$

and a new equation allowing us to determine the resonance magnetic field:

$$\omega^2 = (1 + \alpha_{LL}^2)\frac{\gamma^2}{\mu^2}(E_{\alpha\alpha} E_{\beta\beta} - E_{\alpha\beta}^2). \qquad (7.62)$$

In the low damping limit, $\alpha_{LL} \to 0$ and Eq. (7.62) reduces to Eq. (7.20), see Section 7.3.

In a similar way, one can calculate the linewidth anisotropy and the resonance condition for other equations of damped gyromagnetic precession.

7.5
Superparamagnetic Narrowing of the Resonance Spectra

7.5.1
De Biasi and Devezas model

A simple "intuitive" way to account for this particle volume-dependent narrowing of the SPR spectra, suggested by de Biasi and Devezas [15, see also [42]], consists in "renormalizing" the magnetic parameters by averaging over the motion caused by thermal fluctuations of the magnetic moments:

$$\langle M \rangle = M\langle \cos\alpha \rangle \qquad (7.63)$$

and

$$\langle K_1 \rangle = K_1 \langle P_n(\cos\alpha) \rangle, \tag{7.64}$$

where α is the polar angle of the magnetization vector **M**, $P_n(\cos\alpha)$ is the nth-order Legendre polynomial, and the angular brackets denote averaged quantities.

For an assembly of particles whose magnetic moments are much larger than the Bohr magneton, the partition function can be calculated as an integral over all possible values of α. De Biasi and Devezas have considered the strong magnetic field case where in Eq. (7.4) the Zeeman term $-\boldsymbol{\mu} \cdot \mathbf{B}$ dominates over the remaining terms. In this approximation the probability of the $\boldsymbol{\mu}$ vector pointing in a particular direction is expressed as

$$dW(\alpha) = Z^{-1} e^{x\cos\alpha} d\alpha, \tag{7.65}$$

where α is measured with respect to the direction of **B**, $x = \mu B/kT$, k is the Boltzmann constant, T is the absolute temperature, and the partition function Z is given by

$$Z = \int_0^\pi e^{x\cos\alpha} d\alpha. \tag{7.66}$$

With Eqs. (7.65) and (7.66), one can easily calculate in the case of cubic symmetry, $n = 4$; $P_4(\xi) = \frac{1}{8}(35\xi^4 - 30\xi^2 + 3)$,

$$L_4(x) = \langle P_4(x) \rangle = \frac{\int_0^\pi P_4(\cos\alpha) e^{x\cos\alpha} d\alpha}{\int_0^\pi e^{x\cos\alpha} d\alpha}$$

$$= \frac{35}{x^2} + 1 - \left(\frac{105}{x^3} + \frac{10}{x}\right) L(x), \tag{7.67}$$

where $L(x)$ is the usual Langevin function:

$$L(x) = \langle P_1(x) \rangle = \frac{\int_0^\pi x e^{x\cos\alpha} d\alpha}{\int_0^\pi e^{x\cos\alpha} d\alpha} = \coth x - \frac{1}{x}. \tag{7.68}$$

Similarly, in the case of axial symmetry, $n = 2$; $P_2(\xi) = \frac{1}{2}(3\xi^2 - 1)$, one gets

$$L_2(x) = \langle P_2(x) \rangle = 1 - \frac{3}{x} L(x). \tag{7.69}$$

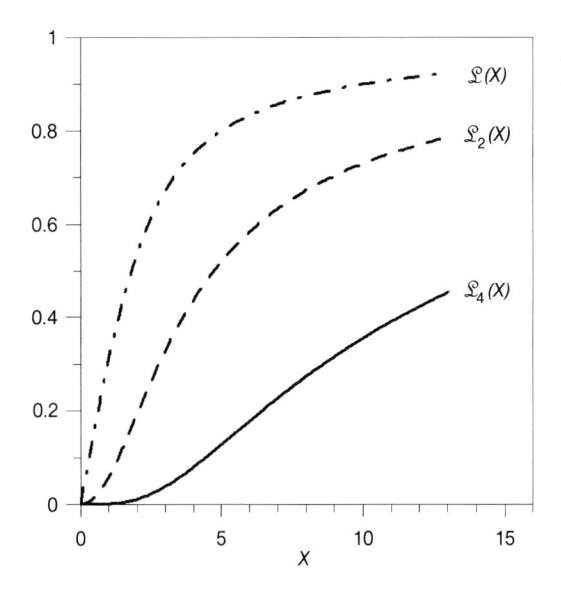

Figure 7.6 The Langevin function family (see the text for details).

Figure 7.6 shows $L_2(x)$ and $L_4(x)$ in comparison with $L(x)$. Note that for $x \to \infty$ all these functions tend to unity while for $x \to 0$ they behave as follows:

$$L(x) \sim \frac{1}{3}x$$

$$L_2(x) \sim \frac{1}{15}x^2$$

$$L_4(x) \sim \frac{1}{945}x^4. \tag{7.70}$$

One can see that in the case of cubic symmetry for very fine nanoparticles the contribution of the magnetocrystalline anisotropy is severely reduced.

In practice, this renormalization is often directly applied to the magnetocrystalline anisotropy field B_a and the demagnetizing field B_d, see Eqs. (7.23)–(7.26). As B_a is proportional to K_1/M, cf. Eqs. (7.24) and (7.25), in order to account for superparamagnetic narrowing of the SPR spectra, it should be multiplied by the following factors:

$$\delta_{\text{cub}} = \frac{\langle L_4(\cos\alpha)\rangle}{\langle L(\cos\alpha)\rangle} = \frac{1 + \frac{35}{x^2}}{L(x)} - \frac{10}{x} - \frac{105}{x^3} \tag{7.71}$$

in cubic symmetry and the factor

$$\delta_{\text{ax}} = \frac{\langle L_2(\cos\alpha)\rangle}{\langle L(\cos\alpha)\rangle} = \frac{1}{L(x)} - \frac{3}{x} \tag{7.72}$$

in axial symmetry.

In their original paper [15], de Biasi and Devezas erroneously stated that, as long as the demagnetizing field B_d is proportional to the magnetization, it is averaged by the usual Langevin function L(x). In fact, B_d derives from the magnetostatic energy term in Eq. (7.4); therefore, e.g., for uniaxial shape anisotropy it should be averaged exactly in the same manner as the magnetocrystalline anisotropy field in the case of axial magnetic symmetry, i.e., in accordance with Eq. (7.72). This fact was first pointed out by Raikher and Stepanov [53].

The model of de Biasi and Devezas can be further improved by taking into account in the partition function (7.66), instead of the Zeeman energy only, the whole anisotropic part of the nanoparticle free energy of Eq. (7.4). A few attempts in this direction have been reported, e.g., see [23, 36].

7.5.2
Raikher and Stepanov Model

In a series of papers, Raikher et al. [20, 21, 26, 53] developed a more elaborate approach to the calculation of the nanoparticle SPR spectra based on the formalism first put forward by Brown Jr. [78]. These authors define the "macroscopic" (averaged) magnetic moment of a statistical assembly of nanoparticles as an average of the "microscopic" moment μ:

$$\langle \mu \rangle = \mu \int u_\mu W(u_\mu, t) du_\mu, \tag{7.73}$$

where $W(u_\mu, t)$ is a normalized time-dependent distribution density of orientations of μ and the unit vector $u_\mu = \mu/\mu$, defined in Section 7.2, stands for the corresponding vectors r^0 in the paper [78] and e in the papers of Raikher et al. $W(u_\mu, t)$ satisfies the Fokker–Planck continuity equation

$$\dot{W} + \nabla \cdot (W \dot{u}_\mu) = 0. \tag{7.74}$$

The "phase velocity" $\dot{u}_\mu = \dot{\mu}/\mu$ in Eq. (7.74) can be expressed from one of the phenomenological equations of damped precession of the magnetization, see Section 7.4. Raikher and Stepanov used the Landau–Lifshitz equation rewritten in terms of u_μ and the nanoparticle free energy E, cf. Eqs.(7.45) and (7.8):

$$\dot{u}_\mu = -\frac{\gamma}{\mu} u_\mu \wedge [\nabla_u + \alpha_{LL}(u_\mu \wedge \nabla_u)] E. \tag{7.75}$$

In order to account for fluctuations of the magnetic moment, they include in E a supplementary term of the form $kT \ln W$ giving rise to a stochastic component of the effective field

$$B_{random} = -\frac{kT}{\mu} \nabla_u \ln W. \tag{7.76}$$

Substituting $\dot{u}_\mu$ from Eq. (7.75) to Eq. (7.74) and introducing the infinitesimal rotation operator $J = u_\mu \wedge \nabla_u$, after some transformations they obtain the following kinetic equation of rotary diffusion (the Brown kinetic equation):

$$2\tau \dot{W} = JWQ\left(\frac{E}{kT} + \ln W\right), \tag{7.77}$$

where $\tau = \mu/2\alpha_{LL}\gamma kT$ is a reference time constant of rotary diffusion of μ and $Q = J + \nabla_u/\alpha_{LL}$. Under steady-state conditions, the solution of Eq. (7.77) has the form of a Gibbs distribution

$$W_0 = Z_0^{-1} e^{-E_0/kT},$$
$$Z_0 = \int e^{-E_0/kT} du_\mu, \tag{7.78}$$

where E_0 is the stationary anisotropic part of the particle free energy. In the presence of an oscillating magnetic field the equilibrium is disturbed, and Eq. (7.77) can be solved to yield the imaginary part of magnetic susceptibility proportional to the superparamagnetic resonance absorption.

Whereas the approach of Raikher and Stepanov is theoretically attractive, in practice, it requires complex calculations while yielding very similar results to those obtained with the simple heuristic model of Biasi and Devezas [15]. Up to now it has been successfully applied only in a few relatively simple cases [20, 21, 26].

7.6
Nanoparticle Size and Shape Distribution

7.6.1
Distribution of Diameters

Most authors agree that for nanoparticle shape close to that of a sphere, the most adequate choice of the size distribution density is the log-normal function [79], e.g., in the case of demixing processes of glasses during heat treatment [80] or for fine magnetic particles in ferrofluids [9–11]. Meanwhile, in the literature concerning this distribution there has been much confusion, so, there is some difficulty in comparing the data given by different authors.

The log-normal distribution density can be described by a number of interrelated expressions [11], the two most currently employed being

$$P(d) = \frac{1}{\sqrt{2\pi}\sigma d} \exp\left(-\frac{1}{2}\frac{\ln^2 d/d_0}{\sigma^2}\right) \tag{7.79}$$

and

$$P(d) = \frac{\exp(-\frac{1}{2}\sigma^2)}{\sqrt{2\pi}\,\sigma\, d_m} \exp\left(-\frac{1}{2}\frac{\ln^2 d/d_m}{\sigma^2}\right). \tag{7.80}$$

Equations (7.79) and (7.80) represent the relative *number* of particles of a given diameter. In Eq. (7.79) the diameter d_0, sometimes erroneously called the mean or median diameter, in fact, is defined as

$$\ln d_0 = \langle \ln d \rangle_n = \int_0^\infty \ln d\, P(d)\, dd, \tag{7.81}$$

where the angular brackets denote mean values and the subscript n indicates averaging over the distribution in the number of particles. So, d_0 should be referred to as the *log-mean* diameter. The diameter d_m in Eq. (7.80) is the *most probable* diameter corresponding to the maximum of $P(d)$, and in both equations σ is the "log-standard deviation," i.e., the standard deviation of $\ln d$.

In order to avoid further ambiguities, below we provide a table showing the interrelation between different characteristics of the log-normal distribution density. In particular, it shows the "real" mean diameter $\langle d \rangle_n$ as well as the corresponding standard deviation σ_{d_n}.

In order to calculate the magnetic resonance intensity, one does not need the relative *number* of particles of a given volume but the *volume fraction* of such particles. The corresponding distribution density for near-spherical particles is defined as

$$f_V(d) = \frac{1}{6}\pi d^3 P(d). \tag{7.82}$$

$f_V(d)$ has the same log-normal form as $P(d)$ [25]; and the characteristics obtained by averaging over $f_V(d)$, designed by the subscript V, are also shown in Table 7.2.

Figure 7.7 illustrates the shape of the log-normal distribution and shows its various characteristics. One can see that in the case of a relatively broad distribution a different choice of the "mean" parameters yields very different numerical results; therefore, an accurate definition of the quoted characteristics is most essential.

7.6.2
Nonsphericity of Nanoparticles: Distribution of Demagnetizing Factors

From the viewpoint of the SPR phenomenon, the basic manifestation of the particle shape anisotropy is the emergence of the demagnetizing field which contributes to the local magnetic field.

7.6 Nanoparticle Size and Shape Distribution

Table 7.2 Characteristics of the log-normal distribution density of the number of particles $P(d)$ and of the corresponding volume fraction $f_V(d)$.

Designation	Notation suggested	Definition	Value
Log-mean diameter	d_0	$\ln d_0 = \int_0^\infty \ln d\, P(d) dd$	d_0
	d_{0V}	$\ln d_{0V} = \int_0^\infty \ln d\, f_V(d) dd$	$d_0 e^{3\sigma^2}$
Most probable diameter	d_m	$\dfrac{dP}{dd}(d_m) = 0$	$d_0 e^{-\sigma^2}$
	d_{mV}	$\dfrac{df_V}{dd}(d_{mV}) = 0$	$d_0 e^{2\sigma^2}$
Mean diameter	$\langle d \rangle_n$	$\int_0^\infty d\, P(d) dd$	$d_0 e^{\frac{1}{2}\sigma^2}$
	$\langle d \rangle_V$	$\int_0^\infty d\, f_V(d) dd$	$d_0 e^{\frac{7}{2}\sigma^2}$
Standard deviation of $\ln d$	σ	$\left[\int_0^\infty \ln^2 d/d_0\, P(d) dd\right]^{1/2}$	σ
		$\left[\int_0^\infty \ln^2 d/d_{0V}\, f_V(d) dd\right]^{1/2}$	
Standard deviation of d	σ_{d_n}	$\left[\int_0^\infty (d-\langle d\rangle_n)^2 P(d) dd\right]^{1/2}$	$d_0 e^{\frac{1}{2}\sigma^2}\sqrt{e^{\sigma^2}-1}$
	σ_{d_V}	$\left[\int_0^\infty (d-\langle d\rangle_V)^2 f_V(d) dd\right]^{1/2}$	$\sigma_{d_n} e^{3\sigma^2}$

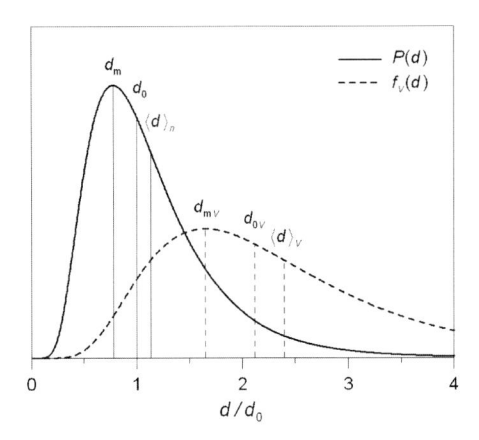

Figure 7.7 Log-normal distribution density of the number of particles and of the corresponding volume fraction.

The calculation is most readily carried out for a uniformly magnetized ellipsoidal particle, in which case the demagnetizing field is uniform and the ratios of the axes can be related to the demagnetizing factors [58]. For the general ellipsoid these relations are rather complicated, involving elliptic integrals of the first and second kind. For an ellipsoid of rotation (spheroid) they reduce to simpler forms. Denoting respectively by $a_\parallel$ and $a_\perp$ the polar and equatorial semiaxes, one gets the following expressions:

For $m = a_\parallel/a_\perp > 1$ (prolate spheroid),

$$N_\parallel = \frac{1}{m^2 - 1}\left[\frac{m \ln\left(m + \sqrt{m^2 - 1}\right)}{\sqrt{m^2 - 1}} - 1\right]$$

$$N_\perp = \frac{1}{2}\frac{m}{m^2 - 1}\left[m - \frac{\ln\left(m + \sqrt{m^2 - 1}\right)}{\sqrt{m^2 - 1}}\right]. \qquad (7.83)$$

For $m = a_\parallel/a_\perp < 1$ (oblate spheroid),

$$N_\parallel = \frac{1}{1 - m^2}\left(1 - \frac{m \arccos m}{\sqrt{1 - m^2}}\right)$$

$$N_\perp = \frac{1}{2}\frac{m}{1 - m^2}\left(\frac{\arccos m}{\sqrt{1 - m^2}} - m\right). \qquad (7.84)$$

The distribution of the demagnetizing factors can be obtained in the following way (see Ref. [29] for further details). The demagnetizing tensor of a particle is expressed as

$$\mathbf{N} = \frac{1}{3}\mathbf{1} + \mathbf{n}, \qquad (7.85)$$

where $\mathbf{1}$ is the unit matrix of order 3, and $\mathbf{n}$ is a symmetric zero-trace tensor with distributed components n_{ij} ($i, j = x, y, z$, $n_{ij} = n_{ji}$, $\mathrm{tr}\,\mathbf{n} = n_x + n_y + n_z = 0$ where the single-subscript components refer to the eigenvalues of the tensor $\mathbf{n}$). In the general case, $\mathbf{n}$ is the sum of two tensors: (i) a fixed one which describes the average distortions of particle shapes from that of a sphere; it vanishes if the particles are spherical on average, and (ii) a random tensor which accounts for random distortions of the individual particle shapes (manifestation of disorder in the particle shapes).

In a general axis system, the tensor $\mathbf{n}$ can be represented by five real quantities v_i, $i = 1, \ldots, 5$, defined as linear combinations of the components of the associated irreducible spherical tensor [81], n_m^l, with $l = 0, 1, 2$ and $m = -l, \ldots, l$:

$$v_1 = n_0^2 = \sqrt{\frac{3}{2}}n_{zz},$$

$$v_2 = \sqrt{\frac{1}{2}}\left(n_2^2 + n_{-2}^2\right) = \sqrt{\frac{1}{2}}\left(n_{xx} - n_{yy}\right),$$

$$v_3 = -i\sqrt{\frac{1}{2}}\left(n_2^2 - n_{-2}^2\right) = \sqrt{2}n_{xy},$$

$$v_4 = i\sqrt{\frac{1}{2}\left(n_1^2 + n_{-1}^2\right)} = \sqrt{2}n_{yz},$$

$$v_5 = \sqrt{\frac{1}{2}\left(n_{-1}^2 - n_1^2\right)} = \sqrt{2}n_{xz}. \tag{7.86}$$

In a macroscopically isotropic system of magnetic nanoparticles, a joint distribution density of v_i, $P(v_1, \ldots, v_5)$, must be invariant with respect to any rotation of the system of coordinates, so that it can depend only on invariants of the tensor $\mathbf{n}$. As such invariants, one can choose the n_z value and the "asymmetry parameter" $\eta = (n_y - n_x)/n_z$. For definiteness, we choose the main axes of the ellipsoid such that $|n_x| < |n_y| < |n_z|$; therefore, η is always confined in the range $0 \leq \eta \leq 1$. The quantities v_i can be expressed in terms of n_z, η, and the set of the three Euler angles relating the principal coordinate system of the tensor $\mathbf{n}$ to a general coordinate system in which the former quantities have been defined.

First, we consider a "totally disordered" nanoparticle system, with no preference for any particular distortion of the particle shapes from that of a sphere. In this case, the most natural (based on the central limit theorem) choice of the joint distribution density function of v_i is a multivariate normal (Gaussian) one, with zero mean values and the same standard deviation σ_v, for each v_i, cf. Ref. [82]:

$$P(v_1, \ldots, v_5) = \frac{1}{(2\pi)^{5/2}\sigma_v^5} \exp\left(-\frac{1}{2}\frac{\Delta^2}{\sigma_v^2}\right), \tag{7.87}$$

where

$$\Delta^2 = \sum_{i=1}^{5} v_i^2 = \sum_{i=1}^{3}\sum_{j=1}^{3} n_{ij}^2. \tag{7.88}$$

Such a choice implies that the different v_i quantities are distributed independently, forming a Gaussian orthogonal ensemble [83]. The relation between the distribution functions of the matrix elements and that of the eigenvalues of $\mathbf{n}$ yields, cf. [83, eq. (3.1.17)],

$$P(n_x, n_y, n_z) = C|n_x - n_y||n_y - n_z||n_x - n_z|\exp\left(-\frac{1}{2}\frac{n_x^2 + n_y^2 + n_z^2}{\sigma_v^2}\right), \tag{7.89}$$

where C is a constant resulting from integration over the Euler angles in accordance with the macroscopic isotropy of the system. As $\mathrm{tr}\mathbf{n} = 0$, the distribution density of Eq. (7.89) can be transformed to the distribution density of n_z and η by applying the following relation of random vector distribution densities:

$$P(n_z, \eta) = \left|\frac{\partial(\mathrm{tr}\mathbf{n}, n_z, \eta)}{\partial(n_x, n_y, n_z)}\right|\delta(\mathrm{tr}\mathbf{n})P(n_x, n_y, n_z) \tag{7.90}$$

where $\partial(\text{tr}\boldsymbol{n}, n_z, \eta)/\partial(n_x, n_y, n_z)$ is the Jacobian determinant and $\delta(\text{tr}\boldsymbol{n})$ is the Dirac function. One gets

$$P(n_z, \eta) = \frac{1}{\sqrt{2\pi}} \frac{n_z{}^4}{\sigma_{n_z}{}^5} \eta \left(1 - \frac{1}{9}\eta^2\right) \exp\left[-\frac{1}{2} \frac{n_z{}^2}{\sigma_{n_z}{}^2}\left(1 + \frac{1}{3}\eta^2\right)\right] \qquad (7.91)$$

where $\sigma_{n_z} = \sqrt{\frac{2}{3}}\sigma_v$. Such form of distribution density has been first obtained by Czjzek et al. [84] for the invariants of the electric field gradient tensor on the nucleus and used to describe the Mössbauer [82, 84, 85] and EPR [86, 87] spectra in disordered systems. The mathematical analogy between the distributions of the tensor $\boldsymbol{n}$, on the one hand, and those of the EFG tensor (and the related tensor $\boldsymbol{P}$) and the quadrupole fine structure tensor $\boldsymbol{D}$, on the other hand, stems from the fact that the magnetostatic energy term in Eq. (7.4) has a form similar to that of the quadrupole fine structure terms in electronic or nuclear spin Hamiltonians, respectively, $H_S = \boldsymbol{S} \cdot \boldsymbol{D} \cdot \boldsymbol{S}$ and $H_I = \boldsymbol{I} \cdot \boldsymbol{P} \cdot \boldsymbol{I}$ [88].

In fact, e.g., magnetostriction or interactions with the environment can impose to the magnetic nanoparticles a preference for some average distortions. This would bring about some ordering on the nanoscopic scale in the statistical assembly of nanoparticle shapes. In order to account for implications of short-range ordering on the components of the $\boldsymbol{P}$ and $\boldsymbol{D}$ tensors, some authors have considered the various tensor components as dependent on a reduced number of new independent random variables. In this case, the power of n_z in the pre-exponential factor of the distribution density, Eq. (7.91), is reduced to $d - 1$, with $d < 5$ the number of "degrees of freedom" of the system [86, 87]. However, the distribution densities for $d < 5$ do not meet the requirement of rotation invariance.

A more accurate way of taking into account a possible nonsphericity of the particles on the average is to assume the following form for the demagnetizing tensor in Eq. (7.85), cf. Refs. [85, 89, 90]:

$$\boldsymbol{n} = \boldsymbol{n}_0 + \boldsymbol{n}', \qquad (7.92)$$

where $\boldsymbol{n}_0$ is a fixed zero-trace tensor which describes the average distortions of particle shapes from that of a sphere and $\boldsymbol{n}'$ is a random tensor which accounts for departures from the average of the shapes of individual particles. No exact analytical form of $P(n_z, \eta)$ can be found in this case, but the marginal distribution density $Q(\Delta)$ of the invariant Δ, see Eq. (7.88), can be derived from the noncentral χ^2 distribution with five degrees of freedom as follows [85, 89, 90]:

$$Q(\Delta) = \frac{\Delta^{5/2}}{\Delta_0^{3/2}\sigma_{n_z}{}^2} I_{3/2}\left(\frac{\Delta\Delta_0}{\sigma_{n_z}{}^2}\right) \exp\left(-\frac{1}{2}\frac{\Delta^2 + \Delta_0{}^2}{\sigma_{n_z}{}^2}\right), \qquad (7.93)$$

where Δ_0 is the mean Δ-value and

$$I_{3/2}(z) = \sqrt{\frac{2}{\pi}} \frac{z \cosh z - \sinh z}{z^{3/2}} \qquad (7.94)$$

is the modified Bessel function. $Q(\Delta)$ can also be represented in a more convenient form [90]:

$$Q(\Delta) = \frac{\Delta \sigma_{n_z}}{\sqrt{2\pi}\Delta_0^3} G\left(\frac{\Delta \Delta_0}{\sigma_{n_z}^2}\right) \exp\left[-\frac{1}{2}\frac{(\Delta - \Delta_0)^2}{\sigma_{n_z}^2}\right], \qquad (7.95)$$

where

$$G(y) = y - 1 + (y+1)e^{-2y}. \qquad (7.96)$$

Figure 7.8 illustrates the aspect of the $G(y)$ function. One gets $G(y) \approx \frac{2}{3} y^3$ for $y \to 0$ and $G(y) \approx y$ for $y \gg 1$, so that the pre-exponential factor in Eq. (7.95) is proportional to Δ^4 or Δ^2, respectively, for $\Delta \Delta_0/\sigma_{n_z}^2 \ll 1$ and $\Delta \Delta_0/\sigma_{n_z}^2 \gg 1$.

7.6.3
Joint Distribution of Diameters and Demagnetizing Factors

In a disordered assembly of magnetic nanoparticles dispersed in a diamagnetic matrix, the particle shape can vary with the particle size, resulting in a *correlated* distribution of the axis lengths and the components of the tensor **N**. In order to somewhat simplify the analysis, we assume that the deviations of the particle shapes from that of a sphere are not large, so that in the first approximation the log-normal distribution of diameters, Eqs. (7.79) and (7.80), still remains valid. This distribution can then be associated with that of the demagnetizing factors to constitute a joint distribution density.

Taking into consideration only axial distortions of the particle shapes, one gets $n_x = n_y$, $n_z = -2n_x = -2n_y$, and $\eta \equiv 0$, and the random vector ξ in

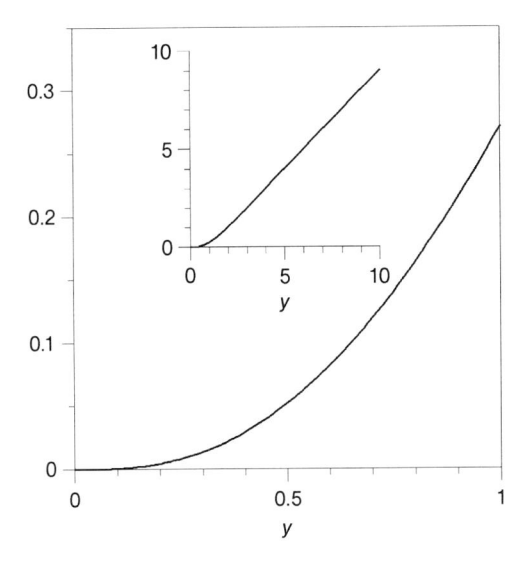

Figure 7.8 The $G(y)$ function (the inset shows this function on a larger scale).

Eqs. (7.1)-(7.3), see Section 7.2, reduces to $\xi = (d, n_z)$. In this case $\Delta = |n_z|$ and the marginal distribution of n_z has the form described by Eq. (7.95).

The correlation between the most probable diameter d_m and n_z can be rendered by including a correlation coefficient ρ in the "Gaussian" part of the joint distribution density $P(d, n_z)$, cf. Ref. [90]:

$$P(d, n_z) \propto |n_z| G\left(\frac{|n_z n_0|}{\sigma_{n_z}^2}\right)$$

$$\times \exp\left\{-\frac{1}{2}\frac{1}{1-\rho^2}\left[\frac{(n_z - n_0)^2}{\sigma_{n_z}^2} - 2\rho\frac{n_z - n_0}{\sigma_{n_z}}\frac{\ln\frac{d}{d_m}}{\sigma} + \frac{\ln^2\frac{d}{d_m}}{\sigma^2}\right]\right\}, \quad (7.97)$$

where n_0 is the mean value of n_z. For $n_0 = 0$, $P(d, n_z)$ simplifies to

$$P(d, n_z) \propto n_z^4 \exp\left\{-\frac{1}{2}\frac{1}{1-\rho^2}\left[\frac{n_z^2}{\sigma_{n_z}^2} - 2\rho\frac{n_z}{\sigma_{n_z}}\frac{\ln\frac{d}{d_m}}{\sigma} + \frac{\ln^2\frac{d}{d_m}}{\sigma^2}\right]\right\}. \quad (7.98)$$

In the corresponding computer code a correlated distribution $P(d, n_z)$ can be readily implicated by, first, generating *uniform* correlated random variables d and n_z and, second, applying the noncorrelated distribution density

$$P(d, n_z) \propto |n_z| G\left(\frac{|n_z n_0|}{\sigma_{n_z}^2}\right) \exp\left\{-\frac{1}{2}\left[\frac{(n_z - n_0)^2}{\sigma_{n_z}^2} + \frac{\ln^2\frac{d}{d_m}}{\sigma^2}\right]\right\}. \quad (7.99)$$

Figure 7.9 illustrates the typical shape of the $P(d, n_z)$ distribution densitiy and the corresponding marginal distribution densities:

$$P(d) = \int_{-\infty}^{\infty} P(d, n_z) dn_z$$

$$P(n_z) = \int_{0}^{\infty} P(d, n_z) dd. \quad (7.100)$$

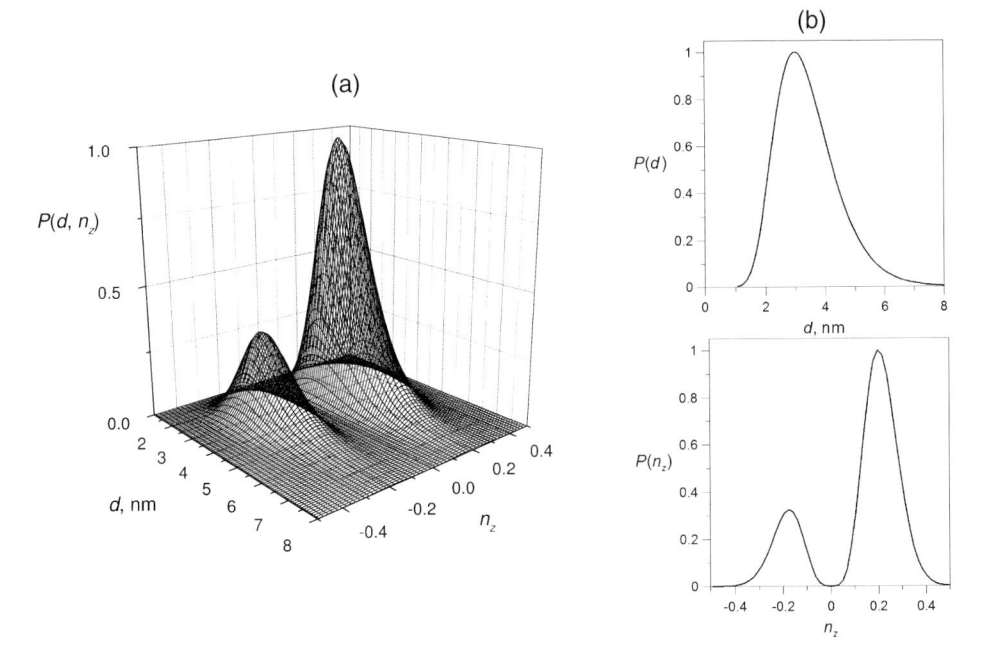

Figure 7.9 The joint distribution density of diameters and demagnetizing factors. (a) The $P(d, n_z)$ density calculated with $d_m = 3.0$ nm, $\sigma = 0.3$, $n_0 = 0.03$ and $\sigma_{n_z} = 0.1$. (b) The marginal distribution densities $P(d)$ (top) and $P(n_z)$ (bottom).

7.7
Superparamagnetic Resonance in Oxide Glasses: Some Experimental Results

7.7.1
Lithium Borate Glass

In order to illustrate the basic ideas described in the previous sections, below we summarize some recent experimental EMR studies of nanoparticles formed in oxide glasses.

Lithium borate glass of the molar composition 0.63 B_2O_3–0.37 Li_2O containing 0.75×10^{-3} mole% of Fe_2O_3 was annealed by repeated stages for 0.5 h at increasing anneal temperatures T_a starting at the glass transition temperature $T_g = 708$ K. Figure 7.10 shows the evolution with T_a of the X-band (ca. 9.5 GHz) room-temperature EMR spectra of these glasses. The as-prepared glass exhibits an asymmetric EPR spectrum with the effective g-factor $g_{eff} \approx 4.3$ characteristic of isolated Fe^{3+} ions. As T_a increases, this spectrum gradually decreases in intensity and finally disappears. Simultaneously, a new resonance emerges at $g_{eff} \approx 2.0$, appearing as a narrow line superposed with a broader one (the "two-line pattern"). The narrow component predominates at lower

T_a and it is progressively replaced by the broader one at higher T_a. Such behavior is indicative of the devitrification process, as confirmed by X-ray diffraction results. Numerical integration of the experimental derivative-of-absorption spectra yields the EMR *absorption* spectra shown in Figure 7.11 and the overall EMR intensity, see Figure 7.12. In the course of the heat treatment this intensity increases at least by two orders of magnitude, indicating the *change of the nature of the resonance* – from the EPR of isolated ions to the SPR of iron-containing magnetic nanoparticles.

Because of the very low iron content, no iron-containing phases could be observed by X-ray diffraction, but nanoparticles arising in annealed borate glasses with higher iron oxide contents were identified by X-ray diffraction as lithium ferrite $LiFe_5O_8$ [91]. Therefore, in computer simulations of the nanoparticle SPR spectra in the borate glass, the magnetic parameters of lithium ferrite were used: $M = 310$ kA m^{-1} and $K_1 = -8.0$ kJ m^{-3} (cubic symmetry). The simulations were carried out using the approach based on the joint distribution density of the particle diameters and demagnetizing factors $P(d, n_z)$, see Section 7.5. Figure 7.13 shows the computer fits as well the corresponding best-fit distribution densities; the simulation parameters are given in Table 7.3.

From an inspection of these data, one can conclude that with the increase of the anneal temperature, the most probable diameter d_m of the magnetic nanoparticles increases while the standard deviation σ_d decreases, so the assembly of nanoparticles becomes more ordered. The absolute values of mean demagnetizing factors n_0 remain small in comparison with the corresponding

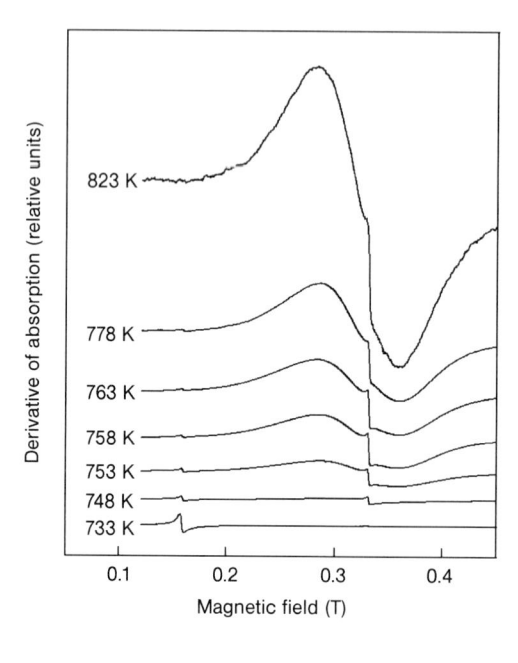

Figure 7.10 X-band room temperature EMR spectra of the 0.63 B_2O_3–0.37 Li_2O–0.75 × 10^{-3} Fe_2O_3 glass annealed at the indicated temperatures during 1/2 hour [25, 34].

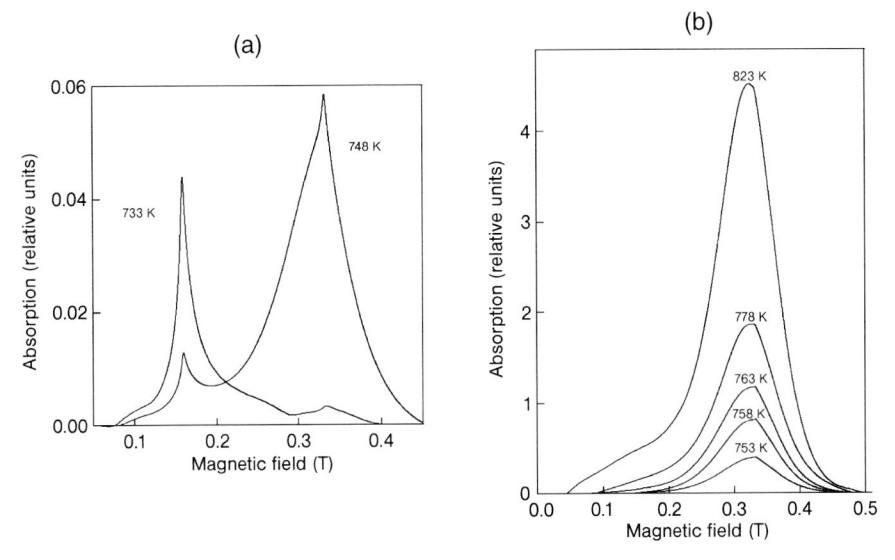

Figure 7.11 Integrated EMR spectra of the 0.63 B_2O_3–0.37 Li_2O–0.75 × 10^{-3} Fe_2O_3 glass for lower (a) and higher (b) anneal temperatures, indicated near the corresponding curves [25].

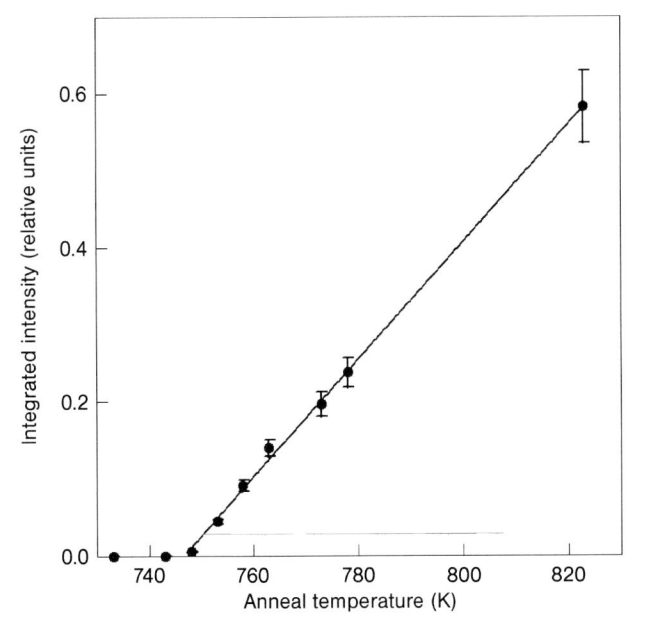

Figure 7.12 Intensity of the EMR spectra of the 0.63 B_2O_3–0.37 Li_2O–0.75 × 10^{-3} Fe_2O_3 glass (obtained by double integration of the experimental curves) versus the anneal temperature. The straight line shows a linear dependence in the 748–823 K range [34].

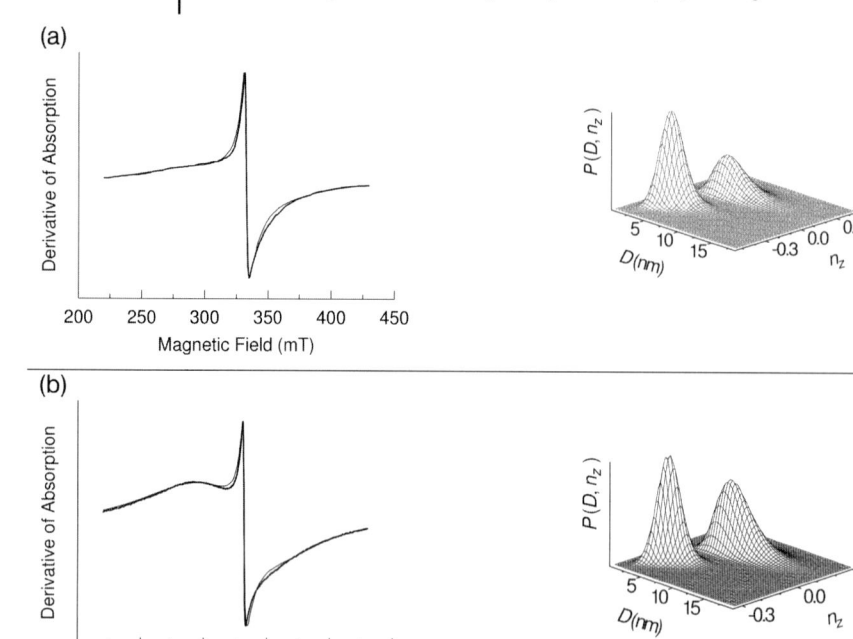

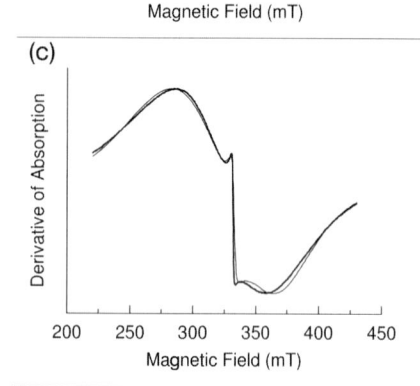

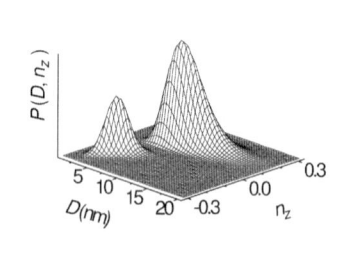

Figure 7.13 Computer fits to the room-temperature X-band SPR spectra of the borate glass annealed at (a) 738 K, (b) 748 K, and (c) 753 K. Left: experimental spectra (full lines) and best-fit computer-generated spectra (dashed lines). Right: reconstructions of the best-fit joint distribution density of diameters and demagnetizing factors of Eq. (7.97). See Table 7.3 for the simulation parameters [29].

distribution widths σ_{n_z}, indicating random distortions of the nanoparticle shapes from that of a sphere with no marked preference for any average distortion.

In spite of the very specific shape of the joint distribution density $P(d, n_z)$, the marginal distribution density of nanoparticle diameters $P(d)$ is unimodal, cf.

Table 7.3 Parameters of the joint distribution density of diameters and demagnetizing factors; see Eq. (7.97) for magnetic nanoparticles in annealed borate glasses [29].

T_a (K)	738	748	753
d_m (nm)	3.4	3.9	4.7
σ	0.39	0.37	0.34
n_0	−0.022	−0.006	0.010
σ_{n_z}	0.095	0.060	0.060
ρ	0.39	0.52	0.60

Section 7.6, Figure 7.9. It might seem surprising that such distribution could well reproduce the "two-line pattern" in the SPR spectrum. In fact, one deals here with a specific example of a very general characteristic of the EMR spectra of disordered systems: singularities (sharp features) occur in such spectra if the resonance magnetic field is stationary with respect to the distributed magnetic parameters [92]. This situation is similar to that of the EPR spectra of diluted Fe^{3+} or Gd^{3+} ions in oxide glasses; namely, a unimodal distribution of the fine structure parameters brought about by disorder inherent in the vitreous state gives rise to several well-defined resonance lines. In the present case, such singularity is personified by the sharp feature at $g_{\text{eff}} \approx 2.0$, resulting from the smaller magnetic nanoparticles for which the magnetocrystalline anisotropy and demagnetizing fields are almost entirely averaged by thermal fluctuations of their magnetic moments.

The mean diameters determined by SPR are in reasonable agreement with those obtained by other methods in annealed glasses [12, 91, 93] and in ferrofluids [9, 10, 94, 95]. The σ values for the magnetic particles in the borate glass are larger than those generally found for ferrofluids (0.2–0.5), indicating a higher degree of disorder in the glass.

Figure 7.14 illustrates the superparamagnetic narrowing of the high-temperature SPR spectra. The left part of this figure (a) shows a series of spectra of the $0.63\ B_2O_3$–$0.37\ Li_2O$–$0.75 \times 10^{-3}\ Fe_2O_3$ glass annealed at 753 K and recorded at measurement temperatures T_m from 300 to 723 K. The corresponding temperature dependence of the peak-to-peak linewidth is shown in Figure 7.14, (b). With an increase in T_m, the SPR spectrum drastically narrows, its peak-to-peak increases and the two-component structure of the $g_{\text{eff}} \approx 2.0$ feature is no longer resolved at higher T_m. These transformations remain quite reversible for $T_m < T_a$; therefore, they are due to dynamic effects and not to a structural change.

The temperature dependence of the intrinsic linewidth can be well described by the following phenomenological expression [34, 39]:

$$\Delta_B = \Delta_T \delta_{\text{cub}}\left(\frac{K_1 V_s}{kT}\right) = \Delta_0 L\left(\frac{MB_{\text{eff}} V}{kT}\right)\delta_{\text{cub}}\left(\frac{K_1 V_s}{kT}\right), \qquad (7.101)$$

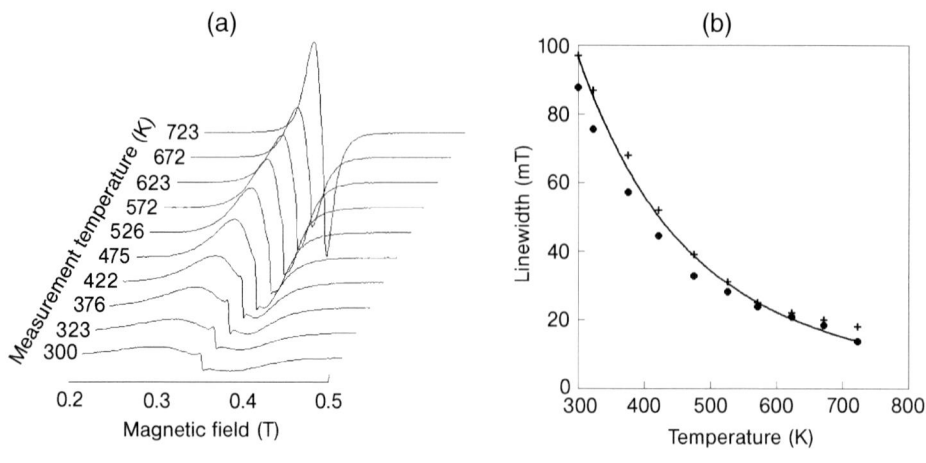

Figure 7.14 (a) EMR spectra of the 0.63 B_2O_3–0.37 Li_2O–0.75 × 10^{-3} Fe_2O_3 glass annealed at 753 K and recorded at different temperatures T_m with a constant gain. (b) Experimental peak-to-peak width (●) and saturation linewidth Δ_T (+) defined in Eq. (7.101) versus T_m. The curve is a fit calculated in accordance with the same equation [34].

where Δ_T is a temperature-dependent saturation linewidth, Δ_0 is a saturation linewidth at 0 K, L is the Langevin function with parameters defined in Section 7.5 The $\delta_{cub}(K_1 V_s/kT)$ function, see Eq. (7.71), accounts for the thermal fluctuation-induced modulation of the energy barrier between easy magnetization axes, K_1 is the first-order anisotropy constant, and V_s designs some reference particle volume.

One can see from Figure 7.14(b) that the apparent peak-to-peak linewidth and Δ_T determined by computer simulations of the SPR spectra have very similar temperature dependences. However, at a given temperature the peak-to-peak linewidth remains smaller than Δ_T because of thermal averaging of the anisotropic magnetic interactions, more pronounced for smaller particles.

7.7.2
Sol–Gel Silica Glass

Below room temperature the X-band SPR spectra of the borate glass broaden beyond the possibility of any meaningful study. The low-temperature behavior of the SPR spectra has been studied for a sol–gel glass subjected to heat treatment at ca. 1250 K during 6 h. Magnetic nanoparticles formed in this glass were identified by X-ray and Mössbauer spectroscopy as maghemite γ-Fe_2O_3 [39]. The magnetic parameters of maghemite used in computer simulations of the SPR spectra are $M = 370$ kA m^{-1} and $K_1 = -4.64$ kJ m^{-3} (cubic symmetry) [96]. The experimental X-band EMR spectra of this glass at

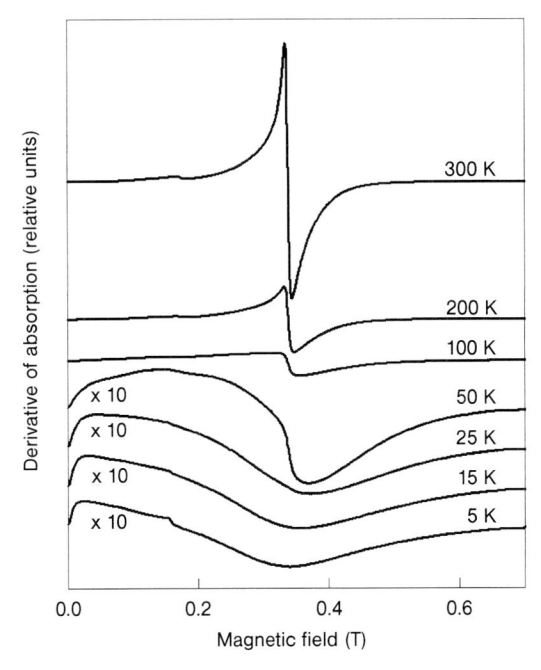

Figure 7.15 Experimental derivative-of-absorption EMR spectra of the sol–gel glass at different temperatures indicated on the right alongside the curves. The spectra at 5 to 50 K are shown with a gain 10 times larger than that of the spectra at 100 to 300 K [39].

different temperatures are shown in Figure 7.15. At room temperature the SPR reduces to a single asymmetric hyper-Lorentzian line with the effective g-factor $g_{\text{eff}} = 2.0$. With a decrease in temperature this line broadens, becomes asymmetric, and shifts toward lower fields. (A weak resonance at $g_{\text{eff}} \approx 4.3$ is due to a small number of Fe^{3+} ions diluted in the glass matrix, as in the borate glass case.) Figure 7.16 shows the temperature dependence of the apparent resonance field B_{max} and the peak-to-peak linewidth.

The results of computer simulations of the spectra recorded at different temperatures are shown in Figure 7.17. The simulations have been carried out following Eq. (7.3) with the Landau–Lifshitz intrinsic lineshape case (ii), i.e., the first derivative of the form given in Eq. (7.49). The particle shapes were assumed as ellipsoids of revolution characterized by the respective demagnetizing factors $N_\parallel$ and $N_\perp$ in the directions parallel and perpendicular to the major axes. The following set of parameters provides the best fits in the whole temperature range of this study:

$$\Delta_0 = 0.2944 \text{ T}, \quad d_{m_V} = 6.8 \text{ nm},$$
$$\sigma = 0.40, \quad V_s = 6370 \text{ nm}^3, \quad N_\perp - N_\parallel = 0.33.$$

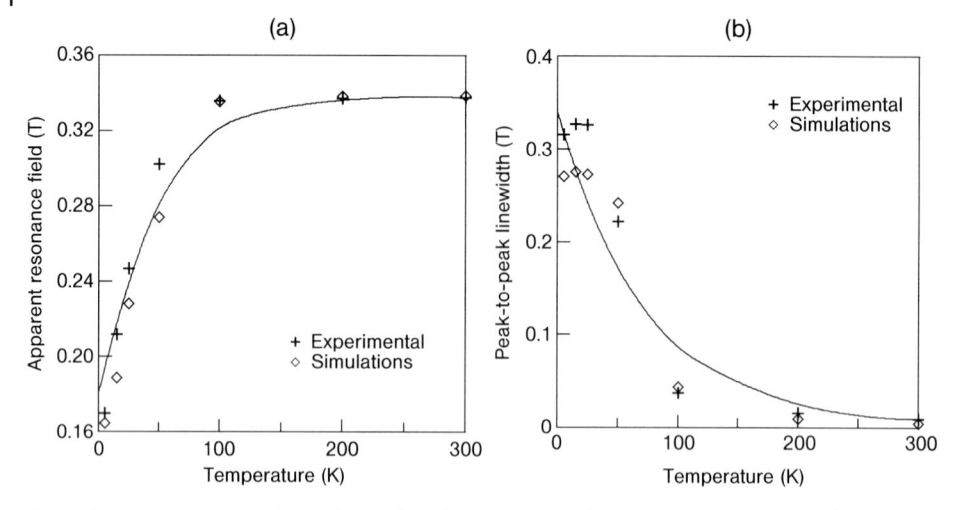

Figure 7.16 Temperature dependence of the apparent resonance field (a) and of the peak-to-peak linewidth (b) in the experimental and computer-simulated spectra. The full curve in the left part is the dependence of the apparent resonance field (B_{max}) given in Section 7.4, Table 7.1 for the Landau-Lifshitz equation, case (ii), and that in the right part shows the theoretical $\Delta_T(T)$ dependence, see Eq. (7.101) [39].

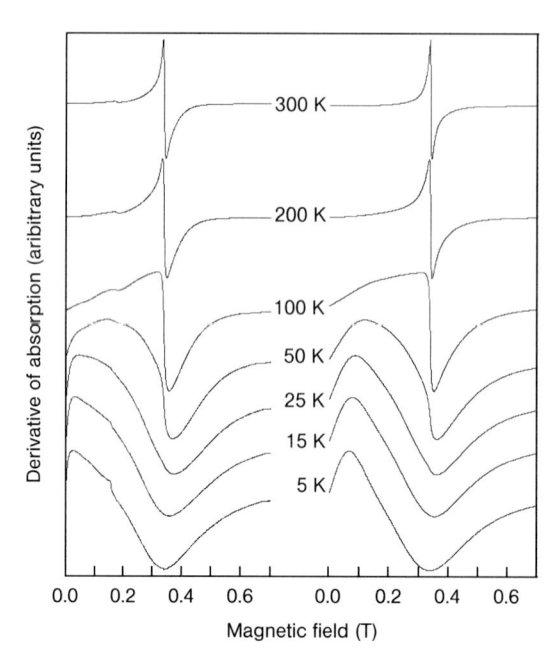

Figure 7.17 Experimental (left) and the corresponding best-fit computer-generated SPR spectra (right) of the sol–gel glass for different measurement temperatures indicated alongside the curves. All spectra are displayed with the same peak-to-peak amplitude [39].

The broadening and concomitant shift of the SPR spectra towards lower fields with the decrease in temperature have been observed in a number of different nanoparticle systems, such as silica supported nickel particles [17], ultrafine Mn–Zn ferrite particles [19], maghemite [26] and ferrite [40] nanoparticles in ferrofluids, maghemite nanoparticles in polyethylene matrix [32], a granular Cu–Co alloy [33], LaSrMnO$_3$ nanoparticles [49, 51], FeOOH [54] and ferrihydrite nanoparticles [56], hematite nanoparticles in Al$_2$O$_3$ matrix [97]. One can see from Figure 7.7 that within the approach put forward in Ref. [39] this behavior is perfectly fitted to; moreover, the whole shape of the SPR spectra at different temperatures is quite well reproduced. (The minor discrepancies between experimental and computed spectra observed in the vicinity of $g_{\text{eff}} \approx 4.3$ are mainly due to the EPR signal of isolated Fe^{3+} ions.)

7.7.3
Potassium-Alumino-Borate Glass

An interesting example of nanoparticle formation in oxide glass is provided in a recent comparative study of as-prepared and thermally treated glasses of the potassium-alumino-borate system 22.5 K$_2$O–22.5 Al$_2$O$_3$–55 B$_2$O$_3$ containing two transition metal oxides: 1.5 mass% of Fe$_2$O$_3$ and 0.4 mass% of MnO over 100 mass% [52]. The glasses were heat treated at 833 K during 2 h.

In the EMR spectra of an as-prepared glass one observes the $g_{\text{ef}} = 4.3$ characteristic feature mainly due to Fe^{3+} and the $g_{\text{ef}} = 2.0$ feature mainly due to Mn^{2+}. The only significant change in the spectra between liquid helium and room temperatures is a decrease in the relative amplitude of the former feature in expense of the latter one, see Figure 7.18(a).

The thermal treatment produces relatively small changes in the EMR spectra at lower temperatures; in contrast, room temperature spectra become very different in comparison with those of the as-prepared samples, cf. the spectra series at the left and at the right of Figure 7.18. As the temperature increases, a new large resonance line, centred at relatively low magnetic field, appears, gradually narrows and shifts to higher fields. The intensity of this resonance does not follow the Curie law, indeed, at room temperature it is several orders of magnitude greater than the resonance intensity due to remaining diluted ions, while the concentrations of ions contributing to these two resonances are comparable. This new resonance is identified as the SPR of magnetically ordered nanoparticles.

From Faraday rotation studies the magnetite Fe$_3$O$_4$ structure of the nanoparticles formed in the glass was inferred, and the corresponding magnetic parameters, $M = 480$ kA m^{-1} and $K_1 = -12.8$ kJ m^{-3} (cubic symmetry) [98] were used to computer simulate the difference spectrum obtained by subtracting the contribution of diluted paramagnetic ions, see Figure 7.19. The joint distribution density shown at the right part of this figure has been obtained

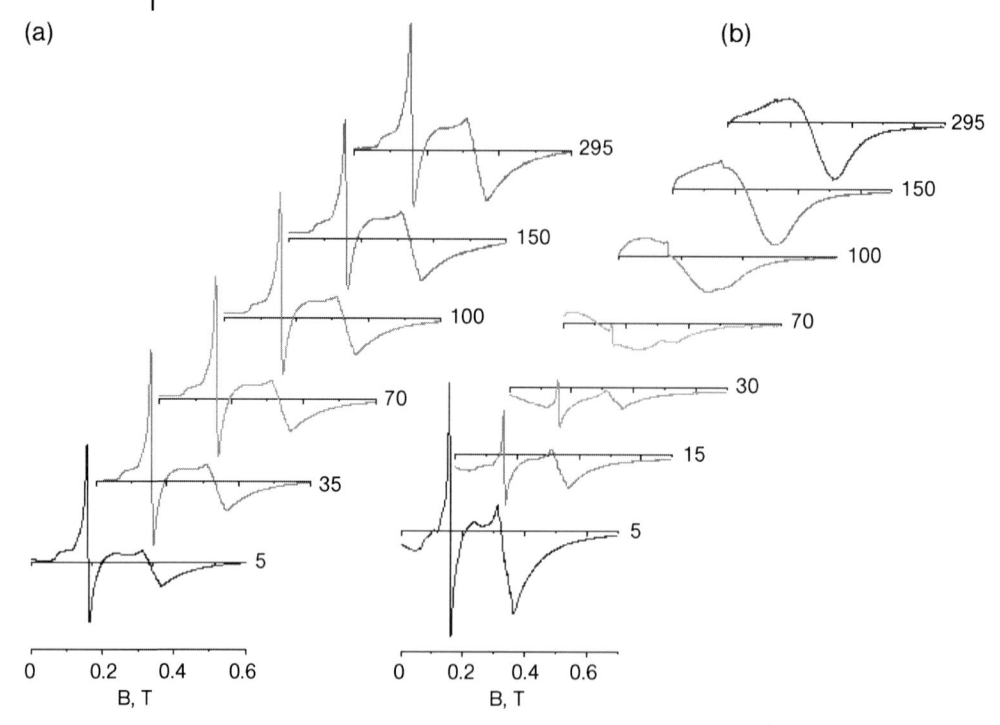

Figure 7.18 Evolution with temperature of the EMR spectra of the 22.5 K$_2$O– 22.5 Al$_2$O$_3$–55 B$_2$O$_3$ glass containing 1.5 mass% of Fe$_2$O$_3$ and 0.4 mass% of MnO. (a): as prepared glass; the intensities are multiplied by absolute temperatures. (b): Thermally treated glass; the intensities are plotted as measured [52].

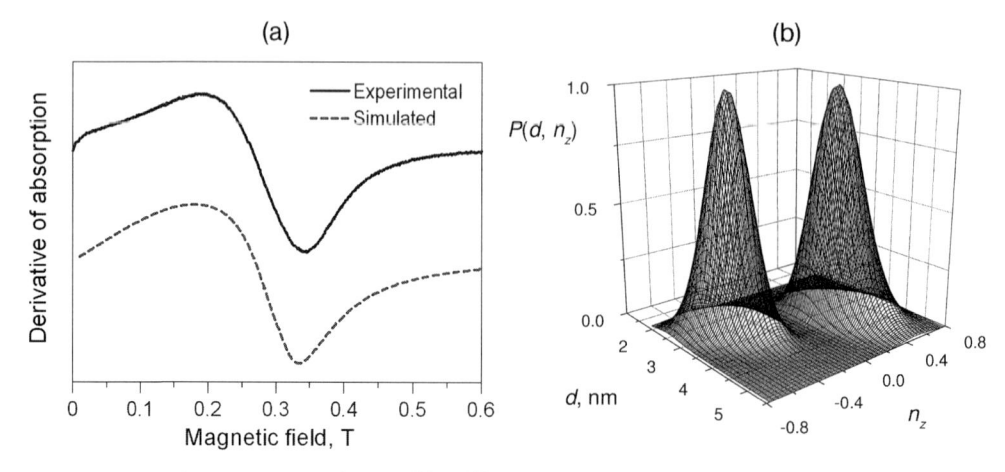

Figure 7.19 Simulation of the difference spectrum of sample 2 at room temperature (left) with the joint distribution density of diameters and demagnetizing factors shown at the right. See the text for the simulation parameters [52].

with $g = 2.20$; the diameter distribution with $d_m = 3.2$ nm and the logarithmic width $\sigma = 0.15$; the demagnetizing factor distribution with $n_0 = 0$ and $\sigma_{n_z} = 0.18$; and the correlation coefficient $\rho = 0.4$. These parameters indicate a relatively large distribution in the nanoparticle sizes and shapes.

7.7.4
Gadolinium-Containing Multicomponent Oxide Glass

The formation of magnetically ordered nanoparticles was evidenced by EMR in gadolinium-containing oxide glasses of the system 20 La_2O_3–22 Al_2O_3–23 B_2O_3–35 ($SiO_2 + GeO_2$) with a part of La_2O_3 substituted by Gd_2O_3 in concentrations from 0.1 (Gd1) to 10 mass% (Gd4) [44, 50]. At low Gd contents the EPR from isolated Gd^{3+} ions is observed, whereas at higher doping levels the overall shape of the EMR spectra shows the presence of clustering. At low temperatures the cluster-related resonance signal is altered in shape, indicating an onset of magnetic anisotropy field.

In order to extract the EMR absorption due to clusters, a numerical analysis of the experimental spectra was used. In a somewhat simplified way, the procedure used can be outlined as follows (see Ref. [44] for details). The EPR spectrum of the Gd1 glass is convoluted with a relatively broad lineshape to produce an "intermediate" spectrum representing the contribution of Gd^{3+} ions diluted in the matrix of the Gd4 glass. By subtracting this spectrum from the total EMR spectrum of Gd4, one gets a "difference" spectrum describing the contribution of clustered Gd^{3+} ions.

The results of fitting to the EMR spectrum recorded at liquid helium temperature are illustrated in Figure 7.20. At low temperatures the "difference" spectrum becomes clearly asymmetric, see Figure 7.20(b). As the actual magnetic parameters of the hypothetical ferromagnetic nanoparticles were not known, in computer simulations the magnetization value of $M = 5 \; 10^5$ A m^{-1} was rather arbitrarily assumed. Under this assumption, the following parameters were deduced from the fitting to the low-temperature underlying resonance: the magnetic anisotropy constant $K = -10$ kJ m^{-3} (including contribution of both the magnetocrystalline anisotropy and the particle shape anisotropy); the most probable diameter $d_m = 1$ nm, the log-normal diameter distribution width $\sigma = 0.2$, and the "modified Bloch" case (i) intrinsic lineshape, see Section 7.4, Eq. (7.37), with the linewidth parameter $\Delta_B = 37.5$ mT.

7.8
Conclusions and Prospective

The importance of consistently using computer simulations in the analysis of the superparamagnetic resonance spectra cannot be overestimated;

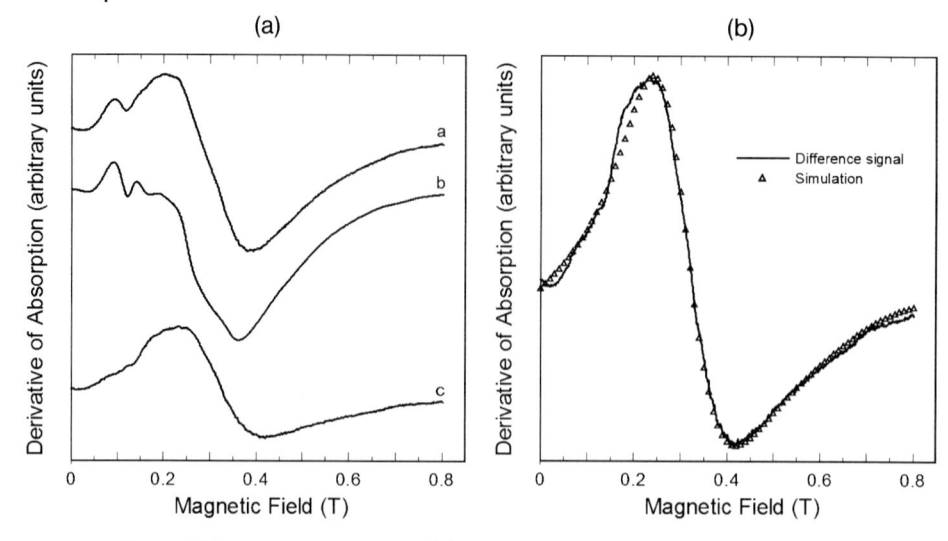

Figure 7.20 (a) Representation of the EPR spectrum of Gd4 at 4.5 K (a) as a linear combination of the liquid helium-temperature spectrum of Gd1 convoluted with the lineshape of Eq. (7.37) for $\Delta_B = 5.8$ mT (b) and an underlying resonance: (c) = (a) -0.55 (b). (b) Fitting to the curve (c) with a superparamagnetic resonance signal (see the text for the simulation parameters) [44].

unfortunately, too many authors prefer the facility of interpreting their experimental results on the basis of a visual inspection. With such an approach, different factors contributing, e.g., to the apparent resonance field, or to the observed "peak-to-peak" resonance width can hardly be reliably separated, so the conclusions drawn in such studies often remain contestable. In contrast, the superparamagnetic resonance assisted by computer simulations has proven its efficiency; in particular, it can be considered as a new reliable technique of morphological analysis of the magnetic nanoparticles.

In various oxide glasses doped with paramagnetic ions superparamagnetic nanoparticle systems are formed, giving rise to characteristic resonance spectra. The computer fits to these spectra based on the joint distribution density of diameters and demagnetizing factors show that the nonsphericity of the nanoparticles may take a prominent part in determining the characteristics of the superparamagnetic resonance spectra and in any case cannot be neglected *a priori*.

Amongst other factors interfering with the SPR of nanoparticles, the surface effects such as surface anisotropy and surface disorder deserve a particular mention. Indeed, because of the smallness of the nanoparticle, its properties are greatly influenced by its surface layer where the environment of magnetic atoms differs from that in a core [1, 6, 26, 35, 47]. Unfortunately, the characteristics of the surface layer are still less easily mastered during the synthesis than those of the core. Different methods of sample preparation and different embedding matrices used result in a large variety of magnetic characteristics observed in nanoparticle systems. Therefore, the surface effects in nanoparticle systems are far from being well understood and such effects have remained beyond the scope of the present chapter.

As a final remark, we would like to mention the following issue. When passing from the classical EPR to FMR and, particularly, to SPR, one is somewhat embarrassed by the necessity to coincidentally pass from the strict quantum mechanical perspective to more loose statistical if not phenomenological approaches. In part, this change of viewpoint is imposed by the physical systems themselves; indeed, the question is, in the first case, of an individual paramagnetic species and, in the second case, of collective phenomena in assemblies of a more or less large number of paramagnetic ions. Nevertheless, there clearly is a need for more rigorous microscopic approaches to the physics of nanoparticles in general and to the description of the SPR phenomenon in particular.

References

1. R.H. Kodama, A.E. Berkovitz, *Phys. Rev. B*, **1999**, *59*, 6321.
2. X. Batlle, A. Labarta, *J. Phys. D: Appl. Phys.*, **2002**, *35*, R15.
3. D. Fiorani, A.M. Testa, F. Lucari, F. D'Orazio, H. Romero, *Physica B*, **2002**, *320*, 122.
4. C.R. Vestal, Z.J. Zhang, *J. Am. Chem. Soc.*, **2003**, *125*, 9828.
5. J. Wang, C. Zeng, Z.M. Peng, Q.W. Chen, *Physica B: Condens. Matter*, **2004**, *349*, 124.
6. O. Masala, R. Seshadri, *Chem. Phys. Lett.*, **2005**, *402*, 160.
7. L Néel, *Ann. Geophys.*, **1949**, *5*, 99.
8. L. Dormann, D. Fiorani, E. Tronc, *Adv. Chem. Phys.*, **1997**, *98*, 283.
9. J.-C. Bacri, F. Boué, V. Cabuil, R. Perzynski, *Colloids, Surfaces A*, **1993**, *80*, 11.
10. J. Popplewell, L. Sakhnini, *J. Magn. Magn. Mater.*, **1995**, *149*, 72.
11. R.V. Upadhyay, G.M. Sutariya, R.V. Mehta, *J. Magn. Magn. Mater.*, **1993**, *123*, 262.
12. C. Estournès, T. Lutz, J. Happich, P. Quaranta, P. Wissler, J.L. Guille, *J. Magn. Magn. Mater.*, **1997**, *173*, 83.
13. M. Jamet, V. Dupuis, P. Mélinon, G. Guiraud, A. Pérez, W. Wernsdorfer, A. Traverse, B. Baguenard, *Phys. Rev.*, **2000**, *B 62*, 493.
14. V.K. Sharma, F. Waldner, *J. Appl. Phys.*, **1977**, *48*, 4298.
15. R.S. de Biasi, T.C. Devezas, *J. Appl. Phys.*, **1978**, *49*, 2466.
16. D.L. Griscom, E.J. Friebele, D.B. Shinn, *J. Appl. Phys.*, **1979**; *50*, 2402.
17. V.K. Sharma, A Baiker, *J. Chem. Phys.*, **1981**, *75*, 5596.
18. J. Dubowik, J. Baszyński, *J. Magn. Magn. Mater.*, **1986**, *59*, 161.
19. K. Nagata, A. Ishihara, *J. Magn. Magn. Mater.*, **1992**, *104–107*, 1571.
20. Yu.L. Raikher, V.I. Stepanov, *Phys. Rev. B*, **1994**, *50*, 6250.
21. Yu.L. Raikher, V.I. Stepanov, *J. Magn. Magn. Mater.*, **1995**, *149*, 34.
22. R. Berger, J.-C. Bissey, J. Kliava, B. Soulard, *J. Magn. Magn. Mater.*, **1997**, *167*, 129.
23. J.F. Saenger, K. Skeff Neto, P.C. Morais, M.H. Sousa, F.A. Tourinho, *J. Magn. Res.*, **1998**, *34*, 180.
24. M. Respaud, M. Goiran, F. Yang, J.M. Broto, T. Ould Ely, C. Amiens, B. Chaudret, S. Askenazy, *Physica B*, **1998**, *246–247*, 580.

25. R. Berger, J. Kliava, J.-C. Bissey, V. Baïeto, *J. Phys.: Condens. Matter*, **1998**, *10*, 8559.
26. F. Gazeau, J.-C. Bacri, F. Gendron, R. Perzynski, Yu.L. Raikher, V.I. Stepanov, E. Dubois, *J. Magn. Magn. Mater.*, **1998**, *186*, 175.
27. I. Hrianca, I. Malaescu, F. Claici, C.N. Marin, *J. Magn. Magn. Mater.*, **1999**, *201*, 126.
28. M. Respaud, M. Goiran, J.M. Broto, F.H. Yang, T. Ould Ely, C. Amiens, B. Chaudret, *Phys. Rev. B*, **1999**, *59*, R3934.
29. J. Kliava, R. Berger, *J. Magn. Magn. Mater.*, **1999**, *205*, 328.
30. R.D. Sánchez, M.A. López-Quintela, J. Rivas, A. González-Penedo, A.J. García-Bastida, C.A. Ramos, R.D. Zysler, S. Ribeiro Guevara, *J. Phys: Condens. Matter*, **1999**, *11*, 5643.
31. J.R. Fermin, A. Azevedo, F.M. De Aguiar, Biao Li, S.M. Rezende, *J. Appl. Phys.*, **1999**, *85*, 7316.
32. Yu.A. Koksharov, S.P. Gubin, I.D. Kosobudsky, G.Yu. Yurkov, D.A. Pankratov, L.A. Ponomarenko, M.G. Mikheev, M. Beltran, Y. Khodorkovsky, A.M. Tishin, *Phys. Rev. B*, **2000**, *63*, 12407.
33. H.K. Lachowicz, A. Sienkiewicz, P. Gierlowski, A. Slawska-Waniewska, *J. Appl. Phys.*, **2000**, *88*, 368.
34. R. Berger, J. Kliava, J.-C. Bissey, *J. Appl. Phys.*, **2000**, *87*, 7389.
35. A.F. Bakuzis, P.C. Morais, *J. Magn. Magn. Mater.*, **2001**, *226–230*, 1924.
36. J.F. Hochepied, M.P. Pileni, *J. Magn. Magn. Mater.*, **2001**, *231*, 45.
37. Yu.A. Koksharov, D.A. Pankratov, S.P. Gubin, I.D. Kosobudsky, M. Beltran, Y. Khodorkovsky, A.M. Tishin, *J. Appl. Phys.*, **2001**, *89*, 2293.
38. L.M. Lacava, B.M. Lacava, R.B. Azevedo, Z.G.M. Lacava, N. Buske, A.L. Tronconi, P.C. Morais, *J. Magn. Magn. Mater.*, **2001**, *225*, 79.
39. R. Berger, J.-C. Bissey, J. Kliava, H. Daubric, C. Estournès, *J. Magn. Magn. Mater.*, **2001**, *234*, 535.
40. R.S. de Biasi, W.S.D. Folly, *Physica B*, **2002**, *321*, 117.
41. J. Kliava, R. Berger, in *Recent Res. Devel. Non-Crystalline Solids*, Transworld Research Network, Kerala, India, **2003**, Vol. 3, p. 41.
42. E. de Biasi, C.A. Ramos, R.D. Zysler, *J. Magn. Magn. Mater.*, **2003**, *262*, 235; E. de Biasi, C.A. Ramos, R.D. Zysler, *J. Magn. Magn. Mater.*, **2004**, *278*, 289.
43. J. Kliava, R. Berger, *Molec. Phys. Rep.*, **2004**, *39*, 130.
44. J. Kliava, A. Malakhovskii, I. Edelman, A. Potseluyko, E. Petrakovskaja, S. Melnikova, T. Zarubina, G. Petrovskii, I. Bruckental, Y. Yeshurun, *Phys. Rev. B*, **2005**, *71*, 104406.
45. J. Kliava, R. Berger, in *Smart Materials for Ranging Systems*, J. Franse (editor), Springer, Berlin, **2006**, p. 27.
46. J. Kliava, R. Berger, A. Potseluyko, I. Edelman, E. Petrakovskaja, T. Zarubina, *Phys. Met. Metallogr.*, **2006**, *102*, S39.
47. D.S. Schmool, R. Rocha, J.B. Sousa, J.A.M. Santos, G. Kakazei, *J. Magn. Magn. Mater.*, **2006**, *300*, e331.
48. E. De Biasi, C.A. Ramos, R.D. Zysler, H. Romero, *Physica B*, **2004**, *354*, 286.
49. V. Krivoruchko, T. Konstantinova, A. Mazur, A. Prokhorov, V. Varyukhin, *J. Magn. Magn. Mater.*, **2006**, *300*, e122.
50. J. Kliava, A. Malakhovskii, I. Edelman, A. Potseluyko, E. Pertrakovskaja, I. Bruckental, Y Yeshurun, T. Zarubina, *J. Supercond. Novel Magnetism*, **2006**, *20*, 149.
51. V.N. Krivoruchko, A.I. Marchenko, A.A. Prokhorov, *Low Temperature Phys.*, **2007**, *33*, 433.
52. J. Kliava, A. Marbeuf, I. Edelman, R. Ivantsov, O. Ivanova, E. Petrakovskaja, S.A. Stepanov, V.I. Zaikovskii, in *XXIst International Congress on Glass, Enlarged Abstracts*, Strasbourg, **2007**, A24.
53. Yu.L. Raikher, V.I. Stepanov, *Sov. Phys. – JETP*, **1992**, *75*, 764.

54. M.M. Ibrahim, G. Edwards, M.S. Seehra, B. Ganguly, G.P. Huffman, *J. Appl. Phys.*, **1994**, *75*, 5873.
55. M. Respaud, *Thesis*, INSA Toulouse, **1997**.
56. A. Punnoose, M.S. Seehra, J. van Tol, L.C. Brunel, *J. Magn. Magn. Mater.*, **2005**, *288*, 168.
57. G.V. Skrotskii, L.V. Kurbatov, in *Ferromagnetic Resonance*, V Vonsovskii (editor), Pergamon, Oxford, **1966**.
58. J.A. Osborn, *Phys. Rev.*, **1945**, *67*, 351.
59. J. Smit, H.G. Beljers, *Philips Res. Rep.*, **1955**, *10*, 113.
60. L. Baselgia, M. Warden, F. Waldern, S.L. Hutton, J.E. Drumheller, Y.Q. He, P.E. Wigen, M. Marysko, *Phys. Rev. B*, **1988**, *38*, 2237.
61. M. Masi, *Am. J. Phys.*, **2007**, *75*, 116.
62. C.E. Patton, in *Magnetic Oxides*, D.J. Craik (editor), Wiley, New York, **1975**, Vol. 2, p. 575.
63. F. Bloch, *Phys. Rev.*, **1946**, *20*, 460.
64. B. Bloembergen, *Phys. Rev.*, **1950**, *78*, 572.
65. R.S. Codrington, J.D. Olds, H.C. Torrey, *Phys. Rev.*, **1954**, *95*, 607.
66. M.A. Garstens, J.I. Kaplan, *Phys. Rev.*, **1955**, *99*, 459.
67. T.L. Gilbert, *Phys. Rev.*, **1955**, *100*, 1243.
68. L. Landau, E. Lifshitz, *Phys. Z. Sowjetunion*, **1935**, *8*, 153.
69. H.B. Callen, *J. Phys. Chem. Solids*, **1958**, *4*, 256.
70. R. Berger, J.-C. Bissey, J. Kliava, *J. Phys.: Condens. Matter*, **2000**, *12*, 9347.
71. R. Kikuchi, *J. Appl. Phys.*, **1956**, *27*, 1352.
72. J.C. Mallinson, *IEEE Trans. Magn.*, **1987**, *23*, 2003.
73. B. Lax, K.J. Button, in *Microwave Ferrites and Ferrimagnetics*, McGraw-Hill, New York, **1962**.
74. M.A. Garstens, *Phys. Rev.*, **1954**, *93*, 1228.
75. Yu.L. Raikher, V.I. Stepanov, in *Adv. Chem. Phys.*, S.A. Rice (editor), **2004**, Vol. 129, p. 419.
76. S. Iida, *J. Phys. Chem. Solids*, **1963**, *4*, 625.
77. A.G. Flores, L. Torres, V. Raposo, L. López-Díaz, M. Zazo, J. Iñiguez, *phys. status solidi (a)*, **1999**, *171*, 549.
78. W.F. Brown, Jr., *IEEE Trans. Magn.*, **1979**, *15*, 1196.
79. C.G. Granquist, R.A. Buhrman, *J. Appl. Phys.*, **1976**, *47*, 2200.
80. J. Zarzycki, F. Naudin, *Phys. Chem. Glasses*, **1967**, *8*, 11.
81. J. Jerphagnon, D. Chemla, R. Bonneville, *Adv. Phys.*, **1978**, *27*, 609.
82. R.A. Brand, G. Le Caër, J.M. Dubois, *J. Phys.: Condens. Matter*, **1990**, *2*, 6413.
83. M.L. Mehta, *Random Matrices* (2nd edition), Academic Press, Boston, **1991**, pp. 39 ff, 55 ff.
84. G. Czjzek, J. Fink, F. Götz, H. Schmidt, J.M.D. Coey, J.-P. Rebouillat, A. Liénard, *Phys. Rev. B*, **1981**, *23*, 2513.
85. G. Le Caër, R.A. Brand, K. Dehghan, *J. Phys.*, **1985**, *46*, C8–169.
86. C. Legein, J.Y. Buzaré, J. Emery, C. Jacoboni, *J. Phys.: Condens. Matter*, **1995**, *7*, 3853.
87. C. Legein, J.Y. Buzaré, C. Jacoboni, *J. Non-Cryst. Solids*, **1995**, *184*, 160.
88. A. Abragam, B. Bleaney, *Electron Paramagnetic Resonance of Transition Ions*, Clarendon, Oxford, **1970**, pp. 151, 166.
89. G. Le Caër, J.M. Cadogan, R.A. Brand, J.M. Dubois, H.J. Güntherodt, *J. Phys. F: Met. Phys.*, **1984**, *14*, L73.
90. M. Maurer, *Phys. Rev. B*, **1986**, *34*, 8996.
91. E. Rezlescu, N. Rezlescu, M.L. Craus, *J. Phys.*, **1997**, *IV-7*, 553.
92. J. Kliava, *phys. status solidi (b)*, **1986**, *134*, 411.
93. S. Roy, B. Roy, D. Chakravorty, *J. Appl. Phys*, **1996**, *79*, 1642.
94. J.L. Dormann, F. d'Orazio, F. Lucari, E. Tronc, P. Prené, J.P. Jolivet, D. Fiorani, R. Cherkaoui, M. Noguès, *Phys. Rev. B*, **1996**, *53*, 14291.
95. N. Feltin, M.P. Pileni, *J. Phys.*, **1997**, *IV-7*, C1–609.
96. E. Schmidbauer, R. Keller, *J. Magn. Magn. Mater.*, **1996**, *152*, 99.

97. R. Zysler, D. Fiorani, J.L. Dormann, A.M. Testa, *J. Magn. Magn. Mater.*, **1994**, *133*, 71.

98. Z. K1kol, J.M. Honig, *Phys. Rev. B*, **1989**, *40*, 9090.

8
Micromagnetics of Small Ferromagnetic Particles
Nickolai A. Usov and Yury B. Grebenshchikov

8.1
Introduction

Fine ferromagnetic particles of a nanometer scale size have important applications in various fields of modern nanotechnology, such as magnetic recording, permanent magnet industry, biomedical applications, etc. From a fundamental point of view, the theory of fine ferromagnetic particle constitutes one of the basic parts of Micromagnetics [1–3], a general theory developed by Landau, Lifshitz, Neel, Brown, Kittel, Kondorsky, Aharoni and many other researchers to describe magnetic properties of various types of ferromagnetic materials. The classical results of Micromagnetics concerning the properties of fine ferromagnetic particles are as follows: (1) the notion of a single-domain radius for ideal ferromagnetic particle of ellipsoidal external shape; (2) the theory of nucleation modes of a single-domain particle under the influence of external uniform magnetic field; and (3) the thermally assisted switching of particle magnetization at elevated temperatures.

Just after the discovery of the quantum mechanical nature of the exchange interaction in ferromagnetic materials [4, 5] it was recognized [6] that a subdivision of a macroscopic ferromagnetic body into ferromagnetic domains is a consequence of long-range magnetic dipolar interactions between ferromagnetic spins. Actually, the magnetostatic energy of a uniformly magnetized body decreases greatly due to domain subdivision. However, there is a critical size for this subdivision because of the increase of the exchange and anisotropy energy contributions to the total energy. Following this idea, first estimations of the characteristic single-domain size for particles of soft and hard magnetic types were made in Refs. [7–11]. Later Brown [12, 13] formulated an exact definition of the single-domain particle of an ideal ellipsoidal shape comparing the total energy of uniform magnetization with that of any nonuniform magnetization distribution $\vec{M}(\vec{r})$. He also gave the lower and upper estimates for single-domain radius of a spherical particle [12, 13]. The behavior of a single-domain particle in external uniform magnetic field was first studied in the famous paper of Stoner and Wohlfarth [14]. However, it was then recognized [15] that they had considered only the simplest uniform rotation mode of a particle. Later Brown [2, 16] and Aharoni [3, 17–21] developed a complete theory of nucleation modes of a single-domain particle. A behavior of a single-domain particle at a finite temperature was first considered by Neel [22]. He pointed out that the magnetic moment direction of single-domain particle experiences appreciable thermal fluctuations at a temperature comparable with the height of its effective energy barrier. Neel gave a simple estimation of a characteristic relaxation time for magnetization of an assembly of noninteracting superparamagnetic particles. A general approach to the problem was developed by Brown [23] based on the Fokker–Plank type equation for the magnetic moment orientations of a single-domain particle.

It is well known that the basic equations of Micromagnetics have very complicated mathematical structure. First of all, they are nonlinear, because the components of the unit magnetization vector, $\vec{\alpha} = \vec{M}/M_s$, where M_s is the saturation magnetization, are subjected to the restriction $\alpha_x^2 + \alpha_y^2 + \alpha_z^2 = 1$. Besides, the demagnetizing field of magnetic charges distributed over the volume and surface of a ferromagnetic body has to be taken into account self-consistently, by means of solution of the Maxwell equations. That is why, at a classical stage of the development, only few interesting micromagnetic problems were actually solved. The recent progress in Micromagnetics is related mainly with the introduction [24–27] of power numerical simulation methods exploiting the increasing capability of contemporary computers (see Refs. [28, 29] and references therein). Before going into details, let us summarize briefly the theoretical problems concerning the properties of fine ferromagnetic particles.

The aim of Micromagnetics is to study stable equilibrium configurations existing in small particles of various sizes, shapes, and phenomenological material parameters (such as saturation magnetization, M_s, anisotropy constant, K, exchange constant, C, etc.) and to describe the behavior of particles in external magnetic fields (both static and alternating), as well as at elevated temperatures. One important problem is the influence of the magnetostatic interactions between the particles in a dense particle assembly, but this is beyond the scope of present discussion. The classical theory states [2, 3] that for an ideal ellipsoidal particle, there is a critical single-domain radius a_c defined so that for particle with radius $R < a_c$ the lowest energy state is that of uniform magnetization, $\vec{\alpha} = $ const, whereas for $R > a_c$ certain nonuniform magnetization distribution has the lowest total energy. However, one has to take into account that uniform magnetization can be stable even at $R > a_c$, where it becomes *metastable*. Similarly, the nonuniform state may exist in some interval of sizes below the single-domain radius, $R < a_c$. Thus, the notion of a single-domain radius refers to the stable magnetization states of a fine particle. Yet, another quantity can be introduced to characterize *dynamics* of the particle magnetization. It can be proved [2, 3] that if the particle radius is small enough,

$R < a_{c1} < a_c$, the changes of the particle magnetization under the influence of external magnetic field or thermal fluctuations occur by means of uniform rotation only. Thus, the quantity a_{c1} can be termed as a critical radius of *truly* single-domain behavior. Within the interval $a_{c1} < R < a_c$, the particle magnetization is generally uniform, but it can switch between various stable uniform magnetization states by means of nonuniform nucleation modes, such as buckling, curling, etc.

One can see that the situation is rather complicated even for an ideal ellipsoidal particle. For this case, the analysis is simpler, because the uniform magnetization is certainly one of the solutions of the equilibrium micromagnetic equation. Therefore, the problem is to investigate the stability of the uniform magnetization as a function of the particle size, under the influence of external magnetic field. However, the model of an ideal ellipsoidal particle is only an approximation to real experimental situation. The purpose of this review is to discuss some of the theoretical problems related to the properties of real ferromagnetic particles, which differ in several aspects from those of idealized classical model.

In Section 8.2, we consider the magnetization distributions existing in fine ferromagnetic particles of nonellipsoidal external shape. The shapes of patterned ferromagnetic elements [30–32] often approach to those of perfect geometrical shapes, such as cube, parallelepiped, flat or elongated cylinder, etc. For this case, strictly uniform magnetization is not a solution of the equilibrium micromagnetic equation. It was shown both by means of numerical simulation and perturbation theory [24, 28, 33–36] that the lowest energy state for particles of perfect geometrical shapes of small enough size is a certain quasiuniform state, the so-called flower state. The fact that the lowest energy state for nonellipsoidal particle is nonuniform leads to the existence of a specific type of magnetic anisotropy, the so-called configurational anisotropy, discovered by Schabes and Bertram by means of numerical simulation [24, 28]. In some sense, perfect geometrical shape can be considered as large regular deviation of an ideal ellipsoidal shape. On the other hand, a ferromagnetic particle of irregular shape can be represented sometimes as an ideal ellipsoid subjected to small irregular shape deviations. In this case, the lowest energy quasiuniform magnetization distribution can also be constructed by means of the perturbation theory [37].

In recent years, the nonuniform micromagnetic configurations existing in particles of various shapes have been comprehensively studied both experimentally and by means of numerical simulation. Two-domain states, as well as longitudinal, transverse, and tilted curling states, were investigated in spherical and ellipsoidal particles [38–42], vortex states in flat and elongated cylinders [43–57], and the so-called C and S configurations in flat cylinders [58–61]. Also, various nonuniform states were studied in fine cubes and parallelepipeds [35, 36, 62–72]. In addition, these investigations showed that nonuniform states compete in energy with a quasiuniform magnetization distribution, which has the lowest possible energy for particle of sufficiently small size. This enables one to generalize the notion of a single-domain particle to a general case of a particle of nonellipsoidal shape [73].

In Section 8.3, we consider a magnetization distribution within a fine ferromagnetic particle having the so-called surface magnetic anisotropy [2, 3]. The surface anisotropy was introduced by Neel [74] to describe the effect of breaking of the translation symmetry of crystal lattice at the particle surface. In Micromagnetics, surface anisotropy can be taken into account by means of special boundary condition [2, 3]. This boundary condition is incompatible, as a rule, with the existence of strictly uniform magnetization within the particle. Similar problem may also exist for composite ferromagnetic particles consisting of a ferromagnetic core and thin outer shell with different phenomenological material parameters [75, 76]. Finally, in Section 8.4, we discuss the thermal relaxation process in single-domain particles with various types of magnetic anisotropy using the numerical simulation results obtained by means of solution of stochastic Landau–Lifshitz equation.

8.2
Particle Morphology and Single-Domain Radius

8.2.1
Quasiuniform States

As we mentioned in Section 8.1, the theory of single-domain particle [2, 3] is applicable, strictly speaking, only to a particle of ideal ellipsoidal shape. On the other hand, the shapes of real ferromagnetic particles and patterned ferromagnetic elements certainly deviate from that of ideal ellipsoid. In this section, we consider quasiuniform magnetization distributions for two characteristic cases: (a) patterned ferromagnetic elements of perfect geometrical shape; and (b) ellipsoidal particle with small irregular shape deviations.

8.2.1.1 Particles of Perfect Geometrical Shape

The famous theorem of classical magnetostatic states [77] that the demagnetizing field within a uniformly magnetized ideal ellipsoid is uniform. However, the demagnetizing field $\vec{H}'(\vec{r})$ within a uniformly magnetized nonellipsoidal particle, $\vec{\alpha} = \vec{\alpha}^{(0)} = \text{const}$, is nonuniform. This field arises due to surface magnetic poles of the magnetization $M_s \vec{\alpha}^{(0)}$ and can be calculated as follows [78]:

$$H_i^{\prime(0)}(\vec{r}) = -M_s \sum_i N_{ij}(\vec{r}) \alpha_i^{(0)}; \quad N_{ij}(\vec{r}) = -\frac{\partial^2}{\partial r_i \, \partial r_j} \int \frac{d\vec{r}'}{|\vec{r}' - \vec{r}|}, \quad (8.1)$$

where $N_{ij}(\vec{r})$ is the demagnetizing tensor of the nonellipsoidal particle, the integration is over the particle volume V, $(i, j = x, y, z)$. The nonuniform demagnetizing field complicates the micromagnetic calculations considerably.

8.2 Particle Morphology and Single-Domain Radius

To avoid this difficulty, Brown and Morrish [79] suggested that the magnetization distribution in a nonideal single-domain particle can be approximately described by a certain initial vector $\vec{\alpha}^{(0)}$, which satisfies the equilibrium micromagnetic equation

$$[\vec{\alpha}^{(0)}, \vec{H}_{ef}^{(0)}] = 0; \quad M_s \vec{H}_{ef}^{(0)} = \frac{\partial w_a}{\partial \vec{\alpha}^{(0)}} + M_s \left(\vec{H}_0 + \langle \vec{H}'^{(0)}(\vec{r}) \rangle \right). \quad (8.2)$$

Here $\vec{H}_{ef}^{(0)}$ is the vector of effective magnetic field, $w_a(\vec{\alpha})$ is the density of the magnetic anisotropy energy, $\vec{H}_0$ is the uniform external magnetic field, and $\langle \vec{H}'^{(0)}(\vec{r}) \rangle$ is the averaged value of the demagnetizing field (8.1) over the particle volume. Equations (8.1) and (8.2) determine the vector $\vec{\alpha}^{(0)}$ completely. This state was used as initial state of the perturbation theory developed [33, 34, 80] to describe quasiuniform magnetization distributions in particles of perfect geometrical shape. Actually, one can consider the difference between the nonuniform demagnetizing field (8.1) and its average value over the particle volume as a source of perturbation

$$\vec{H}'^{(1)}(\vec{r}) = \vec{H}'^{(0)}(\vec{r}) - \langle \vec{H}'^{(0)}(\vec{r}) \rangle; \quad \langle \vec{H}'^{(1)}(\vec{r}) \rangle = 0. \quad (8.3)$$

Assuming that the unit vector $\vec{\alpha}^{(0)}$ satisfies Eqs. (8.1) and (8.2), one may obtain the actual magnetization distribution in a particle as a series

$$\vec{\alpha}(\vec{r}) = \vec{\alpha}^{(0)} + \vec{\alpha}^{(1)}(\vec{r}) + \cdots, \quad (8.4)$$

where $\vec{\alpha}^{(1)}(\vec{r})$ is a small nonuniform deviation of the first order with respect to the perturbation (8.3). Due to the normalization condition, it satisfies the relation $\vec{\alpha}^{(0)} \cdot \vec{\alpha}^{(1)} = 0$. Setting the series (8.4) in equilibrium micromagnetic equation and taking into account that the initial state $\vec{\alpha}^{(0)}$ satisfies Eqs. (8.1) and (8.2) by convention, one can express the first-order magnetization deviation as follows [33]:

$$\vec{\alpha}^{(1)} = M_s \sum_n \frac{\langle \vec{u}_n | \vec{H}'^{(1)}(\vec{r}) \rangle}{\lambda_n - \lambda^*} \vec{u}_n. \quad (8.5)$$

Here the vectors $\vec{u}_n$ constitute a complete set of eigenfunctions of the well-known Brown's boundary value problem [2, 16], corresponding to the eigenvalues λ_n; the parameter $\lambda^* = -M_s H_{ef.z}^{(0)}$. For sufficiently small particle the exchange energy contribution to the eigenvalues λ_n dominates, $\lambda_n \approx \lambda_n^{(0)} \sim C/L^2$, where C is the exchange constant and L is the characteristic particle size. On the other hand, the characteristic value of the perturbation is given by $|\vec{H}'^{(1)}(\vec{r})| \sim M_s$. Therefore, it follows from Eq. (8.5) that the small parameter of the perturbation theory equals $L^2 M_s^2 / C = (L/L_{ex})^2$, where $L_{ex} = \sqrt{C}/M_s$ is the exchange length. Because the deviation $|\vec{\alpha}^{(1)}| \sim M_s^2$, one can use in Eq. (8.5) the eigenvectors $\vec{u}_n^{(0)}$ and eigenvalues $\lambda_n^{(0)}$ calculated in the so-called exchange approximation, disregarding the magnetostatic energy contribution to the total energy of the Brown's modes. For the same reason, the term $M_s \langle H'^{(0)}_z \rangle$ in the parameter λ^*, Eq. (8.5), must be omitted.

These considerations greatly simplify the calculation of the first-order correction (8.5). Consider a right circular cylinder with radius R and thickness L_z. Let the easy anisotropy axis is parallel to the cylinder axis, K_1 being the anisotropy constant. If the initial vector $\vec{\alpha}^{(0)}$ points along the cylinder axis, $\vec{\alpha}^{(0)} = (0, 0, 1)$, the demagnetizing field perturbation (8.3) and the correction (8.5) in the cylindrical coordinates with the z axis parallel to the cylinder axis can be written as follows:

$$\vec{H}'^{(1)}(\vec{r}) = (H'^{(1)}_\rho, 0, H'^{(1)}_z); \qquad \vec{\alpha}^{(1)}(\vec{r}) = (\alpha^{(1)}_\rho, \alpha^{(1)}_\varphi, 0). \tag{8.6}$$

One can see from Eqs. (8.5) and (8.6) that for the case considered, only $\alpha^{(1)}_\rho$ component of the first-order correction is nonzero. The eigenfunctions and eigenvalues of the radial Brown's modes, $(u_\rho, 0, 0)$, in the exchange approximation are given by

$$u^{(0)}_{\rho;i,n} = N_{i,n} J_1\left(\frac{\mu_{1,i}\rho}{R}\right) \cos\frac{\pi n z}{L_z};$$

$$\lambda^{(0)}_{i,n} = 2K_1 + C[(\mu_{1,i}/R)^2 + (\pi n/L_z)^2]. \tag{8.7}$$

Here $N_{i,n}$ is the normalization constant, $J_1(\rho)$ is the Bessel function, $\mu_{1,i}$, $(i = 1, 2, \ldots)$ are the roots of the equation $J'_1(\mu) = 0$, so that $\mu_{1,1} \approx 1.84$; $\mu_{1,2} \approx 5.33$, and so on, the integer $n = 0, 1, \ldots$

Since the perturbation $H'^{(1)}_\rho$ is an odd function of the variable $z - L_z/2$, only the modes (8.7) with odd integers $n = 2k + 1$ $(k = 0, 1 \ldots,)$ contribute to the sum (8.5). For a small ferromagnetic cylinder with $R/L_z \sim 1$, the eigenvalues (8.7) increase rapidly as functions of integers i and n. As a result, the $\alpha^{(1)}_\rho$ component in fine ferromagnetic cylinder is determined mainly by the contribution with the lowest values of $i = n = 1$

$$\alpha^{(1)}_\rho \sim J_1\left(\frac{\mu_{1,1}\rho}{R}\right) \cos\frac{\pi z}{L_z}. \tag{8.8}$$

Consider now a fine ferromagnetic parallelepiped with an easy anisotropy axis parallel to the z axis. Let the parallelepiped is uniformly magnetized along the z axis, $\vec{\alpha}^{(0)} = (0, 0, 1)$ in the lowest approximation of the perturbation theory. For a parallelepiped with length L_z and square cross-section, $L_x = L_y$, the eigenfunctions and eigenvalues of the Brown's modes, $\vec{u}_n = (u_{nx}, u_{ny}, 0)$, in the exchange approximation are given by [34]

$$\vec{u}^{(0)}_{n,1} = \frac{1}{\sqrt{2}}(\varphi_n, \varphi_n, 0); \qquad \vec{u}^{(0)}_{n,2} = \frac{1}{\sqrt{2}}(\varphi_n, -\varphi_n, 0); \tag{8.9}$$

$$\lambda^{(0)}_n = 2K_1 + C[(\pi n_x/L_x)^2 + (\pi n_y/L_x)^2 + (\pi n_z/L_z)^2],$$

where $n = (n_x, n_y, n_z)$ is the combined subscript and $n_i = 0, 1, \ldots$. The functions

$$\varphi_n = \left[\frac{(2-\delta_{n_x,0})(2-\delta_{n_y,0})(2-\delta_{n_z,0})}{L_x L_x L_z}\right]^{1/2} \cos\frac{\pi n_x x}{L_x} \cos\frac{\pi n_y y}{L_x} \cos\frac{\pi n_z z}{L_z}$$

are normalized to unity and satisfied proper boundary conditions at the surface of the parallelepiped. Then, using Eq. (8.5) one arrives at the conclusion that the components of the first-order correction $\vec{\alpha}^{(1)} = (\alpha_x^{(1)}, \alpha_y^{(1)}, 0)$ for the parallelepiped are given by

$$\alpha_i^{(1)} = M_s \sum_n \frac{\langle \varphi_n | H_i'^{(1)} \rangle}{\lambda_n^{(0)} - \lambda^*} \varphi_n; \quad (i = x, y), \tag{8.10}$$

where $\vec{H}'^{(1)}(\vec{r}) = (H_x'^{(1)}, H_y'^{(1)}, H_z'^{(1)})$ is the perturbation of the demagnetizing field in the parallelepiped. Note that the component $H_z'^{(1)}$ does not contribute to the first-order correction (8.10). In can be shown that the demagnetizing field component $H_x'^{(1)}$ is an odd function of variables $x - L_x/2$ and $z - L_z/2$, but it is an even function of the variable $y - L_x/2$. Thus, for the $\alpha_x^{(1)}$ component the summation in (8.10) extends over the integers $n = (2k_x + 1, 2k_y, 2k_z + 1)$; $k_i = 0, 1, \ldots$ Similarly, for the $\alpha_y^{(1)}$ component one can prove that the summation in (8.10) extends over the integers $n = (2k_x, 2k_y + 1, 2k_z + 1)$. Since the eigenvalues $\lambda_n^{(0)}$ grow rapidly as a function of n, for a small parallelepiped with aspect ratio $L_x/L_z \sim 1$ the spatial variation of the first-order correction is determined mainly by the terms of (8.10) with the smallest possible n_i

$$\alpha_x^{(1)} \sim \cos\frac{\pi x}{L_x} \cos\frac{\pi z}{L_z}; \quad \alpha_y^{(1)} \sim \cos\frac{\pi y}{L_x} \cos\frac{\pi z}{L_z}. \tag{8.11}$$

Finally, let us consider a flat cylinder, $L_z/R < 1$, of a soft magnetic type. Then, due to influence of a shape anisotropy, the unit magnetization vector $\vec{\alpha}^{(0)}$, i.e., the solution of Eqs. (8.1) and (8.2), points perpendicular to the cylinder axis. Suppose that the easy anisotropy axis is parallel to the x axis so that in the lowest approximation $\vec{\alpha}^{(0)} = (1, 0, 0)$. Let us superimpose the origins of the Cartesian and the cylindrical coordinates setting $x = \rho \cos\varphi$, $y = \rho \sin\varphi$. Since $\vec{\alpha}^{(0)}$ is parallel to the end faces of the cylinder, there is a magnetic charge distributed at the lateral surface of the cylinder with the density $\sigma = M_s \cos\varphi$. It can be shown that the corresponding demagnetizing field perturbation has the components $H_y'^{(1)}(\vec{r}) \sim \sin 2\varphi$ and $H_z'^{(1)}(\vec{r}) \sim \cos\varphi$; the component $H_x'(\vec{r})$ does not contribute to the first-order deviation $\vec{\alpha}^{(1)} = (0, \alpha_y^{(1)}, \alpha_z^{(1)})$. The components $\alpha_y^{(1)}$ and $\alpha_z^{(1)}$ can be determined as a series (8.5) with respect to the eigenfunctions $\vec{u}_n = (0, u_{ny}, u_{nz})$ of the corresponding Brown's boundary value problem. In the exchange approximation, the eigenvalue

$$\lambda_n^{(0)} = \lambda_{s,m,n_z}^{(0)} = 2K_1 + C\left[(\mu_{m,s}/R)^2 + (\pi n_z/L_z)^2\right], \tag{8.12}$$

where $n = (s, m, n_z)$ is a combined subscript, turns out to be four times degenerate. The proper set of eigenfunctions $u_n^{(0)}$ is given by [34]

$$(0, \chi_n, 0); \quad (0, \psi_n, 0); \quad (0, 0, \chi_n); \quad (0, 0, \psi_n). \quad (8.13)$$

Here

$$\chi_n = N_{s,m,n_z} J_m(\mu_{m,s}\rho/R) \cos \frac{\pi n_z z}{L_z} \sin m\varphi;$$

$$\psi_n = N_{s,m,n_z} J_m(\mu_{m,s}\rho/R) \cos \frac{\pi n_z z}{L_z} \cos m\varphi, \quad (8.14)$$

where N_{s,m,n_z} is the normalization constant. In Eqs. (8.12)–(8.14) the integers m and n_z are $0, 1, \ldots$ The integer $s = 1, 2, \ldots$ numbers the sequential roots of the equations $J'_m(\mu_{m,s}) = 0$.

With the demagnetizing field component $H_y'^{(1)} \sim \sin 2\varphi$, one obtains that the eigenfunctions $(0, \chi_n, 0)$ with $m = 2$ are the only ones that may contribute to the component $\alpha_y^{(1)}$. Similarly, because $H_z'^{(1)} \sim \cos \varphi$, the eigenfunctions $(0, 0, \psi_n)$ with $m = 1$ are the only ones that may contribute to the component $\alpha_z^{(1)}$. It can be shown that the components $H_y'^{(1)}$ and $H_z'^{(1)}$ are even and odd functions of the variable $z - L_z/2$, respectively. Thus, the summation with respect to n_z extends over $n_z = 2k$, ($k = 0, 1, \ldots$) for the component $\alpha_y^{(1)}$ and over $n_z = 2k + 1$ in the case of the component $\alpha_z^{(1)}$. For sufficiently small particle, the first-order correction is mainly determined by the terms (8.5) with the smallest possible integers s and n_z, so that

$$\alpha_y^{(1)} \sim J_2(\mu_{2,1}\rho/R) \sin 2\varphi; \quad \alpha_z^{(1)} \sim J_1(\mu_{1,1}\rho/R) \cos \varphi \cos(\pi z/L_z),$$

(8.15)

where $\mu_{2,1} \approx 3.05$.

The flower state in a small cubic particle was first discovered by Bertram and Schabes [24, 28] by means of 3D numerical simulation. The perturbation theory for flower states in the ferromagnetic elements of perfect geometrical shapes was developed in [33, 34]. The spatial variation of the magnetization deviations (8.8), (8.11), and (8.15) for small ferromagnetic elements with moderate aspect ratio was also confirmed by means of comparison with the numerical simulation data. The magnetization distributions, $\vec{\alpha} = \vec{\alpha}^{(0)} + \vec{\alpha}^{(1)}$, corresponding to Eqs. (8.8), (8.11), and (8.15), are shown in Figure 8.1. Note that in Figure 8.1 the amplitude of the magnetization deviation $\vec{\alpha}^{(1)}$ is increased considerably for the sake of clarity.

8.2.1.2 Particle of Quasiellipsoidal Shape

Consider now a small ferromagnetic particle with a shape close to the ellipsoidal one. Using Eq. (8.5), one can estimate the amplitude of the magnetization perturbation arising within the particle volume due to the shape deviations. To do this, one has to construct first a proper perturbation operator. Let us

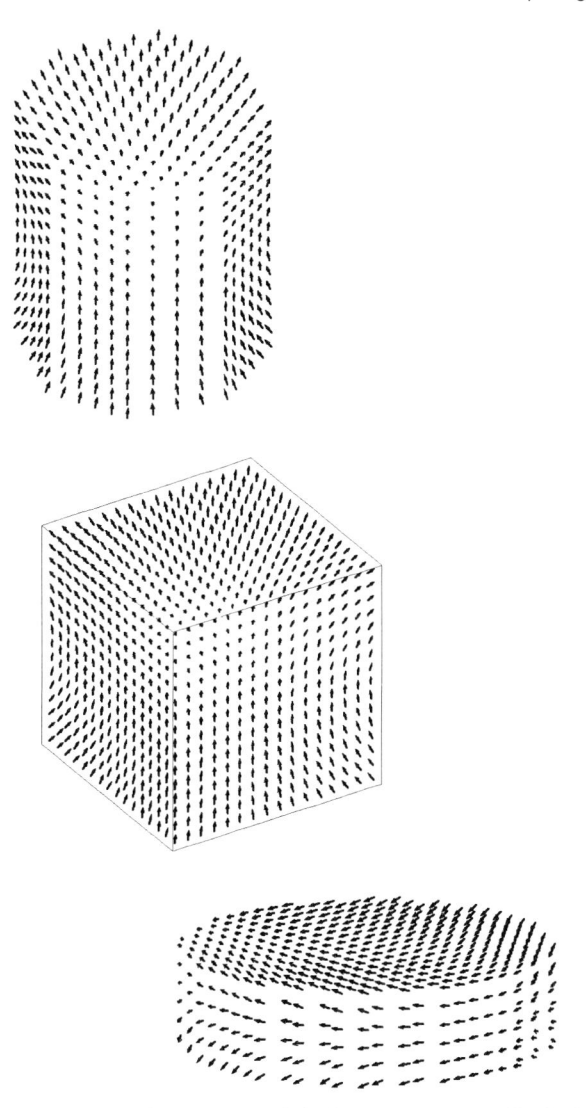

Figure 8.1 Schematic view of the flower states in small cylindrical and cubic particles (see Eqs. (8.8), (8.11), and (8.15)).

encapsulate the particle within an approximate ellipsoid of a smallest possible volume, as shown in Figure 8.2. If we fill the shaded areas in Figure 8.2 by a ferromagnetic material with the same material parameters as that of the particle, a certain stable uniform magnetization state, $\vec{\alpha}^{(0)} = $ const, will exist within the approximate ellipsoid of sufficiently small size. The same uniform magnetization state exists also within the volume of the particle. To construct the perturbation operator and return to the initial particle, it is sufficient to

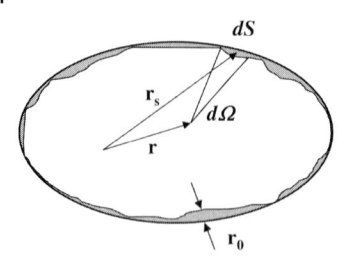

Figure 8.2 Approximate ellipsoid for a small ferromagnetic particle with irregular shape deviations.

fill again the shaded areas in Figure 8.2 with the uniform magnetization, $-M_s\vec{\alpha}^{(0)}$. Evidently, the thin subsidiary layer at the particle surface having the magnetization $\vec{M} = -M_s\vec{\alpha}^{(0)}$ will create within the particle volume a nonuniform perturbation of the demagnetizing field $\delta\vec{H}'(\vec{r})$. The later can be calculated by means of Eq. (8.1), where the integration is now over the volume of the subsidiary layer. The first-order correction $\vec{\alpha}^{(1)}$ to the uniform magnetization $\vec{\alpha}^{(0)}$ can then be calculated by means of Eq. (8.5), where $\vec{u}_n$ and λ_n are the eigenvectors and eigenvalues, respectively, of the Brown's boundary value problem for the nonideal particle. It can be proved [37], however, that if the characteristic amplitude r_0 of the surface deviation is small with respect to the average particle size, $r_0 \ll R$, in the lowest approximation one can use in the series (8.5) the eigenvectors $\vec{u}_n^{(0)}$ and eigenvalues $\lambda_n^{(0)}$ of the Brown's boundary value problem obtained for the approximate ellipsoid.

Of course, for a given particle different approximate ellipsoids can be chosen, but for various ellipsoids the vectors $\vec{\alpha}^{(0)}$ and the corresponding operators $\delta\vec{H}'(\vec{r})$ will be different so that the first-order correction (8.5) remains invariant with respect to the particular ellipsoid chosen. The better choice is to construct the approximate ellipsoid with zero average value of the perturbation operator over the particle volume, $\langle\delta\vec{H}'(\vec{r})\rangle = 0$. In this case, the uniform rotation mode, $\vec{u} = \text{const}$, will not contribute to the series (8.5).

For a particle with small surface deviations, $r_0 \ll R$, one can estimate the characteristic amplitude of the perturbation $\delta\vec{H}'$ using the well-known expression [77] for the potential U of a dipole layer of strength $\tau(\vec{r})$, located at the surface S of the approximate ellipsoid

$$U(\vec{r}) = \int_S \tau(\vec{r}_s) d\Omega. \tag{8.16}$$

Here $d\Omega$ is a solid angle corresponding to the surface element dS (see Figure 8.2). The solid angle has the origin at the point $\vec{r}$ located within the particle volume. Equation (8.16) is valid within the particle volume at the distances from the surface large enough with respect to the amplitude of the surface deviation, $|\vec{r} - \vec{r}_s| > r_0$. One can estimate the strength of the dipole layer as $\tau \approx r_0 M_s$. Then for the characteristic value of the demagnetizing field perturbation, $\delta\vec{H}' = -\vec{\nabla}U$, one obtains

$$|\delta\vec{H}'| \sim 4\pi M_s r_0 l^2/R^3, \tag{8.17}$$

where l is the characteristic length of the perturbation of the particle surface. Therefore, the longest perturbations of the surface with the length $l \sim R$ make the largest contributions to the perturbation of the particle magnetization.

Now to estimate the amplitude of the magnetization perturbation due to correction to the particle demagnetizing field (8.17), let us consider quasispherical particle with averaged radius R of the order of exchange length, $R \sim L_{ex}$. It can be proved [37] that in the exchange approximation, the eigenvalues of Brown's boundary value problem for spherical particle are given by

$$\lambda_{is}^{(0)} = 2K_1 + \mu_{is}^2 \frac{C}{R^2}. \tag{8.18}$$

Here μ_{is} are the successive roots of the derivatives of the spherical Bessel functions, $j_i'(\mu_{is}) = 0$, $i = 0, 1, \ldots$. The integers $s = 1, 2, \ldots$ number the roots of the ith equation. The values μ_{is}^2 grow rapidly as functions of i and s, because for $i, s \gg 1$ one has $\mu_{is} \to \pi(s+i/2)$. Therefore, the basic contributions to the first-order magnetization perturbation within a quasispherical particle make the modes (8.18) with $i = 0, 1$ and $s = 1$. Actually, the values $\mu_{01}^2 = 0$, $\mu_{11}^2 = 4.326$, whereas a minimal value of μ_{is}^2 for $s > 1$ is given by $\mu_{02}^2 > 20$. The value $\mu_{01} = 0$ corresponds to the uniform rotation mode. Its contribution does not lead to the nonuniform perturbation of the particle magnetization. As we discussed above, one can take into account the contribution of this mode choosing properly the initial magnetization state $\vec{\alpha}^{(0)}$. Then, taking into account the contribution of the mode with c $i = s = 1$, one obtains from (8.5), (8.17) and (8.18) the estimation

$$|\vec{\alpha}^{(1)}| \sim \frac{M_s |\delta H'|}{\lambda_{11}^{(0)}} \sim \frac{4\pi}{\mu_{11}^2} \frac{r_0 l^2}{R L_{ex}^2}, \tag{8.19}$$

because in the exchange approximation one can neglect the parameter $\lambda^* \sim M_s^2$ in the denominators of the series (8.5). According to Eq. (8.19), the perturbation of the uniform magnetization within the quasispherical particle is small for particles with the reduced surface deviation $r_0/R \ll 1$.

8.2.2
Nonuniform States

The study of stable nonuniform micromagnetic configurations in various types of small ferromagnetic particles is one of the basic problems of Micromagnetics [2, 3]. These investigations are also important for technical applications at least for two reasons. First, to determine a single-domain radius of a small particle, it is necessary to compare a total energy of a uniform magnetization state with that of the lowest stable nonuniform micromagnetic configuration. Second, in a real particle assembly, a distribution of particle sizes occurs so that there is a certain fraction of particles that exceed the critical size for single-domain behavior. In this section, we discuss the equilibrium micromagnetic states both in ellipsoidal and nonellipsoidal particles and their evolution as function of the particle size and the value of the phenomenological magnetic parameters.

8.2.2.1 Spherical Particle

For a spherical particle, nonuniform magnetization configurations were considered in the early papers that dealt with the estimation of the single-domain radius for a fine ferromagnetic sphere. Kittel [8] compared the energy of a uniform magnetization with that of a sphere containing two domains with opposite magnetization. That calculation gave a certain upper bound for the single-domain radius of a sphere. It can be close to the single-domain radius for magnetically hard particle, but for a magnetically soft particle the Kittel's estimation is certainly far from the exact value. To get a proper estimation of the single-domain radius for a magnetically soft sphere, Kondorsky [11, 81] developed a variational procedure. The Ritz function of Kondorsky turned out to be close to the exact solution, the so-called magnetization curling mode, of Brown's boundary value problem [2, 16]. The latter is the easiest nucleation mode that destroys the uniform magnetization in a spherical particle with $R > a_c$. Later Brown [12, 13] gave the exact upper and lower bounds for the single-domain radius of a spherical particle. These estimations are especially useful for a magnetically soft sphere. Brown's upper and lower estimations were extended by Aharoni [82] to the case of ellipsoidal particle.

For a magnetically hard spherical particle with uniaxial anisotropy, additional results were obtained [38, 39] by means of numerical simulation. Stapper [38] computed a one-dimensional magnetization distribution in a Co sphere subdivided into a large number of thin parallel slices. He showed that the energy of the two-domain state becomes lower than that of the saturated state if the radius of the sphere exceeds some critical value. Aharoni and Jakubovics [39] carried out the two-dimensional calculation of the magnetization distribution in the sphere with the same material parameters under the restriction of cylindrical symmetry. They found that for a cobalt particle, the cylindrically symmetric configuration with two domains separated by the cylindrical domain wall has lower energy than the one of Stapper.

In a more recent three-dimensional (3D) numerical simulations [40], the nonuniform magnetization configurations were studied systematically in spherical particle with the same value of the saturation magnetization, $M_s = 550$ emu/cm^3, but with various values of the uniaxial anisotropy constant: (1) $K_1 = 10^4$ erg/cm^3, (2) $K_1 = 10^5$ erg/cm^3, (3) $K_1 = 5 \times 10^5$ erg/cm^3, and (4) $K_1 = 10^6$ erg/cm^3. This enables us to investigate the evolution of the equilibrium magnetization patterns in spherical particle as a function of the parameter $p = NM_s^2/2K_1$ ($N = 4\pi/3$ is the demagnetizing factor of sphere) in a sufficiently large interval, $0.634 \leq p \leq 63.4$. This parameter characterizes a magnetic softness of a spherical particle. The exchange constant was chosen to be $C = 2 \times 10^{-6}$ erg/cm for all of the cases considered.

In the case $K_1 = 10^4 – 10^5$ erg/cm^3, the Kondorskii–Brown configuration with the axis of the vortex parallel to the easy anisotropy axis was found stable within a certain interval of sizes just above the single-domain radius. However, at a certain critical radius the longitudinal vortex becomes unstable with respect to the rotation of the vortex axis into the plane perpendicular to the easy anisotropy axis. This new state can be termed as a transverse vortex. For larger values of the anisotropy constant, $K_1 = 5 \times 10^5$ erg/cm^3 and $K_1 = 10^6$ erg/cm^3, the longitudinal vortex vanishes, whereas transverse vortex experiences conspicuous elliptic deformation.

Figure 8.3 shows the total energies of various micromagnetic states existing in spherical particles with different values of the uniaxial anisotropy constant as a function of the particle radius. Numerically, the single-domain radii a_c of the particles studied can be determined by means of the intersections of the curves 1–4 in Figure 8.3 with the horizontal line representing the total energy of the uniform magnetization. The latter is very close to the exact value of the magnetostatic energy for the uniformly magnetized sphere, $\varepsilon_0 = 2\pi M_s^2/3 = 6.334 \times 10^5$ erg/cm^3. For magnetically soft particles ($p \gg 1$), the single-domain radii obtained numerically were found to be in reasonable agreement with the exact theoretical bounds [2, 12], whereas for the cases $K_1 = 5 \times 10^5$ erg/cm^3 and $K_1 = 10^6$ erg/cm^3 the gaps between the lower and upper theoretical estimations are too large to determine the single-domain radius with sufficient accuracy.

According to Figure 8.3 there is a striking difference between the cases of magnetically soft, $p \gg 1$ (curves 1 and 2 in Fig. 8.3), and intermediate type,

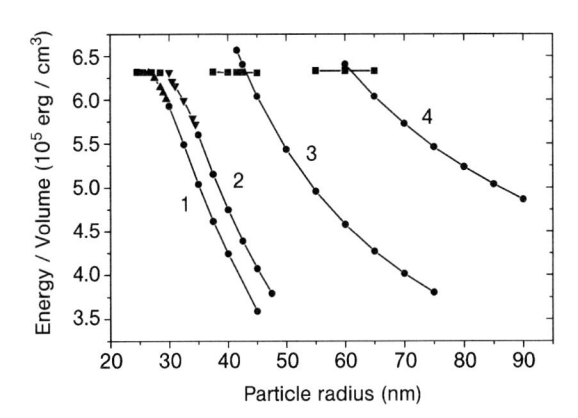

Figure 8.3 Total energy of stable micromagnetic states as a function of radius for spherical particle with different uniaxial anisotropy constants: (1) $K_1 = 10^4$ erg/cm^3; (2) $K_1 = 10^5$ erg/cm^3; (3) $K_1 = 5 \times 10^5$ erg/cm^3; (4) $K_1 = 10^6$ erg/cm^3. Cubes represent uniform magnetization, and triangles and filled circles correspond to longitudinal and transverse vortexes, respectively.

$p \sim 1$, particles. As we mentioned above, for magnetically soft particles there is a longitudinal vortex within certain intervals of sizes close to the single-domain radius. The overall view of the longitudinal vortex in the particle with radius $R = 27.5$ nm, sufficiently close to the single-domain radius, $a_c = 26.5$ nm, is shown in Figure 8.4(a). In the cylindrical coordinates (r, φ, z) the longitudinal vortex in a spherical particle can be fairly well described by means of Kondorskii–Brown's curling mode [2, 11] of a finite amplitude A

$$\alpha_\rho = 0, \qquad \alpha_\varphi = A j_1\left(\gamma_{11}\frac{r}{R}\right) P_1^1(\cos\theta), \qquad \alpha_z = \sqrt{1 - \alpha_\varphi^2}, \qquad (8.20)$$

where $j_1(r)$ is the spherical Bessel function, $P_1^1(x)$ is the associated Legendre polynomial, A is the variational parameter of the model, and $\gamma_{11} = 2.08$ being the minimal root of the equation $j_1''(\gamma) = 0$. The components of the unit magnetization vector in Eq. (8.20) are written in the spherical coordinates, (r, θ, φ). Note that the axis of the vortex in Figure 8.4(a) is parallel to the easy anisotropy axis. However, at certain critical particle radius, the longitudinal vortex becomes unstable with respect to the rotation of the vortex axis into the plane perpendicular to the easy anisotropy axis. As a result of this instability, a transverse vortex appears within the particle. This state is shown in Figure 8.4(b) for the particle with radius $R = 40$ nm. The transverse vortex turns out to be stable up to the largest radii investigated. The total energy of the transverse vortex as a function of the particle radius is shown by filled circles at the curves 1 and 2 in Figure 8.3.

In contrast to the Kondorskii–Brown's configuration, the transverse vortex has only weak dependence of the magnetization distribution on the coordinate parallel to the vortex axis. For the transverse vortex, the plane perpendicular to the easy axis becomes an easy anisotropy plane. Thus, the axis of the vortex may have arbitrary direction within this plane.

In the case of particles with $p \sim 1$ (see curves 3 and 4 in Figure 8.3), the longitudinal vortex is absent, whereas the transverse vortex remains stable even close to the single-domain radius. Furthermore, the transverse vortex retains its stability in a certain interval below the single-domain radius, $R < a_c$, where its total energy exceeds that of the uniform magnetization. Another difference in the case of soft magnetic particles is that the total magnetic moment of the particle being in the transverse vortex turns out to be very small even close to the single-domain radius. As we mentioned above, for particles with $p \sim 1$, transverse vortex experiences appreciable elliptic deformation so that the vortex extends along the easy anisotropy axis and contracts in perpendicular direction. Due to this deformation, the anisotropy energy contribution to the total particle energy decreases. For a particle with large enough radius, the elliptic deformation leads to gradual transformation of the transverse vortex into a nearly flat 180° domain-wall separating two domains with opposite magnetization. Therefore, the Stapper's two-domain configuration [38] develops as a result of gradual

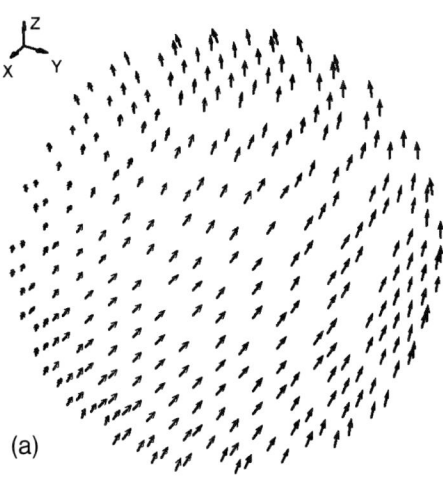

Figure 8.4 Nonuniform magnetization distributions in spherical particles with different radius: (a) Kondorskii–Brown magnetization curling state in the particle with $R = 27.5$ nm, (b) transverse vortex state for the particle with $R = 40$ nm. Saturation magnetization $M_s = 550$ emu/cm^3, uniaxial anisotropy constant $K_1 = 10^4$ erg/cm^3, and the easy anisotropy axis is parallel to the z axis.

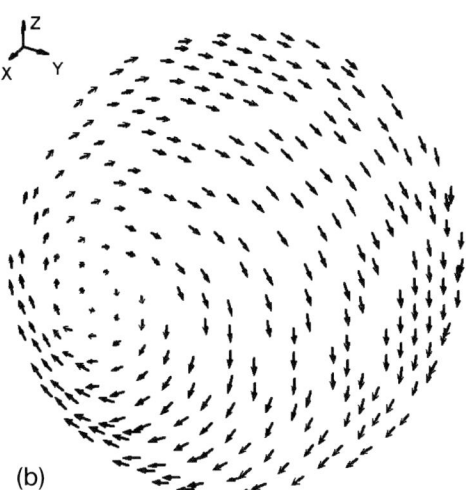

transformation of the transverse vortex with the increase of the particle radius and the anisotropy constant. On the other hand, a stable equilibrium configuration similar to the cylindrical magnetic domain discussed by Aharoni and Jakubovich [39] turned out to be absent even for the case $K_1 = 10^6$ erg/cm^3.

8.2.2.2 Ellipsoidal Particle

The nonuniform magnetization configurations for spheroidal ellipsoids with moderate aspect ratio in the range of $1.5 \leq L_z/D \leq 3$ were studied in Ref. [41]. Here L_z and D are the particle length and the transverse particle diameter, respectively. To compare the data with the case of spherical particle discussed above, the same material parameters, $M_s = 550$ emu/cm^3, $C = 2 \times 10^{-6}$ erg/cm, and two different values of uniaxial anisotropy constant, $K_1 = 10^4$ erg/cm^3, and $K_1 = 10^5$ erg/cm^3, were used in the numerical simulations. To preserve the axial symmetry of the particle, the easy anisotropy axis was taken to be parallel to the largest particle axis. The particle diameters studied were within the range of 60 nm $< D <$ 180 nm.

In the case of ellipsoids with aspect ratio $L_z/D = 1.5$, the situation was found to be similar to that of soft type ($p \gg 1$) spherical particle for both the cases considered, $K_1 = 10^4$ erg/cm^3 and $K_1 = 10^5$ erg/cm^3. Just above the single-domain radius, the longitudinal vortex directed along the long particle axis appears. With increase of the particle diameter, the total particle magnetization decreases rapidly. At certain critical diameter, the longitudinal vortex becomes unstable so that the transverse vortex originates within the particle. The transverse vortex is always elliptically deformed in the plane, perpendicular to the vortex axis even for a soft elliptical particle, because the deformation decreases the density of surface magnetic charge arising at the particle surface. The transverse vortex is in fact a precursor of a two-domain state existing in a spheroidal ellipsoid of sufficiently large size.

For aspect ratios of $L_z/D = 2$ and $L_z/D = 3$, the situation is surprisingly different. After the instability of the longitudinal vortex, the transverse vortex state directed along a short particle diameter does not appear. Instead, there is a transition from the longitudinal vortex to one tilted at an angle to the easy anisotropy axis. One can suppose that the tilted vortex appears in spheroidal ellipsoid with $L_z/D > 2$ as a result of a compromise between the exchange and magnetostatic energy contributions to the total particle energy.

The case of asymmetrical ellipsoidal particles with the same magnetic parameters, but with various aspect ratios, (1) $L_x : L_y : L_z = 2:1:2$; (2) $L_x : L_y : L_z = 2:1:3$; and (3) $L_x : L_y : L_z = 2:1:4$ was studied in Ref. [42]. The uniaxial anisotropy axis was supposed to be parallel either to the longest particle axis (z-axis) or the shortest particle axis (y-axis). The particle sizes studied were within the range of 120–440 nm. For asymmetrical ellipsoidal particles, the transverse and the longitudinal vortexes were obtained for all aspect ratios studied. The tilted vortex observed for spheroidal particles with aspect ratio $L_z/D \geq 2$ is absent for asymmetrical particle. To see clear a difference between spheroidal and asymmetrical ellipsoidal particles, one has to take into account that the longitudinal vortex competes in energy with the uniform magnetization near the single-domain size of spheroidal particle. Besides, the tilted vortex develops in the spheroidal particle at sizes much larger than the single-domain size. In fact, it originates as a result of the longitudinal vortex instability. In contrast to the case of spheroidal particle, the single-domain radius of asymmetrical ellipsoidal particle is determined by the transverse curling state, the axis of the vortex being parallel to the shortest particle's axis. Another difference is that for asymmetrical ellipsoidal particle, the uniform magnetization turns out to be stable far above the single-domain size. It transforms gradually to the longitudinal vortex with axis of the vortex parallel to the longest particle axis for large-enough particle sizes only.

8.2.2.3 Cylindrical Particle

Let us now consider the situation for a particle of a nonellipsoidal external shape, for example, for a fine cylindrical particle. As we discussed in Section 8.2.1, in a cylindrical particle of sufficiently small size the flower state has the lowest total energy. Its average magnetization is either parallel to the cylinder axis or to the end faces of the cylinder, depending on the particle aspect ratio and the easy axis direction. However, a magnetization curling with the axis of the vortex parallel to the cylinder axis is the lowest energy state for soft-type cylindrical particle of large-enough sizes. The magnetization curling in a thin soft-type cylindrical particle with aspect ratio $L_z/R \leq 1$ was studied in detail by means of numerical simulation, [43, 44]. An overall view of the magnetization curling state in a flat cylindrical particle is presented in Figure 8.5(a). It was shown [43] that the vortex structure is rather simple in a thin cylindrical particle with thickness $L_z \sim L_{ex}$. For such a particle in the cylindrical coordinates (r, φ, z), the radial component, α_ρ, of the unit magnetization vector is small as compared with the axial, α_z, or azimuthal, α_φ, components. In addition, for a particle of moderate aspect ratio there is only a weak dependence of the unit magnetization vector on the z-coordinate parallel to the particle axis. Therefore, in cylindrical coordinates, the components of the unit magnetization vector for the vortex state can be approximated as follows [43]:

$$\alpha_r = 0; \quad \alpha_\varphi = f(r); \quad \alpha_z = \pm\sqrt{1 - f^2(r)}. \tag{8.21a}$$

The Ritz function $f(r)$ is given by

$$f(r) = \begin{cases} 2br/(b^2 + r^2), & 0 \leq r \leq b \\ 1, & b < r \leq R \end{cases} \tag{8.21b}$$

where b is the radius of the vortex core. In accordance with Eq. (8.21), within the core region, $r \leq b$, the z-component of the particle magnetization reduces gradually with increasing radial coordinate and vanishes at $r > b$. The value of b can be considered as a variational parameter of the given model. It can be obtained by means of a minimization of the total particle energy [43, 44]. Note that the same magnetization distribution, Eq. (8.21), also describes a transverse vortex in a spherical particle of soft magnetic type with a reasonable accuracy.

Qualitatively different magnetization pattern appears only in cylindrical particle with large aspect ratio, $L_z/D \gg 1$ [45]. As Figure 8.5 shows, in this case a middle part of the cylindrical particle turns out to be uniformly magnetized along the particle axis, whereas the vortices of various chirality exist near to the particle ends.

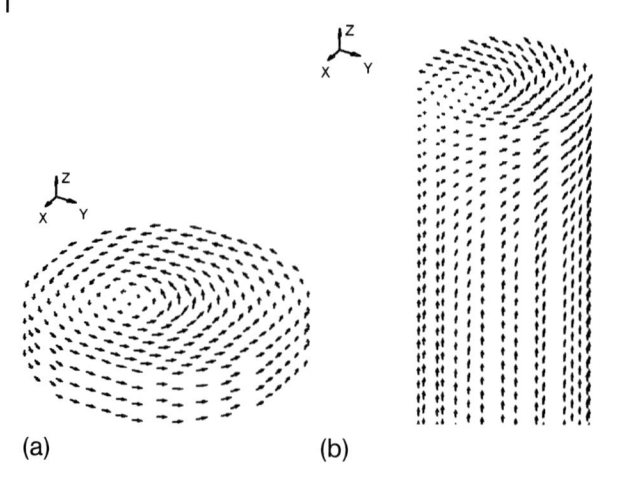

Figure 8.5 Magnetization curling states in Permalloy cylindrical particles ($M_s = 800$ emu/cm^3, $K_1 = 10^3$ erg/cm^3) with various aspect ratios: (a) flat particle with diameter $D = 48$ nm and aspect ratio $L_z/D = 0.25$; (b) elongated particle with $D = 80$ nm and $L_z/D = 5.0$ (only upper half of the particle is shown).

At present, it is well understood that the behavior of the magnetization vortices in thin ferromagnetic elements of soft magnetic type is responsible for their static and dynamic properties. The presence of the vortices in the remnant state of soft magnetic elements was confirmed recently by means of transmission electron microscopy [48, 50, 51], as well as in magnetic-force microscopy investigations [52–54]. The influence of the vortices on the magnetization reversal was observed in thin ferromagnetic elements of circular [49, 56], elliptical [57], or rectangular shape [69, 70]. The structure of the vortex core was investigated by means of spin-polarized scanning tunneling microscopy [55]. Also, the process of vortex nucleation and annihilation under the influence of in-plane applied magnetic field was studied both theoretically and experimentally for submicron cylindrical particles of different thickness and aspect ratio [46, 47, 58, 59]. Interestingly, new nonuniform bending states, the so-called *C* and *S* configurations, were discovered [46, 59] for particle sizes close to the effective single-domain radius (see Section 8.2.3). Evidently, these investigations provide important information about the transformations between various stable micromagnetic states of a cylindrical particle.

Recently [60], detailed numerical simulations were carried out for flat Permalloy cylindrical particles ($M_s = 800$ emu/cm^3, $C = 2 \times 10^{-6}$ erg/cm, uniaxial anisotropy constant $K_1 = 10^3$ erg/cm^3) with different thickness $L_z = 12, 18, 24, 30$ and 36 nm, and for diameters $D \leq 210$ nm. In addition to the flower state, *C* and *S* configurations were revealed for particles of sufficiently large diameter (see also Ref. [61], where the micromagnetic configurations for *C* and *S* states in a flat cylinder are presented). With increasing particle diameter at fixed thickness, the C configuration first originates due to instability of the flower state at certain critical diameter D_{c1}, which depends on the particle thickness. In contrast to the flower state, the in-plane magnetization of the C state diminishes considerably as a function of particle diameter. In turn, the C configuration becomes unstable at another critical diameter D_{c2} due to the large curvature of the magnetization near the lateral side of the particle. This new instability leads to entering of the vortex inside the particle. Therefore, the bending states provide continuous transformation of the quasiuniform magnetization into the vortex pattern with increase in the particle diameter. It was found that the critical diameters D_{c1} and D_{c2} reduce gradually with the increase in the particle thickness. Interestingly, no stable state with appreciable in-plane magnetization was found in cylindrical Permalloy particle with thickness $L_z \geq 36$ nm.

8.2.2.4 Cubic Particle

The flower and vortex states in magnetically soft cubic particle with uniaxial anisotropy were first calculated in a seminal paper by Schabes and Bertram [24]. In Ref. [35], the lowest energy states in small cubic particles with uniaxial anisotropy were thoroughly investigated as a function of the anisotropy constant K_u and a cube size L. The regions for stability of flower, vortex, and double vortex states were determined and a phase diagram in the coordinates (Q, λ), where $Q = 2K_u/M_s^2$ is the reduced anisotropy strength and $\lambda = L/L_{ex}$ is the reduced particle size was constructed. For a soft magnetic particle, $Q \ll 1$, the following stable structures were found depending on the reduced particle size λ: (1) flower state, (2) longitudinal vortex with the axis of the vortex parallel to the easy anisotropy axis, (3) transverse vortex, (4) twisted vortex, whose axis is inclined to the cube sides, and (5) double vortex states of various configurations. For a case of magnetically hard particle, $Q > 1$, the transverse vortex is transformed into two-domain state with nearly flat domain wall, whereas double-vortex states are changed into three-domain states, respectively. It is interesting to note a similarity between stable magnetization states in spherical and cubic particles with comparable sizes and anisotropy strength (see Figure 8.3). Hertel and Kronmuller [36] investigated carefully a stability of the flowers state in a magnetically soft cubic particle. They showed that in this case, a continuous second-order transition exists between the flower and longitudinal vortex states, similar to the case of magnetically soft spherical particle described above (see Figure 8.3). In addition to the simple vortex state, Hertel and Kronmuller [36] also obtained a vortex state with a Bloch point singularity within the vortex core. C and S configurations probably do not exist in a cubic particle, whereas they compete in total energy with the flower state for a square platelet and a parallelepiped [66, 67, 71, 72].

8.2.3
Effective Single-Domain Radius

As we have seen above, the notion of a single-domain ferromagnetic particle can be introduced, strictly speaking, only for a particle of ideal ellipsoidal shape. As for a particle of nonideal shape, Brown assumed [13] that "the nonellipsoidal particle can be either in a state of nearly uniform magnetization, which we can define as a 'single-domain' state, or in various states of drastically nonuniform magnetization with nearly zero resultant moment, which we can define as a 'multidomain' state."

In agreement with Brown's idea, the results of numerical simulations [24, 28, 33–36, 72] show that the flower state is the lowest energy state for particles of different external shape at small-enough sizes. An average magnetization of a particle being in the flower state is very close to the saturation magnetization. On the other hand, as we discussed in Section 8.2.2, the magnetization curling, i.e., a nonuniform state with relatively small average magnetization, is usually the lowest energy state for soft-type particles of larger sizes. In Ref. [73], it was suggested to define an effective single-domain radius for a nonellipsoidal particle as a characteristic length at which a reduced energy of the flower state turns out to be equal to that of the corresponding lowest nonuniform state of the particle. Of course, the effective single-domain diameter of a nonellipsoidal particle may depend on the actual particle shape. In fact, the same notion of the effective single-domain radius has been used in Refs. [35, 36], but without special definition.

In Ref. [73], numerical simulations were made for soft-type parallelepipeds and cylinders with different aspect ratios, $L_z/D = 0.5, 1, 2, 3$, where L_z is the length of the particle and D (or L_x) is the transverse particle's diameter. It was shown that the effective single-domain radii for soft-type cube and cylinder with aspect ratio $L_z/D = 1$ are only $\approx 10\%$ lower than the single-domain radius of an ideal sphere with the same phenomenological magnetic parameters. On the other hand, the effective single-domain diameters of elongated parallelepiped and cylinder may be considerably lower than the single-domain diameter of ideal spheroid with the same aspect ratio due to the influence of the nonuniform demagnetizing field near the end faces of these particles.

8.3
Surface and Interface Effects

It was shown in Section 8.2.1 that a shape deviation of a single-domain particle from that of ideal ellipsoid can only slightly disturb its magnetization if the characteristic particle size L is sufficiently small, $M_s^2 L^2/C \ll 1$. Using similar arguments, one can prove that any kind of internal defects that can be described, for example, by a local perturbation of a particle anisotropy energy density, $w_a(\vec{\alpha}, \vec{r}) = w_a^{(0)}(\vec{\alpha}) + w_a^{(1)}(\vec{\alpha}, \vec{r})$, also makes only

8.3 Surface and Interface Effects

small influence on the particle magnetization distribution provided that $|w_a^{(1)}|L^2/C \ll 1$, where $|w_a^{(1)}|$ is the characteristic amplitude of the anisotropy energy perturbation.

In this section, we consider another source of perturbation of the magnetization in a single-domain particle related with surface anisotropy energy. The latter notion was introduced by Neel [74] to account for the breaking of the translation symmetry in the electronic structure close to the particle surface. For the Neel model, the local surface anisotropy energy has the form

$$w_{sN}(i) = \frac{1}{2} K_{sN} \sum_{j=1}^{z_i} (\vec{S}_i \vec{e}_{ij})^2, \qquad (8.22)$$

where K_{sN} is the surface anisotropy constant, z_i is the number of the nearest neighbors to the spin moment $\vec{S}_i$, and $\vec{e}_{ij}$ is the unit vector from the lattice site i to j. Note that for symmetrical lattices Eq. (8.22) is reduced to a constant for a site i located within the ferromagnetic body. But it depends on the direction of the vector $\vec{S}_i$ if the site i is close to the surface, where the number of the nearest neighbors is reduced.

One can see that Eq. (8.22) defines a *microscopic* quantity that cannot be explicitly used in Micromagnetics. That is why Brown [2] introduced a surface anisotropy energy density of the form

$$w_{sB} = \frac{1}{2} K_{sB} (\vec{\alpha} \vec{n})^2, \qquad (8.23)$$

where $\vec{n}$ is the unit vector of outward normal to the particle surface and K_{sB} is a phenomenological constant of a dimension erg/cm^2 that can be negative or positive depending on the experimental situation. It can be shown [83] that Eq. (8.22) reduces to Eq. (8.23) if the spin directions are nearly parallel to each other close to the surface, at the distances large enough with respect to the lattice constant. This follows from the fact that outward normal is the only preferable direction near the particle surface.

Later Aharoni [84] suggested another expression for the surface anisotropy energy density

$$w_{sA} = \frac{1}{2} K_{sA} (\vec{\alpha} \vec{n}_0)^2. \qquad (8.24)$$

In contrast to Eq. (8.23), the unit vector $\vec{n}_0$ has the same direction in all points at the particle surface. In particular, it can be parallel to one of the particle crystallographic directions. One may assume that Eq. (8.24) takes into account the fact that the spin-orbit interaction is modified near the surface due to reduced crystalline symmetry.

It is worth to be noted that in Micromagnetics, the surface anisotropy, being a surface energy contribution, can be taken into account by means of a proper boundary condition only [2]. Actually, by making a variation of the total particle energy augmented with a surface anisotropy term, one obtains the following boundary conditions [2, 84]:

$$C\frac{\partial \vec{\alpha}}{\partial n} = K_{sB}\vec{\alpha} \cdot \vec{n}[(\vec{\alpha} \cdot \vec{n})\vec{\alpha} - \vec{n}]; \qquad (8.25a)$$

$$C\frac{\partial \vec{\alpha}}{\partial n} = K_{sA}\vec{\alpha} \cdot \vec{n}_0[(\vec{\alpha} \cdot \vec{n}_0)\vec{\alpha} - \vec{n}_0], \qquad (8.25b)$$

for the cases of Eqs. (8.23) and (8.24), respectively.

One can see that uniform magnetization satisfies the boundary condition (8.25b) if the unit magnetization vector $\vec{\alpha} = \vec{n}_0$ or $\vec{\alpha}\vec{n}_0 = 0$. Therefore, Eq. (8.24) is easier to use in the micromagnetic calculations, because the uniform magnetization remains an exact energy state of a single-domain particle at least in the case when the vector $\vec{n}_0$ is parallel to one of the particle easy anisotropy axes. On the contrary, uniform magnetization does not satisfy the boundary condition (8.25a). This means that, strictly speaking, the uniform magnetization is not an eigenstate of a single-domain particle with a surface anisotropy energy density given by Eq. (8.23).

Physically it is evident, however, that the magnetization deviation from the uniform magnetization has to be small for a particle of sufficiently small size L. Actually, the characteristic value of the derivative in the left hand side of Eq. (8.25) can be estimated as $\delta\alpha/L$. Therefore, the perturbation of the particle magnetization is proportional to a small parameter

$$\delta\alpha \sim \frac{K_s L}{C} = \frac{K_s/L}{C/L^2} \ll 1, \qquad (8.26)$$

(here and further we set $K_s = K_{sA}$ or $K_s = K_{sB}$ depending on the situation considered).

8.3.1
Brown's Surface Anisotropy

Let us discuss the influence of the Brown's boundary condition (8.25a) on the magnetization distribution in a small ferromagnetic particle. As an illustrative example, consider first a magnetization pattern in a thin cylindrical particle with radius R and thickness L_z. Supposing that $L_z \ll R$, it is easy to see that the unit magnetization vector is parallel to the particle plane, $\vec{\alpha} = (\alpha_x, \alpha_y, 0)$, and its components satisfy the equilibrium micromagnetic equation

$$\alpha_x \Delta\alpha_y - \alpha_y \Delta\alpha_x = 0. \qquad (8.27)$$

For simplicity, we neglect here the volume anisotropy contribution, as well as the influence of nonuniform demagnetizing field of the cylinder (the latter effect has been considered in Section 2.1). Because the boundary condition (8.25a) is rotationally invariant, without the loss of generality one can construct the magnetization distribution in the particle as a series $\vec{\alpha} = \vec{\alpha}^{(0)} + \vec{\alpha}^{(1)} + \cdots$, where

$$\vec{\alpha}^{(0)} = (1, 0, 0); \qquad \vec{\alpha}^{(1)} = (0, \alpha_y^{(1)}, 0). \tag{8.28}$$

Note that Eq. (8.28) satisfies the usual normalization condition, $\vec{\alpha}^{(0)} \cdot \vec{\alpha}^{(1)} = 0$. It follows from Eqs. (8.25a), (8.26), and (8.27) that in the cylindrical coordinates (r, φ, z) the perturbation $\alpha_y^{(1)}$ satisfies the relations

$$\Delta \alpha_y^{(1)} = 0; \qquad C \left[\frac{\partial \alpha_y^{(1)}}{\partial r} \right]_{r=R} = -\frac{K_s}{2} \sin 2\varphi;$$

$$\left[\frac{\partial \alpha_y^{(1)}}{\partial z} \right]_{z=0} = \left[\frac{\partial \alpha_y^{(1)}}{\partial z} \right]_{z=L_z} = 0. \tag{8.29a}$$

Thus, Eq. (8.29a) has a solution

$$\alpha_y^{(1)} = -\frac{K_s R}{4C} \left(\frac{r}{R} \right)^2 \sin 2\varphi. \tag{8.29b}$$

One can see that the later is just proportional to the small parameter, Eq. (8.26). The magnetization distributions given by Eqs. (8.28) and (8.29) are shown in Figure 8.6 for positive and negative values of the surface anisotropy constant K_s. For the sake of clarity, the amplitude of the magnetization perturbation in Figure 8.6 is set to unity, $K_s R/C = 1$, though in reality it has to be small to validate the correctness of the solution (8.28) and (8.29).

The solutions (8.28) and (8.29) are rotationally invariant within the particle plane, because the initial direction of the uniform magnetization, $\vec{\alpha}^{(0)}$, can be chosen arbitrarily in this plane. Therefore, the energy of the particle is degenerate so that no in-plane magnetic anisotropy appears in the case considered. Similar solution can be obtained for a spherical particle (see also Ref. [85]). But again, due to spherical symmetry of the boundary condition (8.25), in continuous micromagnetic approach the energy of the particle remains degenerate with respect to arbitrary rotation of the unit magnetization vector.

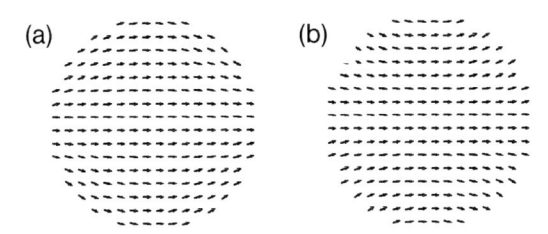

Figure 8.6 Magnetization distribution in a thin cylindrical particle for different signs of the surface anisotropy constant: (a) $K_s > 0$; (b) $K_s < 0$.

However, additional in-plane anisotropy may appear for a whole particle in a nonsymmetrical case, as a consequence of the boundary condition (8.25a). Consider, for example, the case of a rectangular particle with dimensions L_x and L_y and small thickness $L_z \ll L_x, L_y$. Again, for a soft magnetic particle the unit magnetization vector is parallel to the particle plane and satisfies Eq. (8.27). The boundary conditions for the α_x component at the lateral particle surface are given by

$$C\left[\frac{\partial \alpha_x}{\partial x}\right]_{x=L_x} = -K_{sx}\alpha_x\alpha_y^2; \quad C\left[\frac{\partial \alpha_x}{\partial x}\right]_{x=0} = K_{sx}\alpha_x\alpha_y^2; \quad (8.30a)$$

$$C\left[\frac{\partial \alpha_x}{\partial y}\right]_{y=L_y} = K_{sy}\alpha_x\alpha_y^2; \quad C\left[\frac{\partial \alpha_x}{\partial y}\right]_{y=0} = -K_{sy}\alpha_x\alpha_y^2.$$

Here we assume that the surface anisotropy constants have different values, K_{sx} and K_{sy}, for the lateral surfaces perpendicular to x and y axes, respectively. Similarly, for the α_y component

$$C\left[\frac{\partial \alpha_y}{\partial x}\right]_{x=L_x} = K_{sx}\alpha_x^2\alpha_y; \quad C\left[\frac{\partial \alpha_y}{\partial x}\right]_{x=0} = -K_{sx}\alpha_x^2\alpha_y; \quad (8.30b)$$

$$C\left[\frac{\partial \alpha_y}{\partial y}\right]_{y=L_y} = -K_{sy}\alpha_x^2\alpha_y; \quad C\left[\frac{\partial \alpha_y}{\partial y}\right]_{y=0} = K_{sy}\alpha_x^2\alpha_y.$$

On the other hand, $\partial \alpha_x/\partial z = \partial \alpha_y/\partial z = 0$ at the top and bottom surfaces of the particle, i.e., at $z = 0$ and $z = L_z$. It is easy to see that the uniform magnetization is a solution of Eqs. (8.27) and (8.30) when the unit magnetization vector is parallel to either x, $\vec{\alpha} = (1, 0, 0)$, or y, $\vec{\alpha} = (0, 10)$, axis. Then, according to Eq. (8.23), the corresponding total surface anisotropy energies of the particle equal $E_x = L_y L_z K_{sx}$ and $E_y = L_x L_z K_{sy}$, respectively, so that $E_x \ne E_y$, as a rule. This fact leads to an effective in-plane anisotropy of the whole particle.

To prove this statement, suppose for a moment that the unit magnetization vector made a certain angle φ with respect to the x axis, $\vec{\alpha} = (\cos \varphi, \sin \varphi, 0)$, then the total surface anisotropy energy of the particle would be

$$E(\varphi) = \text{const} + L_z(L_x K_{sy} - L_y K_{sx}) \sin^2 \varphi. \quad (8.31)$$

This is a well-known expression for the uniaxial anisotropy energy of the particle with the effective uniaxial anisotropy constant given by $K_{ef} = (K_{sy}/L_y - K_{sx}/L_x)$. Of course, the uniform magnetization is not a solution of Eqs. (8.27) and (8.30) at arbitrary angle $\varphi \ne 0, \pi/2$, etc. Nevertheless, it can be proved [86] that it is the effective anisotropy energy, Eq. (8.31), that determines the *average* direction of the particle magnetization in external in-plane magnetic field $\vec{H}_0 = H_0(\cos \omega, \sin \omega, 0)$, because the effective energy functional of the particle in external magnetic field can be represented as follows:

$$W/V = -M_s H_0 \cos(\varphi - \omega) + K_{ef} \sin^2 \varphi, \quad (8.32)$$

where $V = L_x L_y L_z$ is the particle volume. Actually, to obtain the magnetization distribution within the particle in external magnetic field $\vec{H}_0$, one has to determine first the average direction of the particle magnetization, i.e., the angle φ_0, minimizing the effective functional (8.32) at a given value of H_0 and ω. In this manner, one obtains the lowest approximation of the perturbation theory, $\vec{\alpha}^{(0)} = (\cos\varphi_0, \sin\varphi_0, 0)$. Then, similar to the derivation of Eq. (8.29), it can be shown that the nonuniform perturbation $\alpha^{(1)}$ to the particle magnetization is small, being proportional to the small parameter (8.26).

One can see that the average direction of the particle magnetization in external magnetic field can be determined by means of effective energy functional (8.32), similar to the usual case of a single-domain particle. The same is true for the Aharoni's type of the surface energy density (8.24). Besides, the stationary magnetization of a sufficiently small particle remains nearly uniform in external magnetic field applied at arbitrary direction to the effective easy anisotropy axis. Therefore, the existence of the surface anisotropy is compatible with the notion of the single–domain particle provided that the criterion (8.26) is fulfilled.

Of course, the influence of the boundary conditions (8.25) on the magnetization dynamics and magnetization reversal process of single-domain particle has to be investigated separately. This problem was partly studied in the papers [84, 87–93]. Aharoni showed that the surface anisotropy energy contribution, Eq. (8.24), makes an appreciable influence on the nucleation field of various nucleation modes of a sphere [84, 87], infinite cylinder [88], and prolate spheroid [89]. It also affects the exchange resonance modes of a small sphere with a surface anisotropy [90, 91]. It was shown [92, 93] that the boundary condition (8.25b) leads to a shift of the ferromagnetic resonance frequency of a spherical ferromagnetic particle, proportional to the value of the surface anisotropy constant, K_s. Similarly, the influence of a unidirectional surface anisotropy on the magnetization oscillations in a spherical particle was studied in Ref. [94].

8.3.2
Surface Spin Disorder

It is important to note that both magneto-dipole and spin–orbit interactions have relativistic nature [78]. Therefore, the characteristic anisotropy energy density, w_a, has to be small with respect to the characteristic energy of exchange interaction w_{ex}, because to the order of magnitude, $w_a \sim (v/c)^2 w_{ex}$, where c is the velocity of light and $v \ll c$ is the characteristic velocity of electrons in atoms. Recently it becomes customary [95–100] to study the properties of small ferromagnetic particles using *classical* Heisenberg–Hamiltonian model

$$H = -\sum_{\langle i,j \rangle} J_{ij} \vec{S}_i \vec{S}_j - k_V \sum_i (\vec{S}_i \vec{e}_0)^2 - k_s \sum_l (\vec{S}_l \vec{n}_l)^2. \tag{8.33}$$

Here J_{ij} are the exchange coupling constants between the classical spins $\vec{S}_i$ and $\vec{S}_j$ located in nearest neighbor lattice sites $\langle i, j \rangle$, $\vec{e}_0$ is the easy axis direction (we consider only uniaxial magnetic anisotropy for simplicity), k_V and k_s are the bulk and surface microscopic anisotropy constants, respectively. For the last term of Eq. (8.33), the summation is over the lattice sites belonging to the particle surface, and $\vec{n}_l$ is the unit vector perpendicular to the particle surface near the site l. More precisely [100], the unit vector $\vec{n}_l$ can be defined, for example, through the vectors $\vec{e}_{lj}$ of the Neel surface anisotropy model, Eq. (8.22)

$$\vec{n}_l = \sum_j \vec{e}_{lj} \Big/ \Big| \sum_j \vec{e}_{lj} \Big|,$$

where summation runs over the nearest neighbors of the surface lattice site l.

Note that the microscopic constants J_{ij}, k_V, and k_s in Eq. (8.33) have the dimension of energy. In most of the models published [95–99], the exchange coupling is assumed to be ferromagnetic, $J_{ij} = J > 0$. Then it is easy to see that the Hamiltonian (8.33) is equivalent to the energy functional used in Micromagnetics. Actually, the expression equivalent to Eq. (8.33) arises in the numerical simulation scheme based on the micromagnetic equations [28, 29] if one assumes very fine numerical cell size of the order of the lattice constant a. The direct mapping can be established through the relations [2, 83]

$$C = \xi \frac{JS^2}{a}; \quad K_V = \frac{k_V}{a^3}; \quad K_s = \frac{k_s}{a^2}, \tag{8.34}$$

where the number $\xi \sim 1$ depends on the type of the crystal structure, $C \approx 10^{-6}$ erg/cm is the macroscopic exchange constant, $K_V \sim 10^5-10^8$ erg/cm^3 is the macroscopic bulk anisotropy constant, and $K_s \sim 1$ erg/cm^2 is the macroscopic surface anisotropy constant [83, 101]. Assuming $S \sim 1$ and $a \sim 10^{-8}$ cm, one obtains $J \sim 10^{-14}$ erg, $k_V \sim 10^{-18}-10^{-16}$ erg, and $k_s \sim 10^{-16}$ erg. Thus, the ratio k_V/J is of the order of 10^{-2} even for a magnetic material with very large value of the bulk anisotropy constant, $K_V = 10^8$ erg/cm^3. It follows from the structure of the Hamiltonian (8.33) that the surface anisotropy energy is also a relativistic correction to the exchange energy contribution. Thus, there is no reason for the surface anisotropy constant k_s to be comparable with the exchange coupling constant J. Therefore, the 'hedgehog' and other complicated structures calculated in some papers for very large k_s values (see, for example, [96, 97]) hardly have physical meaning. On the other hand, slightly nonuniform magnetization patterns can be easily calculated under the condition (8.26) in the frame of Micromagnetics, as we demonstrated in Section 8.3.1. In this respect, the influence of surface anisotropy is similar to the other relativistic contribution, i.e., magneto-dipole interaction. As we have seen in Section 8.2.1, the magneto-dipole interaction can only lead to a small correction to the uniform magnetization of a sufficiently small ferromagnetic particle of nonellipsoidal shape.

Although under the constraint $J \gg k_V, k_s$, the magnetization of a sufficiently small ferromagnetic particle is nearly uniform, surface anisotropy can make appreciable contribution to the effective anisotropy energy of the particle if $k_s > k_V$ and the number of the surface spins N_s is not very small with respect to the total number of magnetic spins N_t. Therefore, the actual theoretical problem is to determine an effective anisotropy energy of a small ferromagnetic particle including the surface anisotropy energy contribution (see Eq. (8.32) and [86]), as well as the configurational anisotropy [24, 31, 64] related with the magneto-dipole interactions. Having in hand the effective energy functional of the type of Eq. (8.32), augmented with the configurational anisotropy term, one can use Stoner–Wohlfarth approach [14] or stochastic Langevin equation [23] to study the behavior of the particle in external magnetic field or at elevated temperatures. In fact, the same statement follows from the calculations based on classical Heisenberg Hamiltonian when the authors [85, 99] assume the realistic values of the ratio k_s/J.

On the other hand, for a very small particle with $N_s \sim N_t$, i.e., for the so-called magnetic cluster, there is no sense to separate volume and surface degrees of freedom. Generally speaking, the determination of the effective anisotropy constant of a magnetic cluster is a task for the first principle calculation. The latter has to take into account the quantum mechanical nature of the spin operators, the structural reconstruction of the particle, the interaction of mechanical and magnetic degrees of freedom, etc. In the recent calculation of the electronic structure of small Co clusters [102], the magnetic ground states in all clusters with mixed (bcc–fcc) and pure crystalline structure have been found to be fully polarized, the average magnetic moment per atom $\langle M \rangle$ being equal to $2\mu_B$. This was ascribed to the fact that the large exchange interaction J dominates independently of the assumed geometrical configuration of the cluster. Therefore, in a typical many-body calculation (see, for example Ref. [103]), the total cluster energy is usually determined as a function of the direction of the average cluster magnetization. In fact, it is found [103] that the net cluster anisotropy is a delicate balance between contributions from the interior and the surface of the cluster that generally have opposite signs.

Situation can be *qualitatively* different for the case when the exchange coupling constants may change sign near the surface of the particle due to strong surface disorder. Another example is the case of a small ferrite particle [104–107]. The latter have several magnetic sublattices, the superexchange interaction between various sublattices being antiferromagnetic. In this case one can expect the existence of a spin disorder near the particle surface, because the variations in coordination of surface cations may result in the distribution of positive and negative net exchange fields at the spins located close to the particle surface. As a result, small ferrite particles may show anomalous magnetic properties at low temperatures, such as reduced magnetization, open hysteresis loops and time-dependent magnetization in very large applied magnetic fields [106]. It is clear, however, that the phenomena observed [104–106] have different physical origin, because they are related with the changes in the exchange interaction between the ferromagnetic spins. Evidently, small relativistic corrections have no meaning in this case. Instead, the actual structure of the largest energy term, i.e., exchange interaction, has to be taken into account to describe the phenomenon of surface spin disorder correctly. For this purpose, it seems reasonable to study quantum mechanical Heisenberg Hamiltonian with properly defined exchange coupling constants. This approach is certainly beyond of the scope of Micromagnetics.

8.3.3
Interface Boundary Condition

Now consider the situation when there is a thin layer with different magnetic characteristics at the surface of a small ferromagnetic particle. First of all, in the frame of Micromagnetics one can consider only sufficiently thick layer with thickness d considerably larger than the interatomic distance. Otherwise, the influence of the layer on the particle behavior has to be described by means of a proper boundary condition, similar to Eq. (8.24). Let the surface layer is characterized by the exchange constant $\tilde{C}$, saturation magnetization $\tilde{M}_s$, and anisotropy constant $\tilde{K}_V$. The magnetization distribution within the layer can be represented by means of the unit magnetization vector $\tilde{\vec{\alpha}}$. At the boundary of the surface layer with a nonmagnetic media, one can use the same boundary conditions (8.24), that reduce to the ordinary one, $\partial \tilde{\vec{\alpha}}/\partial n = 0$, if the surface anisotropy constant is zero, $K_s = 0$. On the other hand, at the interface between the particle core and the surface layer the variational principle leads to the following boundary conditions

$$\tilde{\vec{\alpha}}(\vec{r}) = \vec{\alpha}(\vec{r}); \tag{8.35a}$$

$$\left[\vec{\alpha}(\vec{r}), \, C\frac{\partial \vec{\alpha}}{\partial n} - \tilde{C}\frac{\partial \tilde{\vec{\alpha}}}{\partial n}\right] = 0, \tag{8.35b}$$

where $\vec{n}$ is the unit vector of the outward normal to the interface. For simplicity, possible contribution of the surface anisotropy energy is omitted in Eq. (8.35). Also, a magneto-striction interaction does not take into account in Eq. (8.35), though it may be important near the interface due to difference in the mechanical characteristics of the core and shell layers.

For the first time, the influence of a thin surface layer on the behavior of small ferromagnetic particle was correctly considered in the papers [108–112]. Stavn and Morrish [108] studied the influence of a thin ferromagnetic layer at the surface of ellipsoidal particle on the particle hysteresis loop, but they considered only magnetostatic interaction between the core and the outer shell. The nucleation field of the magnetization curling mode in a composite spherical particle having thin surface layer with different phenomenological material parameters was investigated in Refs. [109, 110] as a function of the layer thickness d. It was found that the nucleation field of the curling mode decreases with decreasing layer's parameters $\tilde{C}$ and $\tilde{K}_V$ with respect to the corresponding core values. Under the condition $\tilde{K}_V/\tilde{M}_s \ll K_V/M_s$, the effect was found to be appreciable even for a small layer thickness, $d/R \ll 1$, where R is the particle radius. Similar investigations were carried out by Aharoni [111, 112]. He studied in cylindrical geometry the nucleation modes of γ-Fe$_2$O$_3$ particle coated with a thin layer of a cobalt ferrite. The same values of the saturation magnetization and exchange constant were used for the core and shell layers, whereas the uniaxial anisotropy constant of the surface layer was assumed to be much larger than that of the core. It was found that the nucleation field of the curling mode first increases and then saturates as a function of the ratio of coated particle radius R_2 to the uncoated particle radius R_1.

It is worth mentioning that in Refs. [110–112], the full solution of the Brown's equations within the shell layer, containing a linear combination of Bessel functions of the first and second kind, was used to satisfy both the boundary conditions (8.35a) and (8.35b). Recently [113, 114] the so-called bulging nucleation mode was suggested to describe a nucleation process in a soft spherical or cylindrical particle covered by a very hard surface layer of fixed magnetization. The authors [113, 114] set to zero the amplitude of the nucleation mode at the interface. Thus, they used only boundary condition Eq. (8.35a), which is not enough. Besides, the self-magnetostatic energy of the bulging mode was omitted in the calculations presented [113, 114]. The importance of the boundary condition, Eq. (8.35b), was recently emphasized in Ref. [115], where theoretical analysis of the magnetization reversal process in a bilayer structure with hard and soft magnetic layers was carried out.

One interesting theoretical problem is the nature of the so-called exchange bias effect [116, 117] observed in small ferromagnetic particles covered by a thin antiferromagnetic layer, as well as in antiferromagnetic–ferromagnetic bilayers at a temperature below the Neel temperature of the antiferromagnetic layer (see reviews [118–120] and references therein). The macroscopic description of the phenomenon was given in Refs. [117, 121], where the existence of unidirectional anisotropy energy was postulated at the interface. The effect has certainly a microscopic origin being related with the exchange interaction of spins with ferromagnetic and antiferromagnetic coupling constants at the interface. Nevertheless, for a special case of a perfectly compensated antiferromagnetic interface it was shown [122] that antiferromagnetic–ferromagnetic exchange interaction can be described in terms of interface magnetic anisotropy, the interaction energy being proportional to a square of a scalar product of the unit ferromagnetic and antiferromagnetic vectors at the interface. This enables one to apply micromagnetic approach discussed in this section to study the exchange bias effect in this particular case. Unfortunately, general experimental situation is much more complicated, because the antiferromagnetic–ferromagnetic interface is usually only partly compensated and can hardly be considered as atomically flat.

8.4
Thermally Activated Switching

The study of thermal relaxation rates of magnetic nanoparticles is important for various technical applications, especially for high-density magnetic recording [123]. In principle, the relaxation times of single-domain particles with different types of magnetic anisotropy can be calculated by means of solution of the Fokker–Plank equation for the distribution of magnetic moment orientations. This famous equation was stated by Brown many years ago [23] on the basis of the stochastic Langevin equation. Unfortunately, exact analytical solution of the Brown's equation is absent even for the simplest case of a particle with a uniaxial anisotropy energy. Therefore, numerical and approximate analytical methods were developed [124–142] to solve this equation for a number of important cases. Recently, a new method of numerical simulation of the relaxation processes in nanoparticles was introduced in Refs. [143–146]. It utilizes an explicit solution of the stochastic Langevin equation. Of course, the numerical integration of the Langevin equation is time consuming. Nevertheless, this method has the advantage of great universality. It can be applied both for a dilute assembly of particles of any type of magnetic anisotropy, and for a dense particle assembly with strong magneto-dipole interaction.

The numerical scheme for solving the stochastic Langevin equation is well developed at present [143–146]. However, the calculated relaxation times for uniaxial ferromagnetic particle differ by a factor of 4 [145], or even 30 [144], from the corresponding Brown's analytical estimate [23]. The reason for such a big difference is unclear. Therefore, the calculation of the relaxation times of nanoparticles by means of direct integration of the Langevin equation seems doubtful. However, it is worth mentioning that in the Refs. [144, 145] the relaxation times of nanoparticles were estimated through the number of switching events happened during a large, enough elapsed time. To our opinion, it is not evident that the quantity determined in this way strictly corresponds to the relaxation time of a nanoparticle obtained by means of the solution of the Brown's equation [23]. Furthermore, even the notion of the "switching event" seems unclear for the process under consideration. Actually, in the spherical coordinates the path of the unit magnetization vector on a surface of a unit sphere is very irregular. In particular, the unit magnetization vector may spend some time near the separatrix line of the energy landscape when it is difficult to say which of the potential wells it currently belongs to. We would like to stress that due to a stochastic nature of the Langevin equation, a special procedure for proper interpretation of the numerical simulation data has to be developed. A reasonable approach has been used in the experimental paper [147], where characteristic relaxation time was determined by means of measuring a waiting time histogram of an assembly of independent ferromagnetic nanoparticles. Evidently, similar approach must be used in the numerical simulation. The goal of the Brown's equation [23] is to describe the magnetization relaxation process in a large assembly of identical single-domain particles. To simulate this process numerically, one has to average the time-dependent particle magnetization over sufficiently large number of statistically independent numerical experiments performed at the same initial conditions. Using this approach, we prove in this section [148] that the calculated relaxation times coincide with reasonable accuracy with the corresponding analytical estimates for a number of cases considered.

8.4.1
Analytical Estimates of the Relaxation Time

The fact that the exact solution of the Brown's equation [23] is absent stimulated the development of numerical and analytical methods [124–142] of studying the problem of magnetization relaxation in an assembly of single-domain particles with various types of magnetic anisotropy energy. A symmetrical problem, i.e., the case of a particle with uniaxial anisotropy in external magnetic field parallel to the easy anisotropy axis, was first considered by Brown [23], who gave explicit analytical estimates for the corresponding relaxation time in the limits of high, $\varepsilon = K_1 V/k_B T \gg 1$, and low, $\varepsilon \ll 1$, energy barriers. Here K_1 is the uniaxial anisotropy constant, k_B is the Boltzmann constant, T is the absolute temperature, and V is the particle volume. In the symmetrical case considered by Brown, in the limit of high-energy barrier the relaxation time is determined by the smallest nonvanishing eigenvalue of the one-dimensional Fokker–Planck equation. It was shown by Aharoni [124, 125] by means of numerical solution of the corresponding one-dimensional boundary value problem that the Brown's asymptotic estimate for the high-energy barrier turns out to be good even for a barrier with $\varepsilon \sim 1$. It is important to note that for symmetrical case, the Brown's estimate holds for any value of the damping parameter κ due to the axial symmetry of the energy barrier. Situation becomes more complicated for nonsymmetrical problems, for example, for a particle with cubic type of magnetic anisotropy, or uniaxial particle in external magnetic filed applied at certain angle to the easy anisotropy axis. In this case, even in the limit of high-energy barrier, most interesting for the applications, one can obtain analytical estimates for high- and low-dissipation regimes only.

The case of a particle with cubic type of magnetic anisotropy was first considered in Refs. [126–129] in the intermediate to high damping limit, $\kappa > 1$. Later Klik and Gunther [130, 131], using the theory of first-passage times [142], stated a general formula for the relaxation time of a single-domain particle in the limit of low damping, $\kappa \ll 1$. This approach was successfully used [139] to study a simplest nonsymmetrical problem of uniaxial particle in external magnetic field applied perpendicular to the easy anisotropy axis. For this case, the explicit asymptotic formulae valid for low- and high-dissipation limits were obtained [139]. More general situation, when external magnetic field is applied at an arbitrary angle to the easy anisotropy axis, was studied in detail in Refs. [133, 135–137]. Also, the case of a particle with cubic type of magnetic anisotropy was recently reexamined in Refs. [138, 141]. Therefore, at present there are analytic estimates both for symmetrical and several nonsymmetrical problems that can be used to check and validate the numerical solution of the stochastic Langevin equation.

8.4.2
Stochastic Langevin Equation

To study a relaxation process for a single-domain particle, one can directly use stochastic Landau–Lifshitz equation [23, 143–146]

$$\frac{d\vec{\alpha}}{dt} = -\gamma_1 \vec{\alpha} \times (\vec{H}_{ef} + \vec{H}_{th}) - \kappa \gamma_1 \vec{\alpha} \times (\vec{\alpha} \times (\vec{H}_{ef} + \vec{H}_{th})), \quad (8.36)$$

where $\gamma_1 = |\gamma_0|/(1+\kappa^2)$, γ_0 is the gyromagnetic ratio, $\kappa = |\gamma_0|\eta M_s$ is the dimensionless damping parameter, η is the dissipation constant, $\vec{H}_{ef}$ is the effective magnetic field, and $\vec{H}_{th}$ is the thermal field. The latter is assumed to be a Gaussian random process with the following statistical properties of its components

$$\langle H_{th,i}(t) \rangle = 0; \quad \langle H_{th,i}(t) H_{th,j}(t') \rangle = \frac{2 k_B T \kappa}{|\gamma_0| M_s V} \delta_{i,j} \delta(t-t'), \quad (8.37)$$

where $i = (x, y, z)$. Equation (8.37) follows from the fluctuation–dissipation theorem [23].

The calculations can be carried out for various values of the damping parameter, from the so-called low-damping limit, $\kappa = 0.001$–0.01, up to the intermediate-to-high damping limit, $\kappa \sim 1$. To ensure the accuracy of the simulations performed, we use simple Milshtein scheme [145] and keep the physical time step t_0 lower than $1/500$ of the characteristic particle precession time T_p. The aim of the calculations is to get a quantity that can be explicitly compared with the experimental data on the magnetization relaxation in a dilute assembly of superparamagnetic nanoparticles. To do so, a time-dependent magnetization $\vec{M}(t)$ of an isolated ferromagnetic particle has been calculated according to Eqs (8.36) and (8.37) in a large series of numerical experiments with fixed initial conditions. Because various runs of the calculations are statistically independent, an average magnetization of an assembly of noninteracting particles is given by

$$\langle \vec{M}(t) \rangle = \frac{1}{N_{exp}} \sum_{n=1}^{N_{exp}} \vec{M}_n(t). \quad (8.38)$$

To keep dispersion of the average magnetization at appreciable level, the total number of the experiments, N_{exp}, has to be sufficiently large. It has been found empirically that $N_{\text{exp}} \sim 1000$ is usually sufficient to reduce the dispersion of the average magnetization up to several percent.

Generally speaking, two types of initial conditions can be used in the calculations. To simulate a demagnetized initial state of an assembly, the initial value $\vec{\alpha}(0)$ of the unit magnetization vector can be chosen with equal probability within the different equivalent potential wells of the particle. With this initial state, the process of magnetization of an assembly in a constant applied magnetic field can be studied. On the other hand, if the vector $\vec{\alpha}(0)$ belongs to a certain preferable potential well one can simulate the relaxation process of an assembly initially magnetized in sufficiently high external magnetic field.

From experimental point of view, the case of moderate or large reduced energy barrier, $\varepsilon \gg 1$, is the most interesting. Unfortunately, due to processor's time limitation the numerical simulation turns out to be very time consuming for high-energy barriers, $\varepsilon > 10$. Therefore, the results of the calculations for moderate values of $\varepsilon = 4$–8 are presented below for the number of cases. Note that for $\varepsilon \geq 4$, the initial state $\vec{\alpha}(0)$ can be chosen more or less arbitrary within a particular potential well because the quasiequilibrium distribution for magnetization orientations within the given well is established much faster than the total equilibrium between different wells.

8.4.3
Simple Examples

In this section, the numerical results obtained for nanoparticles with different types of magnetic anisotropy are compared with the corresponding theoretical estimates discussed briefly in Section 8.4.1.

8.4.3.1 Uniaxial Anisotropy

Consider a particle with uniaxial type of magnetic anisotropy having an effective anisotropy constant K_1. The later may include the shape anisotropy contribution if the particle axis of symmetry is parallel to the easy anisotropy axis. The Brown's expression [23] for the mean relaxation time of uniaxial particle in the absence of external magnetic field can be written as follows

$$\tau = \frac{1}{f_0} \exp\left(\frac{K_1 V}{k_B T}\right); \quad f_0 = \frac{4\kappa \gamma_1 K_1}{M_s} \sqrt{\frac{K_1 V}{\pi k_B T}}. \tag{8.39}$$

To compare analytical estimation directly with the numerical simulation data, Eq. (8.38), we define the mean relaxation time, Eq. (8.39), so that the time-dependent magnetization of a particle averaged over a large enough assembly of identical particles is given by

$$M(t)/M_s = \exp(-t/\tau). \tag{8.40}$$

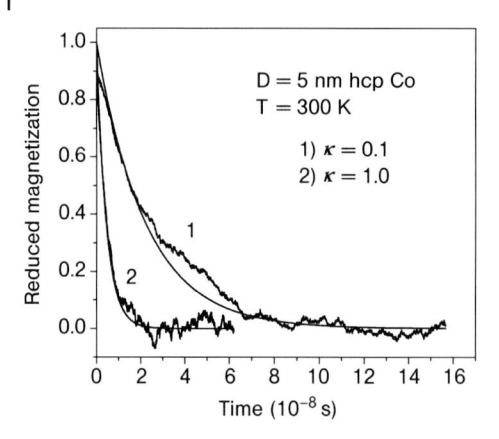

Figure 8.7 The relaxation process in oriented assembly of uniaxial Co-hcp nanoparticles with diameter $D = 5$ nm for two values of the damping parameter: (1) $\kappa = 0.1$; and (2) $\kappa = 1.0$.

The estimation (8.39) is valid, strictly speaking, in the limit of sufficiently high energy barrier, $\varepsilon = K_1 V/k_B T \gg 1$. However, it was shown [124] that it turns out to be good even for moderate barriers, $\varepsilon \sim 1$. As we mentioned above, Eq. (8.39) holds for arbitrary value of the damping parameter κ. This fact is unique for uniaxial single-domain particle [131, 140]. It follows from the axial symmetry of the energy barrier between the equivalent potential wells.

The numerical solution of the stochastic Landau–Lifshitz equation (8.36) is carried out for uniaxial Co-hcp particle with diameter $D = 5$ nm at a room temperature, $T = 300$ K, for various values of the damping parameter κ. The particle anisotropy constant is given by $K_1 = 4.1 \times 10^6$ erg/cm^3, saturation magnetization $M_s = 1400$ emu/cm^3, and the reduced energy barrier being $\varepsilon \approx 6.48$. Supposing that the particle easy anisotropy axis is parallel to the z axis, the initial magnetization state is chosen to be $\vec{\alpha}(0) = (0, 0, 1)$. The irregular curves 1 and 2 in Figure 8.7 show the reduced particle magnetization averaged by means of Eq. (8.38) over $N_{exp} = 1000$ runs of independent simulations. These curves describe the time-dependent magnetization of oriented assembly of identical Co nanoparticles initially magnetized along the easy anisotropy axis. The smooth curves 1 and 2 in Figure 8.7 are drawn in accordance with Eqs. (8.39) and (8.40). Equation. (8.39) gives the mean relaxation times $\tau = 2.23 \times 10^{-8}$ s for the curve 1, and $\tau = 4.41 \times 10^{-9}$ s for the curve 2, respectively. One can see that for both of the cases, numerical and analytical data are in reasonable agreement. The accuracy of the numerical calculations is further characterized by means of the small value of the parameter $t_0/T_p = 1/625$. Thus, for the curves 1 and 2 shown in Figure 8.7 the numbers of the numerical time-steps are given by $N_{step} = 10^7$ and $N_{step} = 2 \times 10^6$, respectively, for every of 1000 runs of the simulations performed.

8.4.3.2 Nonsymmetrical Case

The Brown's estimation (8.39) characterizes the very special case of axially symmetrical energy barrier. Simple nonsymmetrical case corresponds, for example, to uniaxial single-domain particle in external magnetic field H_0 applied perpendicular to the easy anisotropy axis [139]. Let the easy anisotropy axis is parallel to the z axis, whereas the external magnetic field is applied along the x axis. Suppose that initially, at $t = 0$, the particle is uniformly magnetized along the easy axis, $\vec{\alpha}(0) = (0, 0, 1)$, and the external magnetic field is switched on abruptly for $t > 0$. Then the reduced particle magnetization along the easy axis is given by [139]

$$M_z(t)/M_s = \sqrt{1 - h^2} \exp\left(-\frac{t}{\tau}\right);$$

$$\tau = \frac{1}{f_{0h}} \exp\left(\frac{K_1 V}{k_B T}(1 - h)^2\right), \qquad (8.41)$$

where $h = H_0/H_K$ is the reduced magnetic field, and $H_K = 2K_1/M_s$ is the particle anisotropy field. In the low damping limit, $\kappa \ll 1$, the following analytical estimation for the frequency f_{0h} has been obtained [139]

$$f_{0h} = \frac{\kappa \gamma_1 K_1}{\pi M_s} \frac{K_1 V}{k_B T} \sqrt{h(1-h)(1-h^2)} \left(16 - \frac{80}{3}h + \frac{32}{3}h^2\right), \qquad (8.42a)$$

whereas in the high damping limit, $\kappa > 1$

$$f_{0h} = \frac{\gamma_1 K_1}{\pi M_s} \sqrt{\frac{1+h}{h}} \left(\kappa(1 - 2h) + \sqrt{\kappa^2 + 4h(1-h)}\right). \qquad (8.42b)$$

For comparison with the analytical estimations (8.41) and (8.42), the numerical simulations were carried out for uniaxial particle with diameter $D = 5$ nm and material parameters $K_1 = 10^6$ erg/cm³, $M_s = 1400$ emu/cm³ at a temperature $T = 70$ K.

The irregular curves 1–3 in Figure 8.8 are obtained by means of averaging the reduced particle magnetization over $N_{exp} = 400$ runs of simulations made for different values of applied magnetic field. The corresponding smooth curves 1–3 in Figure 8.8 are drawn by means of Eqs. (8.41) and (8.42a). It should be noted that according to Eq. (8.41), the effective energy barrier decreases as a function of H_0. As a result, the mean relaxation times for the cases 1–3 are given by $\tau = 1.29 \times 10^{-7}$ s for the curve 1, $\tau = 5.85 \times 10^{-8}$ s for the curve 2, and $\tau = 3.06 \times 10^{-8}$ s for the curve 3, respectively. One can see that at $\kappa = 0.01$, numerical simulation data are in agreement with the estimations (8.41) and (8.42a). Therefore, one can assume that this value of the damping parameter corresponds to the low-damping limit. It is found, however, that systematic deviations of the numerical data from Eqs. (8.41) and (8.42a) appear with increasing damping parameter.

As an example, Figure 8.9 shows the magnetization relaxation curves in comparison with the corresponding analytical estimations calculated for the case of the damping parameter $\kappa = 0.1$ for two values of applied magnetic field. The relaxation curves are obtained by means of averaging the particle

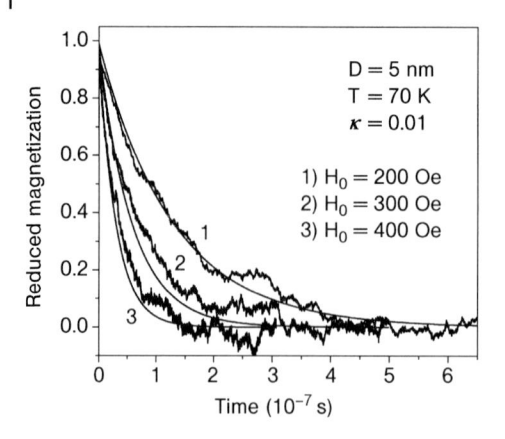

Figure 8.8 The relaxation process in oriented assembly of uniaxial nanoparticles for various values of external magnetic field H_0 applied perpendicular to the easy particle axis. The calculations are made in the low-damping limit, $\kappa = 0.01$.

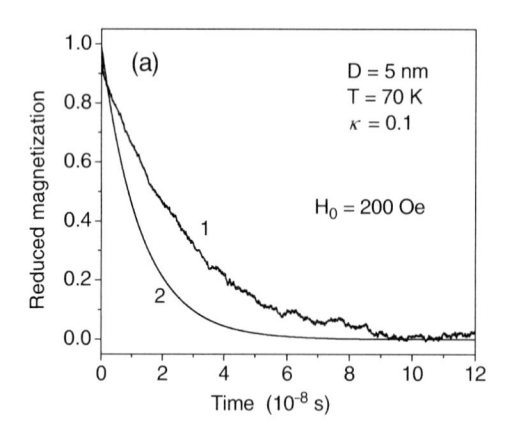

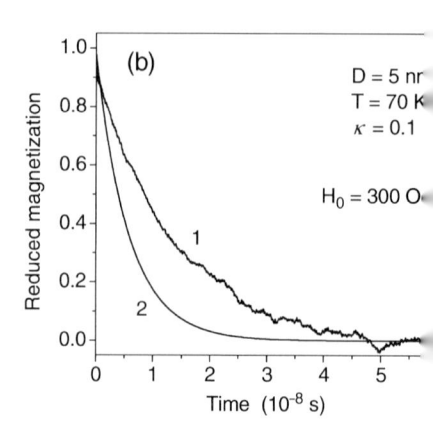

$D = 5$ nm
$T = 70$ K
$\kappa = 0.01$

1) $H_0 = 200$ Oe
2) $H_0 = 300$ Oe
3) $H_0 = 400$ Oe

Figure 8.9 The same as in Figure 8.8, but for larger value of the damping parameter, $\kappa = 0.1$. Curves 1 in the panels (a) and (b) represent the numerical simulation data in comparison with the corresponding analytical estimations (curves 2) given by Eqs. (8.41) and (8.42a).

magnetization over $N_{exp} = 1000$ runs of simulations with small time-step $t_0/T_p = 1/625$. The relaxation times calculated by means of Eqs. (8.41) and (8.42a) are given by $\tau = 1.29 \times 10^{-8}$ s, and $\tau = 5.85 \times 10^{-9}$ s for the panels (a) and (b) of Figure 8.9, respectively. On the other hand, the numerical simulation data in Figure 8.9 can be well approximated by means of Eq. (8.42) with $\tau_{num} = 2.5 \times 10^{-8}$ s, and $\tau_{num} = 1.25 \times 10^{-8}$, correspondingly. These values are approximately twice larger with respect to the analytical estimations. However, as Figure 8.10 shows, in the limit of $\kappa > 1$ the numerical simulation data can be fairly well described by means of the estimations (8.41) and (8.42b).

Therefore, for the given particle it is found that for moderate values of the energy barrier, the regimes of low and high dissipation correspond to the intervals $\kappa \leq \kappa_{small} = 0.01$, and $\kappa > \kappa_{large} = 1.0$, respectively.

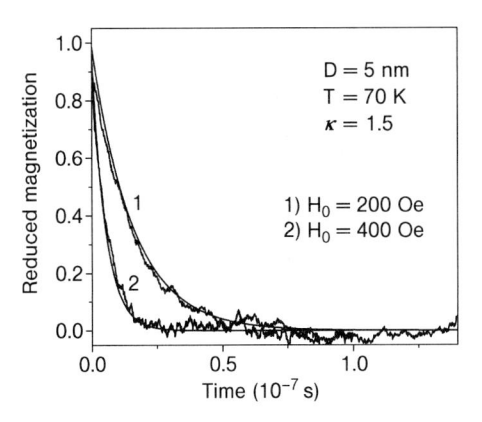

Figure 8.10 The magnetization relaxation curves of the same particle in the high damping limit, $\kappa = 1.5$, for different values of applied magnetic field. Irregular curves represent the numerical simulation data in comparison with the analytical estimations drawn by means of Eqs. (8.41) and (8.42b).

8.4.3.3 Cubic Anisotropy

Finally, let us consider the magnetization relaxation process for a small ferromagnetic particle with cubic type of magnetic anisotropy. For a Fe particle with positive cubic anisotropy constant, $K_{1c} > 0$, in the absence of external magnetic field there are six equivalent potential wells. If the particle is initially magnetized along one of the easy anisotropy axis, its reduced magnetization at $t > 0$ is given by [126–129, 138]

$$M(t)/M_s = \left(1 - \frac{k_B T}{2K_{1c} V}\right) \exp(-t/\tau); \quad \tau = \frac{1}{f_{0c}} \exp\left(\frac{K_{1c} V}{4 k_B T}\right). \quad (8.43)$$

For the frequency f_{0c}, different representations can be obtained in the low, $\kappa \ll 1$

$$f_{0c} = \frac{8\sqrt{2}}{9} \frac{\kappa \gamma_1 K_{1c}}{\pi M_s} \left(\frac{K_{1c} V}{k_B T}\right), \quad (8.44a)$$

and high damping limits, $\kappa > 1$.

$$f_{0c} = \frac{4\gamma_1 K_{1c}(\kappa + 2q)}{\pi M_s} \sqrt{\frac{2(\kappa + 2q)}{\kappa(1 + 12q^2) + 16q^3}};$$

$$q = \frac{\sqrt{9\kappa^2 + 8} - 3\kappa}{8}. \quad (8.44b)$$

The numerical simulations were carried out for spherical iron particle with diameter $D = 9$ nm at a temperature $T = 77$ K. The material parameters of the particle are given by $K_{1c} = 5.5 \times 10^5$ erg/cm^3, $M_s = 1700$ emu/cm^3, so that the reduced energy barrier $K_1 V/4 k_B T \approx 4.94$. The curves 1 in the panels (a)–(d) of Figure 8.11 show the numerical simulation results obtained by means of averaging the particle magnetization over sufficiently large amounts of statistically independent simulations for various values of the damping parameter.

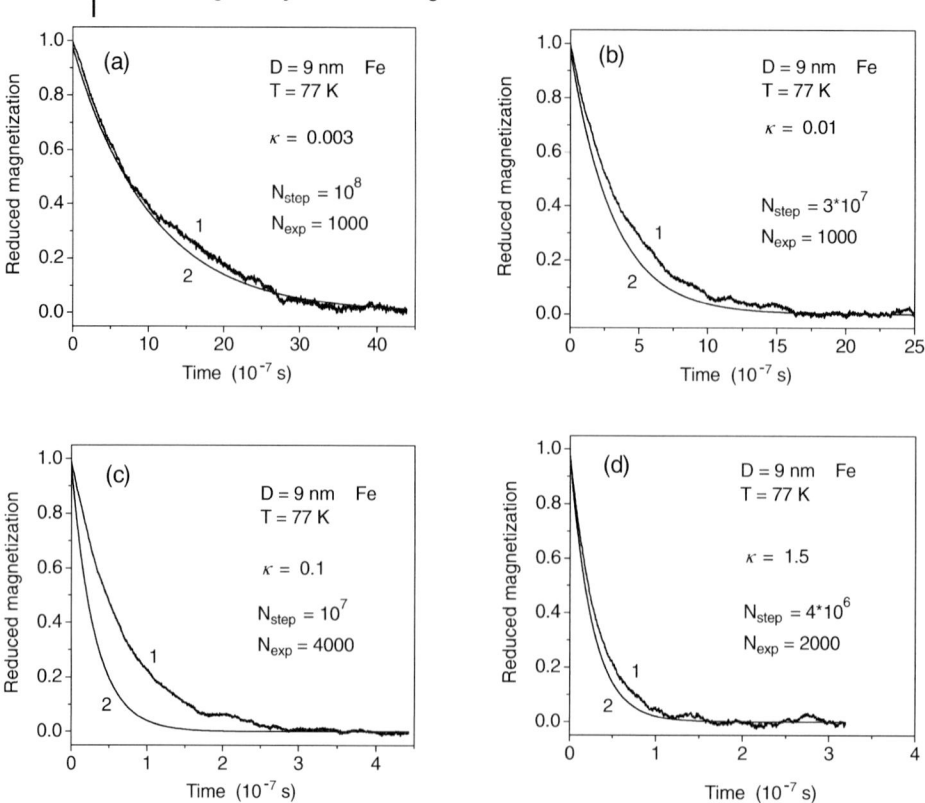

Figure 8.11 The magnetization relaxation process (curves 1) of an assembly of iron nanoparticles with $D = 9$ nm and cubic type of magnetic anisotropy for various values of the damping parameter: (a) $\kappa = 0.003$; (b) $\kappa = 0.01$; (c) $\kappa = 0.1$; and (d) $\kappa = 1.5$. The curves 2 in the panels (a)–(c) are drawn by means of Eqs. (8.43) and (8.44a). The curve 2 in the panel (d) corresponds to the high damping approximation, Eqs. (8.43) and (8.44b).

As Figure 8.11 shows, for iron particle with moderate value of the reduced energy barrier the limit of low dissipation corresponds to rather small values of the damping parameter $\kappa \leq 0.003$, whereas the high damping limit realizes for $\kappa \geq 1.5$ only. The numerical data (curves 1 in Figure 8.11) can be approximated by the relaxation function (8.43) with certain empirical values τ_{num}. They are given in the second row of the Table 8.1. For comparison, the last row of this table represents the corresponding theoretical estimates, τ_{theor}, calculated by mean of Eqs. (8.43) and (8.44a) for the cases $\kappa = 0.003$; $\kappa = 0.01$; and $\kappa = 0.1$, and by mean of Eqs. (8.43) and (8.44b) for the case $\kappa = 1.5$. One can see again that the fairly well agreement of the data is obtained in the limits of low, $\kappa = 0.003$, and high, $\kappa = 1.5$, dissipation regimes. Nevertheless, one notes that even for intermediate range of κ, the numerical and theoretical values differ not more than by factor 2.

Table 8.1 Relaxation times of iron nanoparticles with $D = 9$ nm.

	$\kappa = 0.003$	$\kappa = 0.01$	$\kappa = 0.1$	$\kappa = 1.5$
τ_{num} (s)	1.0×10^{-6}	4.0×10^{-7}	6.8×10^{-8}	3.2×10^{-8}
τ_{theor} (s)	1.034×10^{-6}	3.103×10^{-7}	3.134×10^{-8}	2.60×10^{-8}

Based on the above discussion, one can conclude that the numerical procedure of Section 8.4.2 enables one to investigate the relaxation characteristics of an assembly of noninteracting single-domain particles with *any kind of* magnetic anisotropy in external magnetic field applied at arbitrary direction. However, to get reliable data it is necessary to keep the physical time-step small enough with respect to the characteristic precession time, $t_0/T_p \ll 1$. Besides, it is necessary to make large-enough runs of simulations, $N_{exp} \sim 1000$, with the same initial conditions to keep the dispersion of the assembly magnetization at appreciably low level. It is important that numerical simulation enables one to determine the actual bounds for the low and high dissipation regimes. It is found that for moderate energy barriers, $\varepsilon = 4$–8, the high damping limit occurs generally at $\kappa > \kappa_{large} = 1.0$–$1.5$. However, the regime of low dissipation corresponds usually to sufficiently low values of the damping parameter, $\kappa \leq \kappa_{small} = 0.003$–$0.01$. Interestingly, the numerical simulation shows that the analytical estimation obtained for low damping limit turns out to be not very far from the numerical data even for moderate values of $\kappa \sim 0.1$. Unfortunately, the reliable numerical simulations are possible for the moderate energy barriers, $\varepsilon = 5$–10, only. It generally corresponds to the case of nanoparticles of sufficiently small diameters, $D < 5$–10 nm, in the temperature interval 50–300 K, depending on the value of the particle anisotropy constant and the type of the magnetic anisotropy. As Figures 8.7–8.11 show, the corresponding relaxation times turn out to be very small, $\tau \sim 10^{-8}$–10^{-6} s. For lager barriers, only analytical estimations of the relaxation times for particles with various types of magnetic anisotropy seem possible. However, the conditions for their validity, i.e., the corresponding bounds for the low and high damping regimes, have to be stated.

8.4.4
Nonuniform Modes

While comparing the theoretical results with the experimental data, one has to keep in mind two important limitations of the Brown's equation [23]. Strictly speaking, it describes the relaxation process in an assembly of ideal single-domain particles only. Therefore, the influence of any kind of magnetization deviations from the uniform magnetization discussed in the Sections 8.2 and 8.3 has to be investigated separately. Taking into account the discussion in Sections 8.2 and 8.3 one may hope that particle shape deviations and surface anisotropy contributions may be generally accounted for by introducing certain additional energy terms to the total energy of a single-domain particle, such as configurational [24, 31, 64] or effective anisotropy energy (see Eq. (8.32)). Generally, for small-enough particles the influence of these contributions has to be small, because it is proportional to the parameter $w_p L^2/C \ll 1$, where w_p is the characteristic energy density of the corresponding perturbation.

Another limitation of the Brown's approach is the restriction to the uniform rotation mode for the thermally agitated motion of the unit magnetization vector. This restriction can be justified only for a single-domain particle of sufficiently small size when the thermal excitations of the magnetization curling or other high-order switching modes of single-domain particle have sufficiently small probability. The first experimental observation of the influence of the curling mode on the relaxation time of spherical ferromagnetic particle was made in Refs. [149, 150], where it was found that the relaxation time τ can *decrease* with increasing particle size in certain interval of sizes close to the single-domain radius. This behavior is in evident contradiction with Eqs. (8.39), (8.43). The latter show that the energy barrier associated with the magnetic anisotropy energy increases exponentially as a function of the particle volume both for particles with uniaxial and cubic types of magnetic anisotropy. But this is true for uniform rotation of the particle magnetization only. It was proved [149, 151] that for spherical particle with cubic anisotropy, the energy barrier decreases with volume in certain interval of sizes near the critical size for nucleation by curling if the magnetization reversal proceeds by means of a "rigid" rotation of the magnetization curling distribution. As a result, the relaxation time of such a particle shows anomalous nonmonotonic behavior as a function of the radius. The estimation of the effect in a more rigorous model of "quasirigid" magnetization rotation [152] confirmed the anomalous decrease of τ with increasing particle radius.

Recently, the thermally assisted magnetization reversal in submicron-patterned ferromagnetic elements was studied both experimentally [153, 154] and theoretically [155]. The stability of the element magnetization in external magnetic field at room or elevated temperatures is crucial for the performance of magnetic memory devices and sensors based on patterned ferromagnetic elements. The elements studied in [153, 154] were rectangles with pointed ends or elongated rectangular bars. Such elongated elements of soft magnetic type can support longitudinal magnetization even at submicron in-plane dimensions; however, they show magnetization curling (i.e., S or C configurations) or vortex states near the sample ends in zero applied field. In the experiment [154], a logarithmic decay of remanent magnetization in an array of rectangular elements was measured to estimate the effective energy barrier for magnetization reversal as a function of the element thickness. In the experiment [153], the magnetization reversal process was studied in applied magnetic field slightly below the coercive field of the element. The magnetization reversal occurred by means of domain-wall nucleation at the opposite sides of the sample. The authors [153] observed a peak in a probability of reversal as a function of the time. The effect was explained on the basis of "energy-ladder" model that assumes that the system climbs stochastically, via thermal activation, up the ladder of states, eventually escaping from the metastable energy minimum.

8.5 Conclusions

Micromagnetics is a powerful tool to study nonuniform magnetization distributions such as domain walls, vortices, magnetization nonuniformities near the surfaces, interfaces, and near various types of magnetic defects. In the framework of Micromagnetics, the notion of a single-domain particle can be correctly formulated and the properties of ideal single-domain particles with various types of magneto-crystalline anisotropy can be investigated in detail. However, particles used in real experiments might have shape deviations, as well as a surface anisotropy. One may hope that the standard model of a single-domain particle can still survive due to domination of the exchange interaction at small particle sizes so that the perturbation of the uniform magnetization can be small for a particle of sufficiently small size. Then, the influence of the shape deviation and surface magnetic anisotropy can be described by means of corresponding effective energy contributions. However, for a very small particle, i.e., for a magnetic cluster with the number of the surface spins N_s of the order of the total number of magnetic spins N_t, the separation of the "surface" and "volume" degrees of freedom appears inappropriate. In this case, the total cluster energy as a function of the direction of the total magnetization has to be determined by means of the first principle many-body calculation.

More subtle phenomena, such as exchange bias and spin-torque effect, need a proper generalization of the phenomenological micromagnetic approach. Micromagnetics is not flexible enough to take into account the actual atomic structure of a nanoparticle, which may have several types of magnetic ions in a primitive cell. Also, it is rather restrictive to study composite particles consisting, for example, of ferromagnetic and antiferromagnetic areas being in atomic contact. Recently, a quantum mechanical Hartree–Fock approximation [122, 156,] was used to describe a magnetic state of a nanoparticle and a thin exchange-coupled bilayer. A set of nonlinear equations for averaged values of spin operators in various lattice sites was solved iteratively. The method enables one to consider any type of exchange interactions and to take into account actual spin structure of magnetic ions in different lattice sites. Other variants of this approach are possible. However, the phonon degrees of freedom of magnetic nanoparticle seem necessary to be taken into account [102, 157]. Also, a generalization of the method for dynamical problems is desirable.

References

1. L.D. Landau, E.M. Lifshitz, *Phys. Z. Sowjetunion*, **1935**, *8*, 153.
2. W.F. Brown Jr., *Micromagnetics*, Wiley, New York, **1963**.
3. A. Aharoni, *Introduction to the Theory of Ferromagnetism*, Clarendon Press, Oxford, **1996**.
4. Ya.I. Frenkel, *Z. Phys.*, **1928**, *49*, 31.
5. W. Heisenberg, *Z. Phys.*, **1928**, *49*, 619.
6. Ya.I. Frenkel, Ya.G. Dorfman, *Nature*, **1930**, *126*, 274.
7. C. Kittel, *Phys. Rev.*, **1946**, *70*, 965.
8. C. Kittel, *Rev. Mod. Phys.*, **1949**, *21*, 541.
9. L. Neel, *Compt. Rend.*, **1947**, *224*, 1488.
10. L. Neel, *Compt. Rend.*, **1947**, *224*, 1550.
11. E. Kondorsky, *Izv. Acad. Sci.*, **1952**, *16*, 398. (in Russian).
12. W.F. Brown Jr., *J. Appl. Phys.*, **1968**, *39*, 993.
13. W.F. Brown Jr., *Ann. NY Acad. Sci.*, **1969**, *147*, 461.
14. E.S. Stoner, E.P. Wohlfarth, *Phil. Trans. Roy. Soc.*, **1948**, *240*, 599.
15. E.H. Frei, S. Shtrikman, D. Treves, *Phys. Rev.*, **1957**, *106*, 446.
16. W.F. Brown Jr., *Phys. Rev.*, **1957**, *105*, 1479.
17. A. Aharoni, S. Shtrikman, *Phys. Rev.*, **1958**, *109*, 1522.
18. A. Aharoni, *J. Appl. Phys.*, **1959**, *30*, 70.
19. A. Aharoni, *Phys. Rev.*, **1963**, *131*, 1478.
20. A. Aharoni, *Phys. Stat. Sol.*, **1966**, *16*, 3.
21. A. Aharoni, *J. Appl. Phys.*, **1986**, *60*, 1118.
22. L. Neel, *Ann. Geophys.*, **1949**, *5*, 99.
23. W.F. Brown, Jr., *Phys. Rev.*, **1963**, *130*, 1677.
24. M.E. Schabes, H.N. Bertram, *J. Appl. Phys.*, **1988**, *64*, 1347.
25. Y.D. Yan, E. Della Torre, *J. Appl. Phys.*, **1989**, *66*, 320.
26. D.R. Fredkin, T.R. Koehler, *J. Appl. Phys.*, **1990**, *67*, 5544.
27. D.R. Fredkin. T.R. Koehler, *IEEE Trans. Magn.*, **1990**, *26*, 1518.
28. M.E. Schabes, *J. Magn. Magn. Mater.*, **1991**, *95*, 249.
29. J. Fidler, T. Schrefl, *J. Phys. D: Appl. Phys.*, **2000**, *33*, R135.
30. S.Y. Chou, P.R. Krauss, W. Zhang, L. Guo, L. Zhuang, *J. Vac. Sci. Technol. B*, **1997**, *15*, 2897.
31. R.P. Cowburn, *J. Phys. D: Appl. Phys.*, **2000**, *33*, R1.
32. B.D. Terris, T. Thomson, *J. Phys. D: Appl. Phys.*, **2005**, *38*, R199.
33. N.A. Usov, S.E. Peschany, *J. Magn. Magn. Mater.*, **1994**, *130*, 275.
34. N.A. Usov, S.E. Peschany, *J. Magn. Magn. Mater.*, **1994**, *134*, 111.
35. W. Rave, K. Fabian, A. Hubert, *J. Magn. Magn. Mater.*, **1998**, *190*, 332.
36. R. Hertel, H. Kronmuller, *J. Magn. Magn. Mater.*, **2002**, *238*, 185.
37. N.A. Usov, Yu.B. Grebenschikov, *Fiz. Met. Metaloved.*, **1991**, *6*, 59 (in Russian).
38. C.H Stapper, Jr., *J. Appl. Phys.*, **1969**, *40*, 798.
39. A. Aharoni, J.P. Jakubovics, *Phil. Mag. B*, **1986**, *53*, 133.
40. N.A. Usov, *Proc. of Moscow International Symposium on Magnetism*, Moscow, **1999** 39.
41. N.A. Usov, J.W. Tucker, *Mater. Sci. Forum*, **2001**, *373–376*, 429.
42. N.A. Usov, L.G. Kurkina, J.W. Tucker, *J. Magn. Magn. Mater.*, **2002**, *242–245*, 1009.
43. N.A. Usov, S.E. Peschany, *J. Magn. Magn. Mater.*, **1993**, *118*, L290.
44. N.A. Usov, S.E. Peschany, *Fiz. Met. Metaloved.*, **1994**, *12*, 13 (in Russian).
45. N.A. Usov, *J. Magn. Magn. Mater.*, **1999**, *203*, 277.
46. V. Novosad, K.Yu. Guslienko, H. Shima, Y. Otani, K. Fukamichi, N. Kikuchi, O. Kitakami, Y. Shimada, *IEEE Trans. Magn.*, **2001**, *37*, 2088.
47. K.Yu. Guslienko, V. Novosad, Y. Otani, H. Shima, K. Fukamichi, *Appl. Phys. Lett.*, **2001**, *78*, 3848.
48. R.P. Cowburn, D.K. Koltsov, A.O. Adeyeye, M.E. Welland, D.M. Tricker, *Phys. Rev. Lett.*, **1999**, *83*, 1042.

49. J. Shi, S. Tehrani, T. Zhu, Y.F. Zheng, J.-G. Zhu, *Appl. Phys. Lett.*, **1999**, *74*, 2525.
50. J. Raabe, R. Pulwey, R. Satter, T. Schweinbock, J. Zweck, D. Weiss, *J. Appl. Phys.*, **2000**, *88*, 4437.
51. M. Schneider, H. Hoffmann, J. Zweck, *Appl. Phys. Lett.*, **2000**, *77*, 2909.
52. T. Shinjo, T. Okuno, R. Hassdorf, K. Shigeto, T. Ono, *Science*, **2000**, *289*, 930.
53. T. Pokhil, D. Song, J. Novak, *J. Appl. Phys.*, **2000**, *87*, 6319.
54. T. Okuno, K. Shigeto, T. Ono, K. Mibu, T. Shinjo, *J. Magn. Magn. Mater.*, **2002**, *240*, 1.
55. A. Wachowiak, J. Wiebe, M. Bode, O. Pietzsch, M. Morgenstern, R. Wiesendanger, *Science*, **2002**, *298*, 577.
56. M. Schneider, H. Hoffmann, J. Zweck, *Appl. Phys. Lett.*, **2001**, *79*, 3113.
57. A. Fernandez, C.J. Cerjan, *J. Appl. Phys.*, **2000**, *87*, 1395.
58. K.Yu. Guslienko, V. Novosad, Y. Otani, H. Shima, K. Fukamichi, *Phys. Rev. B*, **2002**, *65*, 024414.
59. M. Schneider, H. Hoffmann, S. Otto, Th. Haug, J. Zweck, *J. Appl. Phys.*, **2002**, *92*, 1466.
60. N.A. Usov, Yu.B. Grebenshchikov, L.G. Kurkina, Ching-Ray Chang, Zung-Hang Wei. *J. Magn. Magn. Mater.*, **2003**, *258–259*, 6.
61. Yu.B. Grebenshchikov, N.A. Usov, K.S. Pestchanyi. *J. Appl. Phys.*, **2003**, *94*, 6649.
62. Y. Zheng, J.G. Zhu, *J. Appl. Phys.*, **1997**, *81*, 5471.
63. A.J. Newell, R.T. Merrill, *J. Appl. Phys.*, **1998**, *84*, 4394.
64. R.P. Cowburn, A.O. Adeyeye, M.E. Welland, *Phys. Rev. Lett.*, **1998**, *24*, 5414.
65. R. Hertel, H. Kronmuller, *Phys. Rev. B*, **1999**, *60*, 7366.
66. H. Kronmuller, R. Hertel, *J. Magn. Magn. Mater.*, **2000**, *215–216*, 11.
67. W. Rave, A. Hubert, *IEEE Trans. Magn.*, **2000**, *36*, 3886.
68. L. Thomas, S.S.P. Parkin, J. Yu, U. Rudiger, A.D. Kent, *Appl. Phys. Lett.*, **2000**, *76*, 766.
69. K.J. Kirk, S. McVitie, J.N. Chapman, C.D.W. Wilkinson, *J. Appl. Phys.*, **2001**, *89*, 7174.
70. K.J. Kirk, M.R. Scheinfein, J.N. Chapman, S. McVitie, M.F. Gillies, B.R. Ward, J.G. Tennant, *J. Phys. D: Appl. Phys.*, **2001**, *34*, 160.
71. D. Goll, G. Schutz, H. Kronmuller, *Phys. Rev. B*, **2003**, *67*, 094414.
72. H. Kronmuller, D. Goll, R. Hertel, G. Schutz, *Physica B*, **2004**, *343*, 229.
73. N.A. Usov, L.G. Kurkina, J.W. Tucker, *J. Phys. D: Appl. Phys.*, **2002**, *35*, 2081.
74. L. Neel, *J. Phys. Radium*, **1954**, *15*, 225.
75. M. Wu, Y.D. Zhang, S. Hui, T.D. Xiao, S. Ge, W.A. Hines, J.I. Budnick, M.J. Yacaman, *J. Appl. Phys.*, **2002**, *92*, 6809.
76. H. Zeng, S. Sun, J. Li, Z.L. Wang, J.P. Liu, *Appl. Phys. Lett.*, **2004**, *85*, 792.
77. I.E. Tamm, *Theory of Electricity*, Nauka, Moscow, **1976** (in Russian).
78. A.I. Akhiezer, V.G. Bar'yakhtar, S.V. Peletminskii, *Spin Waves*, Wiley, New York, **1968**.
79. W.F. Brown Jr., A.H. Morrish, *Phys. Rev.*, **1957**, *105*, 1198.
80. N.A. Usov, Yu.B. Grebenschikov, S.E. Peschany, *Z. Phys. B*, **1992**, *87*, 183.
81. E. Kondorsky, *IEEE Trans. Magn.*, **1979**, *15*, 1209.
82. A. Aharoni, *J. Appl. Phys.*, **1988**, *63*, 5879.
83. G.S. Krinchik, *Physics of Magnetic Phenomena*, Moscow State University Publisher, Moscow, **1985** (in Russian).
84. A. Aharoni, *J. Appl. Phys.*, **1987**, *61*, 3302.
85. D.A. Garanin, H. Kachkachi, *Phys. Rev. Lett.*, **2003**, *90*, 065504.
86. N.A. Usov, Yu.B. Grebenshchikov, *J. Appl. Phys.*, **2008**, *104*, 043903.
87. A. Aharoni, *J. Appl. Phys.*, **1988**, *64*, 6434.

88. A. Aharoni, *J. Magn. Magn. Mater.*, **1999**, *196–197*, 786.
89. A. Aharoni, *J. Appl. Phys.*, **2000**, *87*, 5526.
90. A. Aharoni, *J. Appl. Phys.*, **1991**, *69*, 7762.
91. A. Aharoni, *J. Appl. Phys.*, **1997**, *81*, 830.
92. V.P. Shilov, J.C. Bacri, F. Gazeau, F. Gendron, R. Perzynski, Yu.L. Raikher, *J. Appl. Phys.*, **1999**, *85*, 6642.
93. F. Gazeau, V.P. Shilov, J.C. Bacri, E. Dubois, F. Gendron, R. Perzynski, Yu.L. Raikher, V.I. Stepanov, *J. Magn. Magn. Mater.*, **1999**, *202*, 535.
94. V.P. Shilov, Yu.L. Raikher, J.C. Bacri, F. Gazeau, R. Perzynski, *Phys. Rev. B*, **1999**, *60*, 11902.
95. D.A. Dimitrov, G.M. Wysin, *Phys. Rev. B*, **1994**, *50*, 3077.
96. Y. Labaye, O. Crisan, L. Berger, J.M. Greneche, J.M.D. Coey, *J. Appl. Phys.*, **2002**, *91*, 8715.
97. H. Kachkachi, M. Dimian, *Phys. Rev. B*, **2002**, *66*, 174419.
98. H. Kachkachi, H. Mahboub, *J. Magn. Magn. Mater.*, **2004**, *278*, 334.
99. H. Kachkachi, E. Bonet, *Phys. Rev. B*, **2006**, *73*, 224402.
100. J. Mazo-Zuluaga, J. Restrepo, J. Mejia-Lopez, *Physica B*, **2007**, *398*, 187.
101. S.V. Vonsovskii, *Magnetism*, Wiley, New York, **1974**.
102. R. Guirado-Lopez, F. Aguilera-Granja, J.M. Montejano-Carrizales, *Phys. Rev. B*, **2002**, *65*, 045420.
103. Y. Xie, J.A. Blackman, *J. Phys.: Condens. Matter*, **2004**, *16*, 3163.
104. A.E. Berkowitz, J.A. Lahut, I.S. Jacobs, L.M. Levinson, D.W. Forester, *Phys. Rev. Lett.*, **1975**, *34*, 594.
105. A.H. Morr, K. Haneda, *J. Appl. Phys.*, **1981**, *52*, 2496.
106. R.H. Kodama, A.E. Berkowitz, E.J. McNiff, Jr., S. Foner, *Phys. Rev. Lett.*, **1996**, *77*, 394.
107. R.H. Kodama, A.E. Berkowitz, *Phys. Rev. B*, **1999**, *59*, 6321.
108. M.J. Stavn, A.H. Morrish, *IEEE Trans. Magn.*, **1979**, *15*, 1235.
109. I.I. Krjukov, N.A. Manakov, *Fiz. Met. Metalloved.*, **1983**, *56*, 5. (in Russian).
110. I.I. Krjukov, N.A. Manakov, V.D. Sadkov, *Fiz. Met. Metalloved.*, **1985**, *59*, 455 (in Russian).
111. A. Aharoni, *J. Appl. Phys.*, **1987**, *62*, 2576.
112. A. Aharoni, *J. Appl. Phys.*, **1988**, *63*, 4605.
113. R. Skomski, J.P. Liu, D.J. Sellmyer, *Phys. Rev. B*, **1999**, *60*, 7359.
114. R. Skomski, J.P. Liu, D.J. Sellmyer, *J. Appl. Phys.*, **2000**, *87*, 6334.
115. K.Yu. Guslienko, O. Chubykalo-Fesenko, O. Mryasov, R. Chantrell, D. Weller, *Phys. Rev. B*, **2004**, *70*, 104405.
116. W.H. Meiklejohn, C.P. Bean, *Phys. Rev.*, **1956**, *102*, 1413.
117. W.H. Meiklejohn, C.P. Bean, *Phys. Rev.*, **1957**, *105*, 904.
118. J. Nogues, I.K. Shuller, *J. Magn. Magn. Mater.*, **1999**, *192*, 203.
119. R.L. Stamps, *J. Phys. D: Appl. Phys.*, **2000**, *33*, R247.
120. M. Kiwi, *J. Magn. Magn. Mater.*, **2001**, *234*, 584.
121. D. Mauri, H.C. Siegmann, P.S. Bagus, E. Kay, *J. Appl. Phys.*, **1987**, *62*, 3047.
122. N.A. Usov, S.A. Gudoshnikov, *J. Magn. Magn. Mater.*, **2006**, *300*, 164.
123. D. Weller, A. Moser, *IEEE Trans. Magn.*, **1999**, *35*, 4423.
124. A. Aharoni, *Phys. Rev.*, **1964**, *135*, A447.
125. A. Aharoni, *Phys. Rev.*, **1969**, *177*, 793.
126. D.A. Smith, F.A. de Rozario, *J. Magn. Magn. Mater.*, **1976**, *3*, 219.
127. I. Eisenstein, A. Aharoni, *Phys. Rev. B*, **1977**, *16*, 1278.
128. I. Eisenstein, A. Aharoni, *Phys. Rev. B*, **1977**, *16*, 1285.
129. W.F. Brown, Jr., *IEEE Trans. Magn.*, **1979**, *15*, 1197.
130. I. Klik, L. Gunther, *J. Stat. Phys.*, **1990**, *60*, 473.
131. I. Klik, L. Gunther, *J. Appl. Phys.*, **1990**, *67*, 4505.
132. W.T. Coffey, D.S.F. Crothers, Yu.P. Kalmykov, J.T. Waldron, *Phys. Rev. B*, **1995**, *51*, 15947.

133. W.T. Coffey, D.S.F. Crothers, J.L. Dormann, L.J. Geoghegan, Yu.P. Kalmykov, J.T. Waldron, A.W. Wickstead, *Phys. Rev. B*, **1995**, *52*, 15951.
134. D.A. Garanin, *Phys. Rev. E*, **1996**, *54*, 3250.
135. W.T. Coffey, D.S.F. Crothers, J.L. Dormann, L.J. Geoghegan, E.C. Kennedy, *Phys. Rev. B*, **1998**, *58*, 3249.
136. W.T. Coffey, D.S.F. Crothers, J.L. Dormann, Yu.P. Kalmykov, E.C. Kennedy, W. Wernsdorfer, *Phys. Rev. Lett.*, **1998**, *80*, 5655.
137. W.T. Coffey, D.S.F. Crothers, J.L. Dormann, L.J. Geoghegan, E.C. Kennedy, W. Wernsdorfer, *J. Phys.: Condens. Matter*, **1998**, *10*, 9093.
138. Yu.P. Kalmykov, S.A. Titov, and W.T. Coffey, *Phys. Rev. B*, **1998**, *58*, 3267.
139. D.A. Garanin, E.S. Kennedy, D.S.F. Crothers, W.T. Coffey, *Phys. Rev. E*, **1999**, *60*, 6499.
140. D.J. McCarthy, W.T. Coffey, *J. Phys.: Condens. Matter*, **1999**, *11*, 10531.
141. Yu.P. Kalmykov, *Phys. Rev. B*, **2000**, *61*, 6205.
142. B.J. Matkowsky, Z. Schuss, C. Tier, *J. Stat. Phys.*, **1984**, *35*, 443.
143. J.L. Garcia-Palacios, F.J. Lazaro, *Phys. Rev. B*, **1998**, *58*, 14937.
144. Y. Nakatani, Y. Uesaka, N. Hayashi, H. Fukushima, *J. Magn. Magn. Mater.*, **1997**, *168*, 347.
145. W. Scholz, T. Schrefl, J. Fidler, *J. Magn. Magn. Mater.*, **2001**, *233*, 296.
146. D.V. Berkov, *IEEE Trans. Magn.*, **2002**, *38*, 2489.
147. W. Wernsdorfer, E.B. Orozco, K. Hasselbach, A. Benoit, B. Barbara, N. Demoncy, A. Loiseau, H. Pascard, D. Mailly, *Phys. Rev. Lett.*, **1997**, *78*, 1791.
148. N.A. Usov, Yu.B. Grebenshchikov, *J. Appl. Phys.*, **2009**, to be published.
149. A.M. Afanas'ev, I.P. Suzdalev, M.Ya. Gen, V.I. Goldanskii, V.P. Korneev, E.A. Manykin, *Sov. Phys. – JETP*, **1970**, *31*, 65.
150. A.P. Amuljavichus, I.P. Suzdalev, *Sov. Phys. – JETP*, **1973**, *37*, 859.
151. A.M. Afanas'ev, E.A. Manykin, E.V. Onishchenko, *Sov. Phys. – Solid State*, **1973**, *14*, 2175.
152. I. Eisenstein, A. Aharoni, *Phys. Rev. B*, **1976**, *14*, 2078.
153. R.H. Koch, G. Grinstein, G.A. Keefe, Y. Lu, P.L. Trouilloud, W.J. Gallagher, S.S.P. Parkin, *Phys. Rev. Lett.*, **2000**, *84*, 5419.
154. H.Q. Yin, W.D. Doyle, *J. Appl. Phys.*, **2002**, *91*, 7709.
155. E. Weinan, W. Ren, E. Vanden-Eijnden, *J. Appl. Phys.*, **2003**, *93*, 2275.
156. N.A. Usov, S.A. Gudoshnikov, *J. Magn. Magn. Mater.*, **2005**, *290–291*, 727.
157. F. Dorfbauer, R. Evans, M. Kirschner, O. Chubykalo-Fesenko, R. Chantrell, T. Schrefl, *J. Magn. Magn. Mater.*, **2007**, *316*, e791.

9
High-Spin Polynuclear Carboxylate Complexes and Molecular Magnets with VII and VIII Group 3d-Metals

Igor L. Eremenko, Aleksey A. Sidorov, and Mikhail A. Kiskin

9.1
Introduction

Metal-containing nanosized molecules have attracted attention for many years because of their unique properties, which are already employed in practice. For example, giant clusters containing 561 palladium atoms have unusual catalytic properties [1–3]. Nanosized molecules of polyoxometalate derivatives of V and VI Group d-metals not only exhibit catalytic activity but also hold promise as bioactive compounds or new analytical reagents [4–6]. In recent years, nanosized high-spin polynuclear organometallic and coordination compounds of transition metals and lanthanides, the so-called molecular magnets, have captured the interest of both chemists and physicists [7–11]. This interest is understandable and is associated with new prospects for the synthesis of magnetically active molecular materials with large magnetic moments. Generally, paramagnetic complexes, in which high-spin metal ions acting as carriers of magnetism are enclosed in the ligand environment formed by organic molecules, serve as building blocks for magnetically active molecular materials. The molecular magnetism is, in essence, the supramolecular phenomenon because it is generated by the collective properties of the components having unpaired electrons, and it depends on their relative arrangement in organized ensembles and crystal structures. The engineering of molecular magnetic systems calls for a search of new high-spin metal-containing components and the appropriate polynuclear or supramolecular assembly of these components in such a way so as to achieve an ordered arrangement due to spin–spin exchange interactions [7–11]. A particular intermolecular organization of molecules giving rise to a highly efficient exchange-coupled macro ensemble is the key problem in the design of molecular magnets, including molecular ferri- and ferromagnets. However, methods for the controlled influence on the structures of multispin systems are lacking for most of the known types of molecular magnets. As a rule, the manifestation of ferromagnetism in crystals of high-spin complexes is still

Magnetic Nanoparticles. Sergey P. Gubin
© 2009 WILEY-VCH Verlag GmbH & Co. KGaA, Weinheim
ISBN: 978-3-527-40790-3

accidental, though some correlations between the crystal structure and the magnetic properties have been found.

The dodecanuclear manganese carboxylate cluster $Mn_{12}O_{12}(OOCMe)_{16}(H_2O)_4$ that crystallizes as a solvate with four water molecules and two acetic acid molecules [12–14] is a prominent example of nanosized molecular magnets. Actually, this single molecular magnet (SMM), on the one hand, has properties of a conventional magnetic material (for example, it is characterized by residual magnetization and a hysteresis loop) and, on the other hand, exhibits quantum characteristics typical of subnanosized materials, such as the quantum tunneling of magnetization (QTM) [15, 16]. A rather large number of nanosized molecular magnets containing different metals are known. These compounds have different structures and sizes. For example, the cluster octacation $[Fe_8O_2(OH)_{12}(tacn)_6]^{8+}$ (tacn is 1,4,7-triazacyclononane) [17, 16] contains only eight iron atoms, whereas the cluster $Mn_{30}O_{24}(OH)_8(OOCCH_2CMe_3)_{32}(H_2O)_2(MeNO_2)_4$ [18] has 30 manganese atoms.

Among a huge number of known nanomolecules containing transition-metal atoms, carboxylate complexes and clusters occupy a special place. This is associated not only with a diversity of structural types formed by carboxylate compounds (which is of importance for the development of fundamentals of synthesis and structures of such metal complexes) but also with the particular properties of these compounds enabling their use in industrial processes. In addition, these complexes are convenient starting compounds for the design of the most appropriate models of certain natural enzymes containing dinuclear carboxylate- or carbamate-bridged metal fragments [19–23].

Since carboxylate anions act as weak-field ligands, d-metal ions exist in the high-spin state in most of polynuclear carboxylate complexes, resulting in the unique magnetic properties of these compounds [7–11, 24, 25]. The magnetic behavior of carboxylates can be controlled by varying the structures and composition of these compounds. Hence, the development of general approaches to the synthesis of polynuclear carboxylates with desired structures is an important problem. Synthetic investigations not only include the simple examination of all carboxylate anions with substituents of different nature but also require a detailed study of the influence of the nature of metal centers, the medium, and other conditions of the synthesis and the elucidation of the structuring role of neutral N-, O-, P-, and other donor ligands involved in the formation of carboxylate clusters and complexes with different structures.

In our investigations on the synthesis of carboxylate derivatives of manganese, iron, cobalt, and nickel, we used trimethylacetate ligands containing bulky donor *tert*-butyl substituents, which generally ensure the formation of discrete structures with high-spin metal atoms. In addition, these compounds are readily soluble in conventional organic solvents. The latter fact is convenient for studying the chemical properties and isolating analytically pure compounds as single crystals. The latter are necessary for performing high-quality physicochemical studies, including X-ray diffraction.

9.2
High-Spin 3d-Metal Pivalate Polymers as a Good Starting Spin Materials

The development of efficient methods for the synthesis of convenient starting spin materials is an important problem in the chemical construction of carboxylate molecular magnets with desired properties. Polymeric 3d-metal carboxylates with particular structures and having high reactivity toward various organic donors can serve as the starting compounds for the easy generation of new magnetic molecules with desired structures and compositions. For example, the compositionally similar pivalate coordination polymers [M(Piv)$_2$]$_n$ (M is a 3d metal, Piv is anion of pivalic acid), which are formed by self-assembly with the use of certain metal-to-carboxylic acid ratios, are such compounds.

The most convenient and promising methods for the synthesis of polymeric molecules [M(Piv)$_2$]$_n$ containing high-spin Mn(II) ($S = 5/2$), Fe(II) ($S = 2$), Co(II) ($S = 3/2$), or Ni(II) ($S = 1$) atoms are based on the metathesis of inorganic salts of these metals with potassium pivalate KPiv (**1**). For example, the reaction of MnCl$_2$·4H$_2$O with **1** in a ratio of 1 : 2 in EtOH at 78 °C affords the polymer [Mn(Piv)$_2$(HOEt)]$_n$ (**2**) in 87% yield [26, 27]. Attempts to synthesize iron-containing analog **2** according to this procedure failed; instead, only the tetranuclear iron(II) cluster, Fe$_4$(μ_3-OH)$_2$(μ-Piv)$_4$(η^2-Piv)$_2$(EtOH)$_6$ (**3**), was produced under these conditions [27, 28]. Hence, we used acetonitrile as the solvent. This reaction afforded the polymer [Fe(μ-Piv)$_2$]$_n$ (**4**) in high yield. The reaction of the latter polymer with ethanol (1 mol) gave iron-containing analog of **2**, [Fe(Piv)$_2$(HOEt)]$_n$ (**5**). According to X-ray diffraction data [26, 28], polymers **2**, **4**, and **5** (Figure 9.1) have a chain structure (Mn–O(OOCR) in **2**, 2.078(2)–2.157(2) Å; Fe–O(OOCR) in **4** and **5**, 1.975(2)–1.984(2) Å, and 2.092(2)–2.175(2) Å, respectively).

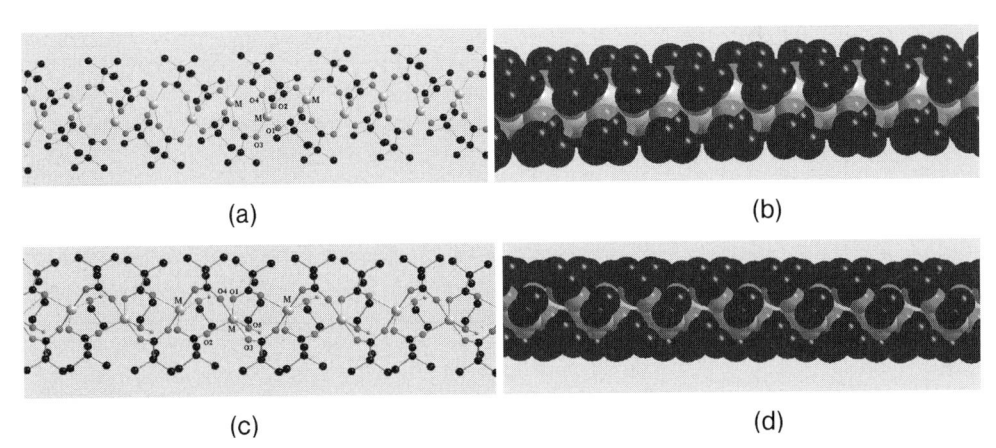

Figure 9.1 Structures of the coordination polymers [M(Piv)$_2$(HOEt)]$_n$ (M = Mn (**2**) or Fe (**5**)) (c) and (d) and [M(μ-Piv)$_2$]$_n$ (M = Fe (**4**)) (a) and (b).

Coordination polymers **2**, **4**, and **5** contain high-spin metal atoms. However, in spite of the fact that their metal carboxylate cores are structurally similar, the magnetic properties of these compounds are substantially different and depend on the nature of the metal centers. For example, manganese derivative **2** proved to be antiferromagnetic (Figure 9.2(a)) [26], whereas its iron-containing analog **5** exhibits ferromagnetic properties at helium temperatures (Figure 9.2(b)) [28, 29].

Iron-containing polymer **4** has even more unusual magnetic characteristics. Thus, the magnetic ordering was observed for this compound at 3.8 K (Figure 9.3).

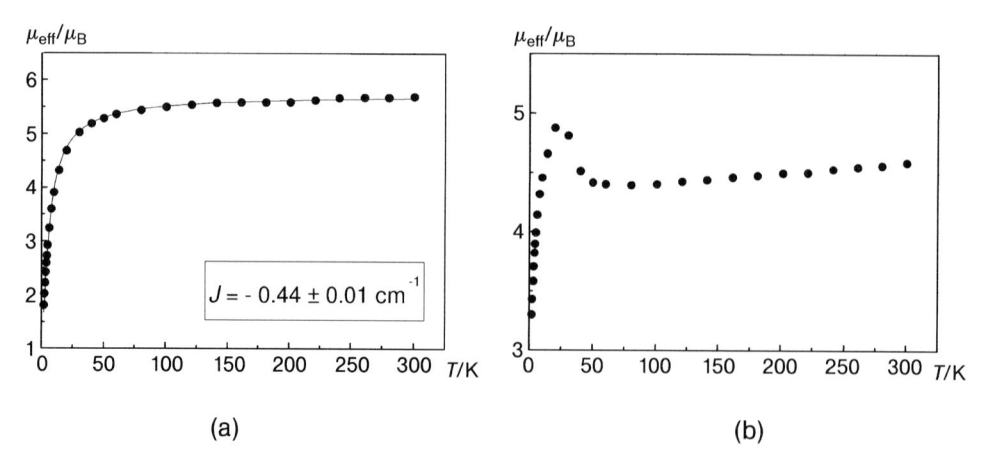

Figure 9.2 Magnetic properties of the polymers $[M(Piv)_2(HOEt)]_n$ (M = Mn (**2**) or Fe (**5**)) (a) and (b).

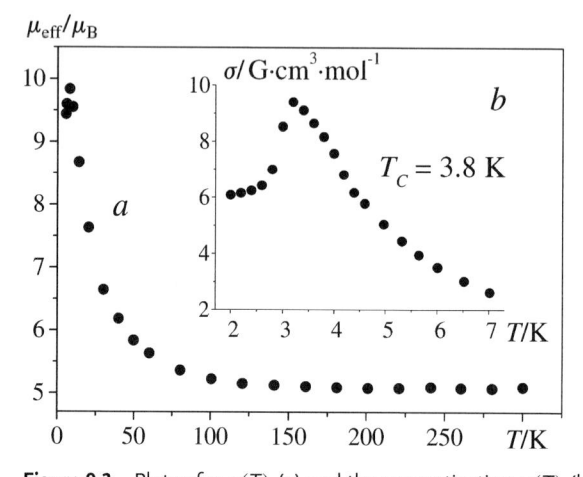

Figure 9.3 Plots of $\mu_{eff}(T)$ (a) and the magnetization $\sigma(T)$ (b) in weak field for **4**.

It should be noted that analogous cobalt(II) and nickel(II) polymers synthesized from hydrated salts of these metals by the metathesis with potassium pivalate (1) in water or by fusion of hydrated acetates (M = Co) with pivalic acid largely correspond to compounds of variable composition $[(HPiv)_xM(OH)_n(Piv)_{2-n}]_m$ (M = Co or Ni) [30, 31], whose structure is unknown. The presence of variable amounts of solvent or coordinated molecules of pivalic acid, as well as of the hydroxo, oxo, or aqua ligands is a considerable obstacle to their use as the starting reagents. It appeared that stable compounds of constant composition $[M(Piv)_2]_n$ containing high-spin atoms M = Co(II) (6) or Ni(II) (7), which are formal analogs of polymeric iron pivalate 4, can be synthesized from various polynuclear structures, including the above-mentioned polymers of variable composition, through mild thermolysis (to 175 °C) in organic solvents [32]. Crystals of cobalt-containing polymer 6 (prepared from different starting compounds) were identified by X-ray powder diffraction. The polymer $[Co(Piv)_2]_n$ was shown to be an isostructural analog of iron-containing polymer 4. The magnetic data (Figure 9.4) indicate that polymer 6, like iron-containing polymer 4, undergoes a magnetic phase transition to the ordered state ($T_c = 3.4$ K, at $H = 1$ Oe) (Figure 9.4(b)) and exhibits properties of a soft magnet (without a hysteresis loop). The magnetization of polymer 6 in strong field ($H = 50$ kOe) reaches $\sigma_{exp} = 200 \pm 5$ G cm^3/mol (Figure 9.4(c)) [32].

The nickel-containing compound of formal composition $[Ni(piv)_2]_n$ (7) can easily be synthesized by thermolysis of the dinuclear complex $Ni_2(\mu\text{-}H_2O)(Piv)_4(HPiv)_2$ in decane under argon at 174 °C as highly air-sensitive brown-yellow crystals. However, the magnetic properties of this compound strongly differ from those of its iron- and cobalt-containing analogs. Complex 7 exhibits antiferromagnetic behavior in the temperature range of 300–2 K ($\mu_{eff} = 3.092$–$2.078\mu_B$).

The X-ray diffraction study (Figure 9.5) of compound 7 gave unexpected results [32]. It appeared that the cyclic hexanuclear complex $Ni_6(\mu_2\text{-}Piv)_6(\mu_3\text{-}Piv)_6$ (7) (Ni...Ni, 3.284(1)–3.325(1) Å; Ni–Ni–Ni angle, 100.33(3)°–103.15(3)°; Ni–O(μ_2-Piv), 1.935(3)–1.983(3) Å; Ni–O(μ_3-Piv), 1.992(3)–2.028(3) Å) is the main structural unit in the crystal structure of the nickel derivative, as opposed to the iron and cobalt derivatives (4 and 6, respectively), whose crystal structures are composed of infinite chains. The outer diameter of molecule 7 (taking into account the C–H bonds) is ca. 15 Å, and the minimal size of the internal cavity is 4.5 Å.

The discovery of magnetic ordering effects in the polymers of simple composition $[M(\mu\text{-}Piv)_2]_n$ (M = Fe (4) or Co(6)) has stimulated research into modifications of such chain structures in an effort to prepare new derivatives of this class exhibiting unusual magnetic behavior. High-spin pivalate polymers can be modified by reactions with certain N-donor organic molecules. For example, the treatment of manganese(II) and iron(II) polymers 2, 4, or 5 with the 1,2-phenylenediamine ligand (L), which generally acts as a chelating agent, in a ratio [M]:L $\geq 3:2$, can lead to redistribution of the bridging pivalate anions

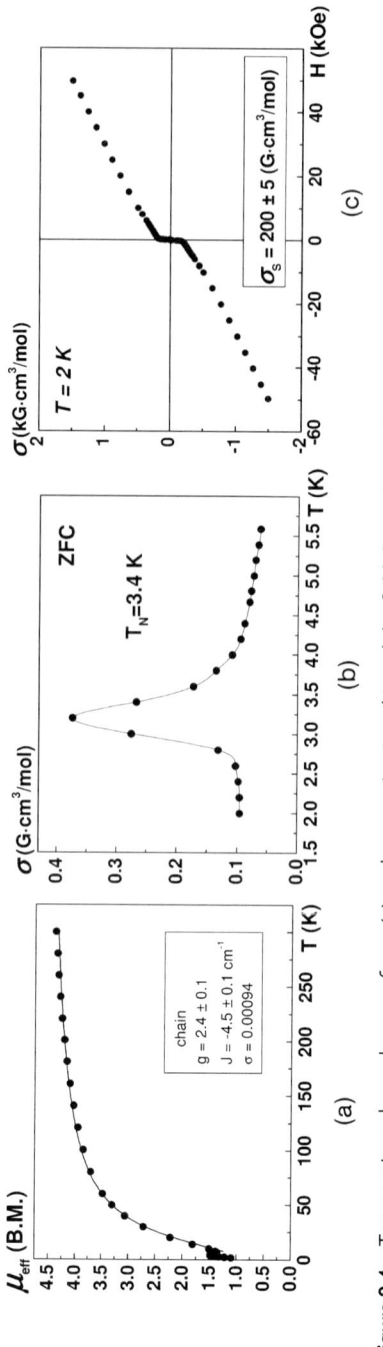

Figure 9.4 Temperature dependences of μ_{eff} (a) and magnetization (b) and the field dependence of magnetization (c) for complex **6**.

in the polymer chain giving rise to the polymers $\{[(\eta^2\text{-}(NH_2)_2C_6H_2R_2)_2M(\mu\text{-}Piv)_2][M_2(\mu\text{-}Piv)_4]\}_n$ (M = Mn (**8**) or Fe (**9**); R = H (**a**) or Me (**b**)) containing the alternating di- and mononuclear fragments, and only the latter fragments contain the *ortho*-phenylenediamine chelate ligands (Figure 9.6) [27, 28].

The magnetic characteristics of the manganese- and iron-containing pivalate polymers, **8** and **9**, modified with *ortho*-phenylenediamines appeared to be substantially different. Thus, the manganese derivatives exhibit antiferromagnetic properties (Figures 9.7(a) and (b)) due to intra- and intermolecular spin–spin exchange [28].

The situation with the magnetic properties of iron-containing analogs **9a** and **9b** is much more complicated [28, 29]. The dependences $\mu_{eff}(T)$ for these complexes are substantially different (Figure 9.8). The magnetic behavior of compound **9a** with unsubstituted (R = H) diamine is unusual in that μ_{eff} is virtually constant in the temperature range of 30–100 K and is close to the theoretical limit of 6.92 μ_B only for two noninteracting spins $S = 2$ with the g factor of 2, although the formally repeated exchange-coupled fragment contains three metal atoms (one metal atom in the mononuclear diaminedicarboxylate fragment and two metal atoms in the dinuclear tetrabridged system (Figure 9.8(a)). It is not inconceivable that μ_{eff} of **9a** decreases in the temperature range of 100–200 K due to phase transitions along with antiferromagnetic exchange interactions.

For polymer **9b** (R = Me), the temperature dependence of μ_{eff} is radically different. Thus, the effective magnetic moment gradually decreases in the range of 300–140 K from 9.16 to 9.03 μ_B, increases to 10.06 μ_B as the temperature is lowered to 30 K, which is indicative of the dominant

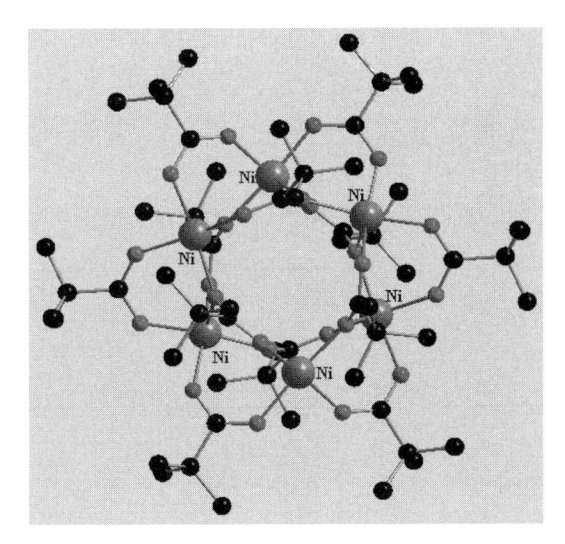

Figure 9.5 Structure of the hexanuclear antiferromagnetic cluster $Ni_6(\mu_2\text{-}Piv)_6(\mu_3\text{-}Piv)_6$ (**7**).

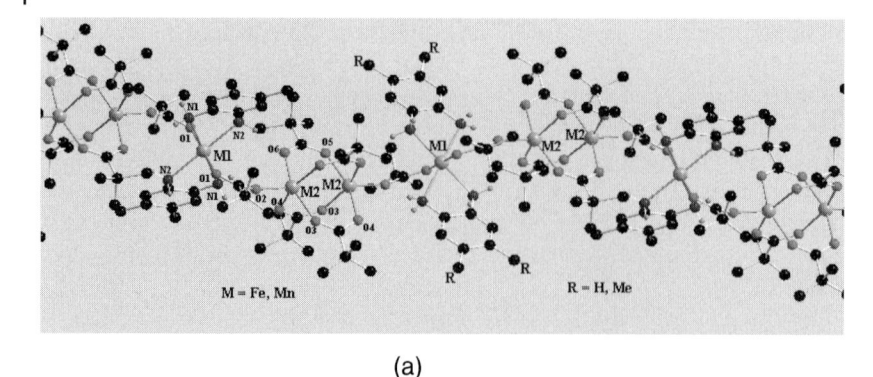

Figure 9.6 Structures of the polymers $\{[(\eta^2\text{-}(NH_2)_2C_6H_2R_2)_2M(\mu\text{-Piv})_2][M_2(\mu\text{-Piv})_4]\}_n$ (M = Mn (**8**) or Fe (**9**); R = H or Me).

ferromagnetic exchange interactions in **9b**, and again decreases to 6.96 μ_B at 2 K due to intermolecular antiferromagnetic interactions. The estimation of exchange interactions in the repeated Fe(1)---Fe(2)---Fe(3) fragments in terms of the isotropic Hamiltonian [25] gave the following parameters: $g_1 = g_2 = 2.09(2)$, $g_3 = 2.00(2)$, $J_{12} = 5.27(9)$ cm^{-1}, $J_{23} = -0.18(7)$ cm^{-1} (J_{12} is the exchange interaction parameter in the dinuclear compound, and J_{23} is the exchange interaction parameter between the dinuclear and mononuclear compounds). The intermolecular exchange parameter is $nJ' = -0.06(2)$ cm^{-1}. It should be noted that the ferromagnetic character of the exchange interactions found for polymeric molecule **9b** has been suggested earlier for the tetrabridged Fe(II) dinuclear compounds, $L_2Fe_2(\mu\text{-OOCR})_4$ (HOOCR is 2,6-di(p-tolyl)benzoic acid, and L is benzylamine or 4-methoxybenzylamine) [33]. In addition, a large ferromagnetic contribution to exchange interactions has been recently substantiated for the dinuclear complex $(2,3\text{-Me}_2C_5H_3N)_2Fe_2(\mu\text{-Piv})_4$ by calculations taking into account the orbital component [34].

The cobalt polymer with methylated diamine $\{[(\eta^2\text{-}(NH_2)_2C_6H_2Me_2)_2Co(\mu\text{-Piv})_2][Co_2(\mu\text{-Piv})_4]\}_n$ (**10**), which was prepared from pivalate polymer **6** or from the polymer of variable composition $[(HPiv)_xCo(OH)_n(Piv)_{2-n}]_m$, exhibits antiferromagnetic properties [35], like manganese-containing analogs **8**. However, an attempt to synthesize an analog of **10** with the unsubstituted ligand,

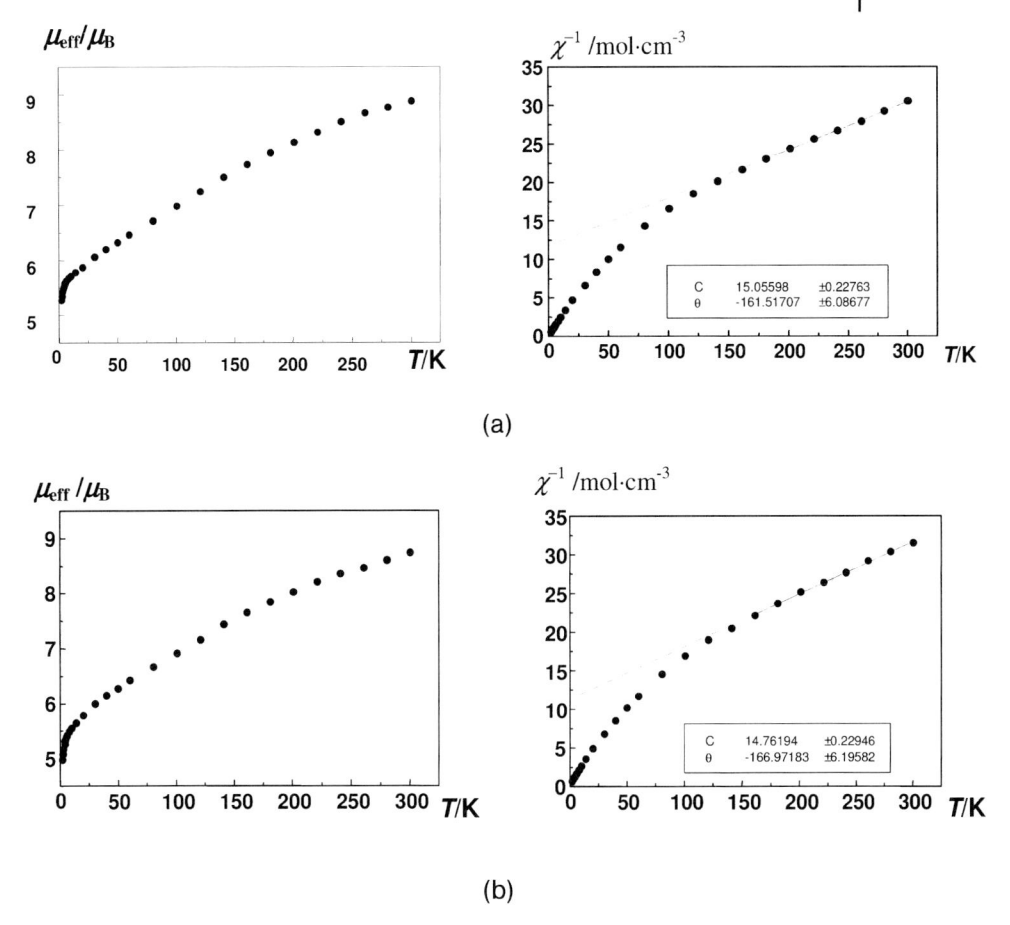

9.2 High-Spin 3d-Metal Pivalate Polymers as a Good Starting Spin Materials | 357

Figure 9.7 Magnetic properties of the coordination polymers $\{[(\eta^2\text{-}(NH_2)_2C_6H_2R_2)_2Mn(\mu\text{-}Piv)_2][Mn_2(\mu\text{-}Piv)_4]\}_n$ (R = H (**8a**) (a) or Me (**8b**) (b)).

1,2-$(NH_2)_2C_6H_4$, unexpectedly led to the formation of the polymeric complex $[Co_2(\mu\text{-}C_8N_2H_7)(\mu\text{-}Piv)_3]_n$ (**11**) containing no diamine ligands as the major product (65% yield). According to X-ray diffraction data, polymer **11** exists as an infinite one-dimensional chain (Figure 9.9) consisting of the dinuclear fragments $M_2(\mu\text{-}Piv)_3$ (Co...Co, 3.253(2) Å; Co–O, 1.949(2)–1.951(2) Å) linked by the N atoms (Co–N, 2.016(8) Å) of the deprotonated 2-methylbenzimidazole molecules, which are assembled through nucleophilic addition of two amino groups of o-phenylenediamine at the C≡N triple bond of the acetonitrile (solvent) molecule followed by elimination of the ammonia molecule [35].

In spite of changes in the structural characteristics of the dicobalt moieties of the chain (in these moieties, two cobalt atoms are linked together by only three carboxylate bridges), compound **11**, like polymer **10**, exhibits antiferromagnetic properties, and the effective magnetic moment of **11**

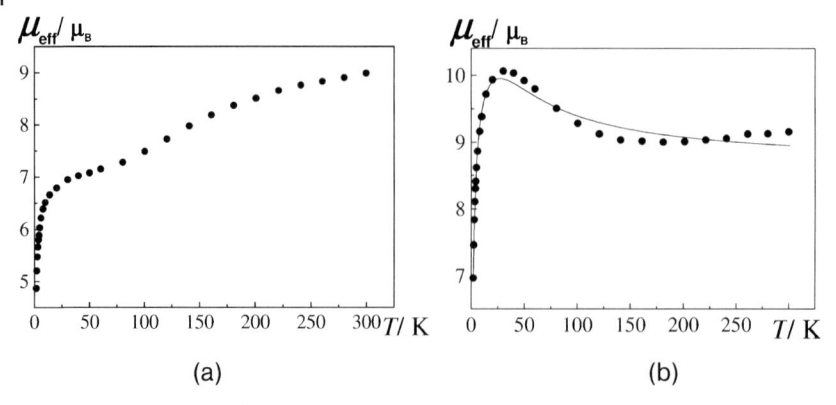

Figure 9.8 Plot $\mu_{\text{eff}}(T)$ for iron-containing polymers **9a** (a) and **9b** (b), the solid line corresponding to theoretical calculations.

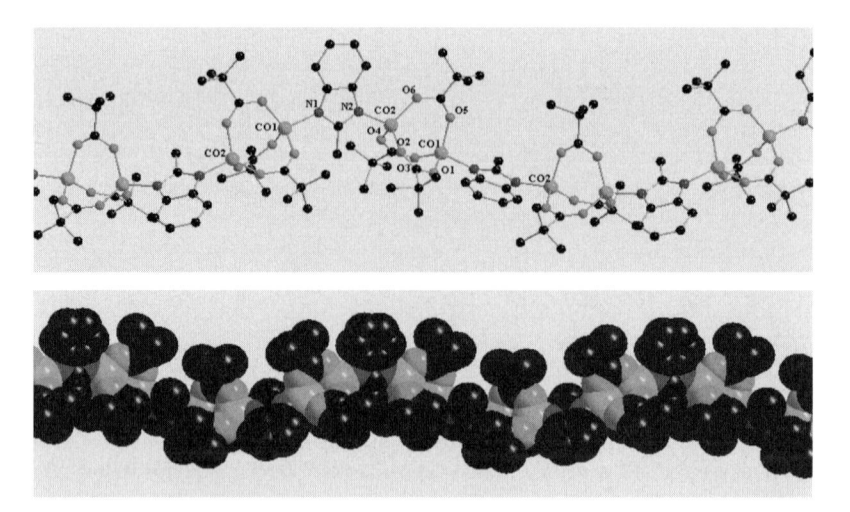

Figure 9.9 Structure of the polymer $[Co_2(\mu\text{-}C_8N_2H_7)(\mu\text{-Piv})_3]_n$ (**11**).

monotonically decreases throughout the temperature range ($\mu_{\text{eff}} = 5.3-0.7\ \mu_B$ (300–2 K)) (Figure 9.10).

The polymers $[M(\mu\text{-Piv})_2]_n$ can also be modified with the use of polydentate pyridine-type N-donors, for example, of pyrimidine or pyrazine. In this case, the ligands serve as bridges between the metal-containing fragments. For example, the reaction of pyrimidine with polymer **6** containing high-spin cobalt(II) atoms affords the 2D polymer $[Co_2(\mu\text{-OH}_2)(\mu\text{-Piv})_2(\text{Piv})_2(\mu\text{-L})_2]_n$ (**12**, L is pyrimidine) (Figure 9.11) [36].

According to X-ray diffraction data, polymeric system **12** consists of the dinuclear $Co_2(\mu\text{-OH}_2)(\mu\text{-Piv})_2(\text{Piv})_2(\mu\text{-L})_4$ moieties. The cobalt atoms in the centrosymmetric unit (Co(1)...Co(1A), 3.630 Å) are bridged by a

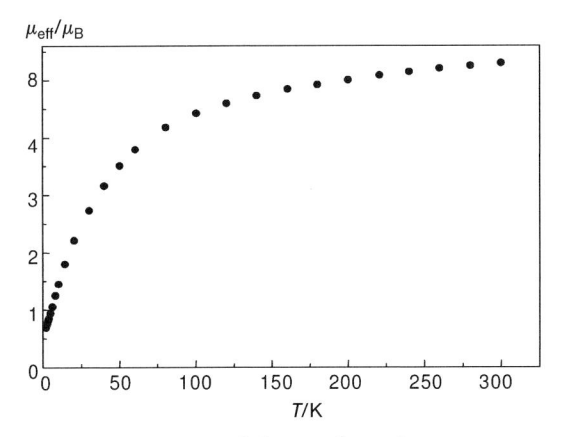

Figure 9.10 Magnetic behavior of complex **11**.

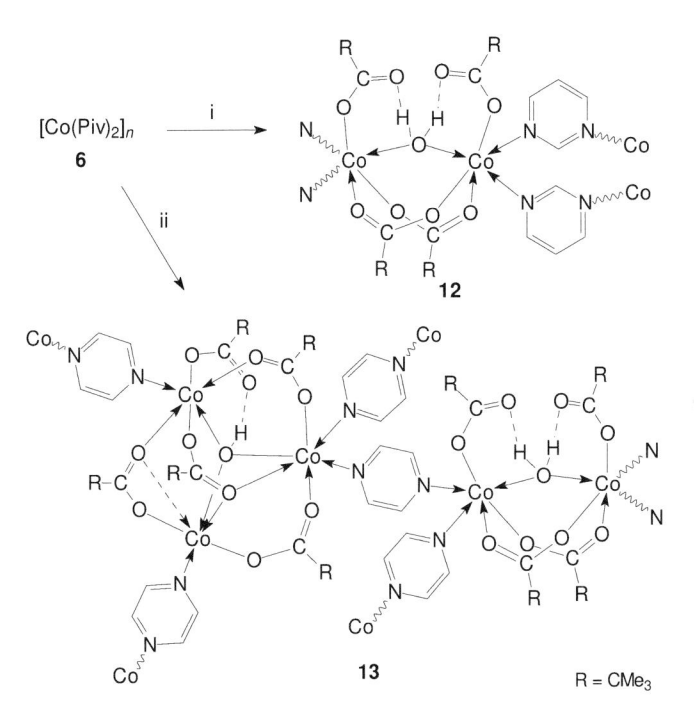

Figure 9.11 Scheme of formation of polymers **12** and **13** [(i), pyrimidine, MeCN, 80 °C; (ii), pyrazine, MeCN, 80 °C].

water molecule and two μ_2-pivalate groups (Figure 9.12). The cobalt atom in compound **12** is in an octahedral ligand environment formed by one monodentate pivalate group and two pyrimidine molecules. The protons of the bridging water molecule are involved in hydrogen bonding with the oxygen atoms of two monodentate pivalate anions. The dinuclear $Co_2(\mu\text{-}OH_2)(\mu\text{-}Piv)_2(Piv)_2$ fragments are bridged by four pyrimidine ligands (two ligands per Co(II) atom in the dinuclear carboxylate moiety) to form a layer of the 2D framework.

Recently, a similar geometry of the dinuclear metal moieties $Co_2(\mu\text{-}OH_2)(\mu\text{-}Piv)_2(Piv)_2(L)_4$ has been observed in the antiferromagnetic dinuclear complexes $Co_2(\mu\text{-}OH_2)(\mu\text{-}Piv)_2(Piv)_2(Py)_4$ and $Co_2(\mu\text{-}OH_2)(\mu\text{-}Piv)_2(Piv)_2(Bpy)_2$ [37, 38].

The reaction of the polymer $[Co(Piv)_2]_n$ (**6**) with pyrazine (L) (Co:L = 1:1) in MeCN gives the polymeric compound $[Co_2(\mu\text{-}OH_2)(\mu\text{-}Piv)_2(Piv)_2(\mu\text{-}L)_4Co_3(\mu_3\text{-}OH)(\mu_3\text{-}Piv)(\mu\text{-}Piv)_3(Piv)]_n$ (**13**) (as a solvate with 0.5 HPiv and L) (Figure 9.12) [36].

According to X-ray diffraction data, coordination polymer **13** consists of the dinuclear moieties $Co_2(\mu\text{-}OH_2)(\mu\text{-}Piv)_2(Piv)_2(\mu\text{-}L)_4$ (**13a**) and the trinuclear moieties $Co_3(\mu_3\text{-}OH)(\mu_3\text{-}Piv)(\mu\text{-}Piv)_3(Piv)(\mu\text{-}L)_4$ (**13b**) bridged by the pyrazine ligands. Dinuclear moiety **13a** is structurally similar to complex **12** (Figure 9.13). The protons of the $\mu\text{-}OH_2$ group and the oxygen atoms of the pivalate groups in moiety **13a** are involved in hydrogen bonding (H(O(1M))...O(6), 1.73 Å; H(O(1M))...O(8), 1.67 Å; the O(1M)...H...O(6) and O(1M)...H...O(8) angles, 142.9° and 143.75°, respectively; O–C, 1.21(5)–1.24(5) Å; the O–C–O angles, 125(4)° and 126(4)°). The cobalt atoms in trinuclear moiety **13b** are bridged by one hydroxo, one μ_3-pivalate, and three μ_2-pivalate groups. The Co_3O fragment in **13b** is nonplanar (the oxygen atom of the OH group protrudes from the Co_3 plane by 0.81 Å). The proton of the μ_3-OH group and the oxygen atom of the pivalate group are involved in hydrogen bonding (H(O(2M))...O(18), 1.94 Å; the O(2M)...H...O(18) angle is 132.6°;

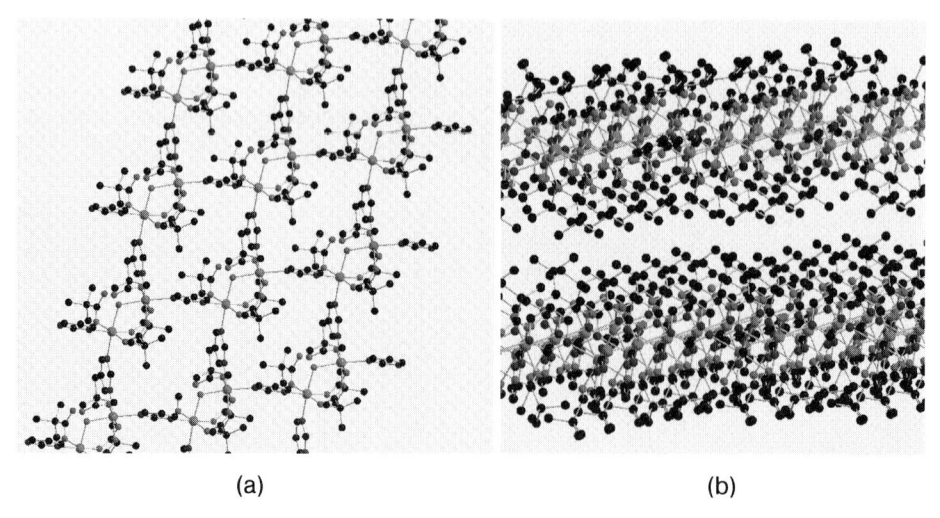

Figure 9.12 Fragment of the polymeric layer of the compound $[Co_2(\mu\text{-}OH_2)(\mu\text{-}Piv)_2(Piv)_2(\mu\text{-}L)_2]_n$ (**12**) (a) and the packing of the layers in the crystal structure (b).

O(17)–C(41), 1.19(7) Å; O(18)–C(41), 1.23(8) Å; O(17)–C(41)–O(18) angle is 126(5)°). The Co(4) and Co(3) atoms are bound to one and two nitrogen atoms of the pyrazine ligands, and both the cobalt atoms are in a distorted octahedral environment. The Co(5) atom is in a distorted trigonal-bipyramidal environment formed by three carboxylate oxygen atoms, one hydroxy oxygen atom, and one nitrogen atom of the pyrazine ligand. In addition, there is a weak interaction with the oxygen atom of the μ_2-pivalate group (Co(5)... O(16), 2.50(4) Å) (Figure 9.14).

In the crystal structure, dinuclear fragments **13a** are linked to two dinuclear moieties and two trinuclear moieties by the bridging pyrazine molecules. Trinuclear moiety **13b** is linked to two dinuclear and two trinuclear moieties.

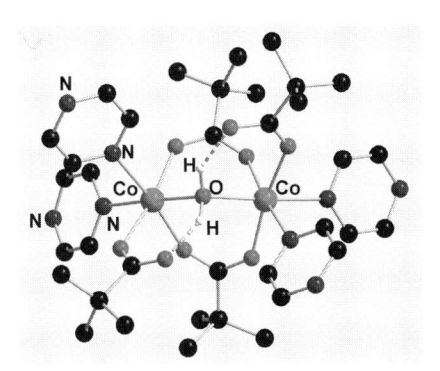

Figure 9.13 Structure of the dinuclear moiety Co$_2$(μ-OH$_2$)(μ-Piv)$_2$(Piv)$_2$(μ-L)$_4$ (**13a**) in polymer **13**.

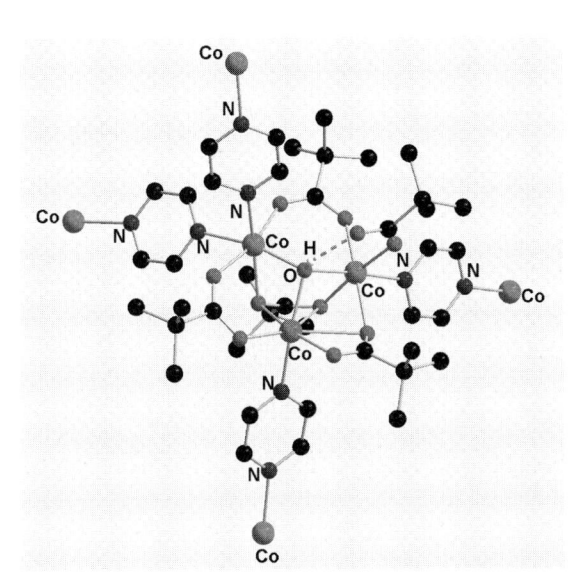

Figure 9.14 Structure of the trinuclear moieties Co$_3$(μ_3-OH)(μ_3-Piv)(μ-Piv)$_3$(Piv)(μ-L)$_4$ (**13b**) in polymer **13**.

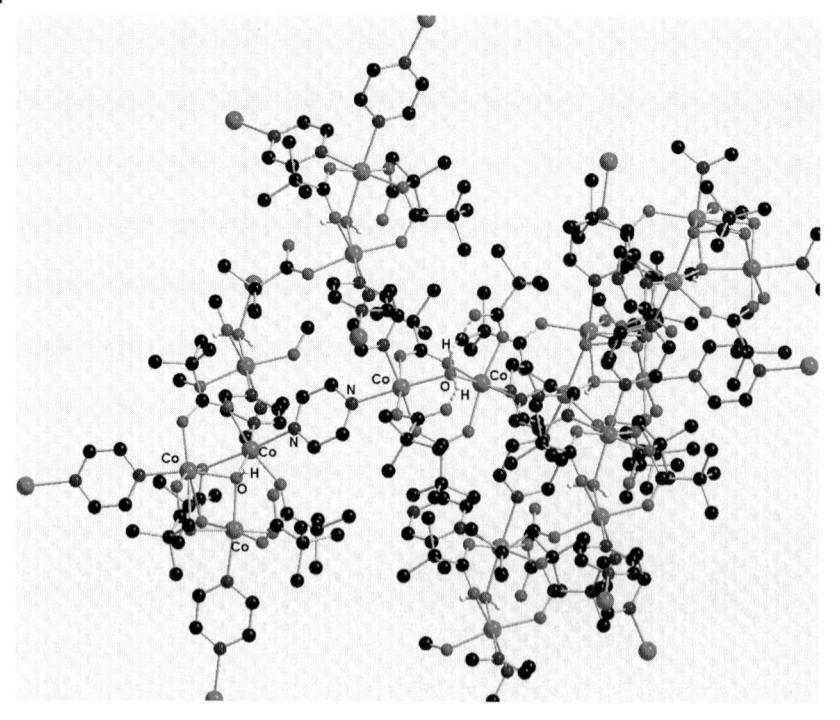

Figure 9.15 Structure of the three-dimensional polymer
$[Co_2(\mu\text{-}OH_2)(\mu\text{-}Piv)_2(Piv)_2(\mu\text{-}L)_4Co_3(\mu_3\text{-}OH)(\mu_3\text{-}Piv)(\mu\text{-}Piv)_3(Piv)]_n$ **(13)**.

This type of bonding gives rise to a 3D framework (Figure 9.15).

Magnetic measurements of compound **12** showed that the effective magnetic moment (μ_{eff}) monotonically decreases from 8.36 to 4.45 μ_B in the temperature range from 300 to 8 K (Figure 9.16(a)). This magnetic behavior indicates that antiferromagnetic spin–spin exchange interactions between the cobalt atoms and spin–orbital interactions are dominant in complex **12**. Below 8 K, the magnetic susceptibility depends on the applied magnetic field. The magnetization isotherms in this region can be defined by the relation $\sigma(H) = \sigma_0 + \chi H$, which is typical of antiferromagnets with a weak ferromagnetic component. However, no hysteresis effects were observed for complex **12**. The data about the temperature dependence of σ_0 show that the Néel point for complex **12** is approximately equal to 4.5 K. Thus, complex **12** is an antiferromagnet having a weak ferromagnetic component with $T_N = 4.5$ K and $\sigma_0(2\text{ K}) = 4000$ G cm^3/mol.

The effective magnetic moment of complex **13** monotonically decreases from 10.89 μ_B at 300 K to 4.33 μ_B at 2 K (Figure 9.17) due probably to antiferromagnetic exchange interactions between the Co(II) ions in an octahedral environment and spin–orbital interactions.

It should be noted that even simple, at first glance, carboxylate 1D polymers can contain very complex repeated metal fragments. A polymeric structure

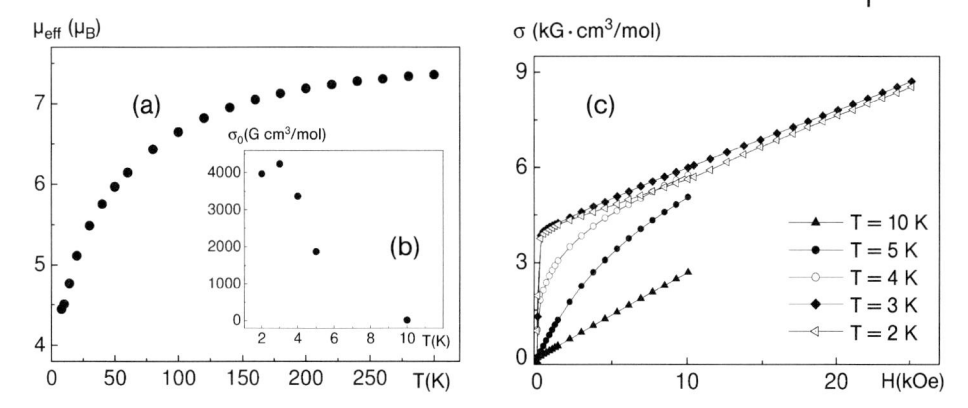

Figure 9.16 Magnetic properties of compound **12**: (a) temperature dependence of magnetic moment (a), temperature dependence of spontaneous magnetization (b), main magnetization curves (c).

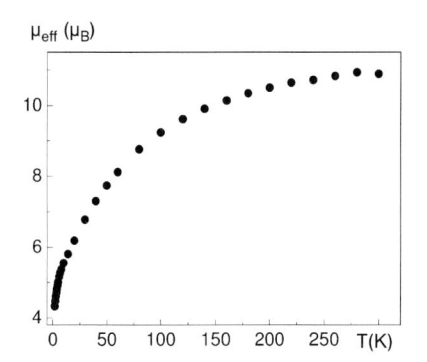

Figure 9.17 Plot $\mu_{eff}(T)$ for compound **13**.

consisting of the repeated decanuclear fragments and at the same time adopting a chain conformation can be cited as an example. The polymer $\{(MeCN)_2(HPiv)_2(H_2O)_2Mn_{10}Cl_2(OH)_2(Piv)_{16} \cdot MeCN\}_n$ (**14**) is formed by the above-mentioned metathesis of Mn(II) chloride with potassium pivalate with the use of a deficient amount of pivalate anions followed by recrystallization of the reaction product from MeCN [29]. In this system, not only pivalate anions but also chlorine atoms serve as bridges (Figure 9.18).

The sizes of the {Mn$_{10}$} units linked together to form an infinite chain, which is formally limited only by the crystal size, are ca. 27 × 6 Å (taking into account all C–H bonds). The magnetic properties of polymeric molecule **14** are similar to those of polymeric pivalate **2**, which does not contain bridging chlorine atoms. A lowering of the temperature leads to a decrease in μ_{eff} of complex **14** to 3.29 μ_B at 2 K (Figure 9.19), which is the evidence that antiferromagnetic interactions are dominant in this complex. The efficiency of these interactions

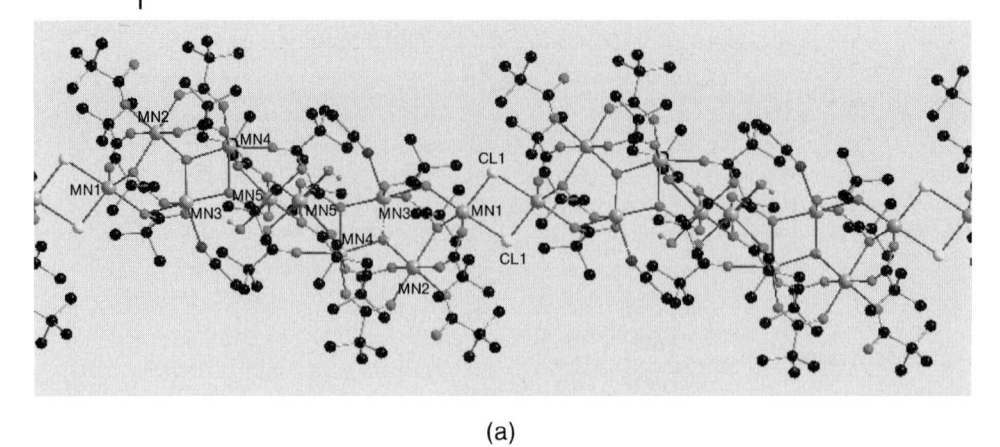

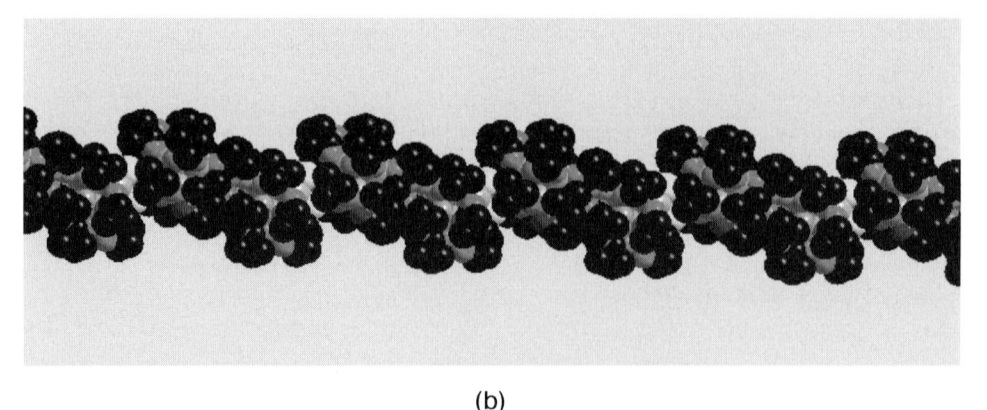

Figure 9.18 Structure of the coordination polymer {(MeCN)$_2$(HPiv)$_2$(H$_2$O)$_2$Mn$_{10}$Cl$_2$(OH)$_2$(Piv)$_{16}$·MeCN}$_n$ (**14**).

is rather high, because μ_{eff} is 12.57 μ_B already at 300 K (the effective magnetic moment was calculated per crystallographically independent formula unit {Mn10}/2) and is close to the pure spin limit (13.2 μ_B) for five weakly interacting Mn(II) ions with the spins $S = 5/2$. In the temperature range of 100–300 K, the magnetic susceptibility obeys the Curie–Weiss law with the parameters $C = 23.2 \pm 0.2$ K cm^3/mol and $\theta = -52.3 \pm 0.2$ K [29].

In principle, the above-considered high-spin pivalate polymers provide the possibility to choose the starting spin materials both in terms of structural features and from the viewpoint of the diversity of magnetic characteristics. In addition, certain polymers are of interest as molecular magnetic materials. A series of other polynuclear structures, e.g., nanosized molecular spin intermediates, such as the antiferromagnetic nonanuclear nickel cluster Ni$_9$(HPiv)$_4$(μ_4-OH)(μ_3-OH)$_3$(μ_n-Piv)$_{12}$ (**15**) [39], can be added to the list of the above-mentioned compounds. Cluster **15** is produced in high yield from

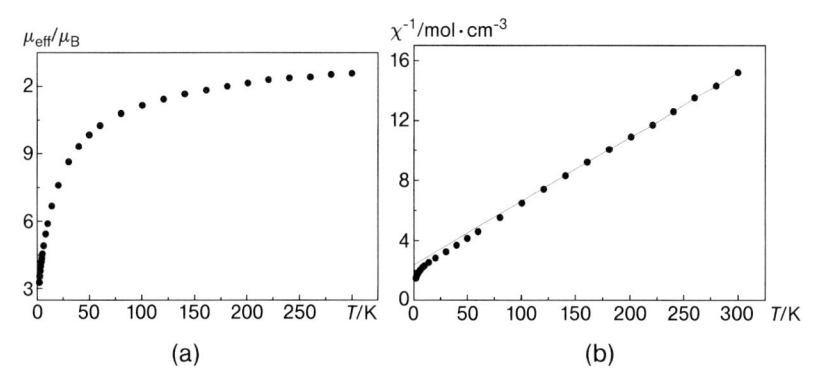

Figure 9.19 Plots $\mu_{eff}(T)$ (a) and $1/\chi(T)$ (b) for polymer **14**.

hydrated nickel chloride and potassium pivalate, followed by the extraction of the reaction product from various hydrocarbon solvents.

In addition to such nanosized molecules, smaller clusters were used in the chemical assembly. The compounds $M_4(EtOH)_6(\mu_3\text{-}OH)_2(\mu_2\text{-}Piv)_4(\eta^2\text{-}Piv)_2$ (M = Co(**16**) or Ni(**17**)) containing high-spin cobalt and nickel atoms [40–42] can be referred to as examples. These molecular units contain coordinated ethanol molecules that are easily eliminated. Under moderate heating, these units generate coordinatively and electronically unsaturated species capable of undergoing further association.

9.3
Chemical Design of High-Spin Polynuclear Structures with Different Magnetic Properties

Among the above-considered pivalate coordination polymers, simple chain polymers are, apparently, most promising as spin sources. Actually, such nanosized building blocks can be cut into various polynuclear fragments employing specific (generally, steric and electronic) characteristics of donor organic ligands that serve as a "cutting tool." The structures of the newly formed polynuclear molecules, i.e., the arrangement of metal atoms (magnetic centers), can be varied as desired. In addition, the reaction and crystallization temperatures or the composition of the medium (either polar or weakly polar solvents) can be used to vary the structures. All these factors often allow the control of the processes giving rise to high-spin molecules. The size and arrangement of magnetic centers in these molecules and, as a result, the physical properties of such compounds can be varied as desired. It should be noted that from the formal point of view, the design of the starting spin materials (polymers) is determined by the bottom-up principle, i.e., the self-assembly of coordination polymeric matrices from coordinatively and electronically unsaturated blocks, which do not necessarily contain only one metal center. These can be fragments consisting of two, three, or more metal centers. In this case, there are wider possibilities to manipulate these systems. In the next step, the top–down principle is employed, e.g., polymers are cut with the use of donor organic molecules. This complex approach allows, in principle, the design of polynuclear structures with any arrangement of metal centers, although detailed studies are obviously necessary to determine the rules of the chemical design in both the first and second steps.

The simplest approach to generate discrete molecular structures from the above-described polymeric pivalates is based on direct reactions of solutions of these compounds in organic solvents with an excess of a donor ligand. In most cases, these reactions afford mononuclear molecules containing high-spin metal ions. For example, the reaction of iron-containing polymer **4** with excess pivalic acid (O-donor) or dimaine (N-donor) gives rise to the mononuclear iron(II) complexes Fe(η^1-Piv)$_2$(η^1-HPiv)$_4$ (5.14 (300 K)–3.38 (2 K) μ_B) and Fe(η^1-Piv)$_2$(η^1-(NH$_2$)$_2$C$_6$H$_4$)$_4$ (5.00 (300 K)–4.66 (4.2 K) μ_B) or Fe(η^1-Piv)$_2$(η^2-(NH$_2$)$_2$C$_6$H$_2$Me$_2$)(η^1-(NH$_2$)$_2$C$_6$H$_2$Me$_2$)$_2$ (5.00 (300 K)–4.12 (2 K) μ_B), in quantitative yields [29]. To control the formation of molecules with desired structures, it is more advantageous to use the steric and electronic effects of donor organic ligands, for example, the known ability of α-substituted pyridines and triethylamine to stimulate the formation of dinuclear tetrabridged transition metal carboxylates LM(μ-OOCR)$_4$ ML [43] and manipulate the metal-to-ligand ratio in the starting reagents. Thus, the reactions of these reagents with polymeric Mn (**2**), Fe (**4** and **5**), and Co (**6**) pivalates and the Ni hexanuclear compound (**7**) afford such antiferromagnetic structures in virtually quantitative yield [26, 28, 29, 44, 45].

To construct molecules containing a large number of magnetic centers, we used tetrazine derivatives, for example, 3-hydroxy-6-(3,5-dimethylpyrazol-1-yl)-1,2,4,5-tetrazine (HL^1) or bis[3,5-(dimethylpyrazolyl)]-1,2,4,5-tetrazine (L^2) containing several donor centers, which can serve as additional bridges and form discrete molecules of a particular shape due to their geometric features. For example, the reaction of HL^1 with cobalt or nickel pivalates (polymers or clusters) affords the pentanuclear complexes M$_5$(μ_3-OH)$_2$(μ-Piv)$_4$(μ-N, N′, N″-3,5-Me$_2$C$_3$HN$_2$C$_2$(O)N$_4$)$_4$(MeCN)$_2$ (M = Co (**18**) or Ni (**19**); M... M, 3.011(1)–3.510(1) Å) (Figure 9.20) [46].

Molecules **18** and **19** can be modified by replacing a part of peripheral carboxylate bridges with bridging chlorine atoms under the action of metal chlorides (in particular, of aqueous NiCl$_2$). The geometric parameters of the resulting clusters M$_5$(μ_3-OH)$_2$(μ-Cl)$_2$(μ-Piv)$_2$(μ-N, N′, N″-3,5-Me$_2$C$_3$HN$_2$C$_2$(O)N$_4$)$_2$(μ-N, N′, N″,O-3,5-Me$_2$C$_3$HN$_2$C$_2$(O)N$_4$)$_2$ (MeCN)$_2$ (M = Co (**20**) or Ni (**21**)) remain mostly unchanged (in **21**, Ni... Ni, 3.080(1)–3.480(1) Å; Ni–Cl, 2.425(2)–2.460(2) Å; based on the unit cell parameters, the cobalt analog is isostructural with the nickel complex).

In spite of the structural analogy, pentanuclear cobalt and nickel clusters **18** and **19** have different magnetic properties [46]. The magnetic behavior of nickel cluster **19** is very unusual. The magnetic moment μ_{eff} monotonically

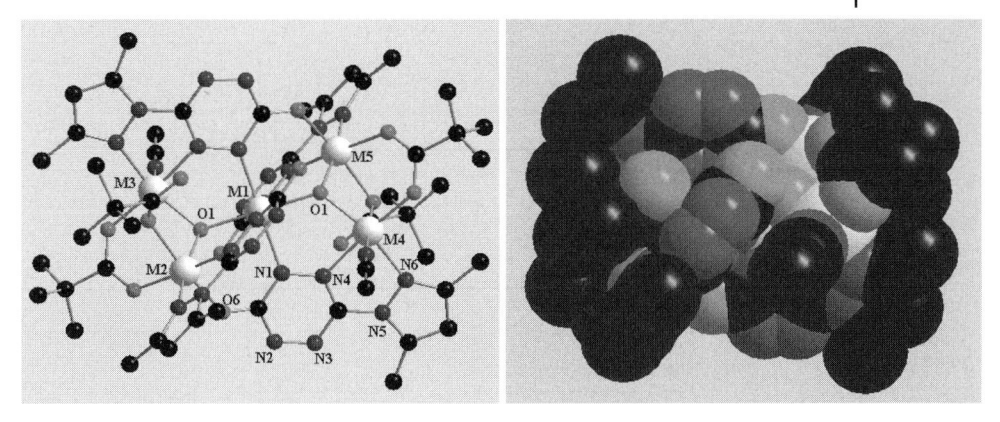

Figure 9.20 Structure of the clusters $M_5(\mu_3\text{-OH})_2(\mu\text{-Piv})_4(\mu\text{-}N, N', N''\text{-}3,5\text{-Me}_2C_3HN_2C_2(O)N_4)_4(C_2NH_3)_2$ (M = Co (**18**) or Ni (**19**)).

decreases ($\mu_{\text{eff}} = 6.4–6.0\ \mu_B$) in the temperature range of 300–125 K, then it passes through a minimum, increases, and reaches a maximum at 15 K ($\mu_{\text{eff}} = 7.3\ \mu_B$). In the temperature range of 15–2 K, a sharp decrease in μ_{eff} is observed apparently due to intermolecular antiferromagnetic spin–spin interactions. This magnetic behavior (the presence of a broad minimum in the curve $\mu_{\text{eff}}(T)$) is typical of ferrimagnetic materials [47] and can be quantitatively interpreted using an approach described in the literature [48]. The calculated exchange parameters ($g = 2.19(6)$, $J_{12} = J_{14} = 31(2)\ \text{cm}^{-1}$, $J_{13} = J_{15} = 56(8)\ \text{cm}^{-1}$, $J_{23} = J_{45} = -11(2)\ \text{cm}^{-1}$, $-2zJ' = 0.129(2)$; $F = 0.01219$, where J_{ij} are the exchange parameters between the Ni$_i$ and Ni$_j$ centers) are indicative of ferromagnetic interactions between the central metal atom and the peripheral metal atoms, whereas interactions between the peripheral metal atoms are antiferromagnetic [46]. Figure 9.21 shows a good correlation between the theoretical curve $\mu_{\text{eff}}(T)$ and experimental data.

The effective magnetic moment μ_{eff} of cobalt-containing pentanuclear compound **18** monotonically decreases ($\mu_{\text{eff}} = 11.6–7.7\ \mu_B$) in the temperature range of 300–2 K apparently due to spin–orbital and antiferromagnetic exchange interactions. Evidently, such substantial differences in the magnetic properties of the pentanuclear clusters are determined primarily by the different electronic nature of the metal centers, which differ by only one electron.

Compared to the starting high-spin cobalt and nickel polymers and oligomers, manganese polymer **2** behaves differently in the reactions with L^1 and L^2 [49]. In the former case, all carboxylate ligands are replaced to form the network high-spin antiferromagnetic 2D polymer $[\text{Mn}(\mu\text{-}N, N', O\text{-}3,5\text{-Me}_2C_3HN_2C_2(O)N_4)_2]_n$ (**22**) (Figure 9.22).

The magnetic behavior of polymer **22** is rather unusual. The effective magnetic moment μ_{eff} of this compound depends slightly only on the temperature in the range of 300–20 K (5.889–5.670 μ_B) and then sharply decreases to

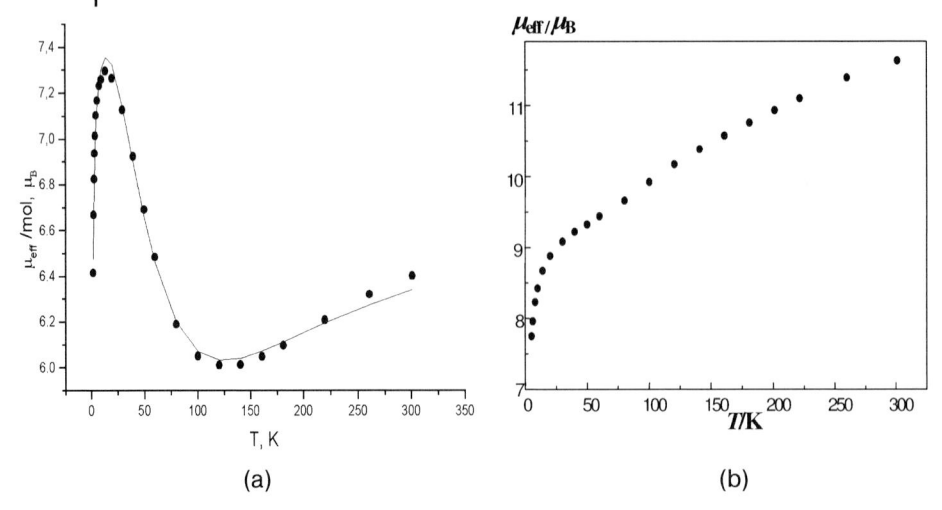

Figure 9.21 Magnetic properties of the pentanuclear clusters M$_5$(μ_3-OH)$_2$(μ-Piv)$_4$(μ-N, N', N''-3,5-Me$_2$C$_3$HN$_2$C$_2$(O)N$_4$)$_4$(MeCN)$_2$ (M = Ni (**19**) (a) or Co(**18**) (b)).

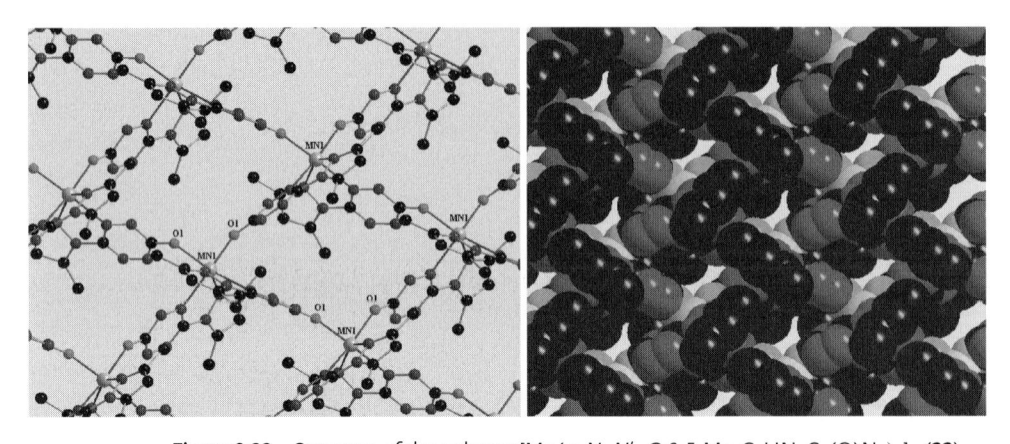

Figure 9.22 Structure of the polymer [Mn(μ-N, N', O-3,5-Me$_2$C$_3$HN$_2$C$_2$(O)N$_4$)$_2$]$_n$ (**22**).

4.639 μ_B at 2 K apparently due to intermolecular antiferromagnetic exchange interactions (Figure 9.23).

The reaction of the starting manganese-containing polymer **2** with L^2 leads to the cleavage of dipyrazolyltetrazine giving rise to the antiferromagnetic heterospin hexanuclear cluster Mn$_6$(μ_4-O)$_2$(μ-Piv)$_{10}$L$_4$ containing dimethylpyrazole molecules as the ligands L (Mn−O(μ_4-O), 1.879(4)−2.219(5) Å; Mn−O(OOCR), 1.950(4)−2.346(5) Å) [49].

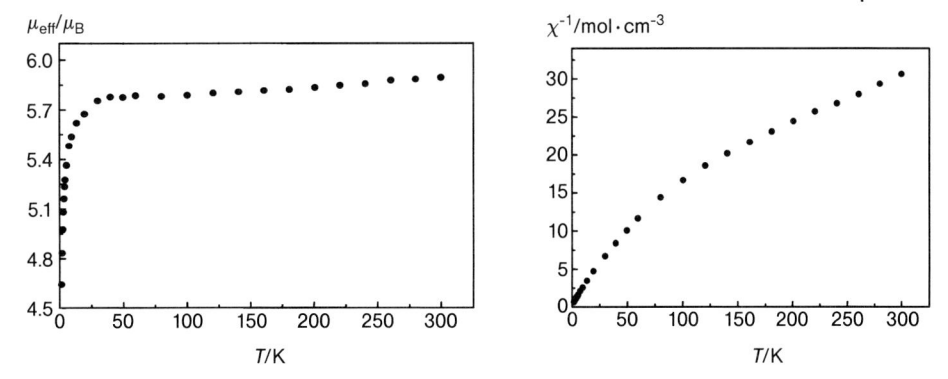

9.3 Chemical Design of High-Spin Polynuclear Structures with Different Magnetic Properties | 369

Figure 9.23 Magnetic characteristics of the polymer [Mn(μ-N, N',O-3,5-Me$_2$C$_3$HN$_2$C$_2$(O)N$_4$)$_2$]$_n$ (**22**).

The use of another polydentate ligand block of the tetrazine series, viz., 3-(3,5-dimethylpyrazol-1-yl)-6-(3,5-diamino-1,3,4-thiadiazolyl)-1,2,4,5-tetrazine (L^3), in the reaction with nonanuclear nickel pivalate **15** made it possible to prepare crystals of the ionic compound [Ni$_8$(μ-OH$_2$)$_4$(μ-Piv)$_4$(η^2-Piv)(Piv)$_9$ (L^3)$_4$]$^+\cdot$Piv$^-$ (**23**, L^3 = N, N', N'', N''', N''''-η^2-N', N'',η^2-N''', N''''-(SC$_2$(NH$_2$)N$_2$)NH(C$_2$N$_4$)(N$_2$(CH)C$_2$(Me)$_2$) (Figure 9.24(a)) [50]. In compound **23**, four ligand molecules L^3 and eight nickel atoms form the cyclic monocationic fragment resembling a box without the bottom and the top (Figure 9.24(b)).

The planar ligands L^3 serve as the walls of the box, whose vertices are formed by nickel atoms (four atoms belong to the inner side of the box and the other

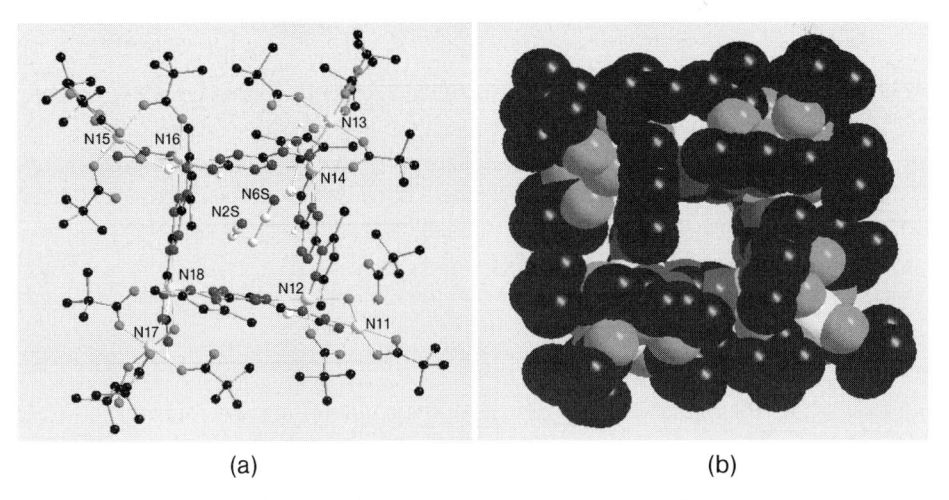

Figure 9.24 Structure of the octanuclear cation [Ni$_8$ (μ-OH$_2$)$_4$ (μ-Piv)$_4$ (η^2-Piv)(Piv)$_9$ (L^3)$_4$]$^+$ with inner acetonitrile solvate molecules (a) and without these molecules (b).

four atoms, to the outer side), and act as bridges between four pairs of nickel atoms (Ni...Ni, 3.325(2)–3.363(1) Å). The inner and outer nickel atoms of each pair are nonequivalent. Hence, the formal scheme of the charge distribution in the octanuclear cation is rather unusual. Each metal center located inside the [Ni(2), Ni(4), Ni(6), Ni(8)] box is coordinated by only one acido ligand (the bridging pivalate group) and is, presumably, positively charged. Three outer atoms located at the periphery of the [Ni(3), Ni(5), Ni(7)] box are coordinated by three pivalate groups (Ni–O, 2.045(7)–2.128(7) Å) each, and the fourth outer metal atom (Ni(1)) is coordinated by only two carboxylate anions, one of which is terminal (Ni–O, 2.092(7) Å) and another is chelate (Ni–O, 2.093(7) Å, 2.138(8) Å). In the latter case, the third carboxylate group (probably belonging to the Ni(1) atom) leaves the metal coordination sphere and becomes the free anion. Therefore, three nickel atoms, Ni(3), Ni(5), and Ni(7), are negatively charged and form zwitterions with the adjacent partially positively charged Ni(4), Ni(6), and Ni(8) atoms. Formally, the Ni(1) atom is neutral, but the presence of the uncompensated positively charged Ni(2) atom in the pair gives rise to a positive charge in octanuclear cation **23** as a whole.

In crystals, nanosized monocationic rings of complex **23** are packed in infinite chains to form a supramolecular ensemble. In spite of the expected repulsion between the positively charged boxes, no disorder is observed. This packing is probably attributed to the arrangement of the sulfur atoms and hydroxy protons of the ligand L^3 (Figure 9.25), which are involved in the five-membered ring and are not formally involved in the binding within the cation, but point in the same direction (for all four ligands in the cation). The presence of lone electron pairs on soft sulfur atoms gives rise to a dipole and, as a result, to dipole–ion interactions with the inner partially positively charged Ni atoms of the next octanuclear monocation. The amino protons of this ring interact with the nitrogen atoms of the acetonitrile solvent molecules (N–H...NCMe, 2.18 Å) located in the inner cavity of the cation (Figure 9.24(a)) [50].

The magnetic properties of this system resemble the behavior of manganese 2D polymer **22**. The effective magnetic moment of complex **23** is virtually temperature-independent in the range of 300–50 K ($\sim$7.45 μ_B). Only in the range of 50–2 K, μ_{eff} monotonically decreases to 4.87 μ_B apparently due, to weak intermolecular antiferromagnetic interactions.

The known examples of the self-assembly of magnetically active manganese-, cobalt-, and nickel-containing clusters illustrate the abilities of tetrazine ligands in the organization of complex and unusual architectures. It is evident that modifications of such systems and their properties based on the replacement of metal centers or carboxylate bridges with other acido ligands hold promise. However, the problem of retention or decomposition of the structure as a whole apparently depends mostly on the nature of metal centers and stability of the structure-forming ligand blocks.

A combination of donor atoms in an organic bridge is one of the main parameters responsible for the strength of binding of bridging ligands to a metal center and the degree of electron density delocalization in M–L–M

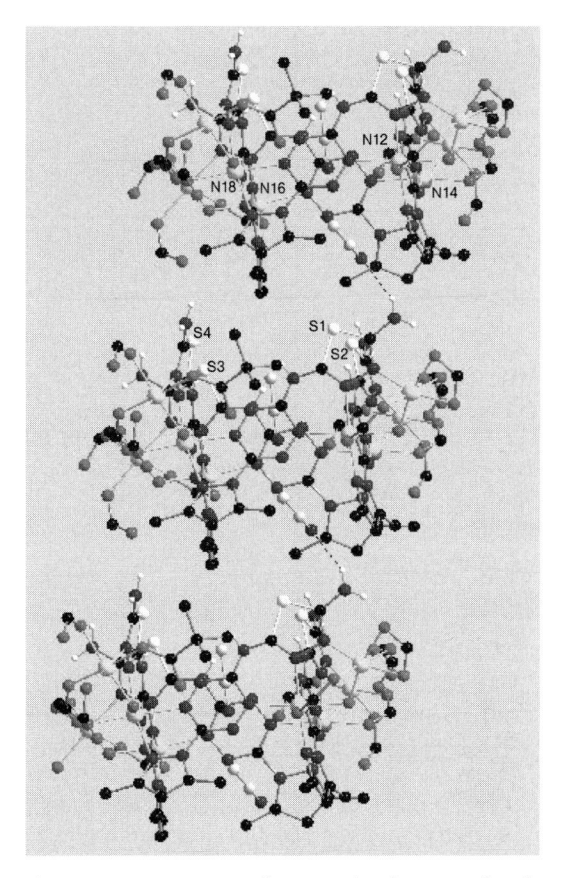

9.3 Chemical Design of High-Spin Polynuclear Structures with Different Magnetic Properties

Figure 9.25 Formation of supramolecular nanotubes from octanuclear nickel cations (*tert*-butyl substituents of the pivalate groups are omitted).

fragments. From this point of view, the carboxylate group −OCO−formally differs from the −NCO−group. Hence, one would expect that the geometric parameters of the resulting molecules or their magnetic properties can be varied by replacing the bridges.

Such structures containing high-spin cobalt or nickel atoms can be prepared by the stepwise replacement of pivalate bridges with 2-hydroxy-6-methylpyridine anions (HL^4) in the starting pivalate polymers or oligomers of these metals. Both cobalt polymer **6** and nickel clusters **7** and **15** react with the HL^4 ligand in MeCN even at room temperature. In both the cases, the metal carboxylate moiety undergoes degradation to form hexanuclear structures (Figure 9.26) [51, 52].

These hexanuclear molecules are different. For example, the cobalt pivalate gives the antiferromagnetic cluster Co$_6$(μ_3-OH)$_2$(η^2, μ_3-L^4)$_2$(μ-Piv)$_8$(HPiv)$_4$ (**24**) containing two bridging deprotonated hydroxypyridine ligands (Co...Co,

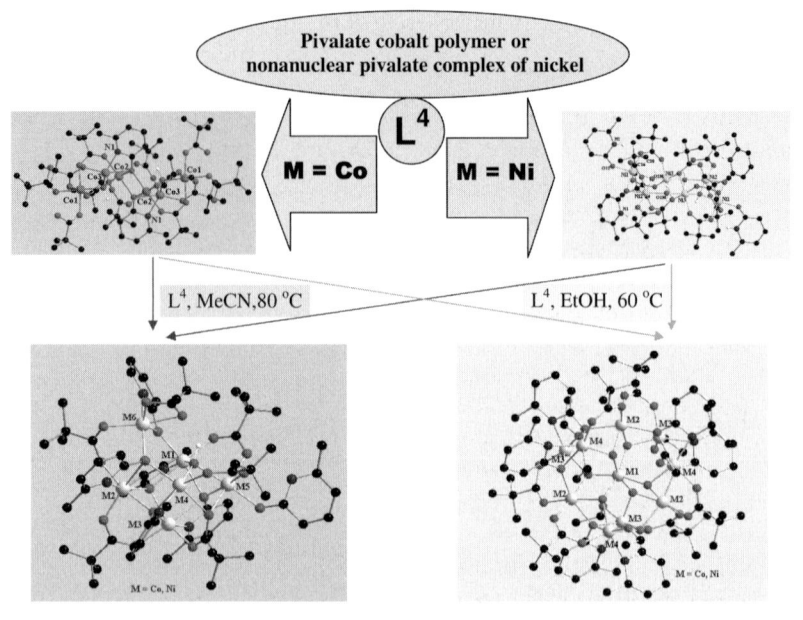

Figure 9.26 Scheme of processes of replacement of pivalate bridges with hydroxypyridine anions in cobalt and nickel derivatives.

3.133(1)–3.385(1) Å; Co–N, 2.032(6) Å; Co–O, 2.390(5)–2.081(5) Å), whereas the nickel derivative forms the antiferromagnetic cluster $(HL)_2(\mu_2\text{-}HL)_2Ni_6(\mu_3\text{-}OH)_2(\mu_2\text{-}H_2O)_2(\mu\text{-}Piv)_8(\eta\text{-}Piv)_2$ (**25**) containing neutral pyridone molecules as ligands, two of which are bridging (Ni–O, 2.135(3)–2.103(3) Å) and the other two are terminal (Ni(1)–O(12), 2.066(3) Å).

An increase in the concentration of the ligand and heating of the reaction mixture in MeCN lead to the further replacement in both hexanuclear molecules (Figure 9.26). The metal core of the identical hexanuclear clusters $(HL^4)M_6(\mu_3\text{-}OH)(\eta^2, \mu_3\text{-}L^4)_3(\eta^2, \mu\text{-}L^4)(\mu_3\text{-}L^4)(\mu_3\text{-}Piv)(\mu\text{-}Piv)_4(\eta^2\text{-}Piv)$ (M = Co (**26**) or Ni (**27**)) formed in both the cases is "crumpled up" compared to the starting structures [51, 52]. From the point of view of the ligand environment, all metal atoms in **26** and **27** are nonequivalent, resulting in the overall asymmetry. In complexes **26** and **27**, the metal atoms are linked by five trimethylacetate bridges, five tridentate bridging hydroxypyridine ligands, and the μ_3-OH group, the distances between the metal atoms being nonbonded (M...M, 3.033(1)–3.769(1) Å).

The changes in the structure of the metal core in clusters **26** and **27** and, as a result, the changes in the number and nature of exchange channels compared to the open starting structures **24** and **25** lead to changes in the magnetic characteristics of these compounds. For example, cobalt compound **26** exhibits antiferromagnetic properties in the temperature range of 300–14 K, and the effective magnetic moment of **26** decreases from 10.88 to 8.06 μ_B.

Then the magnetic moment increases to 8.14 μ_B (6 K) apparently due to intermolecular ferromagnetic exchange interactions and again decreases to 7.25 μ_B (2 K).

The change in the nature of the solvent (the use of ethanol instead of acetonitrile) in the reaction of clusters **24** and **25** with hydroxypyridine (Figure 9.26) results in an expansion of the metal core to form the unusual antiferromagnetic decanuclear complexes $M_{10}(\mu_3\text{-O})_2(\mu_3\text{-OH})_4(\mu\text{-Piv})_6(\mu_3, \eta^2\text{-L}^4)_6(\text{EtOH})_6$ (M = Co (**28**) or Ni (**29**)).

According to X-ray diffraction data, all metal atoms in these isostructural clusters are in an octahedral environment, nine peripheral metal centers being linked to each other to form a closed symmetric system through the μ_3-O atoms of the 6-methyl-2-pyridonate and carboxylate anions. Six of the nine peripheral metal atoms are coordinated by the monodentate ethanol molecule, the oxygen atom of the trimethylacetate anion, and the chelate 6-methyl-2-pyridonate anions (M−O, 2.042(15)−2.245(13) Å; M−N, 2.025(16)−2.043(19) Å). The molecule has C_3 crystallographic symmetry; the threefold axis passes through the central M(1) atom. The coordination environment of this atom is formed by six equivalent oxygen atoms (M_{cen}−O, 2.038(12)−2.084(10) Å). These oxygen atoms form μ_3-O bridges between the M(1) atom and the other nine nickel atoms (M−O, 1.964(11)−2.106(11) Å). Four of these oxygen atoms belong to hydroxy groups and only two of them to oxo bridges. As a result, the protons of four hydroxy groups in the decanuclear molecules are disordered and can formally occupy all six sites (at the oxygen atoms) with occupancy of 2/3.

Based on these data, modifications of carboxylate nanomolecules containing high-spin metal atoms by replacing bridging ligands hold promise, though these processes are less easily controlled. However, this approach can be used to vary the magnetic properties of nanosized molecules. It is important to take into account the specific features of both the carboxylate bridge and the new acido ligand that replaces the bridge.

An attempt to modify manganese-containing chlorine-bridged polymer **14** by the reaction with the chelating 2-benzoylpyridine ligand [53] gave unexpected results. This reaction afforded the ionic complex $[Mn_3(Piv)_5(L)_2(MeCN)]^+[Mn_6Cl(Piv)_{12}]^-$ (L is 2-benzoylpyridine, **30**) containing the unusual hexanuclear anion with the internal hexadentate chlorine atom $\{Mn_6(\mu_6\text{-Cl})\}$ (Figure 9.27). In addition, this anion was prepared as the complex $(NEt_4)^+[Mn_6Cl(Piv)_{12}]^-$ (**31**) by the independent synthesis from polymer **2** and NEt_4Cl in MeCN.

Both ionic compounds **30** and **31** containing the $[Mn_6Cl(Piv)_{12}]^-$ anion, in which the chlorine atom is located inside the octahedron formed by manganese(II) atoms (Mn...Mn, 3.819(8)−3.889(4) Å; Mn−Cl, 2.7086(6)−2.7462(6) Å), exhibit antiferromagnetic properties. The upper values of the effective magnetic moments of **30** and **31** at 300 K (16.653 μ_B and 13.802 μ_B, respectively) are close to the pure spin values corresponding to nine Mn(II) atoms ($S = 5/2$) in **30** (16.26 μ_B) and six Mn(II) atoms in **31** (13.86 μ_B).

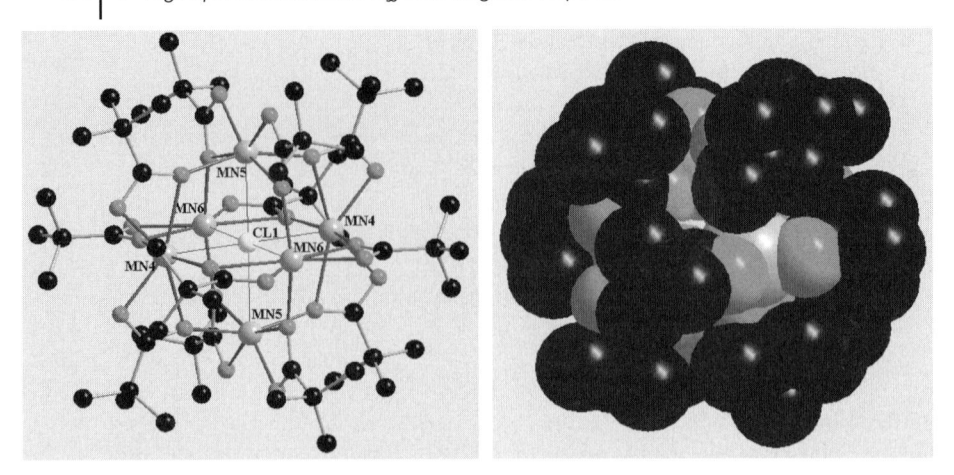

Figure 9.27 Structure of the hexanuclear anion $[Mn_6Cl(Piv)_{12}]^-$ (the diameter of the anion taking into account C–H bonds is ca. 15 Å).

9.4
Pivalate-Bridged Heteronuclear Magnetic Species

The magnetic characteristics of metal-containing molecules can be modified not only by varying the local ligand environment of the metal centers in transition metal complexes but also by partial replacement of magnetic ions with other ions (for example, with ions in another spin state). In this case, it is convenient to use labile metal complexes. Under the corresponding conditions, these complexes generate (in solution) coordinatively unsaturated metal-containing species, which can serve as unusual magnetic ligands with respect to another metal-containing reagent.

The recently synthesized dinuclear antiferromagnetic complex $Co_2(\mu\text{-}Piv)_2(\eta^2\text{-}Piv)_2(Bpy)_2$ (**32**) with two carboxylate bridges is very labile. In this complex, the cobalt atoms contain one extra electron (compared to the saturated 18-electron shell), resulting apparently in a substantial weakening of the bonds between the cobalt atoms and the carboxylate ligands, as evidenced by the length and nonequivalence of the Co–O bonds [54]. As a result, the molecule can, in principle, generate the $Co(Piv)_2(Bpy)$ fragment in the reactions with donors. If the cobalt carboxylate groups $Co(Piv)_2$ from polymer **6** are used as donors, the reaction of **32** with the polymer (in 1:1 ratio with respect to the cobalt atom) in MeCN or benzene affords the dinuclear asymmetric complex $(Bpy)Co_2(\mu_2\text{-}O, \eta^2\text{-}Piv)(\mu_2\text{-}O, O'\text{-}Piv)_2(\eta^2\text{-}Piv)$ (**33**) combining the $(Bpy)Co(OOCR)_2$ (from **32**) and $Co(OOCR)_2$ (from polymer **6**) fragments [56].

According to X-ray diffraction data (Figure 9.28), molecule **33** contains two cobalt atoms at a nonbonded distance of 3.272(1) Å, which is substantially shorter than that in the starting complex **32** (Co.Co, 4.383(1) Å) [54].

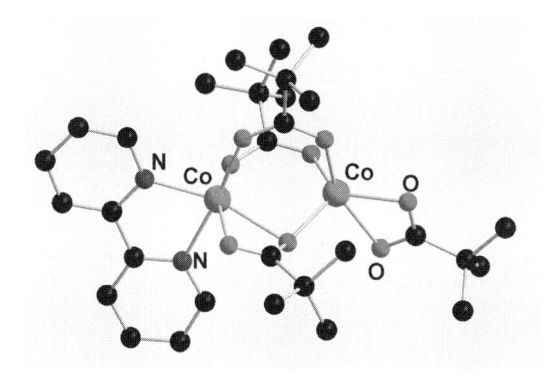

Figure 9.28 Structure of complex **33**.

The dinuclear fragment $Co_2(\mu\text{-Piv})_2(\mu\text{-O}_{OOCR})$ in complex **33** contains two carboxylate bridges with asymmetric Co–O bonds (Co(1)–O, 1.962(2) and 2.014(2) Å; Co(2)–O, 2.017(2) and 2.035(2) Å; C–O, 1.252(4)–1.258(4) Å; the O–C–O angles, 125.7(3) and 124.8(3)°; the angle between the Co_2OCO planes is 95.5°). The bridging oxygen atom (Co(1)–O, 1.987(2) Å; Co(2)–O, 2.259(2) Å; the Co–O–Co angle, 100.59(6)°) belongs to the carboxylate group chelated to another cobalt atom (Co(2)–O, 2.141(2) Å; the O–C–O angle, 119.0(3)°). The dipyridyl ligand (Co(2)–N, 2.075(3) and 2.100(3) Å) is also coordinated to this metal atom, whereas the second metal center is coordinated by the chelate carboxylate group (Co(1)–O, 1.993(2) and 2.270(2) Å, the O–C–O angle, 118.4(3)°).

The magnetic behavior of dinuclear asymmetric complex **33** strongly differs from that of the starting antiferromagnetic symmetric dinuclear compound **32** (Figure 9.29). It appeared that compound **33** exhibits ferromagnetic exchange spin–spin interactions. The effective magnetic moment of **33** increases from 6.51 to 7.62 μ_B (per formula unit) in the temperature range of 300–6 K followed by a decrease to 7.06 μ_B at 2 K. Therefore, the formal removal of one dipyridyl ligand from dinuclear compound **32** leads not only to a substantial rearrangement of the metal carboxylate core in **33** but also to a qualitative change in the magnetic properties of the new compound.

This scheme of transformations with the use of various metal-containing carboxylates bearing the $M(OOCR)_2$ fragment as sources of new donor ligands would be expected to be suitable for the construction of such asymmetric complexes with various combinations of d elements.

The reaction of **32** with the tetranuclear nickel complex $Ni_4(\mu_3\text{-OH})_2(\mu\text{-Piv})_4(Piv)_2(MeCN)_2[\eta^2\text{-}o\text{-}C_6H_4(NH_2)(NHPh)]_2$ (**34**) as a source of metal carboxylate fragments affords heteronuclear cobalt- and nickel-containing complexes in approximately equal yields (40–45%, see Figure 9.30) [55].

The former complex (ICP data, inductive coupled plasma atomic emission spectroscopy) contains cobalt and nickel atoms in a ratio of 1 : 1. The X-ray

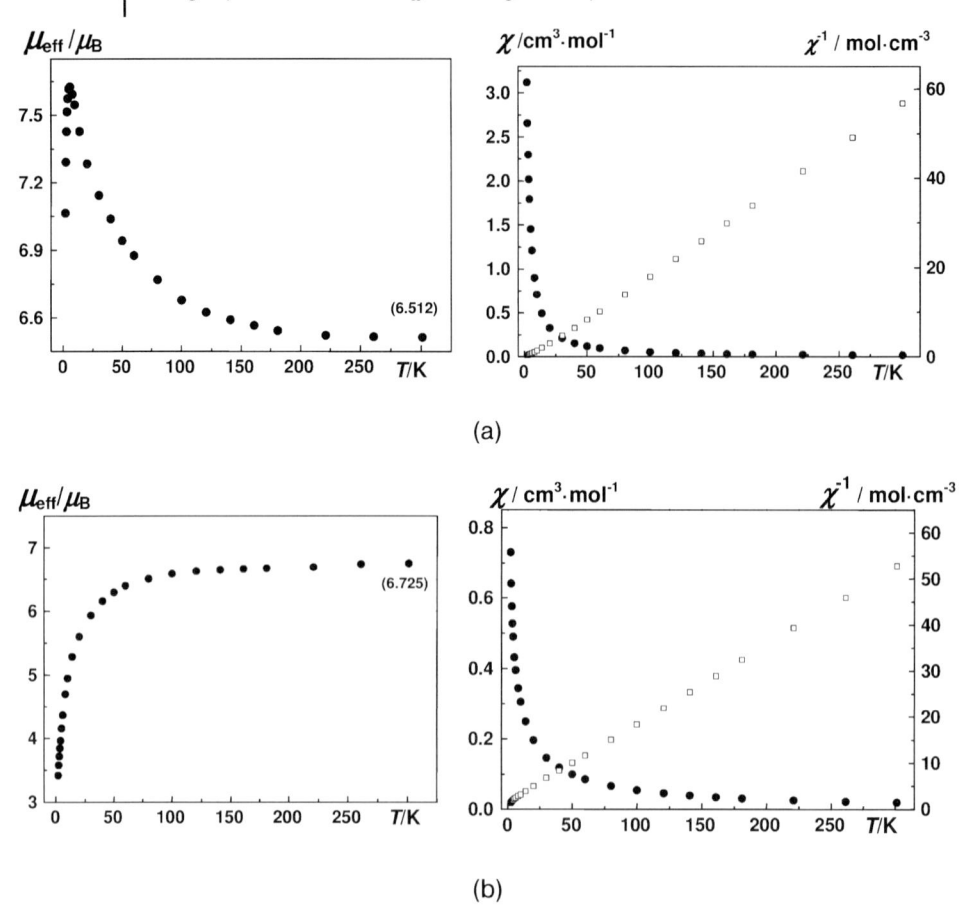

Figure 9.29 Magnetic properties of dinuclear complexes **33** (a) and **32** (b).

diffraction studies showed that this complex is a structural analog of **33**. Although it is very difficult to distinguish between nickel and cobalt atoms based on X-ray diffraction data, it should be noted that the refinement gave the best results for a model, in which the nickel atom occupies a metal site with the coordinated dipyridyl ligand, and the cobalt atom occupies a site in a trigonal-bipyramidal environment. This model is consistent with the known structural data, which provide evidence that the octahedral environment is favorable for nickel atoms in polynuclear pivalates [24, 39, 56–61]. On the other hand, the trigonal-bipyramidal ligand environment of cobalt is observed in structural analogs of **35**, such as complex **33** and the dinuclear anion [$Co_2(\mu_2, \eta^2$-Piv$)(\mu_2$-Piv$)_2(\eta^2$-Piv$)_2$]$^-$ [62]. As a result, compound **35** can, with high probability, be described by the formula (Bpy)Ni(μ_2, η^2-Piv)(μ_2-Piv)$_2$Co(η^2-Piv) (Figure 9.31). In this case, the dipyridyl ligand is transferred from the cobalt atom to the nickel atom in the course of the reaction.

Figure 9.30 Synthesis of heterometallic dinuclear cobalt- and nickel-containing complexes.

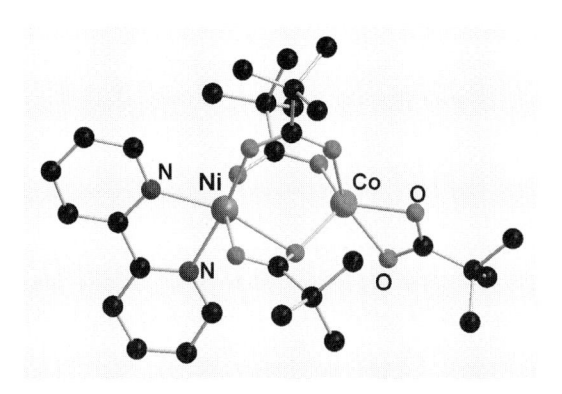

Figure 9.31 Structure of heteronuclear complex **35**.

Like homonuclear analog **33**, compound **35** has ferromagnetic properties (Figure 9.32). However, the temperature dependence of μ_{eff} for **35** is somewhat different from that observed for **33**, and the magnetic moment of the

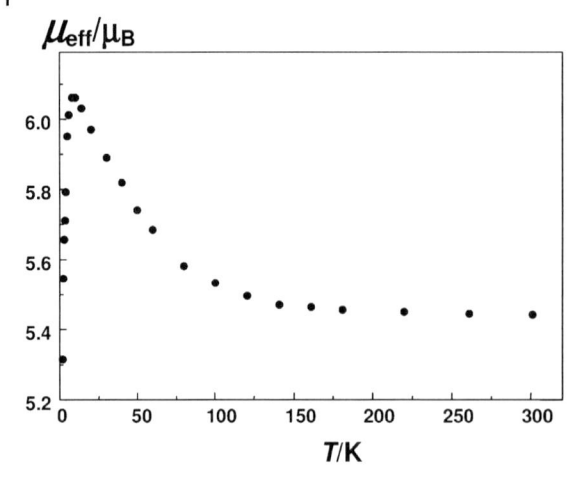

Figure 9.32 Magnetic properties of complex **35**.

heteronuclear complex is substantially smaller in accordance with the change in the spin state of one of the metal centers ($S(Ni) = 1$ instead of $S(Co) = 3/2$).

The reaction affords the dinuclear complex (Bpy)(HPiv)M(μ-OH$_2$)(μ-Piv)$_2$M'(Piv)$_2$[o-C$_6$H$_4$(NH$_2$)(NHPh)] (**36**) as the second product. According to the ICP data, compound **36** contains predominantly nickel atoms (Ni : Co = 1.85 : 0.15). The X-ray diffraction study (Figure 9.33) revealed the presence of the aqua-bridged fragment M(μ-OH$_2$)(μ-OOCR)$_2$M' (M(1)–M(2), Å; M(1)–O(H$_2$O), 2.023(4) Å; M(2)–O(H$_2$O), 2.076(4) Å).

However, the metal atoms in complex **36** are coordinated by different N-donor ligands, such as dipyridyl (M(1)–N, 2.053(5) and 2.071(5) Å) and N-phenyl-o-phenylenediamine bound to the metal atom through the NH$_2$ group (M(2)–N, 2.134(5) Å). The pivalic acid molecule (the hydroxy hydrogen atom was located in a difference Fourier synthesis) serves as the second ligand

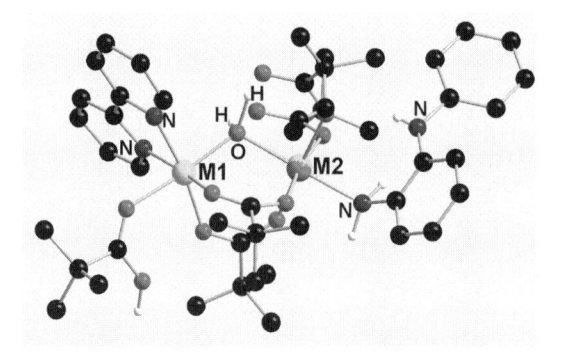

Figure 9.33 Structure of dinuclear complex **36**.

coordinated to the M(1) atom, whereas M(2) is coordinated (in addition to the diamine molecule) by two terminal pivalate anions involved in hydrogen bonding with the bridging water molecule (1.45–1.60 Å).

Since both the metal centers are in a distorted octahedral environment, it is virtually impossible to distinguish between the sites partially occupied by cobalt atoms. It should be noted that the refinement gave the best results for the model characterized by the formula (Bpy)(HPiv)Ni(μ-OH$_2$)(μ-Piv)$_2$Ni$_{0.85}$Co$_{0.15}$(Piv)$_2$[o-C$_6$H$_4$(NH$_2$)(NHPh)] (**36**).

The above results show that the use of labile polynuclear pivalate complexes as a source of coordinatively unsaturated species, which can bind other metal fragments involved in polynuclear counter reagents, is rather efficient for the synthesis of heteronuclear structures. However, this process can be accompanied by deeper transformations, including the formal transfer of N-donor ligands (for example, of dipyridyl) from one to other metal centers, as is observed for both complexes **35** and **36**.

The probability of assembly of heterometallic compounds containing a large number of metal atoms increases in reactions of nickel-containing complexes with ligands that are readily eliminated (for example, with coordinated ethanol molecules). To study this pathway of formation of heteronuclear pivalate systems, we studied the reaction of complex **32** with the tetranuclear compound Ni$_4$(μ_3-OH)$_2$(Piv)$_6$(HOEt)$_6$ (**17**). The reaction of (Bpy)$_2$Co$_2$(Piv)$_4$ (**32**) with **17** in o-xylene at 80 °C affords the trinuclear complex M$_3$(Bpy)$_2$(μ_3-OH)(μ_2-Piv)$_4$(η^1-Piv) (**37**) as the major product (Figure 9.34) [63].

According to the ICP data (inductive coupled plasma atomic emission spectroscopy), nickel and cobalt atoms are present in complex **37** in a ratio of 1.2 : 1. The X-ray diffraction study of blue prismatic crystals of solvate **37**·MeCN (Figure 9.35) showed that this compound contains the metal triangle centered by the hydroxyl group (M(1)–O, 2.085(7) Å; M(2)–O, 2.081(7) Å; M(3)–O, 1.957(6) Å) with nonequivalent nonbonded distances between the M atoms (M(1)...M(2), 3.504(1) Å; M(2)...M(3), 3.318(1) Å; M(1)...M(3), 3.304(1) Å). It should be noted that only two metal atoms, MI(1) and M(2), are coordinated by dipyridyl ligands (M–N, 2.12(1)–2.15(1) Å). Both metal atoms are in an octahedral environment and are linked to each other by two carboxylate groups (M–O, 2.02(1)–2.08(1) Å). The third metal atom is in a tetrahedral environment and is linked to the two other metal centers by only one carboxylate bridge (M(3)–O, 1.972(9)–1.98(1) Å). In addition, the latter atom is coordinated by the terminal carboxylate group (M(3)–O, 2.036(7) Å), which is linked to the tridentate hydroxy bridge by a strong hydrogen bond (O–H...O, 1.78(5) Å).

As a result of this ligand environment, all three metal centers formally have a charge of +2. Although it is virtually impossible to unambiguously distinguish between the sites of the nickel and cobalt atoms in complex **37**, the model, in which the cobalt atom occupies the tetrahedral Co(3) site, seems to be most probable (Figure 9.35). Actually, carboxylate complexes with nickel atoms in a tetrahedral environment formed by carboxylate and

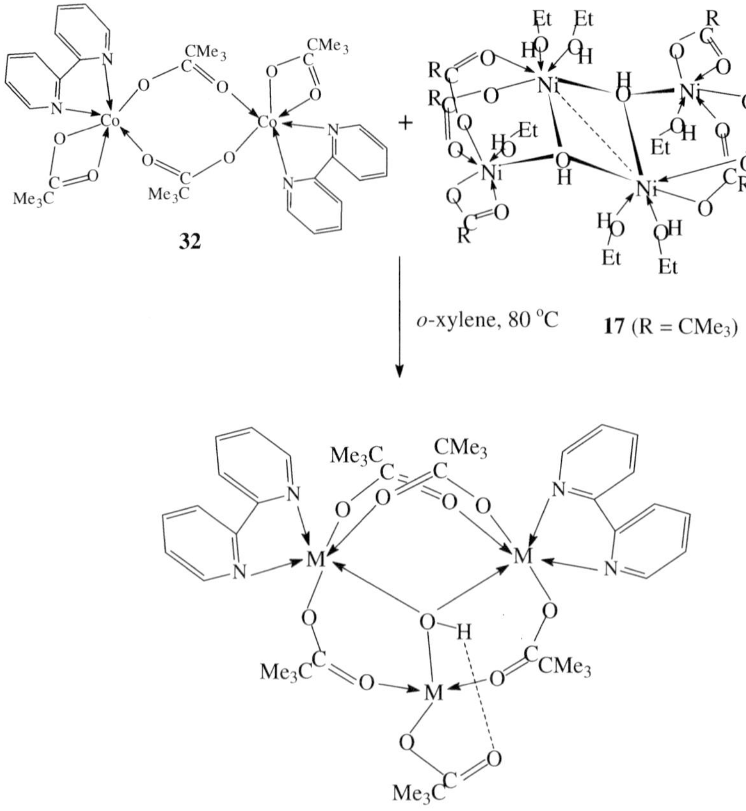

Figure 9.34 Formation of the heterometallic trinuclear complex $M_3(Bpy)_2(\mu_3\text{-OH})(\mu_2\text{-Piv})_4(\eta^1\text{-Piv})$ (**37**).

hydroxy oxygen atoms are unknown. However, the triangular cluster $Co_3L_2(\mu_3\text{-OH})(\mu_2\text{-Piv})_4(\eta^1\text{-Piv})$ (L is 8-amino-2,4-dimethylquinoline) containing the structurally similar metal carboxylate core with the tetrahedral cobalt atom was synthesized by the reaction of 8-amino-2,4-dimethylquinoline with the polymer $[Co(OH)_n(Piv)_{2-n}]_x$ or the tetranuclear cluster $Co_4(\mu_3\text{-OH})_2(\mu\text{-Piv})_4(\eta^2\text{-Piv})_2(EtOH)_6$ (**16**) [65]. In addition, the cobalt atoms in a tetrahedral environment formed by carboxylate and hydroxy oxygen atoms were found, for example, in the hexanuclear pivalate cluster $Co_6(\mu_3\text{-OH})_2(\mu_3\text{-Piv})_2(\mu_2\text{-Piv})_8(HPiv)_4$ [66]. The positions of (Bpy)M in cluster **37** are, apparently, occupied mainly by nickel atoms and are only partially occupied by cobalt atoms. In this case, the formula of **37** can be written as $m[Ni_2Co(Bpy)_2(\mu_3\text{-OH})(\mu_2\text{-Piv})_4(\eta^1\text{-Piv})]\cdot n[Co_3(Bpy)_2(\mu_3\text{-OH})(\mu_2\text{-Piv})_4(\eta^1\text{-Piv})]$, where the coefficients m and n are 9 and 2, respectively, in accordance with the ratio of the metal atoms. It should be noted that the contribution of

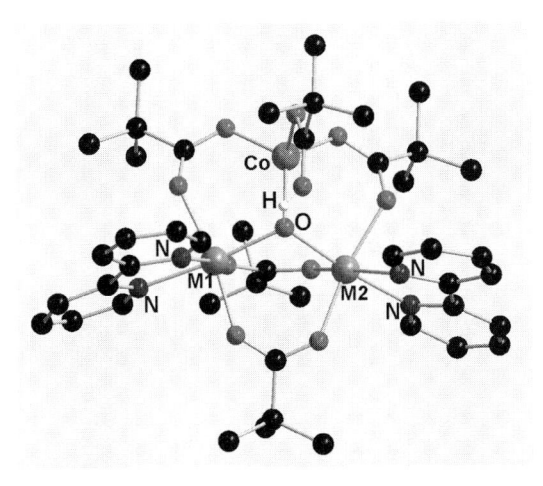

Figure 9.35 Structure of the trinuclear complexes $M_2Co(Bpy)_2(\mu_3\text{-}OH)(\mu\text{-}Piv)_4(Piv)$ (**37**, M_2 = Ni_2; **38**, M_2 = Co_2).

structures containing the $NiCo_2$ core cannot be ruled out. One would expect that the total set of trinuclear clusters would be retained upon dissolution. However, attempts to separate these clusters by fractional crystallization from a solution of **37** in various solvents (benzene, acetonitrile, or CH_2Cl_2) failed, always resulting in isolation of the compound with the above-mentioned composition. Magnetic measurements showed that cluster **37** exhibits antiferromagnetic properties. The magnetic moment per molecular weight of the cluster monotonically decreases with decreasing temperature (Figure 9.36).

Based on magnetic data, the upper value of the effective magnetic moment **37** (per molecule, 6.47 μ_B (300 K)) is somewhat larger than the pure spin moment corresponding to the trinuclear system Ni_2Co ($S_1 = 1$; $S_2 = 1$, $S_3 = 3/2$; $\mu_{\text{eff}} = 5.58\ \mu_B$), which was calculated by an equation published earlier [48,

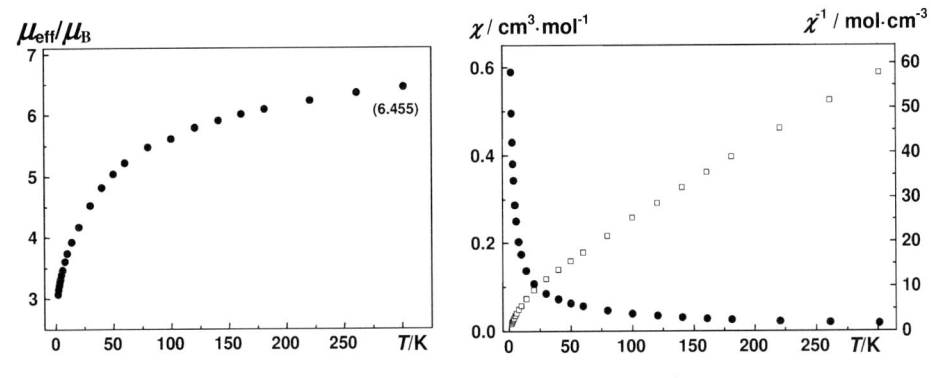

Figure 9.36 Magnetic properties of heteronuclear compound **37**.

67]. This is apparently attributed to the contribution of structures with the Co_3 and $NiCo_2$ cores and a larger total spin, as well as to spin–orbital interactions typical of Co(II) ions. These data confirm the hypothesis that molecules with different metal ratios co-exist in the crystal structure of **37**.

To correctly estimate the possibility that an analogous, at least, homonuclear triangle with the Co_3 core may exist and to adequately compare the magnetic characteristics of **37** and the homonuclear system consisting of three cobalt atoms, we synthesized the homometallic triangular cluster $Co_3(Bpy)_2(\mu_3\text{-}OH)(\mu_2\text{-}Piv)_4(\eta^1\text{-}Piv)$ (**38**). Compound **38** can be synthesized by either the reaction of dipyridyl with $Co_4(\mu_3\text{-}OH)_2(Piv)_6(HOEt)_6$ (MeCN, 50 °C, in a ratio of 2 : 1) or the reaction of Bpy with $Co_8(\mu_4\text{-}O)_2(\mu_n\text{-}Piv)_{12}$ (**39**) (MeCN or benzene, 60–80 °C, in a ratio of 4 : 1).

The X-ray diffraction study of the solvate **38**·MeCN and unsolvated **38** showed that clusters **37** and **38** are virtually isostructural (Figure 9.35) [63]. Crystals of complexes **37** and **38** have different magnetic properties. Thus, these complexes differ in μ_{eff}, and complex **38** is characterized by the steeper temperature dependence of the magnetic moment (Figure 9.37).

The upper value of the effective magnetic moment of **38** calculated per total molecular weight (7.23 μ_B at 300 K) is substantially larger than that found for **37**. This is consistent with an increase in the total spin of molecule **38** containing only cobalt(II) atoms ($S_1 = S_2 = S_3 = 3/2$; $\mu_{eff} = 6.71\ \mu_B$, the pure spin moment without consideration of the spin–orbital contribution).

It is known that thermolysis can lead to an increase in nuclearity of clusters as a result of elimination of weakly coordinated ligands. For example, this is observed upon heating of a solution of tetranuclear pivalate $Co_4(\mu_3\text{-}OH)_2(Piv)_6(HOEt)_6$ (**16**) containing labile molecules of coordinated ethanol in decalin (2 h, 170 °C). This reaction afforded volatile antiferromagnetic octanuclear pivalate $Co_8(\mu_4\text{-}O)_2(\mu_2\text{-}Piv)_6(\mu_3\text{-}Piv)_6$ (**39**), which was isolated as blue-violet prismatic crystals [64]. The formation of **39** from tetranuclear complex **16** is accompanied by the loss of not only all ethanol molecules but also of the water molecule, resulting in the formation of tetradentate bridging

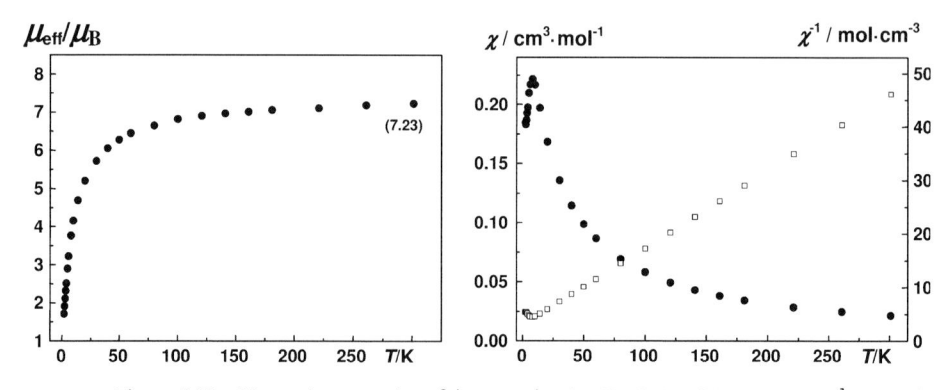

Figure 9.37 Magnetic properties of the complex $Co_3(Bpy)_2(\mu_3\text{-}OH)(\mu_2\text{-}Piv)_4(\eta^1\text{-}Piv)$ (**38**).

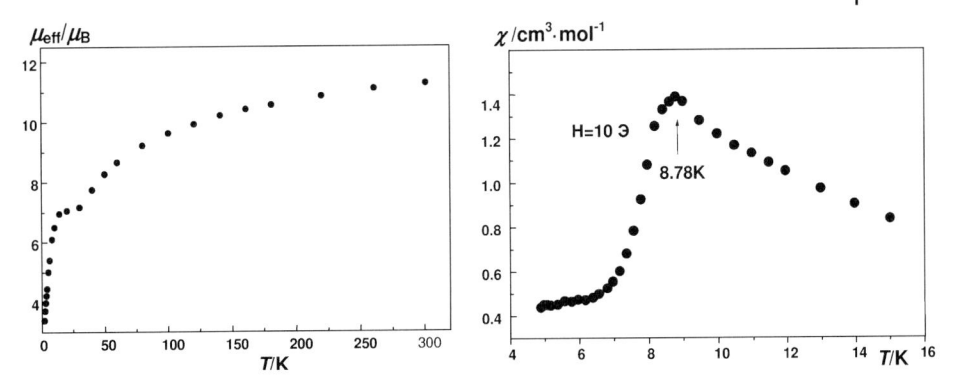

Figure 9.38 Magnetic properties of heterometallic cluster **40**.

oxygen atoms in **39**. As opposed to thermolysis of **16**, thermolysis of its nickel analog, $Ni_4(\mu_3\text{-}OH)_2(Piv)_6(HOEt)_6$ (**17**), under the same conditions affords nonanuclear pivalate **15**. This difference in the chemical behavior of **16** and **17** is apparently attributed to the fact that the reaction performed under these conditions cannot yield octanuclear nickel analog **39** containing metal atoms in different environment formed by oxygen atoms (two metal atoms are in a tetrahedral environment, and the other six atoms are in a distorted trigonal-bipyramidal environment). As mentioned above, unlike cobalt(II) carboxylate (pivalate) derivatives, known nickel pivalates contain metal atoms only in an octahedral and pseudooctahedral environment. However, it cannot be excluded that the reaction affords a stable heteronuclear nickel- and cobalt-containing pivalate cluster isostructural with cobalt derivative **39**, whose stability would be determined by the cobalt-containing moiety of the compound.

It appeared that the simultaneous thermolysis of tetranuclear complexes **16** and **17** (in a ratio of 1:1) in decalin (2 h, 170 °C) gave rise to the heteronuclear cluster $Co_6Ni_2(\mu_4\text{-}O)_2(\mu_2\text{-}Piv)_6(\mu_3\text{-}Piv)_6$ (**40**), which was isolated as blue prismatic crystals in high yield (80%).

The Co-to-Ni ratio in molecule **40** (3:1) was evaluated based on ICP data (inductive coupled plasma atomic emission spectroscopy). The properties of heterometallic cluster **40** differ from those of homonuclear analog **39**. For example, under no conditions does sublimation of **40** proceed, whereas **39** is easily sublimed at 100–150 °C under argon. In addition, the magnetic properties of **40** (Figure 9.38) substantially differ from the behavior of homonuclear cluster **39**. For example, heterometallic cluster **40** exhibits ferromagnetic properties in the temperature range of 8–10 K.

In spite of these differences, **39** and **40** are isostructural (X-ray diffraction data, the crystallographic parameters are identical). Since the structure of **39** has been solved earlier [64], the structure of heterometallic cluster **40** is clear (Figure 9.39). However, it is virtually impossible to correctly distinguish between the sites of the cobalt and nickel atoms in **40**, because molecule **40**

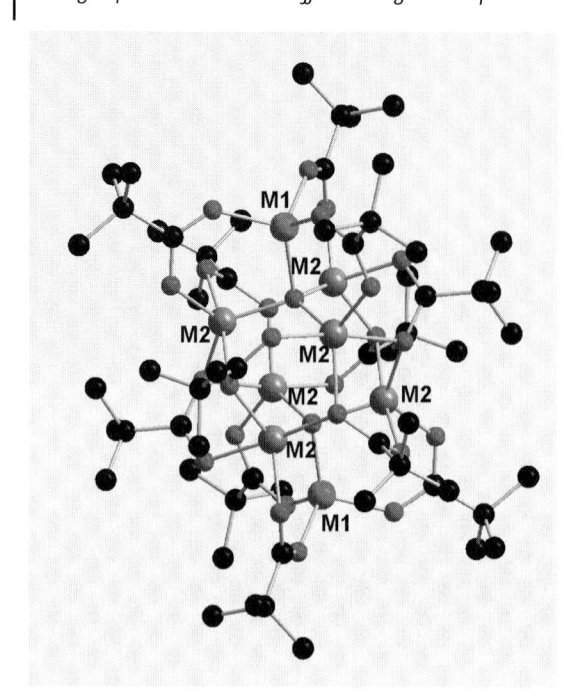

Figure 9.39 Structure of the complex $Co_6Ni_2(\mu_4\text{-}O)_2(\mu_2\text{-}Piv)_6(\mu_3\text{-}Piv)_6$ (**40**).

occupies the crystallographic threefold axis near an inversion center (crystals of **39** and **40** belong to the cubic system) so that only two atoms are independent. One of these metal sites having a tetrahedral environment formed by the oxygen atoms (M(1) or M(1a)) has an occupancy of 1/3. Presumably, this site in the heterometallic molecule is occupied by the cobalt atom, because nickel(II) carboxylate complexes containing metal atoms with coordination number 4 are unknown. In this case, the nickel and cobalt atoms in the triangle located under this tetracoordinated metal atom (Figure 9.40, the M(2), M(2a), and M(2b) atoms and, correspondingly, M(2c), M(2d), and M(2e)) should be disordered with approximate occupancies of 2/3 and 1/3 for Co and Ni, respectively.

Apparently, the presence of a larger number of nickel atoms in a molecule compared to complex **40** is unfavorable because, as mentioned above, the structure contains no metal sites in an octahedral oxygen environment, which is very stable for nickel atoms in polynuclear pivalates [24, 39, 56–61]. As a result, the self-assembly of the heteronuclear molecule stops at the cluster with the Co_6Ni_2 core as the most stable structure regardless of the cobalt-to-nickel ratio in the starting reagents. It should be noted that this reaction produced almost no homonuclear cobalt derivatives (only traces of cluster **39** were detected), which is apparently indicative of the lower stability of the Co_8 core compared to the heterometallic system.

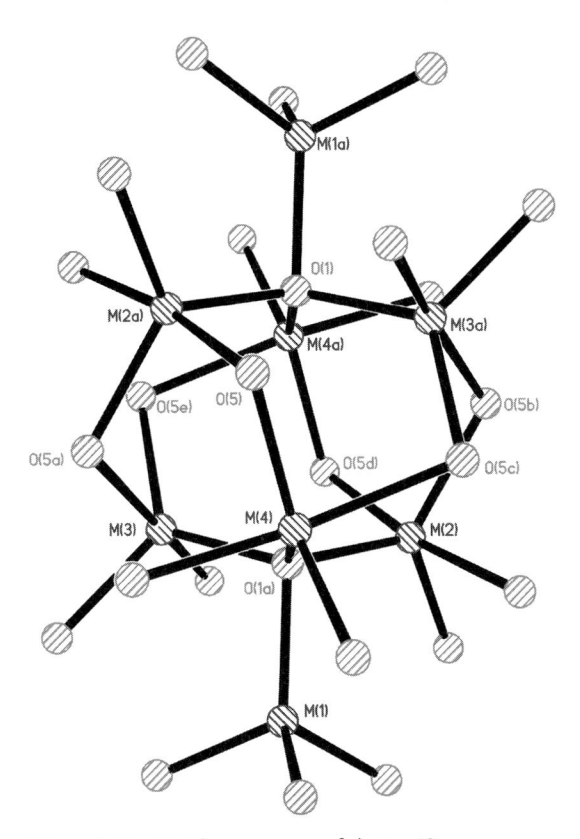

Figure 9.40 Metal oxygen core of cluster **40**.

9.5
Pivalate-Based Single Molecular Magnets

As mentioned above, the ratio of the metal atoms in the starting spin material (e.g., in a chain coordination polymer) to the organic donor ligand that serves a cutting function strongly influences the structures of the newly formed magnetic molecules. However, the crystallization temperature of the reaction products is an additional important factor determining the molecular and crystal structure of the final compounds. This is particularly evident for cobalt pivalate derivatives. Solutions of these compounds presumably contain various molecules with similar energy, which can easily be transformed from one form to another.

For example, crystallization of the reaction product of the starting pivalate polymer **6** with a deficient amount of 4,5-dimethyl-o-phenylenediamine (80 °C, MeCN, Ar, Co_{at}: L = 2 : 1) at room temperature (∼20 °C) affords the above-mentioned polymer $[Co(\eta^2\text{-}(NH_2)_2C_6H_2Me_2)_2(\mu\text{-}Piv)_2Co_2(\mu\text{-}Piv)_4]_n$ (**10**) consisting of alternating mono- and dinuclear metal fragments [35, 69]. Crystals of another product described by the formal formula $\{Co_3(\eta^2\text{-}(NH_2)_2C_6H_2Me_2)_2(\mu\text{-}Piv)_2(\eta^2\text{-}Piv)_2(Piv)_2(HPiv)_2\}\{Co(\eta^2\text{-}(NH_2)_2C_6H_2Me_2)_2(Piv)_2\}_n$ (**41**) can be isolated at 0 °C. Finally, the pentanuclear complex $Co_5(\mu_3\text{-}OH)_2(\mu\text{-}N, N\text{-}1,2\text{-}(NH_2)_2C_6H_2Me_2)_2(\mu\text{-}Piv)_5(Piv)_3(HPiv)$ (**42**) is formed at a crystallization temperature of −5 °C and lower. The yields of these compounds are rather high, varying from 40–50% for **10** and **41** to 80% for **42** [35, 67, 68].

It is noteworthy that this scheme involves gradual transformations of chain coordination polymer **10** through one-dimensional supramolecular chain **41** consisting of the alternating mononuclear $\{Co_1\}$ and trinuclear fragments $\{Co_3\}$ (Figure 9.41) to discrete pentanuclear molecule **42** (Figure 9.43). It should be noted that the metal-to-diamine ratio in the resulting structures (in the case of polymers, for the independent unit of the chain) is 3 : 2 for **10**, 4 : 4 for **41**, and 5 : 2 for **42**, and is unlikely to demonstrate the chemical activity of the diamine toward metal centers; instead, it is more likely attributed to stability of a particular molecular system at a given temperature.

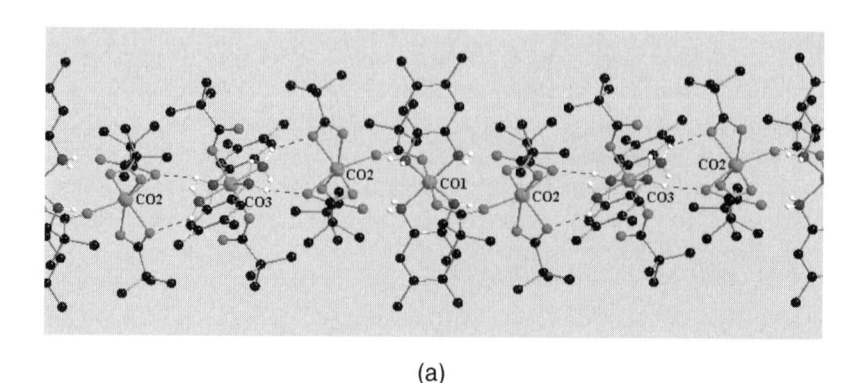

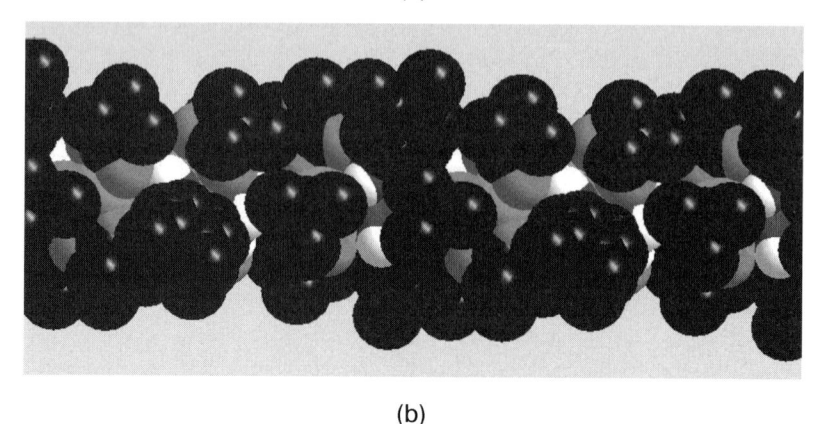

(a)

(b)

Figure 9.41 Structure of the supramolecular chain $\{Co_3(\eta^2\text{-}(NH_2)_2C_6H_2Me_2)_2(\mu\text{-}Piv)_2(\eta^2\text{-}Piv)_2(Piv)_2(HPiv)_2\}\{Co(\eta^2\text{-}(NH_2)_2C_6H_2Me_2)_2(Piv)_2\}_n$ (**41**).

9.5 Pivalate-Based Single Molecular Magnets

Formally, compound **41** exists as co-crystals of the mono- and trinuclear molecules linked together by hydrogen bonding between the NH_2 protons of the ligand of the mononuclear molecule and the O atoms of the chelating carboxylate anions and the C=O groups of the coordinated pivalic acid molecules in the trinuclear complex (N–H...O, 2.107(7) Å, N–H...O, 1.997(5) Å) (Figure 9.41(a)). Apparently, this unusual mode of formation of the chain structure is responsible for the unexpected magnetic properties of compound **41**. As opposed to antiferromagnetic compound **10**, the effective magnetic moment of **41** monotonically decreases from 11.22 to 8.63 μ_B (per independent fragment {Co1 + Co3}) in the temperature range of 300–28 K, then increases to 9.6 μ_B at 20 K, and again decreases to 3.5 μ_B at 2 K, a hysteresis loop with a coercive force of 2 kOe being observed at 5 K (Figure 9.42) [67].

Pentanuclear molecule **42** is a discrete complex containing two vertex-sharing metal triangles (Figure 9.43). The dihedral angle between the planes of these triangles is 69.7° and 70.9° in two independent molecules, and the distances between the cobalt atoms in each triangle vary in the range of 2.74–3.82 Å.

At temperatures below 12 K, complex **42** is transformed into the magnetically ordered state, and the magnetization reaches ~20 000 G cm^3/mol at 2 K; the hysteresis loop is characterized by the large coercive force (5 kOe, see Figure 9.44) [68].

Formally it means that the record magnetic limit is achieved for magnetic materials. Actually, this molecular ferromagnet contains only five magnetic cobalt atoms interacting with each other.

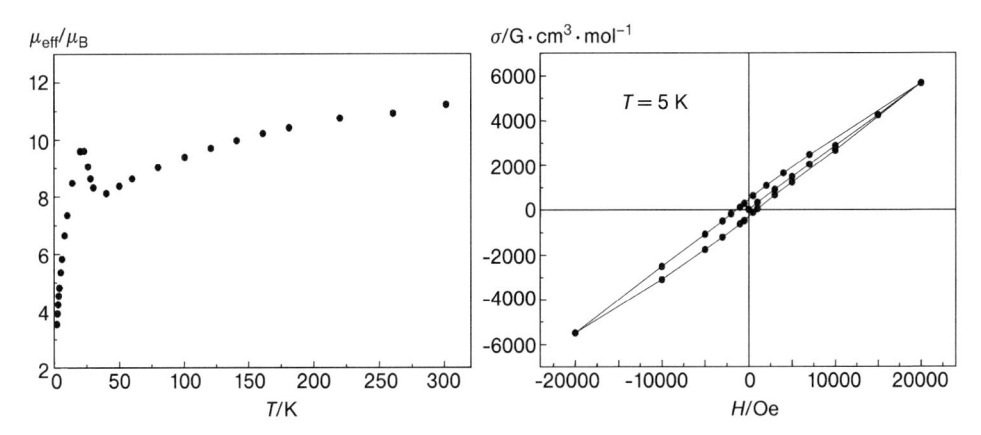

Figure 9.42 Magnetic behavior of {Co$_3$(η^2-(NH$_2$)$_2$C$_6$H$_2$Me$_2$)$_2$(μ-Piv)$_2$(η^2-Piv)$_2$(Piv)$_2$(HPiv)$_2$}{Co(η^2-(NH$_2$)$_2$C$_6$H$_2$Me$_2$)$_2$(Piv)$_2$}$_n$ (**41**).

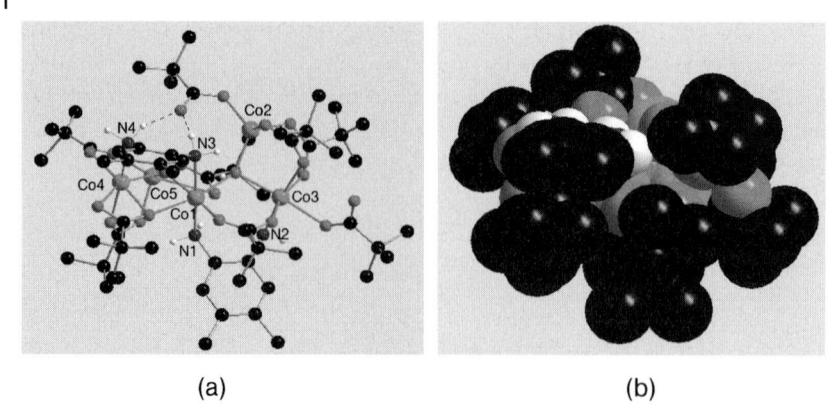

Figure 9.43 Structure of the pentanuclear magnet
$Co_5(\mu_3\text{-}OH)_2(\mu\text{-}N, N\text{-}1,2\text{-}(NH_2)_2C_6H_2Me_2)_2(\mu\text{-}Piv)_5(Piv)_3(HPiv)$ (**42**).

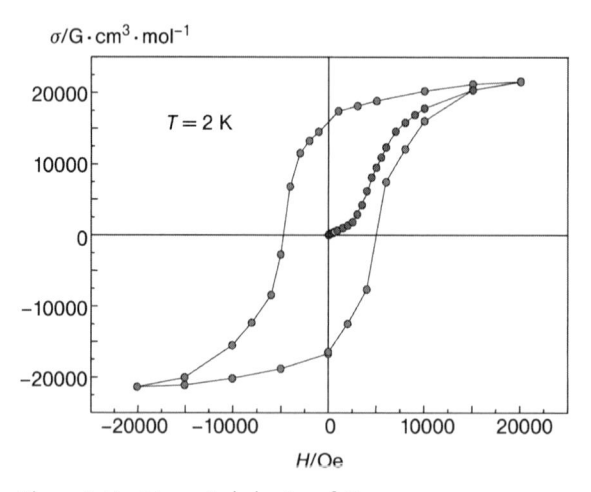

Figure 9.44 Magnetic behavior of **42**.

9.6
Conclusions

The above-described procedures for the construction of high-spin molecules having different magnetic properties demonstrate only a small part of the potential of molecular technologies and are limited to one class of coordination compounds, e.g., polynuclear transition metal carboxylates. Taking into account that organic components of such molecules hold considerable promise as bridging and axial ligands and the possibility of using new combinations of metal centers in a single molecule (the design of heteronuclear or heterospin structures), it is evident that the scope of this field of chemistry is greatly expanded. It is important that chemical technologies of assembly of nanosized molecular structures can easily be controlled, which is particularly advantageous for the design of new molecular nanosized materials with desired properties.

References

1. M.N. Vargaftik, I.I. Moiseev, D.I. Kochubey, and K.I. Zamaraev, *Faraday Discuss*, **1991**, *92*, 13.
2. I.I. Moiseev and M.N. Vargaftik, in *Catalysis by Di- and Polynuclear Metal Cluster Complexes*, R.D. Adams and F.A. Cotton, Wiley-VCH (New York) **1998**, 395.
3. M.N. Vargaftik, V.P. Zagorodnikov, I.P. Stolarov, I.I. Moiseev, D.I. Kochubey, V.A. Likholobov, A.L. Chuvilin, and K.I. Zamaraev, *J. Mol. Catal.*, **1989**, *53*, 315.
4. M.T. Pope and A. Müller, *Angew. Chem., Int. Ed. Engl.*, **1991**, *30*, 34.
5. M.T. Pope, *Heteropoly and Isopoly Oxometallates*, Springer (Berlin) **1983**.
6. S.S. Talismanov and I.L. Eremenko, *Russ. Chem. Rev.*, **2003**, *72*, 627.
7. O. Kahn, *Acc. Chem. Res.*, **2000**, *33*, 647.
8. V.I. Ovcharenko and R.Z. Sagdeev, *Russ. Chem. Rev.*, **1999**, *68*, 345.
9. M. Verdaguer, *Polyhedron*, **2001**, *20*, 1115.
10. G. Christou, D. Gatteschi, D.N. Hendrickson and R. Sessoli, *MRS Bull.*, **2000**, *25*, 66.
11. O. Khan, *Molecular Magnetism*, Wiley-VCH (New York) **1993**.
12. T. Lis, *Acta Crystallogr. Soc. B.*, **1980**, *36*, 2042.
13. D. Gatteschi and R. Sessoli, *Angew. Chem., Int. Ed.*, **2003**, *42*, 268.
14. J. Kortus and A.V. Postnikov, *Molecular Nanomagnets. Handbook of Theoretical and Computational Nanotechnology*, **2005**, *1*, 5.
15. J.R. Freidman, M.P. Sarachik, J. Tejada, and R. Ziolo, *Pys. Rev. Lett.*, **1996**, *76*, 3830.
16. L. Thomas, L. Lionti, R. Ballou, D. Gatteschi, R. Seccoli, and B. Barbara, *Nature*, **1996**, *383*, 145.
17. K. Weighardt, K. Phol, I. Jibril, and G. Huttner, *Angew. Chem.*, **1984**, *23*, 77.
18. M. Soler, W. Wernsdorfer, K. Folting, M. Pink, and G. Christou, *J. Am. Chem. Soc.*, **2004**, *126*, 2156.
19. D. Volkmer, A. Horstmann, K. Grisear, W. Haase, and B. Krebs, *Inorg. Chem.*, **1996**, *35*, 1132.
20. T. Koga, H. Furutachi, T. Nakamura, N. Fukita, M. Ohba, K. Takahoshi, and H. Okawa, *Inorg. Chem.*, **1998**, *37*, 989.
21. S. Uozumi, H. Furutachi, M. Ohba, M. Okawa, D.E. Fenton, K. Shindo, S. Murata, and D.J. Kitko, *Inorg. Chem.*, **1998**, *37*, 6281.
22. K. Yamaguchi, S. Koshino, F. Akagi, M. Susuki, A. Uehara, and S. Suzuki, *J. Am. Chem. Soc.*, **1997**, *119*, 5752.
23. M. Konrad, F. Meyer, A. Jacobi, P. Kircher, P. Rutsch, and L. Zsonali, *Inorg. Chem.*, **1999**, *38*, 4559.
24. I.L. Eremenko, S.E. Nefedov, A.A. Sidorov, and I.I. Moiseev, *Russ. Chem. Bull.*, **1999**, 405.
25. Yu.V. Rakitin and V.T. Kalinnikov, Sovremennaya Magnetokhimiya, Nauka (St-Peterburg) **1994** (in Russian) [Modern magnetochemistry, Science (St.-Petersburg) **1994**].
26. M.A. Kiskin, I.G. Fomina, G.G. Aleksandrov, A.A. Sidorov, V.M. Novotortsev, Yu.V. Rakitin, Zh.V. Dobrokhotova, V.N. Ikorskii, Yu.G. Shvedenkov, I.L. Eremenko, and I.I. Moiseev, *Inorg. Chem. Commun.*, **2005**, *8*, 89.
27. M.A. Kiskin and I.L. Eremenko, *Russ. Chem. Rev.*, **2006**, *75*, 559.
28. I.L. Eremenko, M.A. Kiskin, I.G. Fomina, A.A. Sidorov, G.G. Aleksandrov, V.N. Ikorskii, Yu.G. Shvedenkov, Yu.V. Rakitin, and V.M. Novotortsev, *J. Cluster Sci.*, **2005**, *16*, 331.

29. M.A. Kiskin, G.G. Aleksandrov, Zh.V. Dobrokhotova, V.M. Novotortsev, Yu.G. Shvedenkov, and I.L. Eremenko, *Russ. Chem. Bull., Int. Ed.*, **2006**, *55*, 806.
30. M.A. Golubnichaya, A.A. Sidorov, I.G. Fomina, M.O. Ponina, S.M. Deomidov, S.E. Nefedov, I.L. Eremenko, and I.I. Moiseev, *Russ. Chem. Bull.*, **1999**, *48*, 1751 [*Russ. Chem. Bull., Int. Ed.* (Engl. Transl.)].
31. I.L. Eremenko, A.A. Sidorov, S.E. Nefedov, and I.I. Moiseev, *Russ. Chem. Bull.*, **1999**, *48*, 405.
32. I.G. Fomina, G.G. Aleksandrov, Zh.V. Dobrokhotova, O.Yu. Proshenkina, M.A. Kiskin, Yu.A. Velikodnii, V.N. Ikorskii, V.M. Novotortsev, and I.L. Eremenko, *Russ. Chem. Bull., Int. Ed.*, **2006**, *55*, 1909.
33. S. Yoon and S.J. Lippard, *Inorg. Chem.*, **2003**, *42*, 8606.
34. Yu.V. Rakitin, V.M. Novotortsev, V.N. Ikorskii, and I.L. Eremenko, *Russ. Chem. Bull., Int. Ed.*, **2004**, *53*, 2124.
35. A.E. Malkov, Ph.D. Thesis, N.S. Kurnakov Institute of General and Inorganic Chemistry, Moscow, **2003**.
36. M.A. Kiskin, G.G. Aleksandrov, A.N. Bogomyakov, V.M. Novotortsev and I.L. Eremenko, *Inorg. Chem. Commun.*, **2008**, *11*, 1015.
37. M.A. Golubnichaya, A.A. Sidorov, I.G. Fomina, L.T. Eremenko, S.E. Nefedov, I.L. Eremenko, and I.I. Moiseev, *Russ. J. Inorg. Chem.*, **1999**, *44*, 1401.
38. A.A. Sidorov, Dc.S. Thesis, N.S. Kurnakov Institute of General and Inorganic Chemistry, Moscow, **2002**.
39. I.L. Eremenko, S.E. Nefedov, A.A. Sidorov, M.A. Golubnichaya, P.V. Danilov, V.N. Ikorskii, Yu.G. Shvedenkov, V.M. Novotortsev, and I.I. Moiseev, *Inorg. Chem.*, **1999**, *38*, 3764.
40. A.A. Sidorov, I.G. Fomina, S.S. Talismanov, G.G. Aleksandrov, V.M. Novotortsev, S.E. Nefedov, and I.L. Eremenko, *Koord. Khim.*, **2001**, *27*, 584 [*Russ. J. Coord. Chem.* (Engl. Transl.)].
41. G. Chaboussant, R. Basler, H.-U. Güdel, S. Ochsenbein, A. Parkin, S. Parsons, G. Rajaraman, A. Sieber, A.A. Smith, G.A. Timco, and R.E.P. Winpenny, *Dalton Trans.*, **2004**, 2758.
42. M.A. Golubnichaya, A.A. Sidorov, I.G. Fomina, M.O. Ponina, S.M. Deomidov, S.E. Nefedov, I.L. Eremenko, and I.I. Moiseev, *Russ. Chem. Bull.*, **1999**, *48*, 1751.
43. N.I. Kirilova, Yu.T. Struchkov, M.A. Porai-Koshits, A.A. Pasynskii, A.S. Antsyshkina, L.Kh. Minacheva, G.G. Sadikov, T.Ch. Idrisov, and V.T. Kalinnikov, *Inorg. Chim. Acta*, **1980**, *42*, 115.
44. I.G. Fomina, Zh.V. Dobrokhotova, M.A. Kiskin, G.G. Aleksandrov, O.Yu. Proshenkina, A.L. Emelina, V.N. Ikorskii, V.M. Novotortsev, and I.L. Eremenko, *Russ. Chem. Bull., Int. Ed.*, **2007**, *56*, 1712.
45. I.G. Fomina, Zh.V. Dobrokhotova, G.G. Aleksandrov, M.A. Kiskin, M.A. Bykov, V.N. Ikorskii, V.M. Novotortsev, and I.L. Eremenko, *Russ. Chem. Bull., Int. Ed.*, **2007**, *56*, 1722.
46. A.E. Malkov, I.G. Fomina, A.A. Sidorov, G.G. Aleksandrov, I.M. Egorov, N.I. Latosh, O.N. Chupakhin, Yu.V. Rakitin, G.L. Rusinov, V.M. Novotortsev, V.N. Ikorskii, I.L. Eremenko, and I.I. Moiseev, *J. Mol. Struct.*, **2003**, *656*, 207.
47. A.L. Barra, A. Caneschi, and A. Cornia, *J. Am. Chem. Soc.*, **1999**, *121* 5302.
48. Yu.V. Rakitin, V.T. Kalinnikov, and M.V. Eremin, *Theoret. Chim. Acta (Berl.)*, **1977**, *45*, 167.
49. M.A. Kiskin, A.A. Sidorov, I.G. Fomina, G.L. Rusinov, R.I. Ishmetova, G.G. Aleksandrov, Yu.G. Shvedenkov, Zh.V. Dobrokhotova, V.M. Novotortsev, O.N. Chupakhin, I.L. Eremenko, and I.I. Moiseev, *Inorg. Chem. Commun.*, **2005**, *8*, 524.
50. I.L. Eremenko, A.E. Malkov, A.A. Sidorov, I.G. Fomina, G.G. Aleksandrov, S.E. Nefedov,

G.L. Rusinov, O.N. Chupakhin, V.M. Novotortsev, V.N. Ikorskii, and I.I. Moiseev, *Inorg. Chim. Acta*, **2002**, *10*, 334.
51. A.A. Sidorov, M.E. Nikiforova, E.V. Pahmutova, G.G. Aleksandrov, V.N. Ikorskii, V.M. Novotortsev, I.L. Eremenko, and I.I. Moiseev, *Russ. Chem. Bull., Int. Ed.*, **2006**, *55*, 1920.
52. M.E. Nikiforova, A.A. Sidorov, G.G. Aleksandrov, V.N. Ikorskii, I.V. Smolyaninov, A.O. Okhlobystin, N.T. Berberova, and I.L. Eremenko, *Russ. Chem. Bull., Int. Ed.*, **2007**, *56*, 943.
53. M.A. Kiskin, G.G. Aleksandrov, V.N. Ikorskii, V.M. Novotortsev, and I.L. Eremenko, *Inorg. Chem. Commun.*, **2007**, *10*, 997.
54. M.O. Talismanova, A.A. Sidorov, V.M. Novotortsev, G.G. Aleksandrov, S.E. Nefedov, I.L. Eremenko, and I.I. Moiseev, *Russ. Chem. Bull., Int. Ed.*, **2001**, *50*, 2251.
55. I.G. Fomina, A.A. Sidorov, G.G. Aleksandrov, V.I. Zhilov, V.N. Ikorskii, V.M. Novotortsev, I.L. Eremenko, and I.I. Moiseev, *Russ. Chem. Bull., Int. Ed.*, **2004**, *53*, 114.
56. I.L. Eremenko, M.A. Golubnichaya, S.E. Nefedov, A.A. Sidorov, I.F. Golovaneva, V.I. Burkov, O.G. Ellert, V.M. Novotortsev, L.T. Eremenko, A. Sousa, and M.R. Bermejo, *Russ. Chem. Bull.*, **1998**, *47*, 704.
57. V.M. Novotortsev, Yu.V. Rakitin, S.E. Nefedov, and I.L. Eremenko, *Russ. Chem. Bull.*, **2000**, *49*, 438.
58. A.A. Sidorov, P.V. Danilov, S.E. Nefedov, M.A. Golubnichaya, I.G. Fomina O.G. Ellert, V.M. Novotortsev, and I.L. Eremenko, *Zh. Neorg. Khin.*, **1998**, *43*, 930 [*Russ. J. Inorg. Chem.* (Engl. Transl.)].
59. V. Ovcharenko, E. Fursova, G. Romanenko, I. Eremenko, E. Tretyakov, and V. Ikorskii, *Inorg. Chem.*, **2006**, *45*, 5338. org. Chem. **2006**, 45, 5338.
60. G. Chaboussant, R. Basler, H.-U. Güdel, S. Ochsenbein, A. Parkin, S. Parsons, G. Rajaraman, A. Sieber, A.A. Smith, G.A. Timco, and R.E.P. Winpenny, *Dalton Trans.*, **2004**, 2758.
61. G. Aromí, A.S. Batsanov, P. Christian, M. Helliwell, O. Roubeau, G.A. Timco, and R.E.P. Winpenny, *Dalton Trans.*, **2003**, 4466.
62. I.G. Fomina, A.A. Sidorov, G.G. Aleksandrov, V.N. Ikorskii, V.M. Novotortsev, S.E. Nefedov, and I.L. Eremenko, *Russ. Chem. Bull., Int. Ed.*, **2002**, *51*, 1581.
63. G.G. Aleksandrov, I.G. Fomina, A.A. Sidorov, T.B. Mikhailova, V.I. Zhilov, V.N. Ikorskii, V.M. Novotortsev, I.L. Eremenko, and I.I. Moiseev, *Russ. Chem. Bull., Int. Ed.*, **2004**, *53*, 1200.
64. A.A. Sidorov, I.G. Fomina, G.G. Aleksandrov, M.O. Ponina, S.E. Nefedov, I.L. Eremenko, and I.I. Moiseev, *Russ. Chem. Bull.*, **2000**, *49*, 958.
65. M.A. Golubnichaya, A.A. Sidorov, I.G. Fomina, M.O. Ponina, S.M. Deomidov, S.E. Nefedov, I.L. Eremenko, and I.I. Moiseev, *Russ. Chem. Bull.*, **1999**, *48*, 1751.
66. J.H. Van Vleck, *The Theory of Electronic and Magnetic Susceptibilities*, Oxford University Press (London), **1932**.
67. A.E. Malkov, I.G. Fomina, A.A. Sidorov, G.G. Aleksandrov, V.N. Ikorskii, V.M. Novotortsev, and I.L. Eremenko, *Russ. Chem. Bull., Int. Ed.*, **2003**, *52*, 489.
68. I.L. Eremenko, *Nanotechnologies in Russia*, **2008**, *3*, 6.

10
Biomedical Applications of Magnetic Nanoparticles

Vladimir N. Nikiforov and Elena Yu. Filinova

10.1
Introduction

Nanotechnology is an enabling technology that deals with nanometer-sized objects. It is expected that nanotechnology could be developed for several applications: materials, devices, and systems. At present, the nanomaterial application is the most advanced one, both in scientific knowledge and in commercial applications. A decade ago, nanoparticles were studied because of their size-dependent physical and chemical properties. Now they have entered a commercial exploration period.

Magnetic nanoparticles offer some attractive possibilities in medicine. Living organisms are built of cells that are typically 10 μm in diameter. However, the cell parts are much smaller and in the submicron size domain. First advantage in medicine is that nanoparticles have controllable sizes ranging from a few nanometers up to tens of nanometers, which places them at dimensions that are smaller than those of a cell (10–100 μm), or comparable to size of a virus (20–450 nm), a protein (5–50 nm), or a gene (2 nm wide and 10–100 nm length). This means that they can "get close" to a biological entity of interest. This simple size comparison gives an idea of using nanoparticles as very small probes that would allow us to spy at the cellular machinery without introducing too much interference. Indeed, they can be coated with biological molecules to make them interact with or bind to a biological entity, thereby providing a controllable means of "tagging" or addressing it.

Second, if nanoparticles are magnetic, they can be manipulated by an external magnetic field gradient. This "action at a distance," combined with the intrinsic penetrability of magnetic fields into human tissue, opens up many applications involving the transport and immobilization of magnetic nanoparticles, or of magnetically tagged biological entities. In this way, they can be made to deliver a package, such as an anticancer drug, to a targeted region of the body, such as a tumor.

Magnetic Nanoparticles. Sergey P. Gubin
© 2009 WILEY-VCH Verlag GmbH & Co. KGaA, Weinheim
ISBN: 978-3-527-40790-3

Third, the magnetic nanoparticles can be made to resonantly respond to a time-varying magnetic field, with advantageous results related to the transfer of energy from the exciting field to the nanoparticle. For example, the particle can be made to heat up, which leads to their use as hyperthermia agents, delivering toxic amounts of thermal energy to targeted bodies such as tumors; or as chemotherapy and radiotherapy enhancement agents, where a moderate degree of tissue warming results in more effective malignant cell destruction. These, and many other potential applications, are made available in biomedicine as a result of the special physical properties of magnetic nanoparticles. Understanding of biological processes on the nanoscale level is a strong driving force behind development of nanotechnology.

Nanoparticles have a size (mass) between single molecules and cells, i.e., a size of 10–1000 nm, or 500,000 to 10^{12} g/mol particle mass. The size is between that of large protein complexes (5–10 nm), e.g., ATP-synthase, and cells. The corresponding native biostructures are cellular compartments, i.e., mitochondria, chloroplasts, and the cytosceleton elements, i.e., actin fibers and microtubuli with the associated molecular motor systems, supplying active motion and transport.

There are many instruments that are able to measure nanoparticles sizes. Some system uses dynamic light scattering and can determine particle diameter due to differences in scattering from solid and liquid phases. Figure 10.1 show atomic force microscope (AFM) picture of relative by

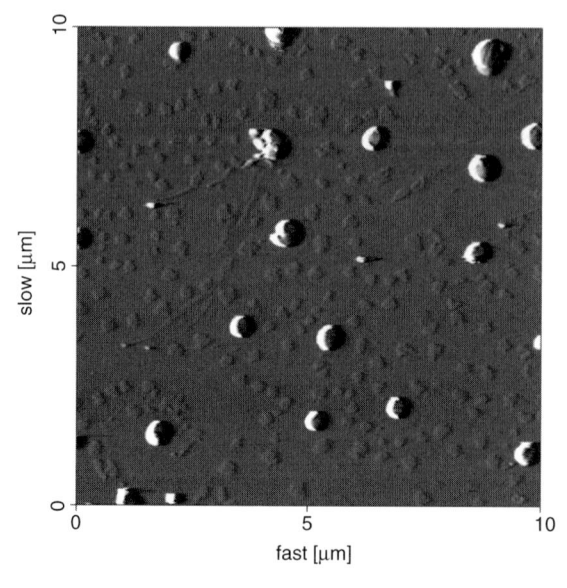

Figure 10.1 AFM imaging of Fe_3O_4 magnetic nanoparticles. Scanning field, 10×10 μm (Author's photo).

large magnetic particles. In case of smaller particles, transmission electron microscopy (TEM) is used.

Magnetic nanoparticles can be a promising tool for several applications *in vitro* and *in vivo*. In medicine, many applications were investigated for diagnostics and therapy and some practical approaches were choosen. Magnetic immunobeads, magnetic streptavidine DNA isolation, cell immunomagnetic separation (IMS), magnetic resonance imaging (MRI), magnetic targeted delivery of therapeutics, or magnetically induced hyperthermia are approaches of particular clinical relevance. Investigations on applicable particles induced a variability of micro- and nanostructures with different materials, sizes, and specific surface chemistry [1].

The nanoparticles for medicine are useful for therapy, imaging, and diagnostics of cancer and other diseases leading an entrapped or bound therapeutic or diagnostic target material to the area of interest, e.g., a tumor. The destination – targeted delivery – may be found by physical forces (magnetic) or with surface-bound antibodies (cell/tissue-specific).

Motile polymers and membranes – a long-term concept for technical application of molecular motion (polymers and chimerical membranes) – are capable of active motion. Nanoparticles are structure components of these motile systems, which can supply the system with the energy required for motion, e.g., by magnetic forces.

The nanoparticles for medical applications as well as motile polymers use the following nanoparticle structure elements as components

1. Magnetic liposomes – liposomes with an internal ferromagnetic iron oxide shell, entrapped magnetic particles or lipid-bound paramagnetic ions. These magnetic target carrier particles can be used for cancer therapy (neutron capture of entrapped boron compounds), magnetic drug targeting (drug entrapped in the liposome lumen), bioanalytics (analytical target signal, imaging), and biophysical experiments (membranes, rheology, cellular traffic, and transport). Magnetic liposomes of 80–250 nm size can be used for targeting *in vivo*, i.e., magnetic drug targeting (MDT), and magnetic radiation targeting for X-rays (photodynamic X-ray therapy (PXT)), neutrons (neutron capture therapy (NCT)), and isotopes (PET).
2. Ferrofluids contain iron oxide nanoparticles (spheres) covered with biocompatible polymers for magnetic drug targeting (cancer therapy), spectroscopy, magnetic imaging (MRI), and technical applications. Biocompatible ferrofluids are water-based and contain only endogenous or bioinert materials. Small ferrofluid particles (usually single-domain) are suitable for hypothermic cancer therapy (overheating by RF application). For biomedical target applications, the magnetic effect of simple ferrofluids is too small. Thus only polyferrofluids of 30–300 nm size, depicting a large macroscopic magnetic moment and magnetic structure generation, can be used for targeting *in vivo*, i.e., magnetic

drug targeting MDT, and magnetic radiation targeting for X-rays (PXT) and neutrons (NCT).

Some present applications of nanomaterials in biology and medicine are fluorescent biological labels [2–4], drug and gene delivery [5, 6], biodetection of pathogens [7], detection of proteins [8], probing of DNA structure [9], tissue engineering [10, 11], tumor destruction via heating (hyperthermia) [12], separation and purification of biological molecules and cells [13], MRI contrast enhancement [14], and phagokinetic studies [15].

As mentioned above, nanomaterials are suitable for biotagging or labeling because they are of the same size as proteins. The other sufficient feature to use nanoparticles as biological tags is their biosusceptibility. In order to interact with a biological target, a molecular linker should be attached to the nanoparticle, acting as a bioinorganic interface. Examples of biological coatings may include antibodies, biopolymers like collagen [16], or molecule monolayers (amino acids, sugars) that make the nanoparticles biocompatible [17]. In addition, as optical detection techniques are wide spread in biological research, it is better if nanoparticles show fluorescence or have other optical features.

Nanoparticles usually form the core of nanobiomaterials. It can be used as a convenient surface for molecular assembly and may be composed of inorganic or polymer materials. It can also be in the form of nanovesicle surrounded by a membrane or a layer. The shape is not automatically spherical but sometimes cylindrical or platelike. Even more complicated shapes are possible. The size and size distribution might be important in some cases, for example, if penetration through a pore structure of a cellular membrane is required. The size and size distribution are becoming extremely critical when quantum-sized effects are used to control material properties. A tight control of the average particle size and a narrow distribution of sizes allow creating very efficient fluorescent probes that emit narrow light in a very wide range of wavelengths. This helps creating biomarkers with many well-distinguished colors. The core itself might have several layers and be multifunctional. For example, by combining magnetic and luminescent layers one can both detect and manipulate the particles.

The core particle is often protected by several monolayers of inert material, for example, silica. Organic molecules that are adsorbed or chemisorbed on the surface of the particle are also used for this purpose. The same layer might act as a biocompatible material. However, more often an additional layer of linker molecules is required to proceed with further functionalization. This linear linker molecule has reactive groups at both ends. One group is aimed at attaching the linker to the nanoparticle surface and the other is used to bind various moieties like biocompatible (for example, dextran), antibodies, fluorophores etc., depending on the function required for the application.

Functionalized magnetic nanoparticles have found many applications including cell separation and probing; these and other applications are discussed next (see below). Most of the magnetic particles studied so far are nearly spherical, which can limit the possibilities to make these nanoparticles multifunctional. Alternative cylindrically shaped nanoparticles can be created by employing metal electrodeposition into nanoporous alumina template [18].

As surface chemistry for functionalization of metal surfaces is well-developed, different ligands can be selectively attached to different parts of nanoparticle surface. It is possible to produce magnetic nanowires by spatially segregated fluorescent parts. In addition, because of the large aspect ratios, the residual magnetization of these nanowires can be high. Hence, weaker magnetic field can be used to drive them. It has been shown that a self-assembly of magnetic nanowires in suspension can be controlled by weak external magnetic fields. This would potentially allow controlling cell assembly in different shapes and forms. Moreover, an external magnetic field can be combined with a lithographically defined magnetic pattern ("magnetic trapping").

10.2
Biocompatibility of Magnetic Nanoparticles

In order to interact with biological target, a biological or molecular coating layer acting as a bioinorganic interface should be attached to the nanoparticle. Examples of biological coatings may include antibodies, biopolymers like collagen [16], or monolayers of small molecules that make the nanoparticles biocompatible [17]. Magnetic particles as carriers for therapeutic agents have been used in experimental animals and clinical applications in humans. Mostly, they have used in combination with either diagnostic imaging procedures, and/or oncological therapeutic regimes. The aim of most of the research is to investigate the possibility of these magnetic particles to be used in clinical applications of musculoskeletal disorders as well (cartilage, joint capsules, bone, tendons and, ligaments).

Biocompatible magnetic nanoparticles *in vitro* experiments insignificantly influence the cell's survive. Biocompatibility is made possible through chemical modification of the surface of the magnetic nanoparticles, usually by coating with biocompatible molecules such as dextran, polyvinyl alcohol (PVA), and phospholipids – all of which have been used on iron oxide nanoparticles [19].

As well as providing a link between the particle and the target site on a cell or molecule, coating has the advantage of increasing the colloidal stability of the magnetic fluid. Specific binding sites on the surface of cells are targeted by antibodies or other biological macromolecules such as hormones or folic acid [20]. As antibodies specifically bind to their matching antigen, this provides a highly accurate way to label cells. For example, magnetic particles coated with immunospecific agents have been successfully bound to red blood cells [19, 21], lung cancer cells [22], bacteria [23], and urological cancer cells [24]. For larger entities such as the cells, both magnetic nanoparticles and larger particles can be used, for example, some applications use magnetic "microspheres"—micron-sized agglomerations of submicron-sized magnetic particles incorporated in a polymeric binder [25].

Generally, the magnetic component of the particle is coated by a biocompatible polymer such as PVA or dextran, although recently inorganic coatings such as silica have been developed. The coating acts to shield the magnetic particle from the surrounding environment and can also be functionalized by attaching carboxyl groups, biotin, avidin, carbodiimide, and other molecules [26]. A common failure in targeted systems is due to the opsonization of the particles on entry into the bloodstream, rendering the particles recognizable by the body's major defense system, the reticuloendothelial system (RES). Magnetic nanoparticles are physiologically well-tolerated, for example, dextran-coated magnetite has nonmeasurable toxicity index LD_{50} [27]. This index shows the margin of safety that exists between the dose needed for the desired effect and the dose that produces unwanted and possibly dangerous side effects. In general, the narrower this margin, the more likely it is that the drug will produce unwanted effects. A quantitative measurement of the relative safety of drugs is the therapeutic index, which is the ratio of the dose that elicits a lethal response in 50% of treated individuals (LD_{50}) divided by the dose that elicits a therapeutic response in 50% of the treated individuals (TD_{50}).

After particles are injected into the bloodstream, they are rapidly coated by components of the circulation, such as plasma proteins. This process, namely, opsonization, is critical in dictating the circumstance of the injected particles [28]. Normally, opsonization renders the particles recognizable by the body's major defense system, the RES. The RES is a diffuse system of specialized cells that are phagocytic, associated with the connective tissue framework of the liver, spleen, and lymph nodes [29]. The macrophage (Kupffer) cells of the liver, and to a lesser extent the macrophages of the spleen and circulation, therefore play a critical role in the removal of opsonized particles. As a result, the application of nanoparticles *in vivo* or *ex vivo* would require surface modification that would ensure that the particles were nontoxic, biocompatible, and stable to the RES.

The particles may be injected intravenously and then be directed to the region of interest for treatment. Alternatively in many cases, the particles' suspension would be injected directly into the general area when treatment was desired. Either of these routes has the requirement that the particles do not aggregate and block their own spread. This leads to questions about the best way to produce a suspension that is stable. Fortunately there is appreciable intracellular space in the body through which nanoparticles can diffuse out of flow. A large proportion of this space is between cells. Brightman found that 9 nm diameter ferritin particles would diffuse rapidly through intercellular spaces to achieve a near uniform distribution in a few minutes [30]. The diffusion to the general mass of tissues was presumably aided by the pressure gradients from the blood vessels (chiefly microcapillaries) to the tissue spaces. Larger particles of 250–300 nm diameter are not being transported this way

10.2 Biocompatibility of Magnetic Nanoparticles

and remained in circulation or attached to the walls of the vascular system. Attaching particles to the vascular walls may be a method of therapy in some instances but carries the risk of thromboses. These considerations suggest that nanoparticles of about 5–10 nm diameter should form the ideal particles for most forms of therapy but that there will also be problems of formulating the particle concentrations and suspending media to obtain best distributions.

Particles that have a largely hydrophobic surface are efficiently coated with plasma components and thus are rapidly removed from the circulation, whereas particles that are more hydrophilic can resist this coating process and are cleared more slowly [31]. This has been used to the advantage when attempting to synthesize RES-evading particles by sterically stabilizing the particles with a layer of hydrophilic polymer chains [32]. The most common coatings described in the literature are derivatives of dextran, polyethylene glycol (PEG), polyethylene oxide (PEO), polyoxamers, and polyoxamines [33]. The role of the dense brushes of polymers is to inhibit opsonization, thereby permitting longer circulation times [34]. A further strategy in avoiding the RES is by reducing the particle size [35]. Despite all efforts, however, complete evasion of the RES by these coated nanoparticles has not yet been possible [31].

Apart from the *in vitro* studies, a first, unsuccessful attempt was made to prove the presence of magnetic particles within the joint by clinical means. In a cadaver limb, particles were injected into the stifle joint of a sheep and CT imaging was performed. Although a defined concentration of particles was injected, no signs of these particles could be noticed on the CT films. Further clinical testings are ongoing for tracking down the particles in an *in vivo* situation.

Magnetic nanoparticles have shown great potential for medical sensors and biomedicine applications. The latter include contrast enhancement agents for MRI and site-specific drug delivery agents for cancer therapies. Usually, Fe, Co, and Ni nanoparticles with controllable sizes and shapes via thermodecomposition, X-ray powder diffraction (XRD), and TEM have been used to characterize the magnetic nanoparticles. Extended X-ray absorption fine structure (EXAFS) and X-ray absorption near-edge structure (XANES) were used to probe the structures of Co nanoparticles.

In order to fully realize the biological applications of these magnetic nanoparticles, one needs to develop the methods for improving their biocompatibility. Also, nanoparticles, superparamagnetic (SPM) at room temperatures, are preferred for medical applications [36]. Furthermore, applications in biology and medical diagnosis and therapy require the magnetic particles to be stable in water at neutral pH and physiological salinity.

The colloidal stability of this fluid depends first on the dimensions of the particles, which should be sufficiently small so that precipitation due to gravitation forces can be avoided, and second on the charge and surface chemistry, which give rise to both steric and coulomb repulsions [37]. Additional restrictions to the possible particles that could be used for biomedical applications strongly depend on whether these particles are going to be used in *in vivo* or *in vitro* applications.

For *in vivo* applications, the magnetic particles must be coated with a biocompatible polymer during or after the synthesis process to prevent the formation of large aggregates, changes from the original structure and biodegradation when exposed to the biological system. The polymer will also allow binding of drugs by covalent attachment, adsorption, or entrapment on the particles [38]. The important factors, which determine the biocompatibility and toxicity of these materials, are the nature of the magnetically responsive component, such as magnetite, iron, nickel, cobalt, neodymium–iron–boron or samarium–cobalt, and the final size of the particles, their core, and the coatings. Iron oxide particles such as magnetite (Fe_3O_4) or its oxidized form maghemite (γ-Fe_2O_3) are by far the most commonly employed for biomedical applications. Highly magnetic materials such as cobalt and nickel are toxic, susceptible to oxidation, and hence are of little interest [39].

Moreover, the main advantage of using particles of sizes smaller than 100 nm (so-called nanoparticles) is their higher effective surface areas (easier attachment of ligands), lower sedimentation rates (high stability), and improved tissular diffusion [40]. Another advantage of using nanoparticles is that the magnetic dipole–dipole interactions are significantly reduced because they scale as $\sim r^6$ (r is the particle radius). Therefore, for *in vivo* biomedical applications, magnetic nanoparticles must be made of a nontoxic and nonimmunogenic material, with particle sizes small enough to remain in the circulation after injection and to pass through the capillary systems of organs and tissues avoiding vessel embolism. They must also have a high magnetization so that their movement in the blood can be controlled with a magnetic field and they can be immobilized close to the targeted pathologic tissue [41].

For *in vitro* applications, the size restrictions are not as severe as in *in vivo* applications. Therefore, composites consisting of SPM nanoparticles dispersed in submicron diamagnetic particles with long sedimentation times in the absence of a magnetic field can be used. The advantage of using diamagnetic matrixes is that the SPM composites can be easily provided with functionality.

Some investigations are devoted to Co@Au and Co@Ag bimetallic (core–shell) nanoparticles. These bimetallic nanoparticles are expected to maintain their magnetic properties of Co, while Au or Ag can improve their biocompatibility. Co@Au and Co@Ag nanoparticles are prepared by growing Au or Ag on the presynthesized Co nanoparticles.

Regarding the choice of magnetic particle, the iron oxides magnetite (Fe_3O_4) and maghemite (γ-Fe_2O_3) are the most studied to date because of their generally appropriate magnetic properties and biological compatibility, although many others have been investigated. Particle sizes less than about 10 nm are normally considered small enough to enable effective delivery to the site of the cancer, either via encapsulation in a larger moiety or suspension in some sort of carrier fluid. Nanoscale particles can be coupled with antibodies to facilitate targeting on an individual cell basis. Candidate materials are divided into two main classes: ferromagnetic or ferrimagnetic (FM) single-domain or multidomain particles, or SPM particles. The heat-generating mechanisms associated with each class are quite different, each offering unique advantages and disadvantages, as discussed later.

Magnetic carriers receive their magnetic responsiveness to a magnetic field from incorporated materials such as magnetite, iron, nickel, cobalt, neodymium–iron–boron, or samarium–cobalt. Magnetic carriers are normally grouped according to size. At the lower end, we have the ferrofluids, which are colloidal iron oxide solutions. Encapsulated magnetite particles in the range of 10–500 nm are usually called magnetic nanospheres and any magnetic particles of just below 1–100 μm are magnetic microspheres (MMS). One can include the magnetic liposomes to that class of the magnetic carriers.

In summary, for biomedical applications, magnetic carriers must be water-based, biocompatible, nontoxic, and nonimmunogenic. The first medical applications directly applied magnetite or iron powder. Improved biocompatibility, however, was reached by encapsulating the magnetic materials. The "shell" material determines the reaction of the body to the microsphere. Matrix materials that have been tested for the MMS include chitosan, dextran, poly(lactic acid), starch, poly(vinyl alcohol), polyalkylcyanoacrylate, polyethylene imine, polysaccharides, gelatin, and proteins.

As promising as these results have been, there are several problems associated with magnetically targeted drug delivery [42]. These limitations include, at the first, the possibility of embolization of the blood vessels in the target region due to accumulation of the magnetic carriers, and at the second, difficulties in scaring up from animal models due to the larger distances between the target site and the magnet. When the drug is released, it is no longer attracted to the magnetic field, and, at last, toxic responses to the magnetic carriers. Recent preclinical and experimental results indicate, however, that it is still possible to overcome these limitations and use magnetic targeting to improve drug retention and address safety issues as well [43].

10.3
Magnetic Separation for Purification and Immunoassay

Historically, the magnetic separation was the first biomedicine applications of magnetic nanoparticles. The basic principle of batch magnetic separation is very simple. Magnetic separation technology, using magnetic particles, is a quick and easy method for sensitive and reliable capture of specific proteins, genetic material, and other biomolecules. The technique offers an advantage in terms of subjecting the analyte to very little mechanical stress compared to other methods. Second, these methods are nonlaborious, cheap, and often highly scalable. Moreover, techniques employing magnetism are more suitable to automation and miniaturization. Now that the human genome is sequenced and tens thousand genes are annotated, the next step is to identify the function of these individual genes, carrying out genotyping studies for allelic variation, ultimately leading to identification of novel drug targets. This magnetic technique is one of the most widespread users of magnetic nanoparticles in biological systems in magnetic cell separation. Magnetic carriers bearing an immobilized affinity or hydrophobic ligand or ion-exchange groups, or magnetic biopolymer particles having affinity to the isolated structure, are mixed with a sample containing target compound. Samples may be crude cell lysates, whole blood, plasma, ascites fluid, milk, whey, urine, cultivation media, wastes from food and fermentation industry, and many others. Following an incubation period when the target compound binds to the magnetic particles, the whole magnetic complex is easily and rapidly removed from the sample using an appropriate magnetic separator. After washing out the contaminants, the isolated target compound can be eluted and used for further work.

10.3.1
Cell Labeling for Separation

In biomedicine, it is often advantageous to separate specific biological entities from their native environment in order that concentrated samples may be prepared for subsequent analysis or other use. Magnetic separation using biocompatible nanoparticles is one way to achieve this. It is a two-step process, involving, in first, the tagging or labeling of the desired biological entity with magnetic material, and, at the second, separating these tagged entities via a fluid-based magnetic separation device. In order to separate nanoparticles embedded a permanent magnet was used for the magnetic separation. The separation principle was based on the balance among the magnetic force, buoyant force, and gravitational force, as described by

$$F = \frac{\chi \cdot B}{\mu_0} \frac{dB}{dz} + \rho_l g - \rho_n g, \tag{10.1}$$

where χ is the magnetic susceptibility of milled materials, μ_0 is the vacuum permeability, B is the magnetic field strength, z depicts the field direction, g is the gravity acceleration, and ρ_l and ρ_n are the density of liquid and nanomaterials, respectively. When $F > 0$, the materials will be attracted to the magnet pole, and consequently, the materials will move downward due to the gravity; thus leading to the materials' separation. Because ρ_l, and ρ_n are approximately same in our case, one may assume that the force F is predominately associated with the spatial distribution of magnetic field and the measurement of volume susceptibility χ. The χ value is closely associated with the content of magnetic component in the nanomaterials. It is reasonable that the materials with high component of Fe will be more subjected to the external magnetic field. Immunoassay is a sensitive and selective approach for low amount of drugs. Magnetic separation immunoassays use magnetic beads to facilitate the separation of bound labeled antigens from free antigens in solution. Tagging is made possible through chemical modification of the surface of the magnetic nanoparticles, usually by coating with biocompatible molecules, such as dextran, polyvinyl alcohol (PVA), and phospholipids – all of which have been used on iron oxide nanoparticles [44–46]. As well as providing a link between the particle and the target site on a cell or molecule, coating has the advantage of increasing the colloidal stability of the magnetic fluid. Specific binding sites on the surface of cells are targeted by antibodies or other biological macromolecules such as hormones or folic acid [47–49]. As antibodies specifically bind to their matching antigen, this provides a highly accurate way to label cells.

The complete separation of mixtures of magnetic particles may be achieved by on-chip free-flow magnetophoresis. The magnetic particles in continuous flow can be deflected from the direction of laminar flow by a perpendicular magnetic field depending on their magnetic susceptibility and size and on the flow rate [50]. The separated particles may be detected by video observation and also by on-chip laser-light scattering. Potential applications of this separation method include sorting of magnetic micro- and nanoparticles as well as magnetically labeled cells.

The magnetically labeled material is separated from its native solution by passing the fluid mixture through a region in which there is a magnetic field gradient that can immobilize the tagged material via the magnetic force of Eq. (10.2). This force needs to overcome the

$$F = 6\pi \eta R \Delta v \tag{10.2}$$

hydrodynamic drag force acting on the magnetic particle in the flowing solution, where η is the viscosity of the medium surrounding the cell (e.g., water), R is the radius of the magnetic particle, and $\Delta v = v_m - v_w$ is the difference in velocities of the cell and the water [51]. There is also buoyancy force that affects the motion, but this is dependent on the difference between the density of the cell and the water, and for most cases of interest in biology and medicine can be neglected. Equating the hydrodynamic drag and magnetic forces, and writing $V_m = 4/3\pi R^3$, gives the velocity of the particle relative to the carrier fluid as

$$\Delta v = \frac{R^2 \Delta \chi}{9\mu_0 \eta} \nabla (B^2); \tag{10.3a}$$

or

$$\Delta v = \frac{\xi}{\mu_0} \nabla (B^2), \tag{10.3b}$$

where ξ is the "magnetophoretic mobility" of the particle – a parameter that describes how manipulable a magnetic particle is. For example, the magnetophoretic mobility of MMS can be much greater than that of nanoparticles due to their larger size. This can be an advantage, for example, in cell separations, where the experimental timeframe for the separations is correspondingly shorter. On the other hand, smaller magnetic particle sizes can also be advantageous, for example, in reducing the likelihood that the magnetic material will interfere with further tests on the separated cells [52].

10.3.2
Magnetic Separator Design

Magnetic separator design can be as simple as the application and removal of a permanent magnet to the wall of a test tube to cause aggregation, followed by removal of the supernatant. However, this method can be limited by slow accumulation rates [53]. It is often preferable to increase the separator efficiency by producing regions of high magnetic-field gradient to capture the magnetic nanoparticles as they float or flow by in their carrier medium. A typical way to achieve this is to loosely pack a flow column with a magnetizable matrix of wire (for example, steel wool) or beads [54] and to pump the magnetically tagged fluid through the column while a field is applied. This method is faster than that in the first case, although problems can arise due to the settling and adsorption of magnetically tagged material on the matrix. An alternative, rapid throughput method, which does not involve any obstructions being placed in the column, is the use of specifically designed field gradient systems, such as the quadrupolar arrangement, which creates a magnetic gradient radially outward from the center of the flow column [55].

As well as separating out the magnetically tagged material, the spatially varying magnitude of the field gradient can be used to achieve fluid flow fractionation [56]. This is a process in which the fluid is split at the outlet into fractions containing tagged cells or proteins with differing magnetophoretic mobility.

The standard methods of magnetic separation have two stages: first, the strong magnet is attached to the container wall of a solution with magnetically tagged and unwanted biomaterials. The tagged particles are gathered by the magnet, and, at the second stage, all rest unwanted supernatant solution is removed. In practice, a solution containing tagged and unwanted biomaterials flows continuously through a region of strong magnetic-field gradient, often provided by packing the column with steel wool, which captures the tagged particles. The central core of the column is made of nonmagnetic material to avoid complications due to the near-zero field gradients there. There after, the tagged particles are recovered by removing the field and flushing through with water magnetophoretic mobilities. In a variant of this, the fluid is static while an applied magnetic field is moved up in the container [57]. The particles move up the container in the resulting field gradient at a velocity dependent on their magnetophoretic mobility. At the top of the container, they enter a removable section and are held here by a permanent magnet. The bottom section of the container moves to the next section, and a magnetic field with different strength to the first is applied, and the process repeats. The result is a fraction of the sample into aliquots of differing magnetophoretic mobility.

10.3.3
The Biomedicine Applications

Magnetic separation has been successfully applied to many aspects of biomedical and biological research. It has proven to be a highly sensitive technique for the selection of rare tumor cells from blood, and is especially well-suited to the separation of low numbers of target cells [58]. This has, for example, led to the enhanced detection of malarial parasites in blood samples either by utilizing the magnetic properties of the parasite [59] or through labeling the red blood cells with an specific magnetic fluid [60]. It has been used as a preprocessing technology for polymerase chain reactions (PCR), through which the DNA of a sample is amplified and identified [61]. Cell counting techniques have also been developed. One method estimates the location and number of cells tagged by measuring the magnetic moment of the microsphere tags [62], while another uses a giant magnetoresistive sensor to measure the location of microspheres, attached to a surface layered with a bound analyte [63].

In another application, magnetic separation has been used, in combination with optical sensing, to perform magnetic enzyme, linked immunosorbent assays [64, 65]. These assays use fluorescent enzymes to optically determine the number of cells labeled by the assay enzymes. Typically the target material must first be bound to a solid matrix. In a modification of this procedure, the MMS act as the surface for initial immobilization of the target material and magnetic separation is used to increase the concentration of the material. The mobility of the magnetic nanoparticles allows a shorter reaction time and a greater volume of reagent to be used than in standard immunoassays where the antibody is bound to its plate. In a variation of this procedure, magnetic separation has been used to localize labeled cells at known locations for cell detection and counting via optical scanning. The cells are labeled both magnetically and fluorescently and move through a magnetic-field gradient toward a plate on which lines of ferromagnetic material have been lithographically etched. The cells align along these lines and the fluorescent tag is used for optical detection of the cells.

10.3.4
The Immunomagnetic Separation

Magnetic separation techniques have several advantages in comparison with standard separation procedures. This process has a few handling stages.

All the steps of the purification procedure can take place in one single test tube or another vessel. There is no need for liquid chromatography systems, centrifuges, filters, etc. The separation process can be performed directly in crude samples containing suspended solid material. In some cases (e.g., isolation of intracellular proteins), it is even possible to integrate the disintegration and separation steps and thus shorten the total separation time [66]. Due to the magnetic properties of magnetic adsorbents (and diamagnetic properties of majority of the contaminating molecules and particles present in the treated sample), they can be relatively easily and selectively removed from the sample. In fact, magnetic separation is the only feasible method for recovery of magnetic particles (diameter from 100 nm to 1 µm) in the presence of biological debris and other fouling material of similar size. Moreover, the power and efficiency of magnetic separation procedures is especially useful at large-scale operations. The magnetic separation techniques are also the basis of various automated procedures, especially magnetic-particle-based immunoassay systems to determine a variety of analytes, among them proteins and peptides. Several automated systems for the separation of proteins or nucleic acids have become available recently.

Magnetic separation is usually very gentle to the target proteins or peptides. Even large protein complexes that tend to be broken up by traditional column chromatography techniques may remain intact when using the very gentle magnetic separation procedure [67]. Both the reduced shearing forces and the higher protein concentration throughout the isolation process positively influence the separation process.

Separation of target proteins using standard chromatography techniques often leads to the large volume of diluted protein solution. In this case, appropriate magnetic particles can be used for their concentration instead of ultrafiltration and precipitation [68].

The necessary materials and equipment for laboratory experiments are mentioned below. Magnetic carriers with immobilized affinity or hydrophobic ligands, magnetic particles prepared from a biopolymer exhibiting affinity for the target compound(s), or magnetic ion-exchangers are usually used to perform the isolation procedure. Magnetic separators of different types can be used for magnetic separations, but many times cheap strong permanent magnets are equally efficient, especially in preliminary experiments.

The magnetic carriers and adsorbents can be either prepared in the laboratory, or commercially available ones can be used. These carriers are usually available in the form of magnetic particles prepared from various synthetic polymers, biopolymers, or porous glass, or magnetic particles based on the inorganic magnetic materials such as surface-modified magnetite can be used. Many of the particles behave like SPM ones responding to an external magnetic field, but not interacting themselves in the absence of magnetic field. This is important due to the fact that magnetic particles can be easily resuspended and remain in suspension for a long time. The diameter of the particles, in most cases, differs from 50 nm to 10 µm. However, also larger magnetic affinity particles, with the diameters up to millimeter range, have been successfully used [69]. Magnetic particles having the diameter larger than 1 µm can be easily separated using simple magnetic separators, while separation of smaller particles (magnetic colloids with the particle size ranging between tens and hundreds of nanometers) may require the usage of high-gradient magnetic separators (HGMS). Commercially available magnetic particles can be obtained from a variety of companies. In most cases, polystyrene is used as a polymer matrix, but carriers based on cellulose, agarose, silica, porous glass, or silanized magnetic particles are also available.

Particles with immobilized affinity ligands are available for magnetic affinity adsorption. Streptavidin, antibodies, protein A, and Protein G are used most often in the course of protein and peptides isolation. Magnetic particles with above-mentioned immobilized ligands can also serve as generic solid phases to which native or modified affinity ligands can be immobilized (e.g., antibodies in the case of immobilized protein A, protein G, or secondary antibodies, biotinylated molecules in the case of immobilized streptavidin).

Also some other affinity ligands (e.g., nitrilotriacetic acid, glutathione, trypsin, trypsin inhibitor, gelatin, etc.) are already immobilized to commercially available carriers. To immobilize other ligands of interest to both commercial and laboratory made magnetic particles, standard procedures used in affinity chromatography can be employed. Usually functional groups available on the surface of magnetic particles such as –COOH, –OH, or –NH_2 are used for immobilization; in some cases, magnetic particles are available already in the activated form (e.g., tosylactivated, epoxyactivated, etc).

The magnetite (or similar magnetic materials such as maghemite or ferrites) particles can be surface modified by silanization. This process modifies the surface of the inorganic particles so that appropriate functional groups become available, which enables easy immobilization of affinity ligands [70]. In exceptional cases, enzyme activity can be decreased as a result of usage of magnetic particles with exposed iron oxides. In this case, encapsulated microspheres, having an outer layer of pure polymer, will be safer.

Biopolymers such as agarose, chitosan, kappa carrageenan, and alginate can be easily prepared in a magnetic form. In the simplest way, the biopolymer solution is mixed with magnetic particles and after bulk gel formation the magnetic gel formed is mechanically broken into fine particles [71]. Alternatively biopolymer solution containing dispersed magnetite is dropped into a mixed hardening solution [69], or water-in-oil suspension technique is used to prepare spherical particles [72].

Basically the same procedures can be used to prepare magnetic particles from synthetic polymers such as polyacrylamide, PVA, and many others [73].

In another approach, used standard affinity or ion-exchange chromatography material was postmagnetized by interaction of the sorbent with water-based ferrofluid. Magnetic particles accumulated within the pores of chromatography adsorbent thus modifying this material into magnetic form [74, 75]. Alternatively magnetic sepharose or other agarose gels were prepared by simple contact with freshly precipitated or finely powdered magnetite [74, 76].

Recently also nonspherical magnetic structures, such as magnetic nanorods, have been tested as possible adsorbent material for specific separation of target proteins [77].

The magnetic separators are necessary to separate the magnetic particles. In the simplest approach, a small permanent magnet can be used, but various magnetic separators employing strong rare-earth magnets can be obtained at reasonable prices. Commercial laboratory scale batch magnetic separators are usually made from magnets embedded in disinfectant-proof material. The racks are constructed for separations in Eppendorf microtubes, standard test tubes, or centrifugation cuvettes; some of them have a removable magnetic plate to facilitate easy washing of separated magnetic particles. Other types of separators enable separations from the wells of microtitration plates and the flat magnetic separators are useful for separation from larger volumes of suspensions (up to approx. 500–1000 ml).

Flow-through magnetic separators are usually more expensive and HGMS are the typical examples. Laboratory scale HGMS is composed from a column packed with fine magnetic grade stainless steel wool or small steel balls, which is placed between the poles of an appropriate magnet. The suspension is pumped through the column, and magnetic particles are retained within the matrix. After removal of the column from the magnetic field, the particles are retrieved by flow and usually by gentle vibration of the column.

For work in dense suspensions, open-gradient magnetic separators may be useful. A very simple experimental setup for the separation of magnetic affinity adsorbents from liter volumes of suspensions was described in [78].

Currently many projects require the analysis of a high number of individual proteins or variants. Therefore, methods are required that allow multiparallel processing of different proteins. There are several multiple systems for high throughput nucleic acid and proteins preparation commercially available. The most often used approach for protein isolation is based on the isolation and assay of 6xHis-tagged (protein purification and assay using 6xHis-tagged biomolecules) recombinant proteins using magnetic beads with Ni–nitriloacetic acid ligand [79].

10.3.5
Basic Principles of Magnetic Separation of Proteins and Peptides

Magnetic separations of proteins and peptides are usually convenient and rapid. Proteins and peptides in the free form can be directly isolated from different sources. Membrane-bound proteins have to be usually solubilized using appropriate detergents. When nuclei are broken during sample preparation, DNA released into the lysate make the sample very viscous. This DNA may be sheared by repeated passage up and down through a 21-gauge hypodermic syringe needle before isolation of a target protein. Alternatively, DNase can be added to enzymatically digest the DNA.

Magnetic beads in many cases exhibit low nonspecific binding of nontarget molecules present in different samples. Certain samples may still require preclearing to remove molecules, which have high nonspecific binding activity. If preclearing is needed, the sample can be mixed with magnetic beads not coated with the affinity ligand. In the case of IMS, magnetic beads coated with secondary antibody or with irrelevant antibodies have been used. The nonspecific binding can also be minimized by adding a nonionic detergent both in the sample and in the washing buffers after isolation of the target.

In general, magnetic affinity separations can be performed in two different modes. In the direct method, an appropriate affinity ligand is directly coupled to the magnetic particles, or biopolymer exhibiting the affinity toward target compound(s) is used in the course of preparation of magnetic affinity particles. These particles are added to the sample, and target compounds then bind to them. In the indirect method, the free-affinity ligand (in most cases an appropriate antibody) is added to the solution or suspension to enable the interaction with the target compound. The resulting complex is then captured by appropriate magnetic particles. In case antibodies are used as free-affinity ligands, magnetic particles with immobilized secondary antibodies, protein A, or protein G are used for capturing the complex. Alternatively, the free-affinity ligands can be biotinylated and magnetic particles with immobilized streptavidin or avidin are used to capture the complexes formed. In both the methods, magnetic particles with isolated target compound(s) are magnetically separated and then a series of washing steps is performed to remove majority of contaminating compounds and particles. The target compounds are then usually eluted, but for specific applications (especially in molecular biology, bioanalytical chemistry, or environmental chemistry) they can be used still attached to the particles, such as in the case of PCR, magnetic ELISA, etc.

These two methods perform equally well, but, in general, the direct technique is more controllable. The indirect procedure may perform better if affinity ligands have poor affinity for the target compound.

In most cases, magnetic batch adsorption is used to perform the separation step. This approach represents the simplest procedure available, enabling to perform the whole separation in one test tube or flask. If larger magnetic particles (with diameters above ca. 1 µm) are used, simple magnetic separators can be employed. In case magnetic colloids (diameters ranging between tens and hundreds of nanometers) are used as affinity adsorbents, HGMS have to be used usually to remove the magnetic particles from the system.

Alternatively magnetically stabilized fluidized beds (MSFB), which enable a continuous separation process, can be used. The use of MSFB is an alternative to conventional column operation, such as packed bed or fluidized bed, especially for large-scale purification of biological products. Magnetic stabilization enables the expansion of a packed bed without mixing of solid particles. High column efficiency, low pressure drop, and elimination of clogging can be reached [80, 81].

Also nonmagnetic chromatographic adsorbents can be stabilized in MSFB if sufficient amount of magnetically susceptible particles is also present. The minimum amount of magnetic particles necessary to stabilize the bed is a function of various parameters including the size and density of particles, the magnetic field strength, and the fluidization velocity. A variety of commercially available affinity, ion-exchange, and adsorptive supports can be used in the bed for continuous separations [82].

Biocompatible two phase systems, composed, for example, from dextran and PEG, are often used for isolation of biologically active compounds, subcellular organelles, and cells. One of the disadvantages of this system is the slow separation of the phases when large amounts of proteins and cellular components are present. The separation of the phases can be accelerated by the addition of fine magnetic particles or ferrofluids to the system followed by the application of a magnetic field. This method seems to be useful when the two phases have very similar densities, the volumetric ratio between the phases is very high or low, or the systems are viscous. Magnetically enhanced phase separation usually increases the speed of phase separation by a factor of about 10 in well-behaved systems, but it may increase by a factor of many thousands in complex systems. The addition of ferrofluids and/or iron oxide particles was shown to have usually no influence on enzyme partitioning or enzyme activity [83, 84].

Proteins and peptides isolated using magnetic techniques have to be usually eluted from the magnetic separation materials. In most cases, bound proteins and peptides can be submitted to standard elution methods such as the change of pH, change of ionic strength, use of polarity reducing solvents (e.g., dioxane or ethyleneglycol), or the use of deforming eluents containing chaotropic salts. Affinity elution (e.g., elution of glycoproteins from lectin-coated magnetic beads by the addition of free sugar) may be both a very efficient and gentle procedure.

10.3.6
Examples of Magnetic Separations of Proteins and Peptides

Magnetic affinity and ion-exchange separations have been successfully used in various areas, such as molecular biology, biochemistry, immunochemistry, enzymology, analytical chemistry, environmental chemistry, etc. [85, 86].

In the case of proteins and peptides purifications, no simple strategy for magnetic affinity separations exists. Various affinity ligands have been immobilized on magnetic particles, or self-magnetic particles have been prepared in the presence of biopolymers exhibiting the affinity for target enzymes or lectins. Immunomagnetic particles, i.e., magnetic particles with immobilized specific antibodies against the target structures, have been used for the isolation of various antigens, both molecules and cells [85, 87] and can thus be used for the separation of specific proteins.

Magnetic separation procedures can be employed in several ways. Preparative isolation of the target protein or peptide is usually necessary if further detailed study is intended. In other cases, however, the magnetic separation can be directly followed (after elution with an appropriate buffer) with electrophoresis. The basic principles of magnetic separations can be used in the course of protein or peptide determination using various types of solid-phase immunoassays. Usually immunomagnetic particles directly capture the target analyte, or magnetic particles with immobilized streptavidin are used to capture the complex of biotinylated primary antibody and the analyte.

Enzyme isolation is usually performed using immobilized inhibitors, cofactors, dyes, or other suitable ligands, or magnetic beads prepared from affinity biopolymers can be used.

Genetic engineering enables the construction of gene fusions resulting in fusion proteins having the combined properties of the original gene products. To date, a large number of different gene fusion systems, involving fusion partners that range in size from one amino acid to whole proteins, capable of selective interaction with a ligand immobilized onto magnetic particles or chromatography matrices, have been described. In such systems, different types of interactions, such as enzyme–substrate, receptor–target-protein, polyhistidines–metal-ion, and antibody–antigen, have been utilized. The conditions for purification differ from system to system, and the environment tolerated by the target protein is an important factor for deciding which affinity fusion partner to choose. In addition, other factors, including protein localization, costs for the affinity matrix and buffers, and the possibilities of removing the fusion partner by site-specific cleavage, should also be considered [88, 89]. As an example, isolation of recombinant oligohistidine-tagged proteins is based on the application of metal chelate magnetic adsorbents [90, 91]. This method has been used successfully for the purification of proteins expressed in bacterial, mammalian, and insect systems.

Aptamers are DNA or RNA molecules that have been selected from random pools based on their ability to bind other molecules. Aptamers-binding proteins can be immobilized to magnetic particles and used for isolation of target proteins.

A new approach for analytical ion-exchange separation of native proteins and proteins enzymatic digest products has been described recently [92]. Magnetite particles were covered with a gold layer and then stabilized with ionic agents. These charged stabilizers present at the surface of the gold particles are capable of attracting oppositely charged species from a sample solution through electrostatic interactions. Au magnetic particles having negatively charged surfaces are suitable probes for selectively trapping positively charged proteins and peptides from aqueous solutions.

10.3.6.1 Perspectives

The first one will be focused on the laboratory-scale application of magnetic affinity separation techniques in biochemistry and related areas (rapid isolation of a variety of both low- and high-molecular-weight substances of various origin directly from crude samples, thus reducing the number of purification steps) and in biochemical analysis (application of immunomagnetic particles for separation of target proteins from the mixture followed by their detection using ELISA and related principles). Such a type of analysis will enable to construct portable assay systems enabling e.g., near-patient analysis of various protein disease markers. New methodologies, such as the application of chip and microfluidics technologies, may result in the development of magnetic separation processes capable of magnetic separation and detection of extremely small amount of target biologically active compounds [93].

In the near future, quite new separation strategies can appear. A novel magnetic separation method, which utilizes the magneto-Archimedes levitation, has been described recently and applied to separation of biological materials. By using the feature that the stable levitation position under a magnetic field depends on the density and magnetic susceptibility of materials, it was possible to separate biological materials such as hemoglobin, fibrinogen, cholesterol, and so on. So far, the difference of magnetic properties was not utilized for the separation of biological materials. Magneto-Archimedes separation may be another way for biological materials separation [94].

It can be expected that magnetic separations will be used regularly both in biochemical laboratories and biotechnology industry in the near future.

In some cases, new methods such as IMS are used. IMS is the process of using small SPM particles or beads coated with antibodies against surface antigens of cells. Using this technique, methods have been described for the efficient isolation of certain eukaryotic cells from fluids such as blood. Additionally, this technique has been shown to be suitable for the detection of prokaryotic organisms such as bacteria and viruses. The technique of IMS is assisted in the fact that bacteria immunologically bound to magnetic beads usually remain viable and can continue to multiply if nutritional requirements are provided. The immunomagnetically isolated fraction can then be washed to remove nonspecifically attached organisms before being placed on suitable growth media. Both polyclonal and monoclonal antibodies have been employed in IMS. These antibodies can be linked to the beads either directly or indirectly, using beads precoated with antimouse or antirabbit antibodies. Several magnetic solid phases in particle form are commercially available for magnetic separation of biological organisms, organelles, or molecules. Common to all of these particles is that specific binding molecules can be attached to them. Most particles are SPM; i.e., they are magnetic in a magnetic field but are nonmagnetic as soon as the magnetic field is removed. This is important because, once separated by a magnet, particles should attach to each other through intermagnetic force but then return directly back into suspension. Physical parameters, i.e., the shape and size of the particles, are also important. In order to perform identically in a suspension, with respect to sedimentation and kinetics of binding to other molecules, identical size and form of the particle are preferred.

The IMS technique has several advantages for microbiologists. When working with samples heavily contaminated with nontarget organisms, IMS facilitates the purification of the target organism. Additionally, larger volumes of samples can be employed and captured target organisms can be concentrated to a volume suitable for analysis. Isolation of specific bacteria by the antigen–antibody reaction has generally been accomplished by inoculating the bead samples to cultivation broths or onto solid media selective for the target bacteria. Identification can then be accomplished by routine or conventional methods. However, increasingly IMS is being chosen as a precursor to a number of downstream detection methods.

There are an increasing number of methods downstream of the IMS process to confirm the presence of target microorganisms. The first application of IMS technology to microbiological science was the separation of bacteria from other nontarget organisms for delivery to liquid or solid culture media. Bacteria do not need to be detached from the beads; as attachment apparently has no effect on their growth. Both solid and liquid media have been used for cultivation of several bacterial species immunologically bound to magnetic beads; however, enumeration must take into account that each colony is not always the product of a single cell – several cells might be attached to a cluster of beads to initiate a single colony. Despite this, IMS has been shown to be a quantitative technique. Both intact bacteria and their soluble antigenic determinants can be detected after magnetic extraction from the test sample, using a second antibody in a sandwich format.

Magnetic separations in biology and biotechnology have diversified in recent years, leading to a wide array of different particles, affinity mechanisms and processes. The most attractive advantage of the magnetic separation technique in biochemistry and biotechnology is the ease of manipulation of biomolecules that are immobilized on magnetic particles. Once target biological cells or molecules are immobilized on magnetic particles, the target biomolecules can be separated from a sample solution, manipulated flexibly in various reagents and transported easily to a desired location by controlling magnetic fields produced from a permanent magnet or an electromagnet. Another advantage is a large surface area of immobilization substrate, which results in a high population of target biological molecules due to a large binding site and high detection signal.

Applications in the nucleic acid realm include products for messenger ribonucleic acid (mRNA) isolation from cells or previously purified total RNA preparations, solid-phase cDNA library construction, double- and single-stranded DNA purification, solid-phase DNA sequencing, and a variety of hybridization-based methodologies.

Magnetic beads are also finding uses in protein purification, immunology, and the isolation of a wide range of specific mammalian cells, bacteria, viruses, subcellular organelles, and individual proteins. There are also products that employ magnetic particles for more conventional isolation and purification methods such as affinity and ion exchange and charcoal trapping of small analytes.

New techniques for magnetic separation are used in work [95]. Magnetic immunoassays utilizing magnetic markers and a high-temperature superconducting interferometer device (SQUID) have been performed. A design of the SQUID was shown there for the sensitive detection of the magnetic signal from the marker, where the spatial distribution of the signal field was taken into account. Using the design, the relationship between the magnetization of the maker and the signal flux detected with the SQUID was obtained. This relationship is important for quantitative evaluation of the immunoassay. The method of the measurements and calculation of the number of magnetic ions in metal–DNA conjugates with the help of SQUID is presented in [96].

Standard liquid column chromatography is currently the most often used technique for the isolation and purification of target proteins and peptides. Magnetic separation techniques are relatively new and still under development. Magnetic affinity particles are currently used mainly in molecular biology (especially for nucleic acids separation), cell biology and microbiology (separation of target cells), and as parts of the procedures for the determination of selected analytes using magnetic ELISA and related techniques (especially determination of clinical markers and environmental contaminants). Up to now, separations in small scale prevail and thus the full potential of these techniques has not been fully exploited.

10.4
Magnetic Nanoparticles in Cancer Therapy

Interest to the nanomedicine has increased dramatically during past years. Facing the known side effects of anticancer drugs, there is a large need for much higher specificity and efficacy. The paradox of high drug concentrations at the target site and almost unharmed normal tissue can be solved by nanoparticles. The second point is the "intelligence" of those nanoparticles, which may release drugs in response to external signals, temperature elevation, or physiological conditions in tumor tissue and cells. Nanoliposomal formulations are already in the market, but they still do not have the features mentioned before. However, as they have shown fewer side effects in clinical studies, this is the right way to go. Finally, there are strong economical arguments for the use of nanotechnology in drug development for cancer. Nanocarrier-based drug targeting systems open a new perspective for conventional anticancer drugs, especially when they are about to loose their patent protection: introducing a new specificity and a new performance by coupling them to nanoparticles brings up completely new products leading to new patent portfolios.

Nanotechnology has the potential to have a revolutionary impact on cancer diagnosis and therapy. It is universally accepted that early detection of cancer (for example, see Chapter 6 – MRI, and positron emission spectroscopy (PET)) is essential even before anatomic anomalies are visible. A major challenge in cancer diagnosis is to be able to determine the exact relationship between cancer biomarkers and the clinical pathology, as well as, to be able to noninvasively detect tumors at an early stage for maximum therapeutic benefit. For breast cancer, for instance, the goal of molecular imaging is to be able to accurately diagnose when the tumor diameter less is 0.3 mm (approximately 1000 cells), as opposed to the current techniques like mammography, which require more than a million cells for accurate clinical diagnosis.

In cancer therapy, targeting and localized delivery are the key challenges. To wage an effective war against cancer, we have to have the ability to selectively attack the cancer cells, while saving the normal tissue from excessive burdens of drug toxicity (see Chapter 5). However, because many anticancer drugs are designed to simply kill cancer cells, often in a semispecific fashion, the distribution of anticancer drugs in healthy organs or tissues is especially undesirable due to the potential for severe side effects. Consequently, systemic application of these drugs often causes severe side effects in other tissues (e.g., bone marrow suppression, cardiomyopathy, neurotoxicity, etc.), which greatly limits the maximal allowable dose of the drug. In addition, rapid elimination and widespread distribution into nontargeted organs and tissues requires the administration of a drug in large quantities, which is often not economical and sometimes complicated due to nonspecific toxicity. This vicious cycle of large doses and the concurrent toxicity is a major limitation of current cancer therapy. In many instances, it has been observed that the patient succumbs to the ill effects of the drug toxicity far earlier than the tumor burden.

10.4.1
Is a Heat Suitable for Health?

The main field of magnetic nanoparticles application in cancer therapy is the magnetic hyperthermia. Hyperthermia is the heat treatment. The temperature of the tissue may be elevated artificially, for example, with the aim of alternating magnetic fields, and that hyperthermia is named "magnetic hyperthermia." Hyperthermia is one of the promising approaches in cancer therapy, and various methods have been employed in hyperthermia [97, 98]. The most commonly used heating method in clinical settings is capacitive heating using a radiofrequency (RF) electric field [99]. However, heating tumors specifically by capacitive heating using an RF electric field is difficult, because the heating characteristics are influenced by various factors such as tumor size, position of electrodes, and adhesion of electrodes at uneven sites. From a clinical point of view, a simple heat mediator is more desirable not only for superficially located tumors but also for deep-seated tumors. Some researchers have proposed inductive heating methods, using magnetic nanoparticles for hyperthermia [100–102], which showed a tenfold higher affinity for the tumor cells than neutral magnetoliposomes [101].

Hyperthermia or the controlled heating of tissue to promote cell necrosis with magnetic nanoparticles has the potential to be a powerful cancer treatment [103]. Magnetic nanoparticles coated in a lipid bilayer, magnetoliposomes, can combine heat therapy with drug delivery to provide a synergistic treatment strategy. Magnetic particles can be injected into a patient and guided to a target site with an external magnetic field and/or specifically bind to target cells via recognition molecules coated onto the surface of the particles. Authors [104] have developed to combine their expertise in opening up the blood-brain barrier, using focused ultrasound with γ-Fe_2O_3 magnetic nanoparticles for the treatment of brain tumors.

For medicine applications purpose, control over the synthesis of the nanoparticles and their surface modification is critical; so in work [104] they have prepared highly uniform, monodisperse, single-crystal maghemite nanoparticles of tailorable size (5–20 nm) via an organometallic decomposition method. A surfactant coats the particles during synthesis and has a crucial role in the nucleation and growth of the particles as well as the resulting particle geometry. The as-synthesized particles are initially hydrophobic. However, coating the particles with a second lipid layer is an effective route to make them hydrophilic. Using well-established principles of lipid self-assembly, authors [104] have successfully coated a second layer of lipid, 1, 2-dipalmitoyl-sn-glycero-3-phosphatidylchocholine (DPPC) on the as-synthesized particles.

Heat generation by SPM particles is highly dependent on particle size, crystallinity, and shape. Heating rates deteriorate quickly with polydispersity of particles. The physics of magnetic heating suitable for biological applications requires considerable investigation, involving the choice of the optimum particle size and magnetic anisotropy and studies of complex magnetic susceptibility, specific heat, and power dissipation.

The process involved in the magnetic hyperthermia, which is based on the known hyper sensibility of tumor cells to heating (hyperthermia), is related to energy dissipation when a ferromagnetic material is placed on an external alternating magnetic field. The magnetic fluid hyperthermia (MFH) is the idea of attaining cytolysis of specific tumor tissues by hyperthermia, through the magnetic losses of subdomain magnetic particles when an alternating (ac) magnetic field is applied.

Clearly, the success of such approach depends critically on the ability to specifically attach a given particle on a certain type of cells, those who are to be killed. This is a quite complex biochemical, biological, and medical issue, though many groups are working on it. Other issues to be solved (depending on the kind of organs to be treated) are to transport to the target, to cheat the body immune system, to minimize the mass of magnetic material, and to detect possible accumulation of magnetic material in other organs.

Regarding the energy loss at the magnetic nanoparticles (MNPs), there are two different effects to be considered: (a) magnetic losses through domain-wall displacements (in multidomain particles) called Néel losses; and (b) energy loss from mechanical rotation of the particles, acting against viscous forces of the liquid medium (Brown losses). For details of these two mechanisms, see [105–107].

The applicability of these mechanisms to oncology has recently been evaluated in phase-2. Additionally, some kind of biochemical reactions of bonding can be detected through magnetic relaxation experiments, suggesting also the potential for monitoring the attachment process of the MNPs to specific tissues. At present, there are at least five phase-3 tests for MFH, involving 100 to 1000 patients.

For clinical applications, the granular materials should present low levels of toxicity, as well as a high-saturation magnetic moment in order to minimize the doses required for temperature increase. In this context, magnetite (Fe_3O_4) is a promising candidate because it presents a high Curie temperature (T_C), high saturation magnetic moment (90–98 emu/g, or ~450–500 emu/cm^3), and has shown the lowest toxicity index in preclinic tests. Although from the point of view of synthesis the material is cheap and can be obtained with high purity from relatively easy routes, the fabrication of MNPs with high structural order and having few nanometer diameters is intrinsically complicated since the high surface/volume ratio makes the surface disorder effects to be important.

One problem with treating cancer successfully is the fact that cancerous cells are very difficult to target specifically. In most respects, they are like normal cells, and even if they are not, they can hide the differences. But malignant cells are reliably more sensitive to heat than normal cells. Raising the temperature of the tumor is one way to selectively destroy cancer cells. It was once believed that hyperthermia damages tumor vasculature. In the 1960s, many researchers confirmed that cancer cells are more vulnerable to heat than their normal counterparts. The hegemony of the three official modalities – surgery, radiation, and chemo – lasted until the 1970s, when hyperthermia was taken off the ACS blacklist. In the late 1970s and early 1980s several trials had shown that hyperthermia combined with radiation produced superior results over radiation alone. However, a US phase-III trial subsequently did not confirm these results, and interest waned. Hyperthermia has since inhabited a strange in-between land of having its value recognized, and being used sporadically in some cancer centers, while ignored or underutilized by most oncologists, and largely unknown.

That situation is beginning to change. It has been admitted that the some study showing negative results in the past was flawed on account of inadequate equipment and quality assurance procedures. And recently, the results of three European and one American phase-III trials have become available. All these trials were well controlled, showing that the use of hyperthermia in combination with radiation therapy results in superior tumor response, tumor control, and survival as compared with radiation therapy alone. Some studies have claimed threefold improvement in results, and positive results have even been noted with very difficult cancers like brain, liver, and advanced kidney. Hyperthermia is particularly suitable in treating small superficial tumors (within 7 cm under the surface).

Hyperthermia can be used by itself and results in impressive shrinkage and even complete eradication (10–15%) of tumors. However, these results usually do not last, and the tumors regrow. (In some animal experiments, cures were affected by hyperthermia. For example, in an animal experiment on transplanted mammary carcinoma, radiation alone produced no cures, heat alone produced 22% cures, and combined modality produced 77% cures.)

The synergistic effects of hyperthermia combined with radiation have been studied too. Hyperthermia has been used for the treatment of resistant tumors of many kinds, with very good results. Combined hyperthermia and radiation has been reported to yield higher complete and durable responses than radiation alone in superficial tumors. In deep-seated tumors, the effect of combined treatment is still under research. The clinical experience has supported hyperthermia as a radiation-enhancing agent. Despite difficulties in increasing human tumor temperatures, recent clinical trials have shown that a combination of hyperthermia with radiation is superior to radiation alone in controlling many human tumors.

So, hyperthermia is one way to improve the therapeutic index of total body irradiation (TBI) by increased killing of neoplastic cells, and also by inhibiting the expression of radiation-induced damage to the normal cell population. It is possible to combine hyperthermia safely with further low-dose radiation in situation where a radical dose has already been delivered.

When it comes to chemotherapy, there are indications that some chemotherapeutic agent's action can be enhanced by hyperthermia. For some agents, hyperthermia increases toxicities and the incidence of damage associated with them at the usual doses, or it can be taken as an advantage in the sense of getting the same results with lower doses of the drug.

Several studies have shown increased apoptosis in response to heat. Hyperthermia damages the membranes, cytoskeleton, and nucleus functions of malignant cells. It causes irreversible damage to cellular perspiration of these cells. Heat above 41.8 °C also pushes cancer cells toward acidosis (decreased cellular pH), which decreases the cells' viability and transplantability.

Hyperthermia activates the immune system. Heat has a well-known stimulatory effect on the immune system causing both increased production of interferon alpha and increased immune surveillance.

Tumors have a tortuous growth of vessels feeding them blood, and these vessels are unable to dilate and dissipate heat as normal vessels do. So tumors take longer to heat, but then concentrate the heat within themselves. Tumor blood flow is increased by hyperthermia despite the fact that tumor-formed vessels do not expand in response to heat. Normal vessels are incorporated into the growing tumor mass and are able to dilate in response to heat, and to channel more blood into the tumor.

Tumor masses tend to have hypoxic (oxygen-deprived) cells within the inner part of the tumor. These cells are resistant to radiation, but they are very sensitive to heat. This is why hyperthermia is an ideal companion to radiation: radiation kills the oxygenated outer cells, while hyperthermia acts on the inner low-oxygen cells, oxygenating them, and so making them more susceptible to radiation damage. It is also thought that hyperthermia's induced accumulation of proteins inhibits the malignant cells from repairing the damage sustained.

It can be hypothesized that hypoxic cells in the center of a tumor are relatively radioresistant but thermosensitive, whereas well-vascularized peripheral portions of the tumor are more sensitive to irradiation. This supports the use of combined radiation and heat; hyperthermia is especially effective against centrally located hypoxic cells, and irradiation eliminates the tumor cells in the periphery of the tumor, where heat would be less effective.

Hyperthermia is considered a modifier of radiation response. Heat selectively kills cells that are chronically hypoxic and nutritionally deficient and have a low pH – characteristics shared by tumor cells in comparison with the better oxygenated and better nourished normal cells. Furthermore, heat preferentially kills cells in the S phase of the proliferative cycle, which are known to be resistant to irradiation.

The researchers are still working out the various practical details – the best machinery, the best way to measure the "dose," and the heat within the tumor, and so on, as well as the optimal sequencing of the treatment and its duration. The cytotoxicity of hyperthermia is dependent on both temperature and time. The hyperthermia at 41.8 °C is now used in several centers.

10.4.2
Basics of Hyperthermia

The preferential killing of cancer cells without damaging normal cells has been a desired goal in cancer therapy for many years. However, the various procedures used to date, including chemotherapy, radiotherapy, or surgery, can fall short of this aim. The potential of hyperthermia as a treatment of cancer was first predicted following observations that several types of cancer cells were more sensitive to temperatures in excess of 41.8 °C than their normal counterparts [108, 109]. In the past, external means of heat delivery were used such as ultrasonic or microwave treatments, but more recently research has focused on the injection of magnetic fluids directly into the tumor body, or into an artery supplying the tumor. The method relies on the theory that any metallic objects when placed in an alternating magnetic field will have induced currents flowing within them. Rosensweig [110] reveals analytical relationships and computations of power dissipation on MNPs in magnetic viscous fluid (ferrofluid) subjected to alternating magnetic field. In the works of Fannin et al. [107, 111], the further development is represented. The nonlinear relaxational properties of a water-based magnetic fluid induce heating effects, often driving the magnetic fluid into the nonlinear region of magnetization.

The amount of current is proportional to the size of the magnetic field and the size of the object. As these currents flow within the metal, the metal resists the flow of current and thereby heats, a process termed "inductive heating." If the metal is magnetic, such as iron, the phenomenon is greatly enhanced. Therefore, when a magnetic fluid is exposed to alternating magnetic field, the particles become powerful heat sources, destroying the tumor cells [112]. The magnetic fluids used are preferably suspensions of SPM particles, prepared much as described for MRI contrast agents, as these produce more heat per unit mass than larger particles [113]. The level heating is simply controlled by the materials Curie temperature, that is, the temperature above which materials lose their magnetic properties and thus their ability to heat [114].

The use of iron oxides in tumor heating was first proposed by Gilchrist et al. [115], and there are two different approaches now. The first is called magnetic hyperthermia and involves the generation of temperatures up to 45–47 °C by the particles. This treatment is currently adopted in conjunction with chemotherapy or radiotherapy, as it also renders the cells more sensitive [116]. The second technique is called magnetic thermoblation and uses temperatures of 43–55 °C that have strong cytotoxic effects on both tumor and normal cells [108, 117]. The reason for this use of increased temperatures is due to the fact that about 50% of tumors regress temporarily after hyperthermic treatment with temperatures up to 44 °C; therefore authors prefer to use temperatures up to 55 °C [117]. The problem of deleterious effects on normal cells is reduced by intratumoral injection of the particles.

The heating power of the particles is quantified as the specific absorption rate (SAR) and describes the energy amount converted into heat per time and mass [118]. Apart from the particle size and shape influencing their magnetic properties, thus consequently their heating power, there is also a dependency between temperature elevation and magnetic field amplitude, which must be considered when comparing experiments with different tissue parameters. On the basis of recent studies, tumors with volumes of approximately 300 mm^3 can be heated and no potential problems were expected with larger tissue volumes (e.g., >1000 mm^3) if there is proper regulation of the magnetic mass used and the intratumoral particle distribution [117]. The frequency should be greater than that sufficient to cause any neuromuscular response, and less than that capable of causing any detrimental heating of healthy tissue, ideally in the range of 0.1–1 MHz [112]. If suitable frequencies and field strength combinations are used, no interaction is observed between the human body and the field.

The susceptibility in ordered materials depends not just on temperature, but also on H, which gives rise to the characteristic sigmoidal shape of the M–H curve, with M approaching a saturation value at large values of H. Furthermore, in ferromagnetic and luidizedng materials one often sees hysteresis, which is irreversibility in the magnetization process that is related to the pinning of magnetic domain walls at impurities or grain boundaries within the material, as well as to intrinsic effects such as the magnetic anisotropy of the crystalline lattice. This gives rise to open M–H curves, called hysteresis loops. The shape of these loops are determined in part by particle size: in large particles (of the order micron size or more) there is a multidomain ground state that leads to a narrow hysteresis loop since it takes relatively little field energy to make the domain walls move; while in smaller particles there is a single domain ground slate which leads to a broad hysteresis loop. At even smaller sizes (of the tens of nanometers or less), one can see superparamagnetism, where the magnetic moment of the particle as a whole is free to fluctuate in response to thermal energy, while the individual atomic moments maintain their ordered slate relative to each other. This leads to the anhysteretic, but still sigmoidal, M–H curve. The underlying physics of superparamagnetism is founded on an activation law for the relaxation time τ of the net magnetization of the particle [105, 106]:

$$\tau = \tau_0 \bullet \exp\left(\frac{\Delta E}{k_B T}\right),$$

where ΔE is the energy barrier to moment reversal, and $k_B T$ is the thermal energy. For noninteracting particles, the preexponential factor is of the order 10^{-10}–10^{-12} s and only weakly dependent on temperature [106]. The energy barrier has several origins, including both intrinsic and extrinsic effects such as the magnetocrystalline and shape anisotropies, respectively; but in the simplest of cases, it has a uniaxial form given by $\Delta E = KV$, where E is the anisotropy energy density and V is the particle volume.

This direct proportionality between ΔE and V is the reason that superparamagnetism – the thermally activated flipping of the net moment direction – is important for small particles, since for them ΔE is comparable to $k_B T$ at room temperature. Superparamagnetism is of great interest as well, being a unique aspect of magnetism in nanoparticles [119]. However, it is important to recognize that observations of superparamagnetism are implicitly dependent not just on temperature, but also on the measurement time τ_m of the experimental technique being used. If $\tau \ll \tau_m$ the flipping is fast relative to the experimental time window and the particles appear to be paramagnetic (PM), while if $\tau \gg \tau_m$ the flipping is slow and quasistatic properties are observed – the so-called blocked state of the system. A "blocking temperature" T_B is defined as the mid-point between these two states, where $\tau = \tau_m$. In typical experiments τ_m can range from the slow to medium timescales of 10^2 s for DC magnetization and 10^{-1}–10^{-5} s for AC susceptibility, through to the fast timescales of 10^{-7}–10^{-9} s for ^{57}Fe Mössbauer spectroscopy.

The different magnetic responses take place for the case of ferromagnetic or ferrimagnectic nanoparticles injected into a blood vessel. Depending on the particle size, the injected material exhibits either a multidomain, single-domain, or SPM M–H curve. The magnetic response of the blood vessel itself includes both a PM response, for example, from the iron-containing hemoglobin molecules, and a diamagnetic (DM) response, for example, from those intravessel proteins that comprise only carbon, hydrogen, nitrogen, and oxygen atoms. It should be noted that the magnetic signal from the injected particles, whatever their size, far exceeds that from the blood vessel itself. This heightened selectivity is one of the important features of biomedical applications of magnetic nanoparticles.

Returning to the hysteresis, which gives rise to the open $M-H$ curve seen for ferromagnets and antiferromagnets, it is clear that energy is needed to overcome the barrier to domain-wall motion imposed by the intrinsic anisotropy and microstructural impurities and grain boundaries in the material. This energy is delivered by the applied field, and can be characterized by the area enclosed by the hysteresis loop. This leads to the concept that if one applies a time-varying magnetic field to a ferromagnetic or ▫luidized▫ng▫ material, one can establish a situation in which there is a constant flow of energy into that material, which will perforce be transferred into thermal energy. The similar energy transfer takes place in SPM nanoparticles, where the energy is needed to align the magnetic moments of particles to achieve the saturation state. These conclusions are physical basis of magnetic hyperthermia treatments.

Self-regulating magnetic hyperthermia can be achieved by synthesizing magnetic nanoparticles with desired Curie temperature [120–126]. The desired range of Curie temperatures 43–44 °C was obtained by doping or combination of melting and ball milling of nickel–copper alloy. The submicron range for possible self controlled magnetic hyperthermia treatment of cancer is possible. It is reported that an increase in tumor temperature decreases the tumor resistance to chemo- and radiation therapies. Self-controlled heating at the tumor site to avoid spot heating is managed by controlling the Curie temperature of the magnetic particles. The process used here is mainly composed of melting of the Cu–Ni mixture and ball milling of the resulted bulk alloy. Both mechanical abrasion and continuous grinding were used to break down the bulk amount into the desired particle size. It was found that the desired alloy is composed of 71% nickel and 29% copper by weight. It was observed that the coarse sand-grinded powder has a Curie temperature of 345 K and the fine ball-milled powder shows a temperature of 319–320 K.

10.4.3
Magnetic Hyperthermia in Cancer Treatment

The possibility of beating cancer by artificially induced hyperthermia has led to the development of many different devices designed to heat malignant cells while sparing surrounding healthy tissue [127–129]. The first investigations of the application of magnetic materials for hyperthermia date back to 1957 when Gilchrist *et al.* [115] healed various tissue samples with 20–100 nm size particles of γ-Fe_2O_3 exposed to a 1.2 MHz magnetic field. Since then, there have been numerous publications describing a variety of schemes using different types of magnetic materials, different field strengths and frequencies, and different methods of encapsulation and delivery of the particles [130–148]. In broad terms, the procedure involves dispersing magnetic particles throughout the target tissue, and then applying an AC magnetic field of sufficient strength and frequency to cause the particles to heat. This heat conducts into the immediately surrounding diseased tissue whereby, if the temperature can be maintained above the therapeutic threshold of 42 °C for half an hour or more, the cancer is destroyed. While the majority of hyperthermia devices are restricted in their utility because of unacceptable coincidental heating of healthy tissue, magnetic particle hyperthermia is appealing because it offers a way to ensure only the intended target tissue is heated.

A number of studies have demonstrated the therapeutic efficacy of this form of treatment in animal models [118, 149, 103]. To date, however, there have been appearing reports of the application of this technology to the treatment of a human patient. The challenge lies in being able to deliver an adequate quantity of the magnetic particles to generate enough heat in the target using AC magnetic field conditions that are clinically acceptable. Most of the laboratory- and animal-model-based studies reported so far are characterized by the use of magnetic-field conditions that could not be safely used with a human patient. In most instances, reducing the field strength or frequency to safer levels would almost certainly lead to such a reduction in the heat output from the magnetic material so as to render it useless in this application. MNPs are promising tools for the minimal invasive elimination of small tumors in the breast using magnetically induced heating. The approach complies with the increasing demand for breast-conserving therapies and has the advantage of offering a selective and refined tuning of the degree of energy deposition allowing an adequate temperature control at the target [150].

Developments by Jordan and Chan led to the current hyperthermia application of single-domain, dextran-coated magnetite nanoparticles in tumors [151, 152, 137]. Magnetic hyperthermia is also possible with larger magnetic particles, as shown by Moroz *et al.* [153]. Their 32-µm plastic particles contain maghemite and embolize the arterial blood supply of the tumor, in addition to the magnetic hyperthermia treatment. In an animal study with 10 rabbits, the tumor volumes decreased by 50–94% within 2 weeks. Ongoing investigations in magnetic hyperthermia are focused on the development of magnetic particles that are able to self-regulate the temperature they reach. The ideal temperature for hyperthermia is 43–45 °C, and particles with a Curie temperature in this range have been described by Kuznetsov *et al.* [120].

It is important, therefore, to understand the underlying physical mechanisms by which heat is generated in small magnetic particles by alternating magnetic fields. Enough heat must be generated by the particles to sustain tissue temperatures at 42 °C for half an hour at least. Calculating the heat deposition rate is complex, due to the presence of blood flow and tissue perfusion. Several authors have analyzed the heat transfer problem whereby a defined volume of tissue is heated by evenly dispersed sources such as microscopic magnetic particles [154–156]. The problem posed by the cooling from discrete blood vessels is generally avoided because of the mathematical complexity and lack of generality of the results. However, an often-used rule of thumb is that a heat deposition rate of 100 mW cm^{-3} of tissue will suffice in most circumstances.

The frequency and strength of the externally applied AC magnetic field used to generate the heating is limited by deleterious physiological responses to high-frequency magnetic fields [157, 158]. These include stimulation of peripheral and skeletal muscles, possible cardiac stimulation, and arrhythmia, and nonspecific inductive heating of tissue. Generally, the useable range of frequencies and amplitudes is considered to be $f = 0.05-1.2$ MHz and $H = 0-15$ kA m^{-1}. Experimental data on exposure to much higher frequency fields come from groups such as Oleson et al. [159] who developed a hyperthermia system based on inductive heating of tissue, and Atkinson et al. [160] who developed a treatment system based on eddy current heating of implantable metal thermoseeds. Atkinson et al. concluded that exposure to magnetic fields H with frequency f, where the multiplication value $H \cdot f$ do not exceed 4.85×10^8 A m^{-1} s^{-1}, is safe and tolerable.

The amount of magnetic material required to produce the required temperatures depends to a large extent on the method of administration. For example, direct injection allows for substantially greater quantities of material to be localized in a tumor than do methods employing intravascular administration or antibody targeting, although the latter two may have other advantages. A reasonable assumption is that ca. 5–10 mg of magnetic material concentrated in each cm^3 of tumor tissue is appropriate for magnetic hyperthermia in human patients.

Regarding the choice of magnetic particle, the iron oxides magnetite (Fe$_3$O$_4$) and maghemite (γ-Fe$_2$O$_3$) are the most studied to date because of their generally appropriate magnetic properties and biological compatibility, although many others have been investigated. Particle sizes less than about 10 nm are normally considered small enough to enable effective delivery to the site of the cancer, either via encapsulation in a larger moiety or suspension in some sort of carrier fluid. Nanoscale particles can be coupled with antibodies to facilitate targeting on an individual cell basis. Candidate materials are divided into two main classes: the ferromagnetic or □luidized□ng□ (FM) single-domain or multidomain particles, or the SPM particles. The heat-generating mechanisms associated with each class are quite different, each offering unique advantages and disadvantages, as discussed later.

The first clinical trail of magnetic delivery of chemotherapeutic drugs to liver tumors using MMS was performed by Lübbe et al. in Germany for the treatment of advanced solid cancer in 14 patients [161]. Their MMS were small, about 100 nm in diameter, and filled with 4-epidoxorubicin. The phase-I study clearly showed the low toxicity of the method and the accumulation of the MMS in the target area. However, MRI measurements indicated that more than 50% of the MMS had ended up in the liver. This was likely due to the particles' small size and low magnetic susceptibility, which limited the ability to hold them at the target organ. There were developed irregularly shaped carbon-coated iron particles of 0.5–5 µm in diameter with very high magnetic susceptibility and used them in a clinical phase-I trial for the treatment of inoperable liver cancer [162]. They have treated 32 patients to date and are able to superselectively (i.e., well-directed) infuse up to 60 mg of doxorubicin in 600 mg MMS with no treatment-related toxicity [163].

10.4.4
Prospective of Clinical Applications

Many researchers are presenting new multidisciplinary research and highlighting important advances in biotechnology, nanotechnology, and defense and homeland security. Investigators are working to determine how to best construct the core–shell structure and learn which shell materials are most ideal for biomedical applications such as magnetodynamic therapy (MDT), or as MRI contrast enhancement agents. In the future it may be possible for a patient to be screened for breast cancer using MRI techniques with engineered enhanced ferrites as the MRI contrast agent. If a tumor is detected, the doctor could then increase the power of the MRI coils and localized heating would destroy the tumor region without damaging surrounding healthy cells. Another promising biomedical application is MDT, which employs MNP that are coupled to the radio frequency of the MRI. This coupling converts the radio frequency into heat energy that kills the cancer cells. European researchers studying MDT have shown that nanoparticles are able to target tumor cells. Because the nanoparticles target tumor cells and are substantially smaller than human cells, only the very few tumor cells are killed, which greatly minimizes damage to healthy cells. The aim of next investigations is to tailor the properties of the nanoparticles to make the use of MDT more universal. The only thing slowing down the development of enhanced ferrites for 100 MHz applications is a lack of understanding of the growth mechanisms and synthesis–property relationships of these nanoparticles. By studying the mechanism for the growth of the enhanced ferrites, it will be possible to create shells that help protect the metallic core from oxidation in biologically capable media. Enhanced ferrites are a class of ferrites that are specially engineered to have enhanced magnetic or electrical properties and are created through the use of core–shell morphology. The core can be a highly magnetic material like iron or iron alloys, while the shell can be a mixed metal ferrite with tailored resistivity. The polymer encapsulated iron oxide particles are used in biomedical applications too. High magnetization of the enhanced ferrite nanoparticles may potentially improve the absorption of the radio frequency, thereby providing better detection of tumor regions and the use of less MRI contrast reagent. The magnetic power of the iron nanoparticles created is about

10 times greater than that of the currently available iron oxide nanoparticles, which translates to a substantial reduction in the amount of iron needed for imaging or therapy.

Cancer researchers have long sought to harness the tumor-targeting ability of monoclonal antibodies with the cell-killing property of radioisotopes, particularly iodine. But clinical results with numerous ^{131}I-antibody formulations have failed to live up to expectations, in large part because the therapy is not specific enough for tumors. In an attempt to remedy that problem, a group led by Jin Chen [164] has added MNPs to the ^{131}I-antibody preparation, and the preliminary results suggest that this approach could be promising for treating human liver cancer.

The investigators describe how they coupled dextran-coated MNPs to an ^{131}I-labeled monoclonal antibody [164] that binds to vascular endothelial growth factor (VEGF), a protein found on the surface of the blood vessels that surround most solid tumors. The idea here was that a focused magnetic field could be used as an initial targeting vector that would concentrate radioactively labeled antibody in the vicinity of a tumor. Once there, the antibody would provide a second level of targeting to the blood vessels surrounding the tumor. Radiation would then kill the neighboring malignant cells. Results in mice with implanted human liver tumors found that a focused magnetic field did indeed concentrate the nanoparticle–antibody formulation as desired, with very little of the formulation accumulating in healthy tissue. In a second experiment, animals in which the nanoparticle–antibody formulation was injected into tumors experienced a marked shrinkage of the tumors, with little toxicity as measured by white blood cell production and weight loss.

Some scientists warn against too much euphoria, saying that the iron oxide nanoparticles used in the treatment could later damage (Figure 10.2) other tissues of the body if they reach the bloodstream. But some authors consider

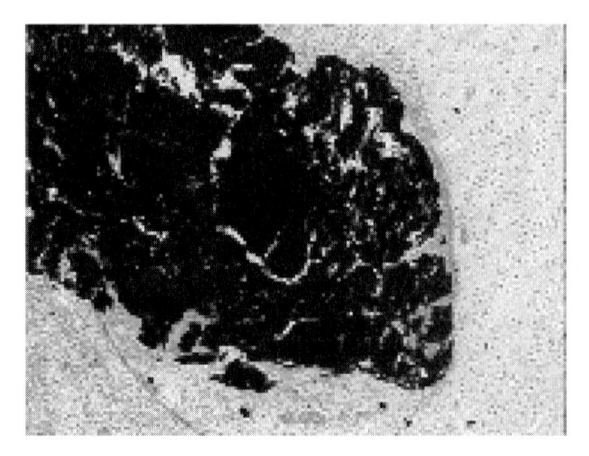

Figure 10.2 Cancer therapy: A tumor saturated with nanoparticles (Author's photo).

10.4 Magnetic Nanoparticles in Cancer Therapy | 427

that as long as the amount of metal injected in the body stays under a certain level, the danger of "nanopoisoning" is relatively low.

Nevertheless, clinical trials on patients in Germany have provided the first verifiable proof that MNPs are capable of completely destroying tumors without surgical intervention [151, 152] and with no side effects, by selectively penetrating and heating cancerous cells.

Thermotherapy (Figure 10.3) will be performed once or twice a week for 60 min each. During treatment, the cardiovascular functions as well as the body temperature and the temperature within the tumor will be controlled.

Magnetic iron oxide nanoparticles are among the most promising nanomaterials being developed as targeted imaging and therapeutic agents for use in detecting and treating cancer. Little is known, however, about how these nanoparticles interact with cancer cells in a living animal; so a team of investigators decided to remedy that knowledge gap [165].

Using thermotherapy with MNPs, many types of cancer have been treated so far, e.g., cancer of the rectum, ovarian, prostate (Figure 10.4), cervix-carcinoma, sarcoma, and different tumors of the brain. The technique of MFH can be used for minimally invasive treatment of prostate cancer. The authors [166] present the first clinical application of interstitial hyperthermia using MNPs in locally recurrent prostate cancer.

Closer examination of individual cells showed that the nanoparticles had accumulated inside cells, where at least some of the nanoparticles formed clusters. Then, there were also detected nanoparticles in the cell nucleus,

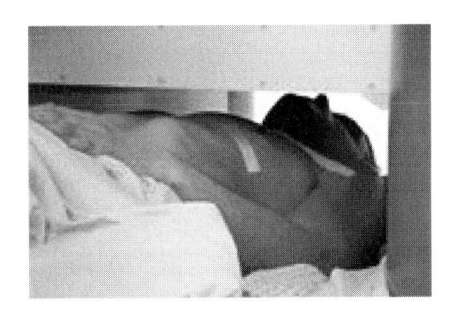

Figure 10.3 For thermotherapy, no fixation or anesthesia of the patient is necessary (Author's photo).

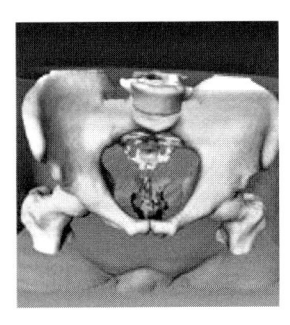

Figure 10.4 Nanoparticles (blue) are instilled into the prostate (green) to treat a prostate carcinoma. Afterward the fluid is heated by an alternating magnetic field [166].

suggesting that these nanoparticles could serve as a tool for delivering anticancer genes and anticancer agents that interact with a tumor cell's genes into a cancer cell's nucleus.

The cytotoxic effect of hyperthermia has been used therapeutically for a long time. Common hyperthermia treatments use different energy sources to achieve higher temperatures within tissues: externally induced electromagnetic waves (e.g., radio frequency or microwave hyperthermia), ultrasonic (externally or interstitial), current flow between two or more electrodes, electrical or magnetic fields between implanted antennas, and electrical or magnetically induced thermo seeds or tubes with warm water.

The major problem of applying hyperthermia treatments in reality is to achieve a homogenous heat distribution in the treated tissue. Failing may either lead to an insufficient supply of tumor areas or damage of neighboring tissue by too high temperatures. The nanotechnology-based new cancer therapy is a special form of interstitial thermotherapy with the advantage of selective heat deposition to the tumor cells. Therefore, it meets the requirement of maximal deposition of heat within the targeted region under maximal protection of the surrounding healthy tissue at the same time. Furthermore, thermotherapy with magnetic nanoparticles enables the physician to select between different treatment temperatures for the first time, after only a single injection of the nanoparticles. He may either choose hyperthermia conditions (up to 45 °C) to intensify conventional therapies like radiation or chemotherapy, or thermoablation by using higher temperatures up to 70 °C.

10.4.5
Conclusions

The MFH is one of techniques used in cancer therapy based on heating tissues for therapeutic purposes. MFH is usually used as an additive therapy with standard treatments (radiotherapy, chemotherapy), and some preliminary studies have showed that the combination "radiation plus hyperthermia" leads to increasing efficiency in tumor regression. The MFH allows reduction of doses of irradiation or drugs used for chemotherapy. It might be particularly helpful in repeated treatment of the target area.

There are several techniques including laser, ionizing radiation, and microwaves to heat up malignant tissues. Although these techniques are capable of rising up the temperature in tumor tissue to the cells death, they may have unwanted bystander effects such as ionization of genetic material (radiation) or lack of selectiveness (microwaves) that affect the surrounding healthy tissues.

A different approach developed mostly for the last decade is the selective thermocytolysis based on the process of magnetic losses. This strategy, called magneto-thermocytolysis or magneto-thermoablation, is a promising technique thanks to the development of precise methods for synthesizing functionalized magnetic nanoparticles.

Magnetic nanoparticles with functionalized surfaces (able to attach with high specificity to a given tissue) are used for hyperthermia treatments seeking their accumulation only in tumor tissue. Depending on the success in solving this biochemical and physiological specificity problem, cancer-specific hyperthermia protocols could be developed.

Thermoablation (hyperthermia of more than 46 °C) alone leads to destruction of pathologically degenerated cells. In case of successful treatment, the tumor is getting smaller in size. Hyperthermia of up to 45 °C intensifies the effectiveness of a radio- and chemotherapy. Herewith also tumor cells, not responding to irradiation or chemotherapy, can be destroyed.

Success of the treatment depends on the sensitivity of the tumor tissue and the achieved temperatures within the tissue. Therefore, a homogeneous distribution of heat in the target area is very important. Using self-controlled magnetic nanoparticles [120–126] can automatize heat control in tissue at alternating electromagnetic fields applied.

Unfortunately, some patients have to be excluded, if there are any irremovable metallic parts in the body within the treatment area (approx. 40 cm distance from the tumor), e.g., arthroplasty, bone nails, and dental prostheses (metal fillings or implants). Also patients with cardiac pace makers or implanted pumps are excluded from thermotherapy treatment.

In the future, nanoparticles can carry drugs and genes specifically into tumor cells, even into the nucleus. By coupling between drug and particles, they can be thermosensitive so that heat releases the drugs within the cells after the magnetic field has been switched on. Using this kind of specific heat activation, transfection rates of vectors, transported by these nanocarriers, are increased.

10.5
Targeted Drug and Gene Delivery

The innovative pharmaceutical treatments obviously require novel modern methods of administration. The possibility of using ferrofluids for drug localization in blood vessels and in hollow organs is a contemporary task for drug development [167]. Pure magnetic particles are not stable in water-based solutions and suspensions; therefore, they cannot be used for medical application without biocompatible coating. The choice of polymers for magnetic nanoparticles coating to prevent them from adhering toxicity occurred to be not an easy one-step task, because these compositions should satisfy both the requirements of biocompatibility and biodegradability.

In the past, pharmaceuticals have been primarily consisted of simple, fast-acting low-molecular chemical compounds or extracted from plants organic compound's mixtures administered orally, as solid pills or liquids, or as injections. During the recent period, however, formulations that control the rate and period of drug delivery (i.e., time-release medications) and target specific areas of the body for treatment have become more complicated.

For example, many drugs' potencies and therapeutic effects are limited or otherwise reduced because of the partial degradation that occurs before they reach a desired target in the body. Once injected, time-release medications deliver treatment continuously rather than provide relief of symptoms and protection from adverse events solely when necessary. Further, injectable medications could be less expensive and easily administered if they could simply be dosed orally. However, this improvement needs safely shepherd drugs that are able to reach specific areas of the body, such as the stomach, where low pH can destroy a medication, or an area where healthy bone and tissue might be adversely affected.

The early biodegradable systems were focused on naturally occurring polymers (collagen, cellulose, etc.) but have recently moved into the area of chemical synthesis. The first synthetic polymers were introduced everywhere in the 1970s and based on lactic acid. However it occurs that low-molecular-weight polymers of lactic acid are having melting temperature point 30–36 °C, and their compositions with magnetic particles cannot keep stable consistence. High-molecular-weight polylactides are toxic and cannot be used for medical applications. Today polymer materials still provide the most important avenues for research, primarily because of their ease of processing and the ability of researchers to control their chemical and physical properties via molecular synthesis. Basically, two broad categories of polymer systems, both known as "microspheres" because of their size and shape, have been studied: reservoir devices and matrix devices.

Examples of such polymers include polyanhydrides, polyesters, polyacrylic acids, poly(methyl methacrylates), and polyurethanes. As a result of extensive experiments with these materials, several key factors have emerged to help scientists design more highly degradable polymers. Specifically, a fast-degrading matrix consists of a hydrophilic, amorphous, low-molecular-weight polymer that contains heteroatoms (i.e., atoms other than carbon) in its backbone and is grown either stepwise or through condensation reactions. Therefore, varying each of these factors allows researchers to adjust the rate of matrix degradation and, subsequently, control the rate of drug delivery.

The goal of all drug delivery systems, therefore, is to deploy medications intact to specifically targeted parts of the body through a medium that can control the therapy's administration by means of either a physiological or chemical trigger. To achieve this goal, researchers are turning to advances in the field of nanotechnology. During the past decade, polymeric microspheres, polymer micelles, and hydrogel-type materials have all been shown to be effective in enhancing drug targeting specificity, lowering systemic drug toxicity, improving treatment absorption rates, and providing protection for pharmaceuticals against biochemical degradation. In addition, several other experimental drug delivery systems show exciting signs of promise, including those composed of biodegradable polymers, dendrimers (so-called star polymers), electroactive polymers, and modified C-60 fullerenes.

During the past two decades, research into hydrogel delivery systems has focused primarily on systems containing polyacrylic acid (PAA) backbones. PAA hydrogels are known for their super-absorbancy and ability to form extended polymer networks through hydrogen bonding. In addition, they are excellent bioadhesives, which mean that they can adhere to mucosal linings within the gastrointestinal tract for extended periods, releasing their encapsulated medications slowly over time.

Dendritic macromolecules make suitable carrier systems because their size and structure can be controlled simply by synthetic means, and they can easily be processed and made biocompatible and biodegradable. However, most of these polymers are poorly suitable for coating of magnetic nanoparticles.

The need for research into drug delivery systems extends beyond ways to administer new pharmaceutical therapies; the safety and efficacy of current treatments may be improved if their delivery rate, biodegradation, and site-specific targeting can be predicted, monitored, and controlled.

10.5.1
Magnetic Targeting Local Hyperthermia

In recent years, nanotechnology has achieved a stage that makes it possible to produce, characterize, and specifically tailor the functional properties of nanoparticles for clinical applications. This has led to various opportunities such as improving the quality of MRI, hyperthermic treatment for malignant cells, site-specific drug delivery, and the manipulation of cell membranes. To this end, a variety of iron oxide particles have been synthesized. A common failure in targeted systems is due to the opsonization of the particles on entry into the bloodstream, rendering the particles recognizable by the body's major defense system, the RES. (see Section 10.2). In some cases, the combination of gene therapy with effects of hyperthermia may be possible. Heat-induced therapeutic gene expression is highly desired for gene therapy to minimize side effects. Furthermore, if the gene expression is triggered by heat stress, combined therapeutic effects of hyperthermia and gene therapy may be possible.

10.5.2
Magnetic Targeting Applications

Drug delivery remains a challenge in the management of cancer and another illness. The focus is on targeted cancer therapy. The newer approaches to cancer treatment not only supplement the conventional chemotherapy and radiotherapy but also prevent damage to normal tissues and prevent drug resistance. Innovative cancer therapies are based on current concepts of molecular biology of cancer. These include antiangiogenic agents, immunotherapy, bacterial agents, viral oncolysis, targeting of cyclic-dependent kinases and tyrosine kinase receptors, antisense approaches, gene therapy, and combination of various methods. Important methods of immunotherapy in cancer involve use of cytokines, monoclonal antibodies, cancer vaccines, and immunogene therapy.

Magnetic drug targeting is a young field. The surgeon Gilchrist published in 1956 on the selective inductive heating of lymph nodes after injection of 20–100-nm-sized maghemite particles into the lymph nodes near surgically removed cancer [115]. Turner and Rand then combined this radiofrequency heating method with embolization therapy [168]. Gilchrist apparently did not, however, envision that these magnetic particles could be magnetically guided and delivered to the target area. In 1963, Meyers [169] described how they were able to accumulate small iron particles intravenously injected into the leg veins of dogs, using a large, externally applied horse shoe magnet. They imagined that it might be useful for lymph-node targeting and as a contrast agent. Hilal then engineered catheters with magnetic ends, and described how they could be used to deposit and selectively embolize arterio-venous malformations with small magnets [170].

The use of magnetic particles for the embolization therapy of liver cancer followed and has recently found renewed interest [171, 172]. More defined spherical magnetic microspheres were made for the first time at the 1979 by Widder *et al.* [173]. Their magnetic albumin microspheres worked well in animal experiments for tumor therapy and as magnet resonance contrast agents, but were not explored in clinical trials [174].

10.5.3
Targeted Liposomal Drug and Gene Delivery

A therapeutic tissue targeting of medicines in the organism with magnetic field force is a doubtful thing. However, one delivery system occurred to be appropriate for the targeting delivery purpose – it is sterically stabilized liposomes platform. In general, magnetic liposomes are also included when speaking about magnetic carriers.

Sterically stabilized (polyethylene glycol-containing) liposomes or PEG-liposomes (SSL) technology has been used to overcome some of the barriers of drug delivery. In general, liposomes consist of phosphatidylcholine (PC), cholesterol (Chol), phosphatydylethanolamine (PEA), and PEG with different molecular weight from 300 to 2000 covalently attached to PEA. They can be used as carriers of magnetic nanoparticles as well as for transportation of antitumor chemotherapeutic agents. Encapsulating anticancer drugs in liposomes (magnetic or nonmagnetic) enables drug delivery to tumor tissues and prevents damage to the normal surrounding tissues. The advantages of administration of sterically stabilized liposomal agents over simple drug forms are well proven in multiple laboratories [175–178] and clinic [179] assays. Liposomes provide time-release of the encapsulated medicines after a single administration together with significant diminishing of systemic toxicity.

For magnetoliposomes, there are not many types of available particles that can be successfully enloaded and stored for a long time in the vesicles. These particles should be small in size up to 10 nm to be naturally passivated in water buffer solution, which is used for lipid film hydration during liposomal membrane formation [175, 180, 181]. Magnetic properties and chemical composition are generally magnetite, Fe_3O_4, and maghemite Fe_2O_3. For the vast majority of these particles, the magnetic component exists in a SPM state. In this case, the force exerted on the particle is a translational force directed along the applied field vector and is dependent on the magnetic properties of the particle and the surrounding medium, the size and shape of the particles, and the product of the magnetic flux density and the field gradient.

Liposomes may be injected directly in tumor or into the bloodstream, and during their circulation guided to the tumor for targeted drug delivery. Monoclonal antibodies (mAbs) can be used for targeted delivery of anticancer payloads such as radionucleotides, magnetic nanoparticles, toxins, and chemotherapeutic agents to the tumors. Antibody-targeted drug delivery systems were developed for modern cancer therapy. ATL systems originally were intended for great enhancement of antitumor effect due to selective activity of the encapsulated chemotherapeutic agent and for keeping the advantages of low total toxicity and better pharmacodynamics (timely drug release) of sterically stabilized liposomal platform [179, 182, 183].

Different possible methods of immunoglobulins coupling to make them built-in in liposomal membranes were described [167]. There are three types presented in Figure 10.5: type A, which was the first attempt to prepare ATL used PEA-bound immunoglobuline added to membrane without PEG, and stabilization theoretically could be made with cardiolipin or ganglioside. Type A did not demonstrate stability of properties and satisfactory results in drug delivery *in vivo*. Type C is sterically stabilized liposomes with long-chain PEG-2000 in which hydrophobic part is PEA and distal end activated with thiolation, carboxylation, or nitride-binding of respectively activated monoclonal antibodies [176]. Thiolation is the most commonly (85–90%) used method for ATL preparation. Type B is a medium variant for ATL development where PEG stabilization was used with first imperfect way of mAbs coupling to PEA [167]. Type C liposomes were admitted to be successful in *in vitro* and *in vivo* assays and all ATL formulations achieved clinical trials belonging to that type. Together with sterical stabilizing of liposomes, PEG-2000 can protect antibodies from protease and RES (macrophages) cleavage if used both in activated for mAbs binding and nonactivated forms during luidizedng film formation.

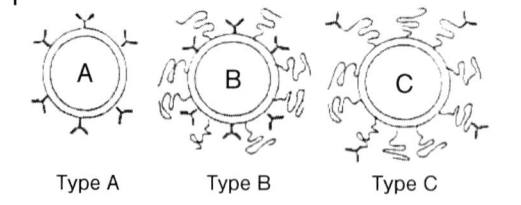

Figure 10.5 Possible types of monoclonal antibody-targeted liposomal constructions [168].

There are several myths or rather romantic believes about antibody-targeting delivery systems that are being adhered every time by doctors and researchers who did not handle them in experiment long enough as "full cycle" beginning from lipids film formation and ending with *in vivo* evaluation. These restrictions also should be taken into account for magnetoliposomes development.

Cytotoxicity of both SSL and ATL increases with the higher concentration of lipids in suspension and lower agent enloading percentage. It proves the fact that for therapeutic applications, concentration of lipids in ready-for-administration suspension should be decreased as much as possible, and for that the percentage of entrapped agent for lipid's weight should be as high as possible [176, 184, 185]. This attitude will diminish nonspecific toxicity effects for targeted liposome vehicles as well (Figure 10.6). The "softness" and "stabilization" balance in complex SS and AT systems is guaranteed with several conditions of their production design, the main of them being the ratio of polyethyleneglycol percentage to total lipids content in liposome's membranes [175, 176, 186] and the length of PEG chain [184]. First, PEG makes lipid liposome's membranes invisible for macrophages and, therefore, provides their circulation in the bloodstream as long as SSLs pass the liver 30–40 times before they are finally destroyed [183, 184, 187]. The second function of PEG is the prevention of leakage of the agent encapsulated in liposomes [181, 182, 186]. The other lipid derivatives like ganglioside, cardiolipine, and ratio of PC and cholesterol in lipid film formation compositions are of help to PEG to achieve these purposes [176, 182].

However, cytotoxicity curves of "empty" ATLs confirm that modified PEG-liposomes are toxic enough for cell cultures (Figure 10.6) and might bring enough additional immunotoxicity effects when administrated to the animals or patients. It is worth mentioning that immunotoxicity of medicine is not a parameter to be tested in nude mice – the common model for ATL study *in vivo* – because in that case immunodeficient animals develop very weak or no immune response to foreign antibodies. At least for ATLs, we cannot prove that toxicity of doxorubicin or another encapsulated cytostatic is being neutralized with the system of antibody-targeted delivery like it happens in sterically stabilized liposomes [176, 182].

The brief analysis of experiments shows that developed ATL system and its ATL modifications are more convenient for delivery of nontoxic pharmaceutical agents. Therefore several years after, many researchers

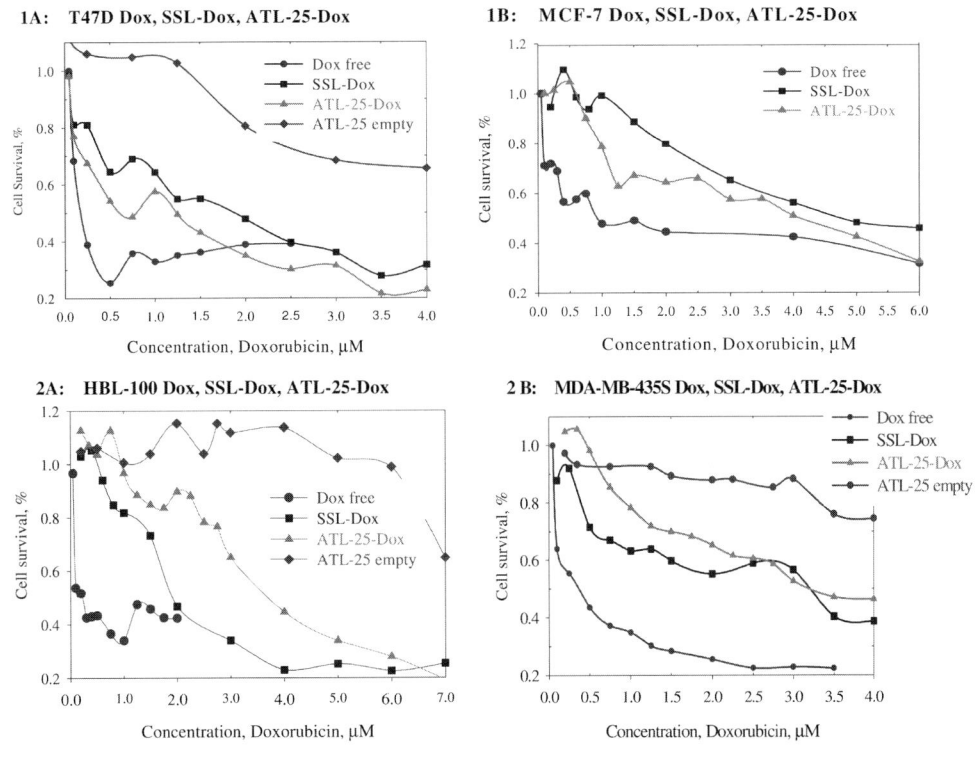

Figure 10.6 Cytotoxicity of antibody-targeted (ATL), sterically stabilized (SSL) enloaded with cytostatic doxorubicine and unloaded (empty) liposomes.

who started with chemotherapeutic agents as anthracyclines [176, 184, 187], amphotericin B, and *cis*-platinum [183] for ATL enloading shifted to gene therapeutic vectors delivery [188] or peptides on base of ATL systems they have developed earlier. Another alternative for diminishing targeting therapy toxicity was the combination of sterically stabilized thermosensitive liposomal encapsulation with local hyperthermia [175, 185, 187, 189, 190]. It is reasonable to use a combination of hydrophilic cytostatics with magnetic particles in fluid for enloading targeted liposomal delivery (ATLs) systems to some nontoxic agents that possess the prospective antitumor activity. This unexpected outcome brought us to consider gene therapeutic agents delivery applications.

Traditional spectrum of methods of antibody-targeted drug delivery systems, efficiency evaluation includes flow cytometry analysis that is trustworthy due to precise quantification and statistical reliability. However, flow cytometry analysis does not provide the possibility to study dynamics of drug entrance and cellular distribution. © first time video-microscopy was tried as quantification method in biomedical application long ago [179].

ATL systems possess a source for enhancement of selective activity due to combination of antigen-specific delivery with reliable accumulation in tumor cells and tissue.

10.5.4
Magnetic Targeting of Radioactivity

Magnetic targeting can also be used to deliver therapeutic radioisotopes [38]. The advantage of this method over external beam therapy is that the dose can be increased, resulting in improved tumor cell eradication, without harm to nearby normal tissue. Different radioisotopes can treat different treatment ranges depending on the radioisotope used – the β-emitters ^{90}Y, for example, will irradiate up to a range of 12 mm in tissue. Unlike chemotherapeutic drugs, the radioactivity is not released, but rather the entire radioactive microsphere is delivered to and held at the target site to irradiate the area within the specific treatment range of the isotope. Once they are not radioactive anymore, biodegradation of the microspheres occurs (and is desired)-Initial experiments in mice showed that intraperitoneally injected radioactive poly(lactic acid)-based MMS could be concentrated near a subcutaneous tumor in the belly area, above which a small magnet had been attached [38]. The dose-dependent irradiation from the β-emitter ^{90}Y-containing MMS resulted in the complete disappearance of more than half of the tumors. Magnetic targeted carriers (MTC; from FeRx), which are more magnetically responsive iron–carbon particles, have been radiolabeled in the last couple of years with isotopes [191] such as ^{188}Re, ^{90}Y, ^{111}In, and ^{125}I and are currently undergoing animal trials. For example, a preliminary *in vivo* investigation of binding stability and localization was performed in normal swine. Eleven millicurie of ^{90}Y-MTC was administered intraarterially to a swine liver via catheterization of the hepatic artery. Blood samples were taken following the administration, which indicated that less than 3% of the total injected activity was circulating 30 min following the administration and decreased over time. Recently DNA-based biocompatible platform to deliver magnetic Gd3 ions was reported [192]. Gd is a potential agent for nuclear capture therapy, so Gd content about 30% in that construction is of prospect.

10.5.5
Other Magnetic Targeting Applications

In all magnetic drug and gene delivery methods, therapeutic drugs or genes are attached to functionalized magnetic particles and injected near the target site. External magnetic fields of varying characteristics (usually produced by rare earth magnets) are applied to the site externally in order to concentrate the particles at the target. In the case of gene delivery, it is envisaged that this method will achieve higher transfection and hence expression rates. In the case of drug delivery, therapeutic drugs are concentrated at the site in the body where they are needed.

Cancer gene therapy is a form of drug delivery for cancer. Various technologies and companies developing them are described. Nucleic acid-based cancer vaccines are also described. For some time, scientists have realized that the antibody–antigen relationship could be a great way to very specifically target pharmaceuticals. If the drug were encapsulated in a material that only certain cells or cell structures could bind with, then only that cell or cell structure would be affected by the drug.

The idea is that drugs are encapsulated in nanoparticles that have antigens that allow them to bind only with tumor tissue. Once in place and the chemical antibody–antigen reaction takes place, the encapsulated drug can kill the cell – removing the need for painful, debilitating, inaccurate and indiscriminate chemotherapy and radiotherapy treatments. Alternatively, the nanoparticle itself can be magnetic and contain no drug. Once bound to the tumor, the magnetic nanoparticles can be used to pinpoint the exact location of the tumor, and through inductive heating from magnets placed outside the body, the tumor can be destroyed.

As well as cancer, this technique could be applicable to many of the worst diseases, including HIV/AIDS. Drugs that require implementation to specific cell structures and cell types can be targeted to only affect those areas. Painkillers, vaccines, and antiviral pharmaceuticals can have their side effects massively reduced while having their primary effect magnified using this technique.

Magnetic drug delivery by particulate carriers is a very efficient method of delivering a drug to a localized disease site. Very high concentrations of chemotherapeutic or radiological agents can be achieved near the target site, such as a tumor, without any toxic effects to normal surrounding tissue or to the whole body. In magnetic targeting, a drug or therapeutic radioisotope is bound to a magnetic compound, injected into a patient's blood stream, and then stopped with a powerful magnetic field in the target area. Depending on the type of drug, it is then slowly released from the magnetic carriers (e.g., release of chemotherapeutic drugs from magnetic microspheres) or confers a local effect (e.g., irradiation from radioactive microspheres; hyperthermia with magnetic nanoparticles). It is thus possible to replace large amounts of freely circulating drug with much lower amounts of drug targeted magnetically to localized disease sites, reaching effective and up to several-fold increased localized drug levels [173].

10.5.6
MRI Contrast Enhancement

Throughout the history of modern medicine, and particularly clinical oncology, important advances in treating illness and injury have usually followed the development of new ways to see better within the body. The advent of CT imaging, for example, provided images of developing tumors in far greater detail than was possible with conventional X-rays, giving oncologists a means of both better localizing tumors before surgically removing them and the first real glimpse of whether a given therapy was causing a tumor to shrink. Similarly, MRI provided greater anatomical detail still, while the development of PET gave both cancer researchers and oncologists the ability to monitor a tumor's metabolic activity, and as a result, an even quicker way of assessing the effectiveness of therapy.

The use of contrast agents and tracers in medical imaging has a long history [193]. Known collectively as imaging contrast agents, these molecules possess physical characteristics that increase the strength of the signal coming out of the body. MRI contrast agents containing the element gadolinium (or iron), for example, do so by altering the magnetic field in the body, which boosts the strength (or reduce) of the MRI signal. They provide important information for diagnosis and therapy, but for some desired applications, a higher resolution is required than can be obtained using the currently available medical imaging techniques.

Acquisition of MR pulse sequences with different weighting (i.e., T1-weighted, T2-weighted, STIR, etc.) allows the detection of diffuse alterations in patients with fatty infiltration or iron overload. The most important diagnostic application of magnetic nanoparticles is as contrast agents for MRI. For the MRI purpose, the ferrite particles of 0.5 μm in diameter were tested *in vivo* for the first time in 1987 [194]. Later, smaller SPM iron oxides (SPIOs) have been developed into nanometer sizes and have, since 1994, been approved and used for the imaging of liver metastases. With the help of a "molecular imaging," SPIO nanoparticles can be effectively used in real-time MRI and PET techniques for target drug delivery control [195, 196]

Molecular and cellular MRI is a rapidly growing field that aims to visualize targeted macromolecules or cells in living organisms. In order to provide a different signal intensity of the target, for example, gadolinium-based MR T1 contrast agents can be employed although they suffer from an inherent high threshold of detectability. SPIO particles can be detected at micromolar concentrations of iron and offer sufficient sensitivity for–T2-weighted imaging. Over the past two decades, biocompatible particles have been linked to specific ligands for molecular imaging. However, due to their relatively large size and clearance by the RES (see Section 10.2), widespread biomedical molecular applications have yet to be implemented and few studies have been reproduced between different laboratories. SPIO-based cellular imaging, on the other hand, has now become an established technique to label and detect the cells of interest. Imaging of macrophage activity was initially, and still is, the most significant application, in particular for tumor staging of the liver and lymph nodes, with several products either approved or in clinical trials. The ability to also label nonphagocytic cells in culture using derivatized particles, followed by transplantation or transfusion in living organisms, has led to an active research interest to monitor the cellular biodistribution *in vivo* including cell migration and trafficking. While most of these studies to date have been mere '"proof-of-principle" type, further exploitation of this technique will be aimed at obtaining a deeper insight into the dynamics of *in vivo* cell biology, including lymphocyte trafficking, and at monitoring therapies that are based on the use of stem cells and progenitors [197].

SPIO was the first tissue-specific contrast agent introduced for liver MRI [194]. It is selectively targeted to the RES, and it accumulates in the Kupffer cells within the liver. The Kupffer cells constitute about 80% of the RES. Thus, for a given dose of SPIO a large part eventually ends up within liver parenchyma. SPIO is an SPM contrast agent. SPIO particles are large enough to act as independent magnetic domains. They create magnetic fields around themselves when placed within an external magnetic field. Therefore, they promote small-field inhomogeneities within the external magnetic field and hence, tissue T2-relaxivity increases due to the rapid dephasing of the spins [198]. Through this so-called susceptibility effect, the signal from normal liver dramatically decreases after SPIO administration. The effective imaging window after a slow SPIO infusion is from approximately hour to several hours. SPIO-enhanced MRI has shown potential in metastases detection [199].

Successful detection of metastases (Figures 10.7–10.9) was done on 6-T MRI with the help of contrast dextran-coated SPIO nanoparticles in live mice [200]. The water-based dextran–ferrite magnetic fluid used in these investigations contained about 28% γ-Fe_2O_3 and Fe_3O_4, 70% dextran, and 2% H_2O.

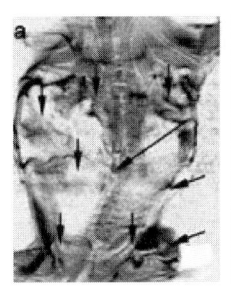

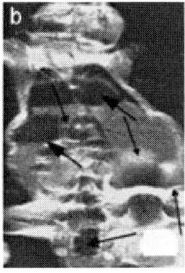

Figure 10.7 Longitudinal dextran ferrite-enhanced MRI: (a) (negative) P388 micro- and macroglobular metastases in the lymph nodes on the left and right sides (small arrows) and in the backbone (large arrow); (b) (positive) Lewis lung carcinoma micro- and macroglobular metastases in the lungs, spleen, and rectum (thick arrows), and in the backbone, liver, and kidney (fine arrows).

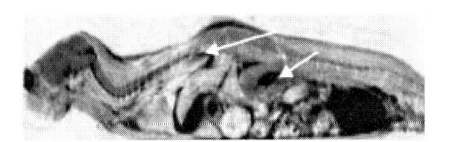

Figure 10.8 Dextran ferrite-enhanced MRI (negative) of an Ehrlich carcinoma: Macrometastases in the backbone and urinary bladder, microglobular metastases in the kidney (arrows).

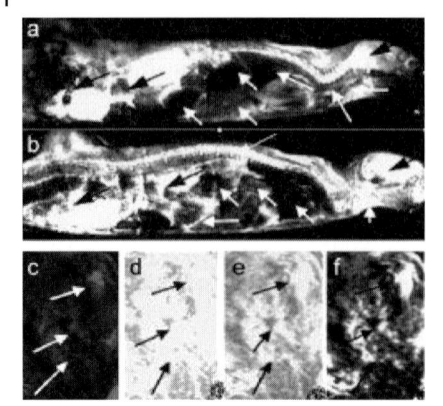

Figure 10.9 Lewis lung carcinoma micro- and macroglobular metastases: (a) dextran ferrite-enhanced MRI for the brain, kidney, and urine bladder (black arrows), throat, lungs, liver, and spleen (white arrows); (b) the same organs and the backbone metastases dextran ferrite-enhanced MRI after 8 days of treatment; (c) nonenhanced MRI for the slice of the liver; (d) Dotarem-enhanced MRI brightness for the same slice, the signal raise indicates DT enrichment; (e) MV-enhanced MRI brightness for the same slice, and (f) DF-enhanced MRI contrast for the liver slice (arrows). The signal loss indicates dextran ferrite enrichment [10].

Ultrasmall SPM iron oxide (USPIO) have mean sizes of 10 nm. The smaller particle size of USPIO compared to SPIO (means sizes up to 100 nm) facilitates much longer plasma half-life, namely, approximately 60 min for USPIO and 6 min for SPIO, because phagocytosis is significantly slower due to the smaller size [201]. Blood pool agents may be utilized in lesion characterization providing information about the vascular pattern of the tumor. SPIO-enhanced T1-weighted MRI was the technique most sensitive to lesion detection and was also helpful in the characterization of liver lesions.

These imaging methods suffer from a common shortcoming – they just are not sensitive enough to accurately find the smallest tumors that are most easily and effectively treated. But increasingly, it appears that nanotechnology may be able to provide that leap in sensitivity that would not only impact today's approach to therapy but could also lead to entirely new pathways for both detecting and treating cancer. The prospect of nanotechnology in cancer imaging is such that we have little doubt that it will lead to far more sensitive and accurate detection of early stage cancer. The best to use novel nanoscale MRI contrast agents are made on the basis of iron or gadolinium, two types of atoms that "resonate" under the influence of magnetic energy. Recently, testing has been done©ing iron oxide nanoparticles to track how dendritic cells move through the body [202]. Dendritic cells are agents for triggering immune responses to kill tumors, but for these cells to perform this way they must first be injected into a patient's lymph nodes. In fact, by labeling dendritic cells with magnetic nanoparticles and tracking them using MRI, the researchers found that interventional radiologists were successful only half the time at injecting these cells into lymph nodes and not into the surrounding tissues. With magnetic nanoparticles, we can use a widely available imaging method, MRI, to ensure that we have accurately delivered therapeutic cells to the exact spot where they can do their job (see Figure 10.10). MRI cell tracking using iron oxides appears clinically safe and well-suited to monitor cellular therapy in humans.

The next step is the multifunctional nanodevices designed to be both imaging agent and anticancer drug. For example, James Baker Jr [203], has been heading a research effort aimed at developing tumor-targeting dendrimers that contain both imaging and therapeutic agents. In a recent paper, Baker's team described its work with a dendrimer linked to a fluorescent imaging agent and paclitaxel, and showed that this agent can identify tumor cells and kill them simultaneously.

The study of USPIO particles [204] shows that MRI enables *in vivo* tracking of SPIO-labeled monocytes longitudinally. Moreover, these data suggest that contrast enhancement after injection of free USPIO does not primarily represent signals from peripherally labeled monocytes that migrated toward the inflammatory lesion. So SPIO-labeled monocytes provide a tool to specifically assess the time window of monocyte infiltration.

SPIO-based colloid has been used clinically as a tissue-specific magnetic resonance contrast agent. In nude mice were tested with a monoclonal antibody A7, conjugated to iron oxides (SPIO), and then examined for the accumulation of this conjugate in xenografted tumors. Mab A7 antibody reacts specifically with human colorectal carcinoma [205].

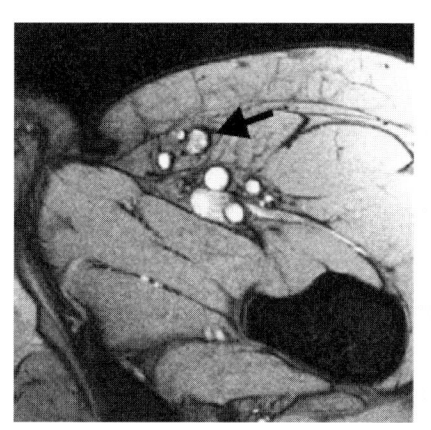

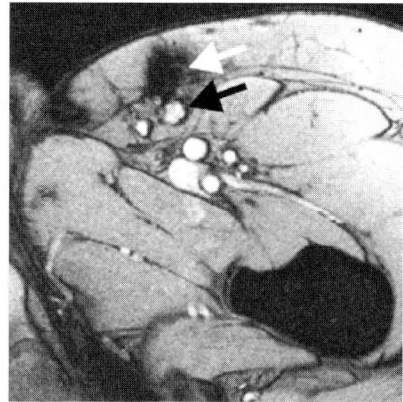

Figure 10.10 Magnetic nanoparticles can be used to monitor the accuracy of delivering therapeutic agents. In this example, the MRI on the left shows a lymph node (black arrow) into which an interventional radiologist wishes to inject dendritic cells, which have been labeled with MNPs. The MRI on the right clearly shows that in this case, the dendritic cells (white arrow) were not injected into the desired target [13].

Authors examined *in vitro* immunoreactivity of Mab A7 coupled to iron oxides and its *in vivo* distribution in nude mice with human colorectal carcinoma. The accumulation levels in normal tissue decreased linearly over time but were lower than levels in tumours from 6 hours (see Figure 10.11). In T2-weighted MRI of the tumour-bearing nude mice, signal intensity was reduced at the margin of the tumour by the injection of A7-iron oxides. Mab A7 coupled to iron oxides is potentially suitable as a magnetic resonance contrast agent for detecting local recurrence of rectal carcinoma.

Targeted nanoparticles show tremendous promise for detecting tumors. In the example shown in Figure 10.12, an SPIO is used to detect a pancreatic tumor in a live mouse [206].

There are new effective diagnostic methods applying bimodal and so on techniques. Combination of optical [207] and MRI images may successfully be applied where both quantum dots and magnetic iron oxide nanoparticles are targeted to tumors. Using one of two targeting molecules, they have shown that they can detect breast tumors in animal models using optical imaging with quantum dots and MRI with the iron oxide nanoparticles.

Moreover, characterization of some focal liver lesions, such as cysts, hemangiomas, and focal fatty infiltration, is often possible. Administration of gadolinium-chelate MR contrast agents may further improve the diagnostic result. These nonspecific MR contrast agents help to assess lesion vascularity during dynamic imaging, similar to that in multiphase helical CT scanning. In general, gadolinium-enhanced MRI does not improve detection of liver metastases, except in the case of primary tumors known to seed hypervascular metastases, but it helps to differentiate between metastases and benign lesions, such as hemangioma.

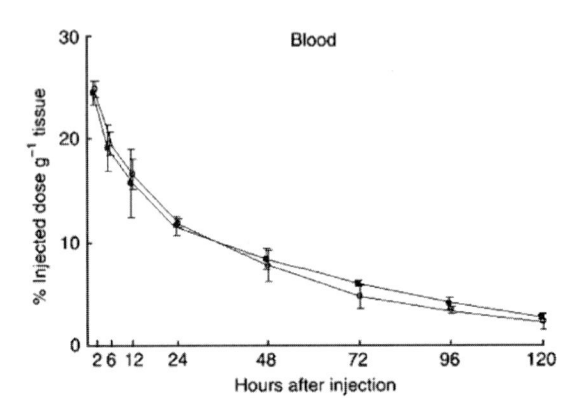

Figure 10.11 Blood concentrations of 125I-labeled A7-Ferumoxides and 125I-labeled normal mouse IgG-Ferumoxides in mice that received an intravenous injection. 125I-labelled A7-Ferumoxides and 125I-labeled normal mouse IgG-Ferumoxides disappeared from blood linearly over time, with similar clearance curves. Hollow circles – A7-Ferumoxides; filled circles – normal mouse IgG-Ferumoxides.

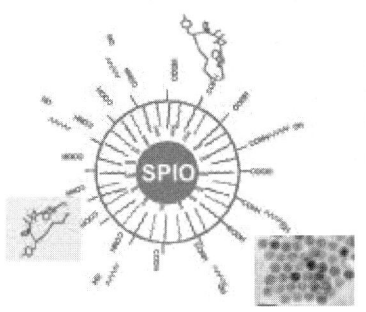

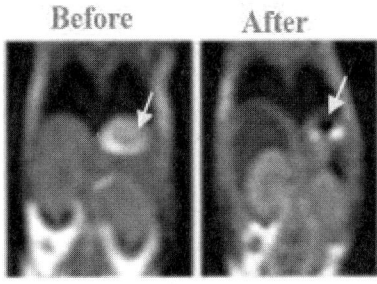

Figure 10.12 The magnetic iron oxide nanoparticle (SPIO) used to detect a pancreatic tumor in a live mouse. The nanoparticle's surface contains a molecule that binds to a receptor found on pancreatic tumors. When injected into the mouse, the nanoparticles accumulate in the tumor, leaving a distinct black void on MRI. The arrows show the location of the tumor in images taken before and after nanoparticle injection.

MRI with liver-specific contrast agents is more accurate than helical CT for detection of liver metastases [208]. Transfer from 1.5 T to more high-resolution (6–8 T) MRI equipment [209] gives ability to see metastases more correctly. Combination of T1- and T2-weighted image contrast for both the range of magnetic fields (1.0–1.5 and 6–8 T) in Fe–Gd nanoparticles [123, 124] is available and shows prospect.

In addition to its well-established role as a diagnostic modality, MRI is becoming increasingly important in therapeutic techniques, reflecting a major trend in healthcare toward cost-effective, minimally invasive procedures. Recently, considerable attention has been focused on the present and future role of MRI in drug delivery, and in the control of local hyperthermia for drug activation and thermotherapy.

10.6
Prospective of MRI

Local drug delivery and activation is a promising future role for MRI. In systemic drug treatment, the drug is, to a greater or lesser extent, dispersed throughout the body. This means that the drug concentration at the disease site may be inadequate, while it may be impossible to increase the dose without endangering healthy tissues. For this reason, there have been many investigations into the possibility of delivering the drug directly to the site where it is required, and then activating it *in situ*. The ideal solution to precise local drug delivery would be to actively target specific cells by molecular vectors attached to small drug carriers such as liposomes or viral particles. Macrophages and lymphocytes have a natural ability to target lesions and inflammation, and may one day be used for both targeting and visualizing local drug delivery. The drug carriers would be injected systemically, and local delivery would be governed by specific antigen distribution, e.g., in response to increased expression of growth factors in tumor cells. The carrier would adhere to the target cell by binding of specific antibodies to the antigens in the cell membrane. Fusion of the carrier and cell membranes would then allow the contents of the carrier to enter the cytoplasm. Here again, the future is very promising, but the technique is not yet fully developed for widespread clinical application.

At present, the most effective method of ensuring precise local delivery is the use of catheter devices. The drugs can then be activated *in situ* by local hypothermia. Liposomes can be used for local drug delivery, as they become "leaky" at the phase-transition temperature. This temperature can be adjusted by modifying the composition of the liposomal membrane. The use of drug-filled liposomes, in combination with local hypothermia, provides significantly higher local drug delivery.

MRI is playing an increasingly important role in local drug delivery and activation. In addition to the excellent visualization and discrimination of various types of tissue, MRI can be used to guide the delivery devices, and for local temperature measurement and temperature mapping. These characteristics seem to make MRI "tailor made" for the purpose. MRI also has the potential to help in monitoring drug uptake, tracking small drug carriers, and evaluating drug efficacy through physiological measurements.

Drug carriers, whether liposomes, viral particles, hydrogels, or modified cells, may be magluidizy labeled with MR contrast agents and followed *in vivo* by MRI. Quantitation of drug release from a hydrogel was shown recently using stoichiometric amounts of contrast agent and drug. Similar methods may be used for tracking microscopic drug carriers and even actively targeting drug delivery vehicles. The development of physiological MRI in recent years has led to powerful tools for the assessment of drug efficacy *in vivo*. Such methods have the potential to give feedback for dosage and regional delivery of drugs.

10.7
Problems and Perspectives

There are several problems associated with magnetically targeted drug delivery. These limitations include the possibility of embolization of the blood vessels in the target region due to accumulation of the magnetic carriers, difficulties in scaring up from animal models due to the larger distances between the target site and the magnet, once the drug is released, it is no longer attracted to the magnetic field, and toxic responses to the magnetic carriers. Recent preclinical and experimental results indicate, however, that it is still possible to overcome these limitations and use magnetic targeting to improve drug retention and also address safety issues.

Ultimately, improvements in the affinity, specificity, and mass production of antibodies will dictate the success or failure of a given technology. Sensitive and specific detection of various agents by immunoassays has improved by several orders of magnitude over the past decades. If recent scientific progress is a fair indicator, the future promises continued improvements in the immunoassays with an ever-increasing array of applications. Magnetic separation techniques are relatively new and still under development. Magnetic affinity particles are currently used mainly in molecular biology (especially for nucleic acids separation), cell biology and microbiology (separation of target cells), and as parts of the procedures for the determination of selected analytes using magnetic ELISA and related techniques (especially determination of clinical markers and environmental contaminants). Till now, separations in small scale prevail and thus the full potential of these techniques has not been fully exploited.

It is a favorable laboratory application of magnetic affinity separation technique in biochemistry and related areas (rapid isolation of a variety of both low- and high-molecular-weight substances of various origin directly from crude samples thus reducing the number of purification steps) and in biochemical analysis (application of immunomagnetic particles for separation of target proteins from the mixture followed by their detection using ELISA and related principles). Such types of analysis will enable to construct portable assay systems enabling, e.g., near-patient analysis of various protein disease markers. New methodologies, such as the application of chip and microfluidics technologies, may result in the development of magnetic separation processes capable of magnetic separation and detection of extremely small amount of target, biologically active compounds.

In the near future, quite new separation strategies can appear. A novel magnetic separation method, which utilizes the magneto-Archimedes levitation, has been described recently and applied to the separation of biological materials. By using the feature that the stable levitation position under a magnetic field depends on the density and magnetic susceptibility of materials, it was possible to separate biological materials such as hemoglobin, fibrinogen, cholesterol, and so on. So far, the difference of magnetic properties was not utilized for the separation of biological materials. Magneto-Archimedes separation may be another way for biological materials separation.

It can be expected that magnetic separations will be used regularly both in biochemical laboratories and biotechnology industry in the near future. Nonsperical magnetic nanoparticles can be used.

In some cases new methods, such as IMS are used. IMS is the process of using small SPM particles or beads coated with antibodies against surface antigens of cells. Most particles are superparamagnetic, i.e., they are magnetic in a magnetic field but nonmagnetic as soon as the magnetic field is removed. This is important because once separated by a magnet, particles should attach to each other through intermagnetic force but then return directly back into suspension. The IMS technique has several advantages for microbiologists. When working with samples heavily contaminated with nontarget organisms, IMS facilitates the purification of the target organism. Magnetic separations in biology and biotechnology have diversified in recent years, leading to a wide array of different particles, affinity mechanisms, and processes. The most attractive advantage of the magnetic separation technique in biochemistry and biotechnology is the ease of manipulation of biomolecules that are immobilized on magnetic particles. Once target biological cells or molecules are immobilized on magnetic particles, the target biomolecules can be separated from a sample solution, manipulated flexibly in various reagents and transported easily to a desired location by controlling magnetic fields produced from a permanent magnet or an electromagnet.

As it stands now, the majority of commercial nanoparticle applications in medicine are geared toward drug delivery. In biosciences, nanoparticles are replacing organic dyes in the applications that require high photostability as well as high multiplexing capabilities. There are some developments in directly and remotely controlling the functions of nanoprobes, for example, driving magnetic nanoparticles to the tumor and then making them either to release the drug load or just heating them in order to destroy the surrounding tissue. The major trend in further development of nanomaterials is to make them multifunctional and controllable by external signals or by local environment, thus essentially turning them into nanodevices.

Although all the components of the body are either dia-, para, superpara-, or ferromagnetic, the magnetic fields required to produce an effect on the body are very large. Even red blood cells with micrograms of the iron protein hemoglobin are relatively unreactive to large fields or steep field gradients. However, there is sufficient iron present for MRI to be possible without adding iron-rich or other contrast enhancing reagents. Many cell types contain magnetite or other iron oxide nanoparticles. It is not clear whether these are generated by some standard biochemical process or are adventitious particles acquired from the environment.

The other natural iron-containing compounds in the body are hemosiderin, ▢luidized▢ngransferrin, and the cytochromes. The question of whether there are hazards or effects from magnetic fields of the magnitudes normally encountered in the environment even in the "built" environment acting on these molecules is unlikely though unresolved. But the electromagnetic fields produced by mobile phones may perhaps be large enough to have pathological effects.

However, the most likely exposures to large fields or field gradients arise from therapy or diagnosis. MRI produces images of both hard and soft organs, and this is useful in the diagnosis and following the course of therapy. MRI contrast reagents are often used, and there are a number of potential compounds for infusion or injection. There are advantages in using strong magnetic nanoparticle contrast reagents, because there is a possibility that the particles can be localized in the desired region by applying local magnetic field gradients. There is also the possibility of producing localized hyperthermia when an organ loaded with nanoparticles is exposed to electromagnetic radiation.

Magnetite or maghemite particles may present an additional risk because the pH in the endolysosomes is low. This final destination is so acidic that ferric ions may be formed, which will damage many biochemical processes. Silica-coated nanopartices may be resistant to this effect.

The drug delivery magnetic nanoparticles offer the possibility of use of external magnetic fields to obtain better localization than could be achieved with nonmagnetic particles. But since magnetic nanoparticles are less easily destroyed or inactivated by cells than many nonmagnetic ones, there is the disadvantage that persistent particles may cause later cell damage and death. The same considerations apply to situations where magnetic nanoparticles are being used for generating hyperthermia by applying external fields.

Abbreviations

AC	alternating current
ATL	antibody targeted liposomes
CT	computer X-ray tomography
DNA	desoxy-ribonucleide acid
ELISA	enzyme-linked immunosorbent assay
IMS	immunomagnetic separation
HGMS	high-gradient magnetic separators
HIV/AIDS	human immunodeficiency virus/acute immunodeficiency syndrome
FM	ferromagnetic
FNH	focal nodular hyperplasia
mAb	monoclonal antibody
MCL	magnetic cationic liposomes
MDT	magnetodynamic therapy
MFH	magnetic fluid hyperthermia
MMS	magnetic microspheres
MRI	magnetic resonance imaging
mRNA	messenger ribonucleic acid
MTC	magnetic targeted carriers
MnDPDP	mangafodipir
MNP	magnetic nanopart–cles
MRX	magnetorelaxometry
MSFB	magnetically st☐luidized fluidised beds
NCT	neutron capture therapy
PAA	polyacrylic acid
PCR	polymerase chain reaction
PEI	polyethyleneimine

PEG	polyethylene glycol
PEO	polyethylene oxide
PET	positron emission spectroscopy
PRF	proton resonance frequency
PVA	polyvinyl alcohol
RES	reticulo-endothelial system
siRNsA	small interfering RNAs
SPIO	superparamagnetic iron oxides
SQUID	superconducting interferometer device
SPM	superparamagnetic
SSL	sterically stabilized liposomes
STIR	short inversion time inversion-recovery
T1	spin-lattice relaxation time, the time constant in the longitudinal z-direction
T2	spin–spin relaxation time, the time constant for the relaxation process in the transverse xy-plane
TBI	total body irradiation
USPIO	ultrasmall superparamagnetic iron oxides
VEGF	vascular endothelial growth factor

References

1. C. Alexiou, R. Jurgons, C. Seliger, H. Iro, *J Nanosci Nanotechnol*, **2006**, *6*, 2762.
2. M. Bruchez, M. Moronne, P. Gin, S. Weiss, A.P. Alivisatos, *Science*, **1998**, *281*, 2013.
3. W.C.W. Chan, S.M. Nie, *Science*, **1998**, *281*, 2016.
4. S. Wang, N. Mamedova, N.A. Kotov, W. Chen, J. Studer, *Nano Letters*, **2002**, *2*, 817.
5. C. Mah, I. Zolotukhin, T.J. Fraites, J. Dobson, C. Batich, B.J. Byrne, *Mol Therapy*, **2000**, *1*, 239.
6. D. Panatarotto, C.D. Prtidos, J. Hoebeke, F. Brown, E. Kramer, J.P. Briand, S. Muller, M. Prato, A. Bianco, *Chem Biol*, **2003**, *10*, 961.
7. R.L. Edelstein, C.R. Tamanaha, P.E. Sheehan, M.M. Miller, D.R. Baselt, L.J. Whitman, R.J. Colton, *Biosens Bioelectron*, **2000**, *14*, 805.
8. J.M. Nam, C.C. Thaxton, C.A. Mirkin. *Science*, **2003**, *301*, 1884.
9. R. Mahtab, J.P. Rogers, C.J. Murphy, *J Am Chem Soc*, **1995**, *117*, 9099.
10. J. Ma, H. Wong, L.B. Kong, K.W. Peng, *Nanotechnology*, **2003**, *14*, 619.
11. A. de la Isla, W. Brostow, B. Bujard, M. Estevez, J.R. Rodriguez, S. Vargas, V.M. Castano, *Mater Res Innovat*, **2003**, *7*, 110.
12. J. Yoshida, T. Kobayashi, *J Magn Magn Mater*, **1999**, *194*, 176.
13. R.S. Molday, D. MacKenzie, *J Immunol Methods*, **1982**, *52*, 353.
14. R. Weissleder, G. Elizondo, J. Wittenburg, C.A. Rabito, H.H. Bengele, L. Josephson, *Radiology*, **1990**, *175*, 489.
15. W.J. Parak, R. Boudreau, M.L. Gros, D. Gerion, D. Zanchet, C.M. Micheel, S.C. Williams, A.P. Alivisatos, C.A. Larabell, *Adv Mater*, **2002**, *14*, 882.
16. V.A. Sinani, D.S. Koktysh, B.G. Yun, R.L. Matts, T.C. Pappas, M. Motamedi, S.N. Thomas, N.A. Kotov, *Nano Lett*, **2003**, *3*, 1177.
17. Y. Zhang, N. Kohler, M. Zhang, *Biomaterials*, **2002**, *23*, 1553.

18. (a) L.G. Gutwein, T.J. Webster, *J Nanoparticle Res*, **2002**, *4*, 231; (b) L.G. Gutwein, T.J. Webster, *Biomaterials*, **2004**, *25*, 4175.
19. (a) R.S. Molday, D. MacKenzie, *J Immunol Methods*, **1982**, *52*, 353; (b) C. Sangregorio, J.K. Wiema'n, ¥.J. O'Connor, Z. Rosenzweig, *J Appl Phys*, **1999**, *85*, 5699; (c) H. Pardoe, W. Chua-anusorn, T.G. St Pierre, J. Dobson, *J Magn Mag Mater*, **2001**, *225*, 41.
20. (a) D. Hogemann, L. Josephson, R. Weissleder, J.P. Basilion, *Bioconjugate Chem*, **2000**, *11*, 041; (b) L. Levy, Y. Sahoo, E.S. Kim, E.J. Bergey, F.N. Prasad, *Chem Mater*, **2002**, *14*, 3715; (c) Y. Zhang, N. Eobl†a, M. Zhang, *Biomaterials*, **2002**, *23*, 1553.
21. A. Tibbe, B. de Grooth, J. Greve, P. Liberti, G. Dolan, L. Terstappen, *Nature Biotechnol*, **1999**, *17*, 1210.
22. Y. Kularatne, P. Lorigan, S. Browne, S.K. Suvarna, M. Smith, J. Lawry, *Cytometry*, **2002**, *50*, 160.
23. S. Morisada, N. Miyala, E. Iwahori, *J MkroMol Methods*, **2002**, *51*, 141.
24. R.E. Zigeuner, R. Riesenberg, H. Pohla, A. Hofstetter, R. Oberneder, *J Vrol*, **2003**, *169*, 701.
25. J. Ugelstad, W.S. Prestvlk, P. Stenstad, L. Kilaas, G. Kvalheim, In. *Medicine*, H. Nowak, Editor, Berlin: Wiley-VCH, **1998**, 471.
26. (a) R.V. Mehla, R.V. Upadhyay, S.W. Charles, ¥.N. Ramchand, *Biotechnol Teclm*, **1997**, *11*, 493; (b) M. Koneracka, P. Kopcansky, M. Antalk, M. Thabo, ¥.N. Ramchand, D. Lobo, R.V. Mehta, R.V. Upadhyay, *J Magn Magn Mater*, **1999**, *201*, 427; (c) M. Koneracka, P. Kopcansky, M. Timko, C.N. Ramchand, A. de Sequeira, M. Trevan, *J Mul Catalysis A Enzymatic*, **2002**, *18*, 13.
27. (a) M. Babincova, D. Leszczynska, P. Sourivong, P. Babinec, *Med Hypoth*, **2000**, *54*, 177; (b) M. Babincova, P. Sourivong, D. Leszczynska, P. Babinec, *Med Hyptoth*, **2000**, *55*, 459.
28. S.S. Davis, *Trend Biotechnol*, **1997**, *15*, 217.
29. (a) J. Kreuter, *Eur J Drug Metab Pharmacokinet*, **1994**, *19*, 253; (b) L. Araujo, R. Lobenberg, J. Kreuter, *J Drug Target*, **1999**, *6*, 373.
30. M.W. Brightman, *Am J Anat*, **1965**, *117*, 193.
31. U. Gaur, S.K. Sahoo, T.K. De, P.C. Ghosh, A. Maitra, P.E. Ghosh, *Int J Pharm*, **2000**, *202*, 1.
32. E. Allemann, J.C. Leroux, R. Gumy, E. Doelker, *Pharm Res*, **1993**, *10*, 1732.
33. L.M. Lacava, *Biophys J*, **2001**, *80*, 2483.
34. (a) T.T. Shen, A. Bogdanov, K. Poss, T.J. Brady, R. Weisleder, *Bioconjugate Chem*, **1996**, *7*, 311; (b) D. Portet, B. Denoit, E. Rump, J.J. Lejeunne, P. Jallet, *J Colloid Interface Sci*, **2001**, *238*, 37.
35. R. Gref, Y. Minamitake, M.T. Peracchia, V. Trubetskoy, V. Torchilin, R. Langer, *Science*, **1994**, *263*, 1600.
36. (a) L.B. Bangs, *Pure Appl Chem*, **1996**, *68*, 1873; (b) J.C. Joubert, *Anales de Quimica Int Ed.*, **1997**, *93*, S70; (c) P.D. Rye, *Biotechnology*, **1996**, *14*, 155.
37. R. Langer, *Science*, **1990**, *249*, 1527.
38. (a) U. Hafeli, W. Schutt, J. Teller, M. Zborowski, *Scientific and Clinical Applications of Magnetic Carriers*, New York: Plenum, **1997**; (b) B. Denizot, G. Tanguy, F. Hindre, E. Rump, J.J. Lejeune, P. Jallet, *J Colloid Interface Sci*, **1999**, *209*, 66.
39. (a) A.E. Merbach, E. Toth, *The Chemistry of Contrast Agents in Medical Magnetic Resonance Imaging*, Chichester, UK: Wiley, **2001**; (b) K. Wormuth, *J Colloid Interface Sci*, **2001**, *241*, 366.
40. D. Portet, B. Denizot, E. Rump, J.J. Lejeune, P. Jallet *J Colloid Interface Sci*, **2001**, *238*, 37.
41. A. Jordan, R. Scholz, K. Maier-Hauff, M. Johannsen, P. Wust, J. Nadobny, H. Schirra, H. Schmidt, S. Deger, S. Loening, W. Lanksch, R. Felix, *J Magn Magn Mater*, **2001**, *225*, 118.

42. (a) A.S. Lubbe, C. Bergemann, J. Brock, D.G. McClure, *J Magn Mater*, **1999**, *194*, 149; (b) U.I. Hafeli, G.J. Pauer, *J Magn Magn Mater*, **1999**, *194*, 76.
43. (a) A.S. Lubbe, C. Bergemann, W. Huhnt, T. Fricke, H. Riess, J.W. Brock, D. Huhn, *Cancer Res.*, **1996**, *56*, 4694; (b) J.M. Gallo, U. Hafeli, *Cancer Res*, **1997**, *57*, 3063.
44. R.S. Molday, D. MacKenzie, *J Immunol Methods*, **1982**, *52*, 353.
45. C. Sangregorio, J.K. Wiema'n, ¥.J. O'Connor, Z. Rosenzweig, *J Appl Phys*, **1999**, *85*, 5699.
46. H. Pardoe, W. Chuaanusorn, T.G. St Pierre, J. Dobson, *J Magn Magn Mater*, **2001**, *22*, S41.
47. D. Hogemann, L. Josephson, R. Weissleder, J.P. Basilion, *Bioconjugate Chem*, **2000**, *11*, 041.
48. L. Levy, Y. Sahoo, E.S. Kim, E.J. Bergey, N. Prasad, *Chem Mater*, **2002**, *14*, 3715.
49. Y. Zhang, N. EŒbl†a, M, Zhang, *Biomaterials*, **2002**, *23*, 1553.
50. N. Pamme, J.C.T. Eijkel, A. Manz, *J Magn Magn Mater*, **2006**, *307*, 237.
51. *Physics of Magnetic Cell Sorting Scientific and Clinical Applications of Magnetic Carrier*, M. Zborowski, Editor, New York: Plenum, **1997**, 205.
52. J.P. Hancock, J.T. Kemsbead, *J Immunol Methods*, **1993**, *164*, 51.
53. ¥.S. Owen, *Magnetic Cell Sorting Cell Separation: Methods and Selected Applications*, New York: Academic, **1983**.
54. T. Rheinlander, R. Kotitz, W. Weilschies, W. Semmler, *J Magn Magn Mater*, **2000**, *219*, 219.
55. L. Moore, A. Rodriguez, P. Williams, B. McCloskey, M. Nakamura, J. Chalmers, M. Zborowski, *J Magn Magn Mater*, **2001**, *225*, 277.
56. T. Rheinlander, D. Roessner, W. Weilschies, W. Scmniler, *Comparison of Size-Selective Techniques for the Fractionalion of Magnetic Nanospheres*, Presented at Frontiers in Magnetism (Stockholm, Sweden), **1999**.
57. P. Todd, R. Cooper, J. Doyle, S. Dunn, J. Vellinger, M. Denser. *J Magn Magn Mater*, **2001**, *225*, 294.
58. P.A. Liberti, ¥.G. Rao, L. Terstappen, *J Magn Magn Mater*, **2001**, *225*, 301.
59. F. Paul, D. Melville, S. Roaih, D. Warhurst, *IEEE Trans Magn*, **1981**, *17*, 2822.
60. N. Seesod, P. Nopparat, A. Hedrum, A. Holder, S. Thaithong, M. Uhlen, J. Lundeberg, *Am J Tropical Med Hygiene*, **1997**, *56*, 322.
61. W.-K. Hofmann, S. de Vos, M. Komor, D. IIoelzer, W. Wachsman, H.P. Kooflÿer, *Blood*, **2002**, *100*, 3553.
62. C. Delgratta, S. Dellapenna, P. Battista, L. Didonato, P. Vitullo, G. Romani, S. Diluzio, *Phys Med Biol*, **1995**, *40*, 671.
63. R.L. Edelstein, C.R. Tamanaha, P.E. Sheehan, M.M. Mille, D.R. Basel, I.J. Whitman, R.J. Colton, *J Biosens Bioelectron*, **2000**, *14*, 805.
64. M. Kala, K. Hajaj, S. Sinha, *Anal Diochem*, **1997**, *254*, 263.
65. S.P. Yazdankhah, A.-L. Ilellemann, K. Ronningen, E. Olsen, *Veterinary Microbiol* **1998**, *62*, 17.
66. M. Schuster, E. Wasserbauer, C. Ortner, K. Graumann, A. Jungbauer, F. Hammerschmid, G. Werner, *Bioseparation*, **2000**, *9*, 59.
67. I. Hofmann, M. Schnolzer, I. Kaufmann, W.W. Franke, *Mol Biol Cell*, **2002**, *13*, 1665.
68. J.D. Alche, K. Dickinson, *Protein Expr Purif*, **1998**, *12*, 138.
69. S. Teotia, M.N. Gupta, *Appl Biochem Biotechnol*, **2001**, *90*, 211.
70. H.H. Weetall, M.J. Lee, *Appl Biochem Biotechnol*, **1989**, *22*, 311.
71. I. Safarik, M. Safarikova, *J Biochem Biophys Methods*, **1993**, *27*, 327.
72. M. Safarikova, I. Roy, M.N. Gupta, I. Safarik, *J Biotechnol*, **2003**, *105*, 255.
73. D. Tanyolac, A.R. Ozdural, *React Funct Polym*, **2000**, *43*, 279.
74. L. Nixon, C.A. Koval, R.D. Noble, G.S. Slaff, *Chem Mater*, **1992**, *4*, 117.
75. K. Mosbach, L. Andersson, *Nature*, **1977**, *270*, 259.
76. B.L. Hirschbein, G.M. Whitesides, *Appl Biochem Biotechnol*, **1982**, *7*, 157.

77. K.B. Lee, S. Park, C.A. Mirkin, *Angew Chem – Int Edit*, **2004**; *43*, 3048.
78. I. Safarik, L. Ptackova, M. Safarikova, *Biotechnol Lett*, **2001**, *23*, 1953.
79. F. Schafer, U. Romer, M. Emmerlich, J. Blumer, H. Lubenow, K. Steinert, *J Biomol Tech*, **2002**, *13*, 131.
80. C.H. Lochmuller, C.S. Ronsick, L.S. Wigman, *Prep Chromatogr*, **1988**; *1*, 93.
81. M.A. Burns, D.J. Graves, *Biotechnol Progr*, **1985**, *1*, 95.
82. A.S. Chetty, M.A. Burns, *Biotechnol Bioeng*, **1991**, *38*, 963.
83. P. Wikstrom, S. Flygare, A. Grondalen, P.O. Larsson, *Anal Biochem*, **1987**, *167*, 331.
84. P.O. Larsson, *Meth Enzymol*, **1994**, *228*, 112.
85. (a) I. Safarik, M. Safarikova, In *Scientific and Clinical Applications of Magnetic Carriers*, U. Hafeli, W. Schutt, J. Teller, M. Zborowski, Editors, New York, London: Plenum, **1997**, 323; (b) M. Safarikova, I. Safarik, *Magn Electr Sep*, **2001**, *10*, 223.
86. Z.M. Saiyed, S.D. Telang, C.N. Ramchand, *BioMagn Res Technol*, **2003**, *1*, 2.
87. I. Safaric, M. Safarikova. *J Chromatogr Biomed Sci Appl*, **1999**, *722*, 33.
88. J. Nilsson, S. Stahl, J. Lundeberg, M. Uhlen, P.A. Nygren, *Protein Expr Purif*, **1997**, *11*, 1.
89. G. Kobs, *Cell Notes.*, **2004**, *9*, 2.
90. V. Gaberc-Porekar, V. Menart, *J Biochem Biophys Methods*, **2001**, *49*, 335.
91. A. Frenzel, C. Bergemann, G. Kohl, T. Reinard, *J Chromatogr B*, **2003**, *793*, 325.
92. C.H. Teng, K.C. Ho, M.Y.S. Chen, *Anal Chem*, **2004**, *76*, 4337.
93. M. Brzeska, M. Panhorst, P.B. Kamp, J. Schotter, G. Reiss, A. Puhler, A. Becker, H. Bruckl, *J Biotechnol*, **2004**, *112*, 25.
94. P. Duncanson, D.R. Wareing, O. Jones, *Lett Appl Microbiol*, **2003**, *37*, 144.
95. (a) K. Enpuku, K. Inoue, K. Soejima, K. Yoshinaga, H. Kuma, N. Hamasaki,, *Appl Superconductivity, IEEE Trans*, **2005**, *15*, 660; (b) K. Enpuku, K. Soejima, T. Nishimoto, T. Matsuda, H. Tokumitsu, T. Tanaka, K. Yoshinaga, H. Kuma, N. Hamasaki, *Appl Supercond, IEEE Trans*, **2007**, *17*, 816.
96. V.N. Nikiforov, V.D. Kuznetsov, Yu.D. Nechipurenko, V.I. Salyanov, Yu.M. Yevdokimov. *JETP Lett*, **2005**, *81*, 264.
97. P.R. Stauffer, T.C. Cetas, R.C. Jones. *Natl Cancer Inst Monogr*, **1982**, *61*, 483.
98. I.A. Brezovich, W.J. Atkinson, M.B. Lilly. *Cancer Res*, **1984**, *44*(10), 4752s.
99. N. Ikeda, O. Hayashida, H. Kameda, *Melanoma Res*, **2003**, *13*, 129.
100. M. Shinkai, M. Yanase, H. Honda, T. Wakabayashi, J. Yoshida, T. Kobayashi, *Jpn J Cancer Res*, **1996**, *87*, 1179.
101. M. Yanase, M. Shinkai, H. Honda, T. Wakabayashi, J. Yoshida, T. Kobayashi. *Jpn J Cancer Res*, **1997**, *88*, 630.
102. M. Yanase, M. Shinkai, H. Honda, T. Wakabayashi, J. Yoshida, T. Kobayashi. *Jpn J Cancer Res*, **1998**, *89*, 463.
103. V.N. Nikiforov, *Russ Phys. J*, **2007**, *50*, 913.
104. M. Gonzales, K.M. Krishnan, *J. Magn Magn Mater* **2005**, *293*, 265.
105. L. Neel, *Ann, Geophys*, **1949**, *5*, 99.
106. W.F. Jr. Brown, *Phys, Rev*, **1963**, *130*, 1677.
107. P.C. Fannin, C.N. Marin, I. Malaescu, N. Stefu, *Physica B: Cond Matter*, **2007**, *388*, 87.
108. A. Jordan, P. Wust, R. Scholz, B. Tesche, H. Fahling, T. Mitrovics, T. Vogl, J. Cervos-Navarro, R. Felix, *Int J Hyperthermia*, **1996**, *12*, 705.
109. I.S. Neilsen, M. Horsman, J. Overgaard, *E J Cancer*, **2001**, *37*, 1587.
110. R.E. Rosensweig, *J Magn Magn Mater*, **2002**, *252*, 370.
111. C.M. Oireachtaigh, P.C. Fannin, *J Magn Magn Mater*, **2008**, *320*, 871.

112. (a) M. Babincova, D. Leszczynska, P. Sourivong, P. Babinec, *Med Hypoth*, **2000**, *54*, 177; (b) M. Babincova, P. Sourivong, D. Leszczynska, P. Babinec, *Med Hyptoth*, **2000**, *55*, 459.
113. M. Mitsumori, M. Hiraoki, T. Shibata, Y. Okuno, Y. Nagata, Y. Nishimura, M. Abe, M. Hasegawa, H. Nagae, Y. Ebisawa, *Hepatogastroenterology*, **1996**, *43*, 1431.
114. J. Rehman, J. Landman, R.D. Tucker, D.G. Bostwick, C.P. Sundaram, R.V. Dayman, *J Endourol*, **2002**, *16*, 523.
115. R.K. Gilchrist, R. Medal, W.D. Shorey, R.C. Hanselman, J.C. Parrot, C.B. Taylor, *Ann Surg*, **1957**, *146*, 596.
116. I. Hilger, K. Fruhauf, W. Andra, R. Hiergeist, R. Hergt, W.A. Kaiser, *Acad Radiol*, **2002**, *9*, 198.
117. I. Hilger, W. Andra, R. Hergt, R. Hiergeist, H. Schubert, W.A. Kaiser, *Radiology*, **2001**, *218*, 570.
118. P. Moroz, S.K. Jones, B.N. Gray, *Int J Hyperthermia*, **2002**, *18*, 267.
119. S.V. Vonsovskii, *Magnetism*, New York: Wiley, **1974**.
120. A.A. Kuznetsov, O.A. Shlyakhtin, N.A. Brusentsov, O.A. Kuznetsov, *Eur Cells Mater*, **2002**, *3*, 75.
121. M. Bettge, J. Chatterjee, Y. Haik, *BioMagn Res Technol*, **2004**, *2*, 4.
122. R.V. Upadhyay, R.V. Mehta, K. Parekh, D. Srinivas, R.P. Pant, *J Magn Magn Mater*, **1999**, *201*, 129.
123. V.D. Kuznetsov, T.N. Brusentsova, N.A. Brusentsov, V.N. Nikiforov, M.I. Danilkin, *Russ Phys J*, **2005**, *48*, 156.
124. T.N. Brusentsova, N.A. Brusentsov, V.D. Kuznetsov, V.N. Nikiforov. *J Magn Magn Mater*, **2005**, *293*, 298..
125. Yu.A. Koksharov, V.N. Nikiforov, V.D. Kuznetsov, G.B. Khomutov. *Microelectron Eng*, **2005**, *81*, 169.
126. O.A. Shlyakhtin, V.G. Leontiev, Young-Jei Oh, A.A. Kuznetsov. *Smart Mater Struc*, **2007**, *16*, 35.
127. J. Van der Zee, *Ann Oncol*, **2002**, *13*, 1173.
128. P. Wust, B. Hildebrandt, G. Sreenivasa, B. Rau, J. Gellermann, H. Riess, R. Felix, P.M. Schlag, *Lancet Oncol*, **2002**, *3*, 487.
129. P. Moroz, S.E. Jones, A.N. Gray, *J Surg Oncol*, **2001**, *77*, 259.
130. J.A. Mosso, R.W. Rand, *Ann Surg*, **1972**, 663.
131. R.W. Rand, M. Snyder, D.G. Elliott, H.D. Snow, *Bull Los Angeles Neurol Soc*, **1976**, *41*, 154.
132. R.T. Gordon, J.R. Hines, D. Gordon, *Med Hypotheses*, **1979**, *5*, 83.
133. R.W. Rand, H.D. Snow, D.G. Elliott, M. Snyder, *Appl Biochem Biotechnol*, **1981**, *6*, 265.
134. N.F. Borrelli, A.A. Luderer, J.N. Panzarino, *Phys Med Biol*, **1984**, *29*, 487.
135. M. Hase, M. Sako, S. Hirota, *Nippon-Igaku-Hoshasen-Gakkai-Zasshi*, **1990**, *50*, 1402.
136. S. Suzuki, K. Arai, T. Koike, E. Oguchi, *J Jpn Soc Cancer Therapy*, **1990**, *25*, 2649.
137. D.¥.F. Chan, D.A. Kirpotin, P.A. Jr. Bunn, *J Magn Magn Mater*, **1993**, *122*, 374.
138. H. Matsuki, T. Yanada, T. Sato, E. Murakami, S. Minakawa, *Mater Sci Eng* **1994**, *A181/A182*, 1366.
139. M. Mitsumori, *Int J Hyperthermia*, **1994**, *10*, 785.
140. M. Suzuki, M. Shinkai, M. Kamihira, T. Kobayashi, *Biotechnol Appl Biochem*, **1995**, *21*, 335.
141. M. Mitsumori, M. Hiraoka, T. Shibata, Y. Okuno, Y. Nagata, Y. Nishimura, M. Abe, M. Hasegawa, H. Nagae, Y. Ebisawa, *Hepato-Gastroenterology*, **1996**, *43*, 1431.
142. A. Jordan, R. Scholz, P. Wust, H. Fahling, J. Krause, W. Wlodarczyk, B. Sander, T. Vogl, R. Felix, *Int J Hyperthermia*, **1997**, *13*, 587.
143. M. Shinkai, M. Yanase, M. Suzuki, H. Honda, T. Wakabayashi, J. Yoshida, T. Kobayashi, *J Magn Magn Mater*, **1999**, *194*, 176.

144. A. Jordan, R. Scholz, P. Wust, H. Fahling, R. Felix, *J Magn Magn Mater*, **1999**, *201*, 413.
145. T. Minamimura, H. Sato, S. Kasaoka, T. Saito, S. Ishizawa, S. Takemori, E. Tazawa, E. Tsukada, *Int J Oncol*, **2000**, *16*, 1153.
146. P. Moroz, S.K. Jones, J. Winter, A.N. Gray, *J Surg Oncol*, **2001**, *78*, 22.
147. S.K. Jones, J.W. Winter, A.N. Gray, *Int J Hyperthermia*, **2002**, *18*, 117.
148. W.A. Kaiser. *Acad Radiol.*, **2002**, *9*, 198.
149. O.A. Kuznetsov, N.A. Brusentsov, A.A. Kuznetsov, *J Magn Magn Mater*, **1999**, *194*, 83; N.A. Brusentsov, V.V. Gogosov and T.N. Brusentsova., *J Magn Magn Mater*, **2001**, *225*, 113; N.A. Brusentsov, L.V. Nikitin, T.N. Brusentsova, Anatoly.A. Kuznetsov, F.S. Bayburtskiy, L.I. Shumakov, N.Y. Jurchenko, *J Magn Magn Mater*, **2002**, *252*, 378; N.A. Brusentsov, T.N. Brusentsova, E.Yu. Filinova, V.D. Kuznetsov, L.I. Shumakov, N.Y. Jurchenko, *J Magn Magn Mater* **2005**, *293*, 450.
150. I. Hilger, R. Hergt, W.A. Kaiser, *IEEE Proc – Nanobiotechnol*, **2005**, *152*, 33.
151. A. Jordan, P. Wust, H. Fabling, W. Johns, A. Hinz, R. Felix, *Int J Hyperthermia*, **1993**, *9*, 51.
152. A. Jordan. *J Magn Magn Mater*, **2001**, *225*, 118.
153. P. Moroz, S.K. Jones, B.N Gray, *J Surg Oncol*, **2002**, *80*, 149.
154. Y. Rabin, *Int J Hyperthermia*, **2002**, *18*, 194.
155. A.M Granov, I.V. Muratov, V.F. Frolov, *Theor Foundations Chem Eng*, **2002**, *36*, 63.
156. V Craciun, G Calugaru, V Badescu, *Czechoslovak J Phys*, **2002**, *52*, 725.
157. J.R. Oleson, T.¥. Cetas, P.M. Corry, *Radial Res*, **1983**, *95*, 175.
158. J.P. Reilly Ami, *New York Acad Sci*, **1992**, *649*, 96.
159. J.R. Oleson, R.S. Heusinfcveld, M.R. Manning, *Int J Radiat Oncol Phys*, **1983**, *9*, 549.
160. W.J. Atkinson, I.A. Brezovich, D.P. Chakraborty, *IEEE Trans Blamed Eng B ME*, **1984**, *31*, 70.
161. (a) A.S. Lübbe, C. Bergemann, H. Riess, *Cancer Res*, **1996**, *56*, 4686; (b) A.S. Lübbe, C. Alexiou, C. Bergemann, *J Surg Res*, **2001**, *95*, 200.
162. S. Goodwin, *Oncol News Int*, **2000**, *9*, 22.
163. J. Johnson, T. Kent, J. Koda, C. Peterson, S. Rudge, G. Tapolsky, *Eur Cells Mater*, **2002**, *3*, 12.
164. J. Chen, H. Wu, D. Han, C. Xie, *Cancer Lett*, **2006**, *231*, 169.
165. J. Zhou, C. Leuschner, C. Kumar, J.F. Hormes, W.O. Soboyejo, *Biomaterials*, **2006**, *27*, 2001.
166. M. Johannsen, U. Gneveckow, L. Eckelt, A. Feussner, N. Waldofner, R. Scholz, S. Deger, P. Wust, S.A. Loening, A. Jordan, *Int J Hyperthermia*, **2005**, *21*, 637.
167. E.K. Ruuge, A.N. Rusetski, *J Magn Magn Mater*, **1993**, *122*, 335.
167. K. Maruyama, *Biol Pharm Bull*, **2000**, *23*, 791.
168. R.D. Turner, R.W. Rand, J.R. Bentson, J.A. Mosso, *J Urol*, **1975**, *113*, 455.
169. P.H. Meyers, F. Cronic, C.M. Nice, *Am J Roentgenol Radium Ther Nucl Med*, **1963**, *90*, 1068.
170. S.K. Hilal, W.J. Michelsen, J. Driller, E. Leonard, *Radiology*, **1974**, *113*, 529.
171. C.B. Wu, S.L. Wei, S.M. He, *J Clin Pharm Sci*, **1995**, *4*, 1.
172. S.K. Jones, J.G. Winter, *Phys Med Biol*, **2001**, *46*, 385.
173. K.J. Widder, A.E. Senyei, D.F. Ranney, *Adv Pharmacol Chemother*, **1979**, *16*, 213.
174. (a) K.J. Widder, W.L. Greif, R.R. Edelman, T.J. Brady, *Am J Roentgenol*, **1987**, *148*, 399; (b) K.J. Widder, R.M. Morris, G.A. Poore, D.P. Howards, A.E. Senyei, *Eur J Cancer Clin Oncol*, **1983**, *19*, 135.
175. M.-S. Martina, V. Nicolas, C. Wilhelm, C. Ménager, G. Barratt, S. Lesieur, *Biomaterials*, **2007**, *28*, 4143.

176. C. Allen, N. Dos Santos, R. Gallagher, G.N.C. Chiu, Y. Shu, W.M. Li, S.A. Johnstone, A.S. Janoff, L.D. Mayer, M.S. Webb, M.B. Bally, *Biosci Rep*, **2002**, *22*, 225.
177. C.L. Hattrup, S.J. Gendler, *Breast Cancer Res*, **2006**, *8*, R37.
178. E.H. Moase, W. Qi, T. Ishida, Z. Gabos, B.M. Longenecker, G.L. Zimmermann, L. Ding, M. Krantz, T.M. Allen, *Biochim Biophys Acta*, **2001**, *1510*, 43.
179. J.W. Park, *Breast Cancer Res*, **2002**, *4*, 95.
180. S.C. De Pinho, R.L. Zollner, M. De Cuyper, M.H. Santana, *Colloids Surf B: Biointerfaces*, **2008**, *63*, 249.
181. J. Giri, S.G. Thakurta, J. Bellare, N.A. Kumar, D. Bahadur, *J Magn Magn Mater*, **2005**, *293*, 62.
182. N. Emanuel, E. Kedar, E.M. Bolotin, N.I. Smorodinsky, Y. Barenholz, *Pharm Res*, **1996**, *13*, 861.
183. W.L. Lu, X.R. Qi, Q. Zhang, R.Y. Li, G.L. Wang, R.J. Zhang, S.L. Wei, *J Pharmacol Sci*, **2004**, *95*, 381.
184. D.C. Drummond, O. Meyer, K. Hong, D.B. Kirpotin, D. Papahadjopoulos, *Pharmacol Rev*, **1999**, *51*, 691.
185. O. Ishida, K. Maruyama, H. Yanagie, M. Eriguchi, M. Iwatsuru, *Jpn J Cancer Res*, **2000**, *91*, 118.
186. K.J. Harrington, M. Mubashar, A.M. Peters, *Q J Nucl Med*, **2002**, *46*, 171.
187. D. Needham, G. Anyarambhatla, G. Kong, M.W. Dewhirst, *Cancer Res*, **2000**, *60*, 1197.
188. C.R. Dass, P.F. Choong, *J Contr Release*, **2006**, *113*, 155.
189. M. Shinkai, M. Yanase, H. Honda, T. Wakabayashi, J. Yoshida, T. Kobayashi, *Jpn J Cancer Res*, **1996**, *87*, 1179.
190. M. Shinkai, M. Yanase, M. Suzuki, H. Honda, T. Wakabayashi, J. Yoshida, T. Kobayashi, *J Magn Magn Mater*, **1999**, *194*, 176.
191. U.O. Häfeli, in *Microspheres, Microcapsules & Liposomes: Magneto- and Radio-Pharmaceuticals*, R. Arshady, Editor; London: Citus Books, **2001**, 559.
192. V.N. Nikiforov, V.D. Kuznetsov, A.V. Ruchkin, V.I. Salyanov, Yu.M. Yevdokimov, *Proceedings on Moscow International Symposium on Magnetism*, Moscow, **2005**, 249.
193. B. Gleich, J. Weizenecker, *Nature*, **2005**, *435*, 1214.
194. S. Saini, D.D. Stark, P.F. Hahn, J. Wittenberg, T.J. Brady, J.T. Ferrucci, *Radiology*, **1987**, *162*, 211.
195. A. Hengerer, J. Grimm, *Biomed. Imaging Interv J*, **2006**; *2*, 238.
196. A. Bogdanov, Jr, L. Matuszewski, C. Bremer, A. Petrovsky, R. Weissleder, *Mol Imaging*, **2002**, *1*, 16.
197. J.W.M. Bulte, D.L. Kraitchman, *NMR Biomed*, **2004**, *17*, 484.
198. P. Gillis, S.H. Koenig, *Magn Reson Med*, **1987**, *5*, 323.
199. E. Rummeny, S. Saini, D.D. Stark, R. Weissleder, C.C. Compton, J.T. Ferrucci, *Am J Roentgenol*, **1989**, *153*, 1207.
200. N.A. Brusentsov, T.N. Brusentsova, E.Yu. Filinova, N.Y. Jurchenko, D.A. Kupriyanov, Yu.A. Pirogov, A.I. Dubina, M.N. Shumskikh, L.I. Shumakov, E.N. Anashkina, A.A Shevelev, A.A. Uchevatkin, *J Magn Magn Mater*, **2007**, *311*, 176.
201. P. Reimer, N. Jähnke, M. Fiebich, W. Schima, F. Deckers, C. Marx, N. Holzknecht, S. Saini, *Radiology*, **2000**, *217*, 152.
202. I. Jolanda, M. de Vries, W.J. Lesterhuis, J.O. Barentsz, P. Verdijk, J. Han van Krieken, O.C. Boerman, W.J.G. Oyen, J.J. Bonenkamp, J.B. Boezeman, G.J. Adema, J.W.M. Bulte, T.W.J. Scheenen, C.J.A. Punt, A. Heerschap, C.G. Figdor, *Nature Biotechnol*, **2005**, *23*, 1407.
203. (a) A. Myc, I.J. Majoros, T.P. Thomas, J.R. Baker, Jr., *Biomacromolecules*, **2007**, *8*, 13; (b) S. Hong, P.R. Leroueil, I.J. Majoros, B.G. Orr, J.R. Baker, Jr, H.M.M. Banaszak, *Chem Biol*, **2007**, *14*, 107; (c) X. Shi, S. Wang, H. Sun, J.R. Baker, Jr., *Soft Matter*, **2007**, *3*, 71; (d) I.J. Majoros, T.P. Thomas,

K.A. Candido, M.T. Islam, S. Woehler, C.B. Mehta, A. Kotlyar, Z. Cao, J.F. Kukowska-Latallo, J.R. Baker, Jr., *Biokemia*, **2007**, *31*, 9; (e) X. Shi, S.-H. Wang, S. Meshinchi, M.E. Van Antwerp, X. Bi, I. Lee, J.R. Baker, Jr., *Small*, **2007**, *3*, 1245.

204. R.D.O. Engberink, E.L.A. Blezer, E.I. Hoff, S.M.A. van der Pol, A. van der Toorn, R.M. Dijkhuizen, H.E. de Vries, *Nature Biotechnol*, **2005**, *23*, 1372.

205. A. Toma, E. Otsuji, Y. Kuriu, K. Okamoto, D. Ichikawa, A. Hagiwara, H. Ito, T. Nishimura, H. Yamagishi, *Br J Cancer*, **2005**, *93*, 131.

206. L. Yang, Z.H. Cao, Y.M. Lin, W.C. Wood, C.A. Staley, *Cancer Biol Ther*, **2005**, *4*, 561.

207. (a) M.K. So, C. Xu, A.M. Loening, S.S. Gambhir, J. Rao, *Nat Biotechnol*, **2006**, *24*, 339; (b) Y. Wang, M. Iyer, A. Annala, L. Wu, M. Carey, S.S. Gambhir, *Physiol Genomics*, **2006**, *24*, 173.

208. (a) E. Seneterre, P. Taourel, Y. Bouvier, J. Pradel, B. Van Beers, J.-P. Daures, J. Pringot, D. Mathieu, J.-M. Bruel, **1996**, *200*, 785; (b) P. Soyer, D.A. Bluemke, R.H. Hruban, J.V. Sitzmann, E.K. Fishman, *Radiology*, **1994**, *193*, 71; C. Valls, E. Andia, A. Sanchez, A. Guma, J. Figueras, J. Torras, T. Serrano, *Radiology*, **2001**, *218*, 55; (c) J. Ward, K.S. Naik, J.A. Guthrie, D. Wilson, P.J. Robinson, *Radiology*, **1999**, *210*, 459.

209. R.M. Koch, G.A. Christoforidis, W.T.C. Yuh, M. Yang, P. Schmalbrock, S. Sammet, N.A. Mayr, J.C. Grecula, M.V. Knopp, *Int J Rad Oncol Biol Phys*, **2007**, *69*, S261.

Index

a

acicular magnetic nanoparticles   153
– alternate   131
– stepwise   131
AFM   130 ff., 155, 158
aggregates   144, 165
– ring-like   165
alloys   61, 89
– cobalt–boron   61
  – coercive force   61
– trimetallic   89
alumosilicate matrix   60
anisotropic nanoparticles   117, 145, 147, 163
anisotropic reaction systems   150
anisotropy   3, 12, 143 ff., 152, 175, 201, 204, 207, 208, 211, 215, 216
– effective volume anisotropy   216
– interface   238
– magnetic anisotropy   3, 176
– magnetocrystalline   12, 204, 207, 211, 215, 239
  – constant   211, 224, 231
– shape   208
– surface   215, 231
  – Néel surface anisotropy   215
– uniaxial   204
– unidirectional   221
anodic alumina as matrix   150
anomaly magnetism   237
antibodies   403, 409
antigen   403
antioxidant   171
applications   145, 183, 241, 405
– biomedicine   405
– ferrofluids   241
  – magneto-optic   243

– mechanical   241
– medicine   393
  – chemical therapy   394
  – chemotherapy   418
  – delivery   429, 443
  – DNA isolation   394
  – gene therapy   431
  – magnetic hyperthermia   394, 415, 416, 422
  – magnetic resonance imaging   394
  – magnetic separation   394, 401, 402, 411
  – magnetic targeted delivery   394, 401
  – magnetic targeting of radioactivity   436
  – magnetodynamic therapy   425
  – radiation therapy   394, 417
  – targeted drug   429, 443
  – thermoablation   429
  – magnetic fluid hyperthermia   416
– nanobiotechnological   183
– spintronics   244, 245
  – spin field-effect transistor   244
approximate ellipsoid   312 ff.
atomic layer deposition technique   162

b

bacteria   170
– magnetotactic   170
band narrowing   219
barium hexaferrite   89, 165
binding   137, 141
biochemistry   410
– biochemical analysis   412
biocolloids   145
biocompatibility   145, 167, 171
biodegradable systems   169, 430

biogenic process 174
biomedical applications 163, 171
biomineralization 172
biomolecules 137, 413
biopolymers 396, 407
Bloch's law 228
– Bloch exponent 229
– Bloch's constant 228
blocking temperature 152, 421
bottom-up 118, 119
boundary condition 306, 323, 325, 327, 330, 331
Brown's boundary value problem 307, 309, 313
Brown's modes 308

c

C and S configurations 305, 320, 321
capsid 157
capsules nanofilm 134, 135, 136 ff.
carbon nanotubes 134, 150, 156, 157, 162
carriers 135
CdTe 167
– luminescent 167
cell labeling 402, 404
chain arrays 173
chain structures 153
chains 145, 153, 154, 159, 166
– of nanoparticles 145
chemical binding 139, 144 ff.
chemical binding energy 140, 142 ff.
chemical deposition 162
chemical nanoreactor 150
chemical stability 172
circle structures 129, 179–181 ff.
cluster 2, 30, 39, 41, 43
– molecular magnetic 2
"cluspol" technique 89
Co 122, 151, 154, 157, 162, 166, 220
– nanotubes 162
– colloidal crystal 121
– Fe_2O_3 106, 108
$Co_{80}Ni_{20}$ 230
coating 145
cobalt 149, 161
– molecular complex 356, 358, 360, 365, 366, 372–374, 376, 378–380, 382, 383, 386, 397
– polymeric complex 353, 356–358, 360, 380, 385, 386
cobalt ferrite 125
cobalt nanoparticles 156
cobalt nanowires 152
coercive force 95, 145, 202, 203, 229, 231

$CoFe_5$ 108 ff.
coherent nanocrystals 237
collective behavior 182
colloid 117, 137, 145
colloidal nanoparticles, stabilizing agent 30, 38, 39, 42, 44, 134
– 1,2-diols 35
– alcohols 35
– aluminum alkyls 36, 51
– capping ligand(s) 32, 34, 37, 39, 46, 50
– electric double-layer(s) 35, 41, 45
– hydroxyacids 35
– ligands 42
– micelles 51
– microemulsions 51
– organoaluminum compounds 37, 38
– polymer 37, 42, 48, 50
– supersolvent(s) 31, 45, 48
– tetraalkylammonium salts 30, 34
columns 148, 153
composite magnetic microcapsules 134
concentrated magnetic semiconductors 235
configurational anisotropy 305
containers 135
CoO 223
coordination numbers 141
CoPt 152, 157, 239
– nanowires 157
$CoPt_3$ 166
Core-shell nanoparticles 39, 61, 161
cotton fiber 171 ff.
Coulomb energy 138
crystalline superlattices 166
cube 141
curling mode 331, 342
cylindrical nanoparticles 151
cylindrical pores 161

d

damping parameter 334, 337
data recording density 238
dead magnetic layer 219
defects 214
defectless 145
delivery 135, 145
demagnetizing factors(s) 259, 283, 283, 286, 287, 289, 292 ff., 297 ff., 294, 298
– distribution 281, 283, 286, 298
– diameter 287
demagnetizing factors shape 299
demagnetizing field 262, 263, 278, 279, 281, 282, 292, 306, 308, 310, 312
demagnetizing tensor 259, 283, 285

diameter 281, 287, 289, 292, 298
– most probable 281, 287, 289
– distribution 285, 287, 298
– mean 281, 292
dicobalt octacarbonyl 154
diluted magnetic semiconductors 233
dimensionality 142 ff.
dipole magnetic moment 146
dipole–dipole energies 179
dipole–dipole interactions 146, 153
dipole–dipole magnetic field 179 ff.
disk 129, 141, 149
distribution 5, 12, 203, 210, 227
– Boltzmann 210
– composition 12
– energy barrier 224
– equilibrium 203
– size 5, 15, 227
  – log-normal 227
DNA 153, 157, 160 ff., 161, 172, 173, 175
DNA/magnetite 165
dots, quantum 2
double hydroxides 152
double vortex 321
drug delivery 160, 163, 171
1D 145, 163
1D polymer 351, 353, 355–358, 362, 363, 367, 371, 385, 386
1D structures 153
2D arrays 120, 124, 132
2D ensembles 122
2D magnetism 124
2D polymer 358, 367, 370
– 2D framework 360
2D reaction system 150
3D polymer 360, 362
3D structures 142

e

effect magnetic viscosity 212
effective anisotropy energy 326, 329, 342
effective magnetic field 259 263, 268, 279
effective single-domain radii 322
effects 198, 208, 214, 220, 418
– collective 226
– exchange anisotropy 223
– finite-size 198, 205
– interparticle interaction 223, 227
– magnetic viscosity
  – viscosity 241 242
– matrix 220, 229
  – antiferromagnetic 220

– paramagnetic 220
– shape 208
– surface 198, 205, 214, 219
– synergistic 418
eigenfunctions 308, 310
eigenvalue(s) 307–308
eigenvectors 307
electrodeposition 152
electron magnetic resonance (EMR) 255, 260, 289, 290, 292–294, 296–298
electrospinning 169
electrostatic energy 132, 142, 143 ff.
emulsions magnetic 168
energetic balance 145
energy 6, 139, 180, 200, 201, 211
– activation 211
– anisotropy 178, 204
– barrier(s) 211, 333, 342
– dipolar 204
– electrostatic energy 139
– exchange 204
– magnetic anisotropy energies 180
– magnetostatic 200
– surface 6
– thermal 211
ensembles 117, 118, 183
– nanoparticles of 117, 118, 183
EPR 272, 285, 289, 292, 296, 298, 299 ff., 300
equilibrium micromagnetic equation 307
exchange 200, 207, 209, 216, 224
– constant 207, 216
– energy 201
– force 200
– length 204, 209
– Ruderman–Kittel–Kasuya–Iosida– mechanism 224
exchange bias 220, 232
external magnetic field 134, 147 ff.

f

Faraday rotation 296
Fe 151
Fe_2C_5 94
Fe_2O_3 89, 94, 101, 148, 169, 422
Fe–Co alloy 90, 149, 157
Fe–Ni alloy 157
Fe–Sm 90
FeO 112
FePt 120, 126, 127, 132, 152, 157, 164, 165, 167, 169, 230, 239
– nanoparticles 169
– nanowires 157

ferrihydrite 232
 –coercive force 62
ferrimagnetic oxides 219
ferrite 133, 153, 289, 296
ferritin 4, 123, 172, 173
ferroelectric 65
 –particles 65
  –microwave adsorption materials 65
ferrofluids 4, 117, 257, 280, 292, 296
ferrofluids, glasses 255
ferromagnetic 60, 61, 64, 156, 161, 207, 209
 –cube 209
 –ellipsoid 209
 –nanocomposites 60
 –particles 64
  –supermagnetism 65
 –resonance 61
  –block copolymers 62
  –interpolymeric complex 62
 –sphere 207
ferromagnetic exchange interactions 352, 356, 367
 –ferromagnetic exchange spin–spin interactions 375
 –ferromagnetic properties 377, 383
ferromagnetic semiconductor(s) 234, 236
 –high-temperature 234
ferromagnetism 233, 235, 237
 –associated with oxygen vacancies 237
 –d^0 235
 –hole-mediated 233
field cooling 82, 103 ff., 220
films 124, 135, 137
flower state 310, 319, 321
flux-closed state 180
flux-closure rings 166
FMR 108, 110, 112
 –g-factor 110
 –linewidth 110
fractal-like 76
 –structures 76
  –metal-containing monomer 76
free energy 137
free-floating 135, 137
free-standing sheet-like nanofilm 138 ff.

g
gadolinium 124, 440
gels
 –magnetic 169
gene delivery 171
genetic engineering 411
geomagnetic field 170

giant magnetoresistive effect 223, 245
glasses 257–259, 274, 280, 289, 290, 292–299
glasslike carbons 168

h
hematite 232, 296
high damping limit 340
high- and low-dissipation regimes 333
high-spin 79
 –iron 79
  –hyperfine fields 79
  –Mössbauer spectra 80 ff.
  –magnetite 79
hyperthermia 171
hysteresis 95, 99, 212, 203 ff.

i
immobilization 407
 –functional groups 407
immune system 418
interactions 139, 174, 210, 219, 223, 224, 411
 –competing 219
 –dipolar 174, 210, 224, 226
 –electrostatic interaction 139
 –exchange 216, 224
  –Heisenberg-type 216
 –interparticle 223
intercalated 60
 –matrix 60
 –MCM-41 60
interfacial 128
interparticle interactions 182
 –Zeeman 175
iron
 –molecular complex 350, 351, 353, 356, 366, 386, 397
 –polymeric complex 351, 353, 355
iron homeostasis 170
iron metabolism 174
iron nanowires 150, 151
iron oxide 122, 126, 128 ff., 129, 132, 135, 136, 138, 145, 148, 152, 154, 157, 159, 160, 167, 169, 173
 –FeO 126
 –nanoparticles 128, 146
 –amorphous 146
iron oxide nanotube 162
iron pentacarbonyl 127, 146, 154, 157
irreversible 82
 –magnetization 82

–metal-containing monomers   83
–solid-phase polymerization   82

j

joint distribution density   284, 286, 287, 289, 292 ff., 296, 297, 299

l

Langevin function   102, 277, 278 ff., 279, 293
Langmuir monolayer   146, 147 ff., 157, 173
Langmuir–Blodgett film(s)   51, 52, 158, 160, 164
Langmuir–Blodgett Technique   124, 125, 130
lattice softening   229
layer-by-layer   63
–nanoparticles   63
  –ferroplastics   63
  –magnetotactic bacteria   63
layer-by-layer Assembly   130, 137, 167
ligand(s)   137, 145, 148, 153, 407, 409
–affinity   409
–polyfunctional   137
linear   159
–aggregates   159
linear arrays   145
–of nanoparticles   145
lineshapes   258, 263–265, 266 ff., 268, 270–273, 275, 276, 294, 298 299 ff.
–Landau–Lifshitz   274, 275, 294
–Lorentzian   272, 275, 294
–normalize   271, 272, 268
linewidth   257–259, 265, 266 ff., 270, 271, 273, 275, 292–294, 295, 298
–angular dependence   275, 276
lipid   120
lipid tubules   162
lithography   163, 165
–block copolymer   165
–nanosphere   165
living organisms   174
log-normal distribution   280, 281, 282 ff., 286
long-range order   165
longitudinal and transverse vortexes   305, 315 ff., 316, 318, 321
low and high dissipation   338
low damping limit   341
low-dimensional systems   246
low-spin   79
–iron   79

m

Mössbauer spectroscopy   7, 212, 255, 285, 293
maghemite   125, 156 ff., 169, 293, 296, 400, 424, 433
magnet
–ferromagnetic   362
–hysteresis loop   353
–magnetic ordering   352
–magnetization isotherms   362
–Neel point   362
magnetic   59, 61, 63, 163, 170, 197, 199, 211
–complex   163, 170
–MRI   59
–coatings   63
  –coercive force   64 ff.
  –Langmuir–Blodgett films   64
–colloids   240
–fluids   240, 395
  –ionic   241
  –surfaced   241
–grains   238
–magnetic materials   199
  –antiferromagnetic   199
  –diamagnetic   199
  –ferrimagnetic   199
  –ferromagnetic   199
  –paramagnetic   199
  –bit-patterned media   239
  –granular magnetic composites   245
  –self-organized magnetic array   239
  –thin film   238
–magnetic resonance imaging   59
–matrices   63
–magnetic moment   199, 211
  –effective   211
  –uncompensated   232
–ordering   199
–properties   197, 199
  –blocking state   212
  –bulk   205
  –classification   199
  –enhanced coercivity   221
  –intrinsic   197, 234
  –rheological   242
  –superparamagnetic state   212
–sensors   59
–susceptibility   61 ff.
magnetic 1D structures   152
magnetic anisotropy   175
magnetic behavior   122
–collective   122

magnetic detection   171
magnetic field   122, 175
–external   122
–local magnetic fields   177
magnetic field flux   166
magnetic fluids   117
magnetic gels   171
magnetic liposomes   395, 432
magnetic microspheres   134
magnetic moment(s)   257, 261, 263, 270, 277, 279, 292
–thermal fluctuations   279, 292
magnetic nanocomposite films   132
magnetic nanowires   150
magnetic properties   174, 182
–ensembles   174, 182
–nanoparticles   174, 182
magnetic random access memory   240
magnetic recording   238
magnetic recording media   163
magnetic semiconductor   127
magnetic separation   170, 171
magnetic susceptibility   258, 264, 265, 280
magnetite   80, 125, 132, 134, 135, 139, 144, 159, 161, 166, 167, 170, 171, 236, 296, 398, 400, 417, 423, 424, 433
–dextran-coated   398, 423
–phase   80
  –coercivity   81, 82 ff.
  –hysteresis loops   80, 81
  –Mössbauer spectra   81
magnetization   151, 199, 200, 206, 212, 213, 219, 228, 257, 258, 260, 263–265, 271, 277, 279, 298
–axes   293
–configuration   206
  –curling   207
  –flower   209
  –uniform   207, 209
  –vortex   207
–curve   200
–easy   293
–easy axis   260
–enhancement   219
–reduction   219
–remanent   202
–reversal of   213
–saturation   200, 202, 228
–spontaneous   199
–temperature dependence   212, 228
magnetization curling   319, 342
magnetization curling mode   314
magnetization curling state   317

magnetization distribution(s)   305, 307
magnetocrystalline anisotropy   259, 262, 278, 279, 298
–axial   259, 263, 278, 279
–constant(s)   258, 259, 263, 293, 298
–cubic   259, 263, 278
–field   262, 263, 278, 279, 292
magnetophoresis   403
–magnetophoretic mobility   404
magnetoresistance   174
magnetorheological fluids   168
magnetostatic energie   262
magnetostatic energy   259, 279, 285
manganese
–molecular complex   368, 37
–polymeric complex   351, 355, 363, 364, 367
manganite   162
–ferromagnetic   162
materials   227, 137, 170
–nanofilm   170
–single-crystal   205
matrices
–rigid   4
matrix-stabilized   67
–nanoparticles   67
  –crystal structure disintegration   75
  –effective activation energies   71
  –metal-containing monomers   67
  –solid-phase polymerization   69
  –specific surface   75
matrix-stabilized magnetic   66
–nanoparticles   66
measuring time   211
membranes   150
metal nanophase   151
metal-containing   78
–phase   78
metal-oxide   78
–phase   78
  –Mössbauer spectra   79
metallopolymeric   59, 65
–magnets   65
–nanocomposites   59
microcapsules   134
–nanocomposite   134
microorganisms   413
microspheres   398, 401, 430
–albumin   432
microwave   133, 135
microwave absorption   109 ff.
minimization procedures   177

Mn_{12} cluster   166
model   224, 234
– carrier-induced ferromagnetism   236
– carrier-mediated ferromagnetism   234
– Dormann–Bessais–Fiorani   225
– impurity band exchange   236
– Mörup and Hansen   225
– Neel–Brown   224
– Shtrikmann–Wohlfarth   225
– Stoner–Wohlfarth   212
– atomic-scale   198
molecular biology of cancer   432
molecular clusters   213
monodisperse   121, 123
– nanoparticles   123
monodomain   64
– particles   64
monolayer(s)   120, 124
Monte Carlo   216, 218, 219
morphology   145
– self-assembled   145
multidomain   64
multifunctionality   133, 135, 172
multilayer   124, 131

n
Nafion® 117   61
– membranes   61
nanoalloy(s)   5
– FeCo   11
– CoPt   13
– Fe–Ni   12
– Fe–Pt   12, 13
nanobiomaterials   396
– biocompatible   396
nanobioreactors   168
nanocomposite   137, 167, 169
– luminescent   167
nanocrystalline   77
– structure   77
nanocrystallites   2
nanocrystals formation   37
– agglomeration   35, 37, 44
– burst nucleation   37
– cluster   41, 43
– coarsening   37
– condensation   41, 43
– core-shell particles   39
– crystal growth   35, 37, 38, 48, 50, 52
– growth   37
– mixture of diamond   46
– nanodisk   38
– nanorods   45
– nucleation and growth   37, 44, 49

– Ostwald ripening   46, 48
– seed particles   39, 41, 47, 48
– shape variation   32, 41, 46–48, 52
nanofibers   170
– nanocomposite   170
nanofibriles   150
nanofilm(s)   119, 137, 145, 171 ff.
– sheetlike   137, 145
nanofilm material(s)   145
nanomagnetism   3
nanomaterials   182
– compact materials   1
– hybrid   182
– integrated   182
– multicomponent   182
– multifunctional   182
– nanodispersions   1
nanomedicine   414
– cancer therapy   414, 429
nanoparticle   1, 258, 259
– anticancer agents   428
– antiferromagnetic   232
– applications   4
   – information storage   13
– assembly   210
   – randomly oriented   212
– biocompatible   397
– classification   2 ff.
– Co   10, 104, 106, 220
– composition   5, 7
– core–shell   220, 400
– critical radii   207, 208, 214 ff., 230
– critical size   206
– domain free   4
– ellipsoidal   259
– functionalized   396, 429
– iron   6, 7
   – iron oxide   10
– magnetization   3, 6
– metal-containing   87
– metallic   6
– multidomain   203, 206, 231
– nickel   8
– nonspherical   210
– oxidation   10
– preparation   5
– rare earth   9
– separation   5
– shape factor   210
– single   4
– single-domain   3, 17, 203, 206, 246, 257–259
– SPIO   438
– superparamagnetic   399

nanoparticle *contd.*
−synthesis 5
−thermal stability 223
nanoparticle core 167
−magnetic 167
−nanocomposite 167
nanoparticle structures 140 ff.
nanoparticle surface chemistry 42
nanoparticles 137, 149, 164, 165, 167, 171
−biogenic 171
−cobalt 8, 165
−core−shell 167
−FePt 165
−iron 149
−iron oxide 164
−magnetic 171
−magnetite 137
−superparamagnetic 171
nanoparticles surface chemistry 45, 52
nanoparticulate structures 142
−2D and 1D structures 142
−planar and linear structures 142
nanopoisoning 426
nanoring(s) 129, 130 ff., 165, 166
nanorods 145
nanosheets 137
nanoshell 161
nanosphere(s) 165, 168, 217 ff. 401
−magnetic 168
nanostructured Media 148, 153
nanostructures 163, 170, 174
−bioinorganic 170, 174
−composite 163, 170
−patterned 163, 170
−self-organized 163, 170
nanostructures 182
−organized 182
nanosystems 182
−hybrid 182
−integrated 182
−multicomponent 182
−multifunctional 182
nanotechnology 1, 183
nanotubes 145, 150, 160, 161, 163
−Fe_3O_4 161
−magnetic 160
−multilayer 161
nanowires 145, 150
needle-like 154
neurodegenerative diseases 170
Ni 149, 151, 157, 162, 168
−molecular complex 353, 364, 365, 366, 369, 372, 373, 375, 376, 378−380, 383
nonspecific binding 409

nonuniform magnetization configurations 314, 317
nonuniform magnetization distributions 343
nonuniform micromagnetic configurations 305

o

"onion"-like structure 181
opsonization 398
ordered arrays 163
−nanoparticles, 163
oxide surface layer 228
oxides 8, 13
−cobalt
  −Co_3O_4 19
  −CoO 8, 10
−iron 13, 14
  −ferrihydrite 18
  −ferrites 16
  −goethite 14, 18
  −hematite 14
  −lipidocrokite 18
  −maghemite 14
  −magnetite 13, 14
  −oxyhydroxides 18
  −wustite 17
−NiO 9, 19

p

particles 168
−conductive magnetic 168
patterned ensembles 163
patterning 163
percolation threshold 235
permeability 135
phase separation 166
photochemical 127, 146
planar 124, 137, 145
plate-like 150
polyacrylic acid 431
polyamine(s) 137, 144, 173
polycarbonate 152
polycarbonate membranes 161
polyelectrolyte 131, 134
polyethylene 62, 87
−matrix, 62
polyfunctional 182
polymer 59
−1D polymer 351, 353, 355−358, 362, 363, 367, 371, 385, 386
−2D polymer 358, 367, 370
−3D polymer 360, 362

–increased thermal stability 92
–matrix 59
　–ferroplastics 59
polymer-bonded 66
–magnets 66
polymerase chain reactions 405
polyphosphate anions 149
pores 150, 160, 161
–cylindrical 160
porous anodic alumina 239
porous membranes 162
positron emission spectroscopy 415
PP 62
–matrices 62
precession 260, 263, 271, 273 275, 279
–Bloch–Bloembergen Equation 265, 268, 272
–Callen Equation 272, 273
–Gilbert Equation 270
–Landau–Lifshitz Equation 270, 272, 279, 295 ff.
–Modified Bloch Equation 268, 270
precursor 5
procedures 176
–energy minimization 176
pseudoatoms 3
–magnetic 3

q
quantum dots 167

r
radioisotopes 436
reductant 172
relaxation 96, 211, 213, 224, 421
–mechanism 213
　–Brown–Neel 213
　–Brownian 213
　–quantum tunneling 213
–time 96, 211, 421
relaxation time 332, 333, 335, 337, 342
remnant magnetization[tab] 95
remote control 145
RES 398
resistance 3
–giant magnetic 3
resonance field 258–260, 262, 265, 275, 292, 294, 295 ff., 299
–apparent 257, 273, 274 ff., 275, 294, 295 ff., 299
RF electric field 415
ring structures 166, 175
ring-shaped memory element 240

s
selected-area electron diffraction 123 ff.
self-assembly 120, 121, 124,137,144,148,163–169
self-organization 129, 137, 140, 144, 153, 154, 165
self-organized 62
–nanoparticles, 62
semiconductors 233
–II–VI 233
–III–V 233
separation 145, 160
shape 145, 257, 258, 280, 281, 283–286
–distribution 286
–ellipsoidal 263, 282, 283, 294
–spherical 257, 260, 280, 281, 283–285
shape anisotropy 145
sheet-like 145
shell 167
signal-to-noise ratio 238
silica 151
silicon channel 151
single molecular magnet 350
–{Co1 + Co3} 387
–[$Fe_8O_2(OH)_{12}(tacn)_6$]$^{8+}$ 350
–$Co_6Ni_2(\mu_4\text{-}O)_2(\mu_2\text{-Piv})_6(\mu_3\text{-Piv})_6$, 383
–hysteresis loop 350, 387
–magnetization 350
–$Mn_{12}O_{12}(OOCMe)_{16}(H_2O)_4$ 350
–$Mn_{30}O_{24}(OH)_8(OOCCH_2CMe_3)_{32}(H_2O)_2(MeNO_2)_4$ 350
–molecular magnets 349
–quantum tunneling of magnetization 350
single-domain radius 304, 313–315, 320
soft lithography 153, 164
solid substrates 124
sonication 149
specific absorption rate 420
sperm cells 171
spermine 137, 171 ff., 172, 173
spherical 77
–clusters 77
spin canting 219
spin configuration 177
spin-coating 120
spin-glass 226
–disorder 225
–droplet model 226
–hierarchical model 227
–spin-glass-like behavior 226
spindle 148
spinel nanotubes 162
spintronics 118, 174

SPR   257–259, 274, 278, 279, 281, 289, 292, 293, 295 ff., 296, 299, 300
SQUID   414
sterically stabilized liposomal platform   433
string   141
structure   135, 200, 206, 217
– "onion"-like structure   175, 178
– circle   178 ff.
– domain   200, 202 ff.
– magnetization   206
– planar   2
– real   202
– spin   217, 218
  – "hedgehog"-type   175, 218
  – "throttled"-type   217
  – collinear   217
  – vortex-type   218
– structural defects   246
– anisotropic   156
– fiber-like   169
– flux-closed vortexes   175
– ring-like   174, 175, 182
– vortex structures   175
styrene copolymer   61
– matrix   61
super-moment   206, 215
superconductivity   235
superlattice   121
superparamagnetic   65, 80, 157
– particles   65
  – block copolymers   65
– nanoparticles   65
– particles   80
  – magnetite   80
superparamagnetic limit   238
superparamagnetic narrowing   278, 292
superparamagnetic resonance   273, 280, 298, 299 ff.
superparamagnetism   210, 213, 255, 421
surface anisotropy energy   328
surface layer   6
surface magnetic anisotropy   306, 323, 327
synthesis   146, 148, 351
– atomic beam   5
– dry methods   16
– inversed micelle   8
– mechanochemical   5, 16
– sonochemical   6
– thermal decomposition   6, 12, 15
– thermal evaporation   5
system dimensionality   142
system geometry   143 ff.

t

TEM   129, 136 ff.
temperature   5, 203, 212, 213
– blocking   5, 212
  – increase   223
– Curie   203, 213, 219, 233, 420
– Neel   213, 223
template   157, 161
templated structures   156–159
theory   198, 208
– micromagnetic   198, 208
thermodynamic   206, 210, 212, 226
– equilibrium   206, 210
– nonequilibrium   212
– phase transition   226
tilted curling states   305
tilted vortex   318
top-down   118, 119
toxicity   434
transverse vortex   316, 318, 319, 321
treatment   136 ff.
– microwave   136 ff.
– thermal   136 ff.
twisted vortex   321
two-domain configuration   316
two-domain state(s)   305, 314, 321

u

uniform magnetization   305, 315, 324, 341
uniform rotation mode   304, 342
unit magnetization vector   304, 332
UV illumination   128, 129, 155 ff.

v

virus   152
volume fraction   258, 281, 282 ff.
vortex   179, 180, 182
vortex core   319–321
vortex states   305, 342
vortex structure   177

z

Zeeman   175, 179
– interactions   175
zeolites   168
zero-field cooling   82, 103 ff., 220